Clinical Bacteriology

Clinical Bacteriology

Editor: Robert Jones

MURPHY & MOORE
www.murphy-moorepublishing.com

www.murphy-moorepublishing.com

MURPHY & MOORE

Cataloging-in-Publication Data

Clinical bacteriology / edited by Robert Jones.
p. cm.
Includes bibliographical references and index.
ISBN 978-1-63987-111-7
1. Diagnostic bacteriology. 2. Diagnostic microbiology. 3. Medical bacteriology. I. Jones, Robert.
QR67.2 .C55 2022
616.014--dc23

Murphy & Moore Publishing,
1 Rockefeller Plaza,
New York City, NY 10020, USA

ISBN 978-1-63987-111-7 (Hardback)

Contents

Preface

Every book is initially just a concept; it takes months of research and hard work to give it the final shape in which the readers receive it. In its early stages, this book also went through rigorous reviewing. The notable contributions made by experts from across the globe were first molded into patterned chapters and then arranged in a sensibly sequential manner to bring out the best results.

The specialty of biology that deals with the study of bacteria and various aspects related to them such as their genetics, morphology, biochemistry and ecology is termed as bacteriology. It also involves the characterization, classification and identification of bacterial species. Clinical bacteriology refers to the study of bacteria and their relation to medicine. It also encompasses the diagnosis of infectious diseases. There has been a significant role of bacteriology in the control of several fatal diseases. It has resulted in the discovery of antibiotics as well as effective vaccines such as tetanus toxoid and diphtheria toxoid vaccines. There are also some vaccines that are not much effective and have side-effects such as typhoid vaccine. Clinical bacteriology is an upcoming field of science that has undergone rapid development over the past few decades. This book brings forth some of the most innovative concepts and elucidates the unexplored aspects of this field. It will help the readers in keeping pace with the rapid changes in this field.

It has been my immense pleasure to be a part of this project and to contribute my years of learning in such a meaningful form. I would like to take this opportunity to thank all the people who have been associated with the completion of this book at any step.

Editor

1

Bacterial Therapy of Cancer: Promises, Limitations, and Insights for Future Directions

M. Gabriela Kramer[1,2], Martín Masner[1†], Fernando A. Ferreira[1,2] and Robert M. Hoffman[3,4]*

[1] Department of Biotechnology, Instituto de Higiene, Facultad de Medicina, Universidad de la República, Montevideo, Uruguay, [2] Laboratory of Carbohydrates and Glycoconjugates, Department of Organic Chemistry, Facultad de Química, Universidad de la República, Montevideo, Uruguay, [3] AntiCancer, Inc., San Diego, CA, United States, [4] Department of Surgery, University of California, San Diego, San Diego, CA, United States

Edited by:
Chew Chieng Yeo,
Universiti Sultan Zainal Abidin, Malaysia

Reviewed by:
Jung-Joon Min,
Chonnam National University, South Korea
Kaushlendra Tripathi,
University of Alabama at Birmingham, United States
Lay-Hong Chuah,
Monash University Malaysia, Malaysia

****Correspondence:***
M. Gabriela Kramer
mgkramer@higiene.edu.uy

Spontaneous tumors regression has been associated with microbial infection for 100s of years and inspired the use of bacteria for anticancer therapy. Dr. William B. Coley (1862–1936), a bone- sarcoma surgeon, was a pioneer in treating his patients with both live bacterial-based and mixture of heat-killed bacteria known as "Coley's toxins." Unfortunately, Coley was forced to stop his work which interrupted this field for about half a century. Currently, several species of bacteria are being developed against cancer. The bacterial species, their genetic background and their infectious behavior within the tumor microenvironment are thought to be relevant factors in determining their anti-tumor effectiveness *in vivo*. In this perspective article we will update the most promising results achieved using bacterial therapy (alone or combined with other strategies) in clinically-relevant animal models of cancer and critically discuss the impact of the bacterial variants, route of administration and mechanisms of bacteria-cancer-cell interaction. We will also discuss strategies to apply this information using modern mouse models, molecular biology, genetics and imaging for future bacterial therapy of cancer patients.

Keywords: bacterial-based therapies, Coley's toxins, antitumor effect, immune response, bactofection, combined therapies, *Salmonella enterica* serovar Typhimurium (*S.* Typhimurium), animal models of cancer

BACK TO THE CONTROVERSIAL FUTURE

The use of microorganisms, in particular live bacteria, for prophylactic vaccination and cancer therapy have been used in humans for long periods in the past and have been a matter of controversy (Payette and Davis, 2001; Hoption Cann et al., 2003). Dr. William B. Coley in the 19th century at the New York Hospital, later to become the Memorial Sloan Kettering Cancer Center (McCarthy, 2006; Hoffman, 2016a), observed and reported spontaneous tumor regression in patients with streptococcal infections (principally erysipelas, known to be caused by *Streptococcus pyogenes*). In 1891 Dr. Coley started to treat his cancer patients with streptococcal living cultures and observed that inducing a fever was crucial for tumor regression; however such a strategy also caused some fatalities (McCarthy, 2006). Coley then generated a variety of "anti-tumor vaccines" mixing heat-killed bacteria, combining *S. pyogenes* with *Serratia marcescens*. In this way he could stimulate the symptoms of an infection (for example, inflammation, chills, fever) without the risks of a bacteremia. These vaccines became known as "Coley's toxins" and were administered to patients with sarcomas, carcinomas, lymphomas, melanomas, and mielomas. Despite the cures and remarkable improvements obtained in patients treated with Coley's bacterial-based therapeutics (Nauts et al., 1946; Nauts, 2004; McCarthy, 2006), his boss, the renowned pathologist James Ewing, forced Dr. Coley's to end all projects involving bacteria-based treatments alleging Coley's inconsistent data and pronouncing himself in favor of radiotherapy, which rapidly took over the market of cancer therapeutics.

Sixteen different preparations of "Coley's toxins" have been used since the method was introduced in 1892, of which three were considerably more potent than the rest (particularly the Buxton's Type VI formula). However, the only preparation available in the United States since 1921 seemed to be weaker compared to the used in the early years (Nauts et al., 1946). Coley's work gradually fell out of favor and by 1962 the Food and Drug Administration (FDA) refused to acknowledge "Coley's toxins" as an approved drug, making it illegal to prescribe them outside of clinical trials. Since then, several small clinical trials have been conducted with mixed results.

To date, Bacillus Calmette-Guerin (BCG), is the only bacterial agent approved by the FDA and it is employed for the treatment of superficial, non-muscle invasive bladder cancer (NMIBC) since the late 1970s (Gontero et al., 2010). BCG is an attenuated strain of *Mycobacterium bovis* obtained at the Pasteur Institute in the early 1900s. Patients typically receive repeated instillations of live bacteria into the bladder. BCG is recommended as the standard of care for high-risk NMIBC and remains the most effective intravesical treatment for this disease, although the response predictor factors of BCG are unknown (Kim and Steinberg, 2001; Zlotta et al., 2009; Gontero et al., 2010).

In the last decades a resurgence of the field has taken place and contemporary investigators demonstrated the efficacy of a number live attenuated bacteria to destroy cancer cells *in vitro*, to selectively accumulate, replicate within and destroy tumors in rodents, to induce an immune-mediated anti-tumor response and to target small metastatic nodules spread in the organism and inhibit their growth (Yu et al., 2004; Adkins et al., 2012; Hoffman, 2012b). Promising results were obtained using modern methods of bacterial genetics, cancer cell and molecular biology, and *in vivo* imaging (Min et al., 2008a,b; Uchugonova et al., 2012; Hoffman, 2015). The mechanism of action of bacterial therapy of cancer and toxicity *in vivo* is not yet clearly understood and the potential acquisition of antibiotic-resistance or mutations that would revert the bacteria attenuated phenotype could be a real risk for the patients. Therefore, the building of a broad integrated picture requires a critical scientific and medical vision, for moving forward.

HIGHLIGHTS BUT STILL MANY QUESTIONS

Bacteria display a number of different characteristics that could be relevant in the therapy against cancer. The direct and immune-mediated anticancer properties derive from biological interactions between the bacteria and the host tumor microenvironment. Important features of the bacteria such as motility, tumor chemotaxis, invasive capacity, cytotoxic potential, pathogen-associated molecular patterns (PAMP) composition/abundance, among others, vary between strains and may affect how they trigger the anti-tumor response (Dang et al., 2001; Cheadle and Jackson, 2002; Hoffman, 2011; Adkins et al., 2012; Kim et al., 2015; Phan et al., 2015). Although the mechanism of bacterial tumor tropism is poorly understood there is evidence indicating that irregular organization of blood vessels within the tumor tissue that often leads to the development of hypoxic and/or necrotic regions and/or an immune-suppressive microenvironment inside the tumor mass may facilitate survival and growth of attenuated auxotrophic bacteria by providing them with nutrients and immune-protection (Forbes et al., 2003; Wouters et al., 2003; Yu et al., 2004). Moreover, niche-specific genes involved in the process of preferential tumor colonization after systemic bacteria delivery, were also identified (Silva-Valenzuela et al., 2014).

Different variants from the genera *Bifidobacterium*, *Clostridium*, *Lactococcus*, *Shigella*, *Vibrio*, *Listeria*, *Escherichia*, and *Salmonella* have been assayed in animal models of cancer (Yazawa et al., 2000; Cheadle and Jackson, 2002; Oelschlaeger, 2010; Patyar et al., 2010; Hoffman, 2012b). Obligate anaerobes such as *Bifidobacterium longum* and a *Clostridium novyi* strain devoid of its lethal toxin (*C. novyi*-NT) have shown preferential localization in low oxygenated necrotic areas of implanted tumors in mice after systemic administration, inducing tumor regression in some cases, although they were unable to grow in viable tumor tissue due to high oxygen tension, a fact that may have limited their efficacy as mono-therapy (Dang et al., 2001; Hoffman, 2012a). However, intra-tumor (i.t.) administration of *C. novyi*-NT has shown objective responses in canine tumors, which are more like those of humans because they are naturally occurring in animals with heterogeneous genetic backgrounds (Roberts et al., 2014). On the other hand, attenuated auxotrophic mutants of the facultative anaerobe *Salmonella enterica* serovar Typhimurium (*S.* Typhimurium) have been shown to invade and destroy a broad number of cancer cell types *in vitro*, as well as to replicate in hypoxic and oxic tumor regions *in vivo*, being the most efficient anti-tumor bacteria assayed in experimental models of cancer thus far (Pawelek et al., 1997; Leschner and Weiss, 2010; Nguyen et al., 2010; Hoffman, 2011, 2016b,c). Among them, *S.* Typhimurium VNP20009, attenuated by the lipid A (*msbB*) deletion and purine (*purI*) auxotrophic mutations, has shown anti-tumor efficacy in mice and swine and was safely administrated to patients with metastatic melanoma and renal carcinoma in a Phase I clinical trial; however, efficacy was not observed, perhaps due to over-attenuation (Toso et al., 2002).

A more tumor-virulent variant and less toxic against the host is *S.* Typhimurium A1-R (Zhao et al., 2006). Unlike VNP20009, the A1-R variant was obtained by successive passages from re-infected human tumor xenografts in nude mice treated with the *S.* Typhimurium A-1 auxotrophic (Leu- Arg-dependent) parental bacteria (Zhao et al., 2005). This selection procedure may account for A1-R's particular tumor-specificity and stronger anti-tumor activity (Zhao et al., 2006). A comparative study between VNP20009 and A1-R in nude mice showed that mice

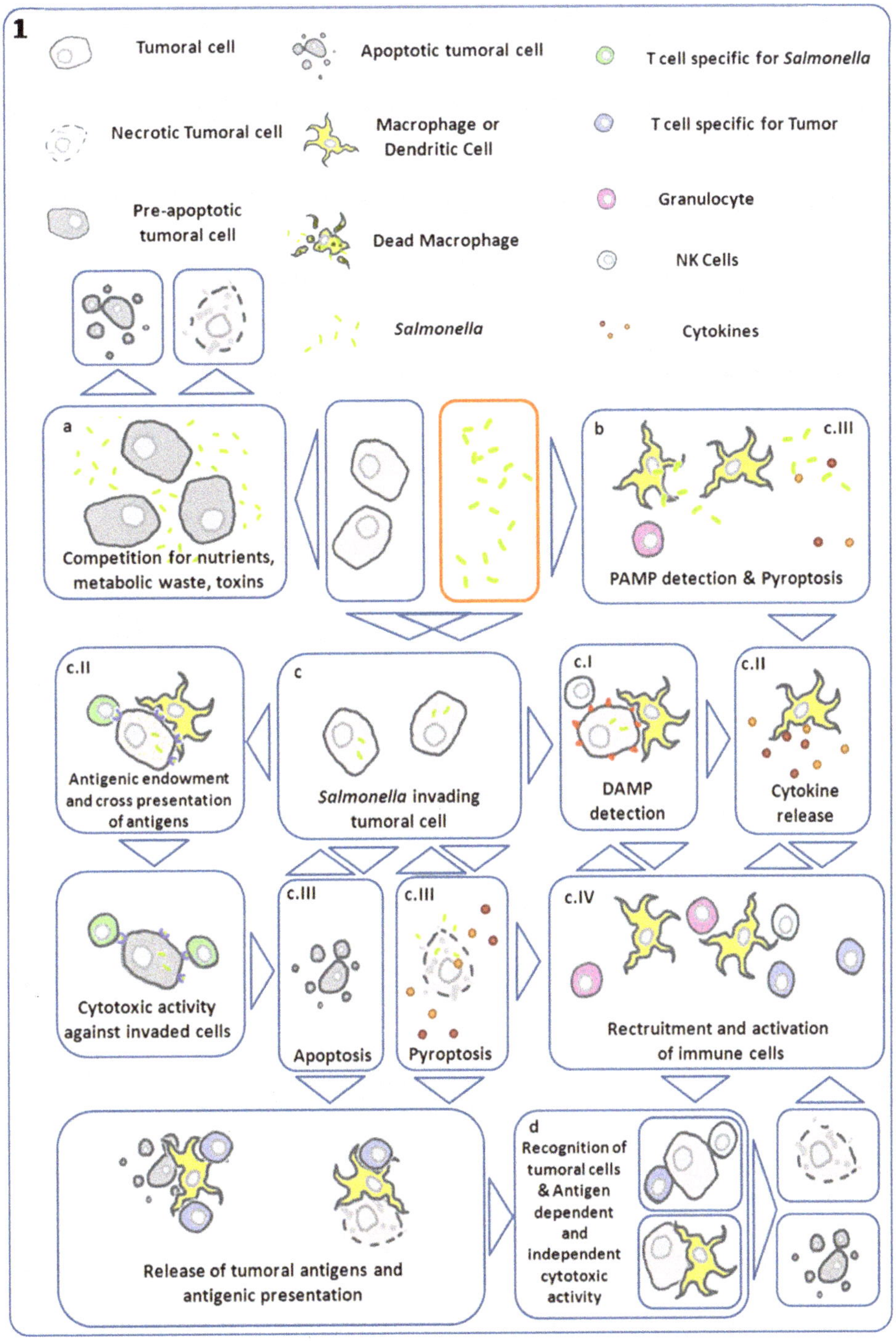

FIGURE 1 | Diagram showing main antitumor mechanisms induced by *S.* Typhimurium (*Salmonella*). Links are established between direct cytotoxicity induced by bacteria and indirect tumor cell death triggered by the immune system. (a) Bacterial infection within the tumor microenvironment results in inhibition of tumor growth and cell death. (b) Detection of bacterial pathogen-associated molecular patterns (PAMP) by immune cells, trigger cytokine release and recruitment of leukocytes capable of initiating anti-tumor immune responses (Patyar et al., 2010). (c) Using their Type III secretion system, *S.* Typhimurium can introduce bacterial factors in cancer cells allowing its internalization and intra-cellular replication (Avogadri et al., 2005; Knodler et al., 2010). (cI) Invasive *Salmonella* induces cell stress responses through danger-associated molecular patterns (DAMP), which are interpreted as damage signals by the immune system. (cII) Simultaneously, this same process can lead to cytokine expression and the transfer of antigens from the bacteria to the cancer cell, enabling the adaptive immune system to recognize and target the invaded cancer cell as infected and bearer of exogenous antigens (Avogadri et al., 2005). Gap junctions are concomitantly induced in the invaded cell and enable cross presentation of antigens to antigen presenting cells (Saccheri et al., 2010). Both processes can give rise to antigen-dependent elimination of infected cancer cells. (cIII) *Salmonella* can lead to the death of the infected cell, by inducing apoptosis or pyroptosis. The later is a programmed inflammatory cell death, characterized

(Continued)

FIGURE 1 | Continued
by activation of caspase 1, activation of the inflamosome, and IL-1B and IL-18 secretion, as well as cell rounding and detachment, cytoskeleton reorganization, nucleus deformation and rupture of the cell membrane, resulting in the release of inflammatory signals (Fink and Cookson, 2005, 2007; Knodler et al., 2010; Wang et al., 2013). This mechanism can result in cancer-cell death and immune-cell activation. Pyroptosis was first described in macrophages, which die quickly as a result of this process, and is of particular interest in cancer immunotherapy, as tumor-associated macrophages have been shown to have immune-suppressive proprieties. Reducing their number could be another component of the *S.* Typhimurium anti-tumor effect. Cancer cell death leads to tumor-antigen liberation, and the released bacteria can infect surrounding cancer cells. (cIV) In the process of pyroptosis, pro-inflammatory cytokines IL1-B and IL-18 can trigger recruitment and activation of immune cells (Knodler et al., 2010; Zhao et al., 2012; Wang et al., 2013). (d) Various mechanisms enhance and converge to enable tumor-antigen recognition and activation of cytotoxic responses both in an antigen-dependent and -independent manner. *S.* Typhimurium proteins injected into the cancer cell cytosol are subject to proteasomal degradation, resulting in bacterial peptides that can be presented through MHC I to cytotoxic lymphocytes (Avogadri et al., 2005; Saccheri et al., 2010).

tolerated *S.* Typhimurium A1-R to at a least twofold higher dose than VNP20009 when the bacteria were administered intravenously (i.v.). In addition, A1-R showed higher tumor targeting and inhibited the Lewis lung carcinoma to a greater extent than VNP20009, with less body weight loss (Zhang et al., 2015). In addition, *S.* Typhimurium A1-R mono-therapy has shown to be effective against primary and metastatic human prostate, breast, and pancreatic cancer as well as osteosarcoma, fibrosarcoma, and glioma in clinically-relevant mouse models (Hoffman, 2016c and references therein). Tumors with a high degree of vascularity were more sensitive to A1-R and vascular destruction appears to play a role in A1-R anti-tumor efficacy (Liu et al., 2010). In addition, A1-R was shown to induce stem-like and non-stem cancer-cell death *in vivo*, indicating that A1-R could be used to kill chemo-resistant cancer stem-like cells (Hiroshima et al., 2013). Together these results suggest that *S.* Typhimurium A1-R may have a greater clinical potential than VNP20009 (Zhang et al., 2015) and that not only the bacterial species, but also their genetic background needs to be taken into account when searching for improvements in bacteria-based therapies.

Salmonella Typhimurium defective in the synthesis of ppGpp (ΔppGpp: depletion of *relA* and *spoT*), showed 10^5 to 10^6-fold attenuation compared with WT strain (Na et al., 2006). This attenuated strain showed very high tumor targeting and stimulation of regional tumor immunity (Kim et al., 2015; Phan et al., 2015; Zheng et al., 2017).

In this regard, high-throughput screenings for *Salmonella* avirulent mutants can identify variants with reduced fitness in normal tissues but unchanged fitness in tumors for potential use as cancer therapeutics (Arrach et al., 2010). As an example, a reported genetically-engineered *S.* Typhimurium *aroA aroD* double mutant harboring the Flt3 Ligand, used to treat melanoma in mice resulted in 50% tumor regression (Yoon et al., 2007). However, *aroA* and *aroD* were later identified by Arrach et al. (2010) as Class 2 mutants which show reduced fitness in tumors compared to Class 1 mutants, increasing the probability that a different avirulent mutant that grows better in tumors might have resulted in a more complete anti-tumor response. In a competitive fitness assay in human prostate tumors growing in mice, Class 1 mutant *STM3120* not only had a fitness advantage over Class 2 mutants, but also effectively targeted tumors after intragastric delivery, suggesting an oral route as an option for bacterial cancer therapy (Arrach et al., 2010). The ability to screen thousands of candidates and evaluate individual mutants in parallel using high-throughput sequencing offers a clear advantage over conventional screening methods. Mutants that retain tumor-targeting while being poor colonizers of normal tissue, are desirable for cancer therapeutics.

The patient-derived xenograft (PDX) mouse models of cancer are emerging as an important component of personalized cancer therapy (Cho et al., 2016). PDX models are generated by implanting sectioned patient tumor fragments into immunodeficient mice, subcutaneously or orthotopically (into the organ or tissue of the cancer origin). Patient-derived orthotopic xenografts (PDOX) have the additional advantage that they usually metastasize as in the patient (Hiroshima et al., 2016). These models retain the histologic characteristics, heterogeneity of cancer cells and genomic signature of the patient tumor enabling the identification of effective individualized therapy (Cho et al., 2016). *S.* Typhimurium A1-R has shown to be effective against osteosarcoma in a PDX model (Murakami et al., 2017) and soft-tissue sarcoma, pancreatic cancer and melanoma in PDOX models (Hiroshima et al., 2014; Murakami et al., 2016; Yamamoto et al., 2016). Although these models need to be immunocompromised in order to allow human tumor engraftments and therefore do not allow evaluation of the immune-mediated bacterial activity, we believe that studies that employ PDOX models would allow the selection of the best-suited bacteria for individual tumors and prediction for its effectiveness in patients. "Humanized" PDOX models (Zitvogel et al., 2016) will be used to determine tumor-immunology effects of bacteria.

Figures 1, 2 show some of the complex net of events that are involved in promoting bacterial anti-tumor efficacy. However, in most models bacteria mono-therapies are not sufficient to eliminate a primary tumor or the metastatic burden. Combined therapies including chemotherapy (Dang et al., 2001; Yamamoto et al., 2016; Yano et al., 2016), radiotherapy (Jiang et al., 2010), traditional herbal medicine (Zhang et al., 2013), anti-angiogenic and/or immunotherapy (Binder et al., 2013; Kramer et al., 2015) or the use of bacteria carrying plasmids coding for anti-tumor genes (reviewed in Moreno et al., 2010; Nguyen and Min, 2017) have shown enhanced results. Based on the use of eukaryotic gene-expression systems it has been suggested that bacteria can act as vector systems for plasmid transfer to mammalian cancer cells, a process known as "bactofection" (Weiss and Chakraborty, 2001; Baban et al., 2010; Othman et al., 2013). However, this trans-kingdom gene delivery assumption is still a matter of controversy (Gahan et al., 2009 and **Figure 2A**). Therefore, for the best performance of a bacteria + plasmid combination, the determination of

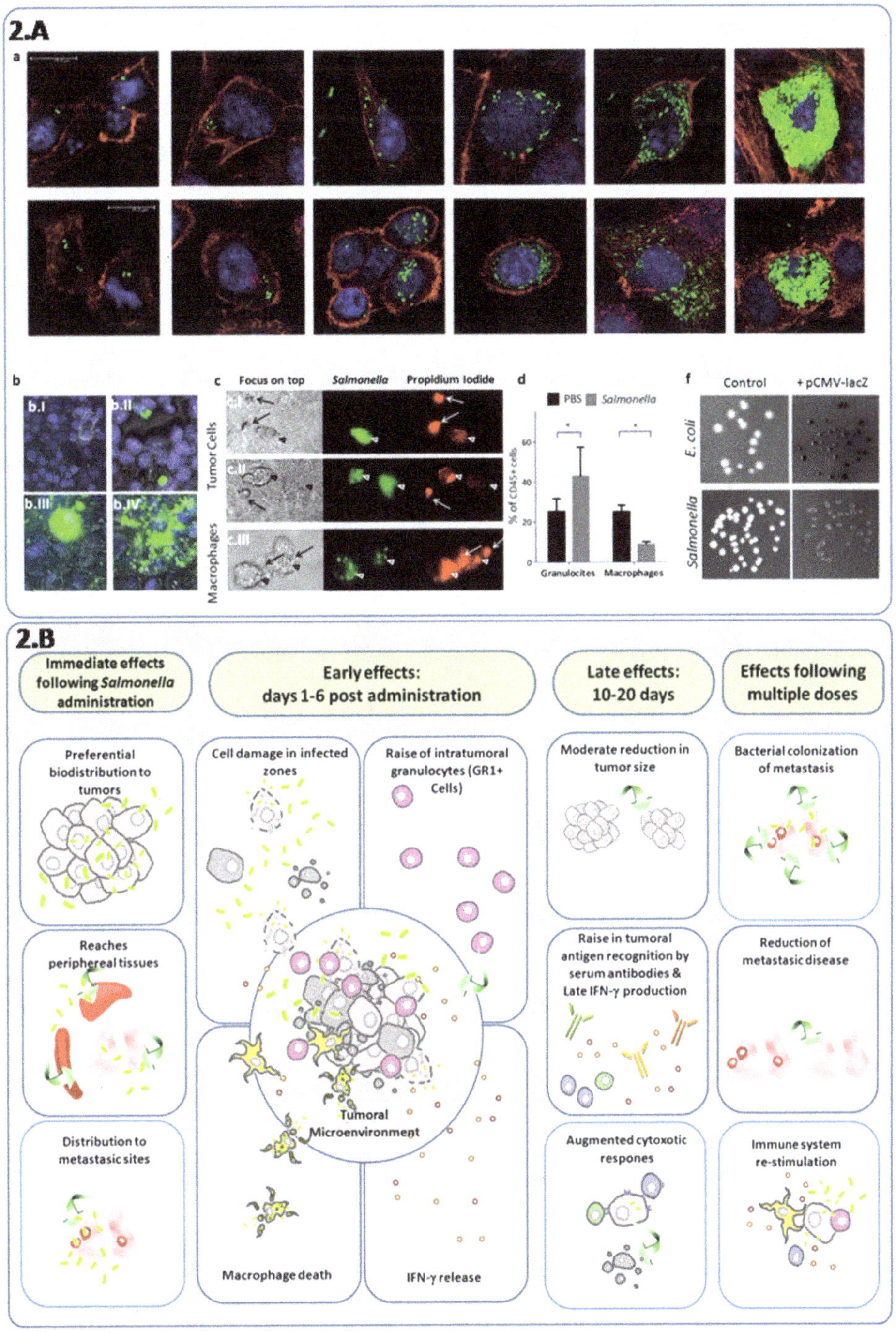

FIGURE 2 | Direct and synergystic anti-tumor effects of attenuated *S.* Typhimurium integrating cellular and systemic immune responses. **(A)** Induction of cell death and granulocyte recruitment associated with intracellular replication of attenuated *S.* Typhimurium LVR01 (*Salmonella*), which previously showed a modest antitumor effect in the 4T1 metastatic breast cancer model (Kramer et al., 2015). (a) Confocal microscopy indicates bacteria invasion and replication in breast cancer cell lines: 4T1 (ATCC-CRL2539) (upper line) and NMU (ATCC-CRL1743) (lower line) in a time-course experiment. Cell cultures were grown in glass coverslips, infected with *Salmonella* expressing the GFP gene and sampled at 2, 12, 24, or 48 h to follow progression of intracellular replication. Specimens were fixed in paraformaldehyde 4%, washed in PBS and stained with DAPI and Phalloidin-Alexa555 (Invitrogen™). After the staining, the coverslips were washed with PBS, mounted using Pro Long Gold (Invitrogen™) and sealed with nail polish. This three color fluorescence pattern allowed the 3D analysis of the infected cultures, by simultaneously visualizing the bacteria, the nucleus and the F-actin cytoskeleton. Intracellular/extracellular determination of bacteria was possible due to the delimited borders of the actin cytoskeleton which are close to the cell membrane. Images were obtained with a LEICA®TCS SP5 II spectral confocal microscope and processed with the software Leica®LAS AF. As observed, bacterial invasion progresses, showing intracellular cytoplasmic hyperreplication over time. (b) Epifluorescence microscopy of 4T1 monolayers infected cells. Cancer cells were infected with *Salmonella*-GFP for 2 h and observed at different time points. Specimens were washed in PBS, fixed

(Continued)

FIGURE 2 | Continued
in paraformaldehyde 4%, and stained with DAPI (Invitrogen^{TM}). After 5 min staining, invaded cultures were washed and observed in a Nikon®Ti-S epifluorescence inverted microscope. At 2 h few peri-nuclear bacteria could be seen (b.I) At 24 h (b.II) bacteria replicated in the cytoplasm and some infected cells appear rounded and extruded. At 48 h (b.III) densely-infected cells were similar, and eventually burst and release their cellular contents (b.IV). (c) Live infected cultures were observed either intact or in the presence of propidium iodide (500 nM) to assess intracellular bacterial mobility and cell viability, respectively. Monolayers of mammary cancer cells: 4T1 (c.I) and NMU (c.II), as well as macrophage cells J774.A (c.III) were infected with *Salmonella*-GFP. At 24 h post-infection, infected cells (green) die as indicated by propidium iodide staining (red). Macrophages died at earlier time points (2–16 h). Arrows point the extruded cells. (d) Flow cytometry of intratumor immune cells at 6 days after *Salmonella* inoculation of 4T1 tumors *in vivo*. As observed, the intra-tumor granulocyte/myeloid-derived-suppressor cell (Ly6G+CD11b+) levels increase and macrophage (F4/80+CD11b+) levels decreased after bacteria administration among total leukocytes (CD45+ cells). (f) X-gal agar plates were used to seed the untransformed bacteria (control) or bacteria transformed with a plasmid containing the *lacZ* gene under the control of the eukaryotic cytomegalovirus (CMV) promoter (pCMV-lacZ). As observed, the *lacZ* gene product β-galactosidase was detected, indicating that the CMV promoter was active in prokaryotic cell species. **(B)** *In vivo* effects of attenuated *S.* Typhimurium (*Salmonella*) in mice bearing metastatic cancer. This integrative diagram shows the anti-tumor effects of attenuated variants of *Salmonella* evaluated as mono-therapy. The bacteria inoculation by different routes (systemic or intratumoral) results in its biodistribution to most organs, but with a marked preference for tumors, including metastasic sites (Pawelek et al., 1997; Low et al., 1999; Forbes et al., 2003; Yu et al., 2004; Hoffman, 2016a). In tumors, bacterial infection is associated with tumor-tissue architecture deterioration, a rise in granulocytic cells and INF-γ induction and a decrease of intra-tumor macrophages (Avogadri et al., 2005; Westphal et al., 2008; Zheng et al., 2017). Late effects (10–20 days after bacteria administration) are characterized by a moderate decrease in tumor size, adaptive immune responses including INF-γ production, antibody recognition of tumor antigens, and cytotoxic immune activities (Avogadri et al., 2005; Kramer et al., 2015; Masner et al., unpublished results). Repeated administration of attenuated bacteria could result in a better targeting of metastases (Zhao et al., 2012), while stimulating immune responses that enhance cancer-cell elimination.

the actual location of transgene expression would allow the right selection of the gene, promoter, and secretion system (if required) to achieve optimized therapy (Forbes, 2010; Zheng et al., 2017). In addition, since the bacteria usually induce death of infected cells within few hours, the rational to use bacteria as a gene delivery system (vector) to immune and/or tumor cells needs to be re-evaluated if medium- or long-term persistence of therapeutic gene expression is necessary *in vivo*.

In terms of combined therapies, a remarkable example of a neoadjuvant (pre-operatory) synergistic efficacy was observed using *S.* Typhimurium *aro C* mutant LVR01 in combination with interleukin 12 (IL-12) expressed from the alfaviral eukaryotic gene vector SFV-IL-12 (Kramer et al., 2015). This approach was evaluated in an immunocompetent mouse model of locally-advanced breast cancer and resulted in a highly effective anti-metastasic therapy, leading to 90% disease free mice, while either mono-therapy was not effective. Moreover, the efficacy of this combined therapy depended on the order in which both agents were administered (Kramer et al., 2015). An initial anti-angiogenic effect associated with a T helper-cell-1-primed response that was timely induced seemed to account for the main global effect. However, the underlying mechanisms of this combination and timing of both factors raised various questions that remain un-answered.

Other relevant questions to be answered for bacteria-based cancer therapy optimization are related to the dose, schedule, and route of administration. A dose-dependent effect of attenuated *S.* Typhimurium was observed, as well as multiple dosing are more efficient than mono-doses (Hayashi et al., 2009; Nagakura et al., 2009; Grille et al., 2014), although the range needs to be determine to avoid toxicity (Zhao et al., 2012). The efficacy and safety of three different routes of *S.* Typhimurium A1-R administration: oral (p.o.), i.v. and i.t. was compared in nude mice with orthotopic human breast cancer indicated that the p.o. route was safer, and the i.v. route was more effective (Zhang et al., 2012). However, such experiments may need to be performed for each type of tumor, since it was also shown in a model of disseminated human ovarian cancer treated with i.v. and intraperitoneal (i.p.) *S.* Typhimurium A1-R, that i.p. treatment was less toxic than i.v. administration (Matsumoto et al., 2015).

Although useful in many approaches, human xenografted tumors into immunodeficient mice limit our knowledge about the range of effects that certain bacterial strains can exert. In this regard, studies in immunocompentent animals are more representative of the complex spectrum of interactions between the bacteria and the tumor microenvironment, thereby enabling immune effects that are otherwise absent in immunocompromised mice. This could be crucial for "tunning" the bacteria to the right degree of immunogenicity/attenuation, avoiding shock while promoting adjuvant effects (Yu et al., 2004). Moreover, toxicity issues regarding immunotherapies are a main concern today. From acute shock to autoimmune diseases, we could gain a better understanding of the risk of side effects of bacteria therapy of cancer from preclinical models that include all the functional branches of the immune system. Undesirable attenuated bacterial infection can be in theory treated with antibiotics; however, long-term clinical trials in humans are required to evaluate toxicity in detail, since the chance of septic shock and/or tumor lysis syndrome could be a fact. In addition, we believe that, there is still a considerable need of work to evaluate bacteria for natural acquisition of antibiotic-resistant genes and/or reversion of attenuation mutations, as well as comparing the anti-tumor efficacy and secondary effects of bacteria or bacterial products *versus* conventional therapies. Moreover, we cannot rule out the possible clearance of bacteria by the immune system before reaching the tumor site in a patient-dependant manner, resulting in treatment failure.

THE FUTURE OF WHAT Dr. Coley BEGAN

Each of the 16 "Coley's toxins" that have been used might have a complex and variable composition, including components of the culture media, products released by the bacteria in the medium, components relevant by bacteria lysis (and autolysis). The inactivation method used to prepare the vaccine and the inclusion, or not, of a filtration step in the preparation of the toxins will affect the final products. The i.v. administration of a suspension of inactivated bacteria cells may mimic a nanodrug, and the number of particles, their size, shape, charge, and surface molecules may affect the immune system response (van Riet et al., 2014).

Both *Streptococcus pyogenes* and *Serratia marcescens* produce exotoxins. *S. pyogenes* produces the pyrogenic exotoxins SpeA, SpeB, and SpeC which have the capacity to unspecifically stimulate CD4+ lymphocytes, leading to a strong secretion of different cytokines (Babbar, 2015). *S. marcescens*, produces prodigiosin, a low-molecular weight red pigmented heterocyclic tripyrrolic toxin with anti-tumor activity (Elahian et al., 2013). The toxins, together with other components of the formulation, result in generation of fever and potential anti-tumor response. The administration route may also influence the efficacy of Coley's toxins including i.v., i.p., direct injection in the tumor, or subcutaneous or intramuscular administration (Nauts et al., 1946).

A chemical description of "Coley's toxins" can be assessed using the analytical tools currently used for proteomic and metabolomic studies (Wishar, 2016). Nuclear magnetic resonance (NMR) and mass spectrometry (MS) methods for the analysis of high- and low-molecular weight components of complex mixtures or their derivatives (Alonso et al., 2015) could be used in combination with multivariate analysis to identify the components responsible for anti-tumor activity. The identification of the active components and their mode of action, would allow the selection of more active and better-defined vaccines, as well as the design of tailored formulations capable of producing the right amount of systemic or tumor-localized fever (Noe, 2016) for optimal stimulation of the host immune system and cytokine secretion to achieve best anti-tumor efficacy.

AUTHOR CONTRIBUTIONS

MGK received the invitation to contribute in this special issue, designed and wrote most of the sections and supervised the artwork. MM made all the drawings, acquired and analyzed the data and participated in the text writing. FAF wrote the final section and was essential in motivating the team work. RMH inspired the main ideas of this perspective article, supplied relevant literature and critically revised the manuscript.

ACKNOWLEDGMENTS

The authors thank Dr. Fernando Gonzalez (Department of Biophysics, UdelaR) for assistance with confocal microscopy and image processing; Dr. Patricia Berasain (Parasitology Unit, UdelaR) for help with informatics; Dr. Teresa Freire (Department of Inmunobiology, UdelaR) for NMU cells; Dr. Lucia Veiga and Dr. José A. Chabalgoity (Department of Biotechnology, UdelaR) for LVR01-GFP; Dr. Helen Bauer (German Research Centre for Biotechnology, Germany) for pCMV-lacZ and Rodrigo Gonzalez (undergraduate student supervised by MGK) for X-gal staining of transformed bacteria.

REFERENCES

Adkins, I., Holubova, J., Kosova, M., and Sadilkova, L. (2012). Bacteria and their toxins tamed for immunotherapy. *Curr. Pharm. Biotechnol.* 13, 1446–1473. doi: 10.2174/138920112800784835

Alonso, A., Marsal, S., and Julià, A. (2015). Analytical methods in untargeted metabolomics: state of the art in 2015. *Front. Bioeng. Biotechnol.* 3:23. doi: 10.3389/fbioe.2015.00023

Arrach, N., Cheng, P., Zhao, M., Santiviago, C. A., Hoffman, R. M., and McClelland, M. (2010). High-throughput screening for *Salmonella* avirulent mutants that retain targeting of solid tumors. *Cancer Res.* 70, 2165–2170. doi: 10.1158/0008-5472.CAN-09-4005

Avogadri, F., Martinoli, C., Petrovska, L., Chiodoni, C., Transidico, P., Bronte, V., et al. (2005). Cancer immunotherapy based on killing of *Salmonella*-infected tumor cells. *Cancer Res.* 65, 3920–3927. doi: 10.1158/0008-5472.CAN-04-3002

Baban, C. K., Cronin, M., O'Hanlon, D., O'Sullivan, G. C., and Tangney, M. (2010). Bacteria as vectors for gene therapy of cancer. *Bioeng. Bugs* 1, 385–394. doi: 10.4161/bbug.1.6.13146

Babbar, A. (2015). *Streptococcal Superantigens, SpringerBriefs in Microbiology.* Cham: Springer, 1–41. doi: 10.1007/978-3-319-22455-8_1

Binder, D. C., Engels, B., Arina, A., Yu, P., Slauch, J. M., Fu, Y.-X., et al. (2013). Antigen-specific bacterial vaccine combined with anti-PD-L1 rescues dysfunctional endogenous T cells to reject long-established cancer. *Cancer Immunol. Res.* 1, 123–133. doi: 10.1158/2326-6066.CIR-13-0058

Cheadle, E. J., and Jackson, A. M. (2002). Bugs as drugs for cancer. *Immunology* 107, 10–19. doi: 10.1046/j.1365-2567.2002.01498.x

Cho, S.-Y., Kang, W., Han, J. Y., Min, S., Kang, J., Lee, A., et al. (2016). An integrative approach to precision cancer medicine using patient-derived xenografts. *Mol. Cells* 39, 77–86. doi: 10.14348/molcells.2016.2350

Dang, L. H., Bettegowda, C., Huso, D. L., Kinzler, K. W., and Vogelstein, B. (2001). Combination bacteriolytic therapy for the treatment of experimental tumors. *Proc. Natl. Acad. Sci. U.S.A.* 98, 15155–15160. doi: 10.1073/pnas.251543698

Elahian, F., Moghimi, B., Dinmohammadi, F., Ghamghami, M., Hamidi, M., and Mirzaei, S. A. (2013). The anticancer agent prodigiosin is not a multidrug resistance protein substrate. *DNA Cell Biol.* 32, 90–97. doi: 10.1089/dna.2012.1902

Fink, S. L., and Cookson, B. T. (2005). Apoptosis, pyroptosis, and necrosis: mechanistic description of dead and dying eukaryotic cells. *Infect. Immun.* 73, 1907–1916. doi: 10.1128/IAI.73.4.1907

Fink, S. L., and Cookson, B. T. (2007). Pyroptosis and host cell death responses during *Salmonella* infection. *Cell. Microbiol.* 9, 2562–2570. doi: 10.1111/j.1462-5822.2007.01036.x

Forbes, N. S. (2010). Engineering the perfect (bacterial) cancer therapy. *Nat. Rev. Cancer* 10, 785–794. doi: 10.1038/nrc2934

Forbes, N. S., Munn, L. L., Fukumura, D., and Jain, R. K. (2003). Sparse initial entrapment of systemically injected *Salmonella typhimurium* leads to heterogeneous accumulation within tumors. *Cancer Res.* 63, 5188–5193.

Gahan, M. E., Webster, D. E., Wesselingh, S. L., Strugnell, R. A., and Yang, J. (2009). Bacterial antigen expression is an important component in inducing an immune response to orally administered *Salmonella*-delivered DNA vaccines. *PLOS ONE* 4:e6062. doi: 10.1371/journal.pone.0006062

Gontero, P., Bohle, A., Malmstrom, P.-U., O'Donnell, M. A., Oderda, M., Sylvester, R., et al. (2010). The role of bacillus Calmette-Guérin in the treatment of non-muscle-invasive bladder cancer. *Eur. Urol.* 57, 410–429. doi: 10.1016/j.eururo.2009.11.023

Grille, S., Moreno, M., Bascuas, T., Marqués, J. M., Muñoz, N., Lens, D., et al. (2014). *Salmonella enterica* serovar *Typhimurium* immunotherapy for B-cell lymphoma induces broad anti-tumour immunity with therapeutic effect. *Immunology* 143, 428–437. doi: 10.1111/imm.12320

Hayashi, K., Zhao, M., Yamauchi, K., Yamamoto, N., Tsuchiya, H., Tomita, K., et al. (2009). Systemic targeting of primary bone tumor and lung metastasis of high-grade osteosarcoma in nude mice with a tumor-selective strain of *Salmonella typhimurium*. *Cell Cycle* 8, 870–875. doi: 10.4161/cc.8.6.7891

Hiroshima, Y., Maawy, A., Zhang, Y., Zhang, N., Murakami, T., Chishima, T., et al. (2016). Patient-derived mouse models of cancer need to be orthotopic in order to evaluate targeted anti-metastatic therapy. *Oncotarget* 7, 71696–71702. doi: 10.18632/oncotarget.12322

Hiroshima, Y., Zhao, M., Maawy, A., Zhang, Y., Katz, M. H. G., Fleming, J. B., et al. (2014). Efficacy of *Salmonella typhimurium* A1-R versus chemotherapy on a pancreatic cancer patient-derived orthotopic xenograft (PDOX). *J. Cell. Biochem.* 115, 1254–1261. doi: 10.1002/jcb.24769

Hiroshima, Y., Zhao, M., Zhang, Y., Maawy, A., Hassanein, M. K., Uehara, F., et al. (2013). Comparison of efficacy of *Salmonella typhimurium* A1-R and chemotherapy on stem-like and non-stem human pancreatic cancer cells. *Cell Cycle* 12, 2774–2780. doi: 10.4161/cc.25872

Hoffman, R. M. (2011). Tumor-seeking Salmonella amino acid auxotrophs. *Curr. Opin. Biotechnol.* 22, 917–923. doi: 10.1016/j.copbio.2011.03.009

Hoffman, R. M. (2012a). Bugging tumors. *Cancer Discov.* 2, 588–590. doi: 10.1158/2159-8290.CD-12-0227

Hoffman, R. M. (2012b). The preclinical discovery of bacterial therapy for the treatment of metastatic cancer with unique advantages. *Expert Opin. Drug Discov.* 7, 73–83. doi: 10.1517/17460441.2012.644534

Hoffman, R. M. (2015). "Back to the future: Are tumor-targeting bacteria the next-generation Cancer therapy?," in *Gene Therapy of Solid Cancers. Methods in Molecular Biology*, Vol. 1317, eds W. Walther and U. Stein (New York, NY: Humana Press), 239–260. doi: 10.1007/978-1-4939-2727-2_14

Hoffman, R. M. (ed.) (2016a). "Preface: a brief history of bacterial therapy of cancer," in *Methods in Molecular Biology*, (New York, NY: Humana Press), doi: 10.1007/978-1-4939-3515-4

Hoffman, R. M. (ed.) (2016b). "Tumor-Targeting *Salmonella typhimurium* A1-R: an Overview," in *Methods in Molecular Biology*, (New York, NY: Humana Press), 1–8. doi: 10.1007/978-1-4939-3515-4_1

Hoffman, R. M. (ed.) (2016c). "Future of bacterial therapy of cancer," in *Methods in Molecular Biology*, (New York, NY: Humana Press), 177–184. doi: 10.1007/978-1-4939-3515-4_15

Hoption Cann, S. A., van Netten, J. P., and van Netten, C. (2003). Dr. William Coley and tumour regression: a place in history or in the future. *Postgrad. Med. J.* 79, 672–680.

Jiang, S. N., Phan, T. X., Nam, T. K., Nguyen, V. H., Kim, H. S., Bom, H. S., et al. (2010). Inhibition of tumor growth and metastasis by a combination of *Escherichia coli*-mediated cytolytic therapy and radiotherapy. *Mol. Ther.* 18, 635–642. doi: 10.1038/mt.2009.295

Kim, J. C., and Steinberg, G. (2001). The limits of bacillus Calmette-Guerin for carcinoma *in situ* of the bladder. *J. Urol.* 165, 745–756. doi: 10.1016/S0022-5347(05)66518-4

Kim, J. E., Phan, T. X., Nguyen, V. H., Dinh-Vu, H. V., Zheng, J. H., Yun, M., et al. (2015). *Salmonella typhimurium* suppresses tumor growth via the pro-inflammatory cytokine interleukin-1beta. *Theranostics* 5, 1328–1342. doi: 10.7150/thno.11432

Knodler, L. A., Vallance, B. A., Celli, J., Winfree, S., Hansen, B., Montero, M., et al. (2010). Dissemination of invasive *Salmonella* via bacterial-induced extrusion of mucosal epithelia. *Proc. Natl. Acad. Sci. U.S.A.* 107, 17733–17738. doi: 10.1073/pnas.1006098107

Kramer, M. G., Masner, M., Casales, E., Moreno, M., Smerdou, C., and Chabalgoity, J. A. (2015). Neoadjuvant administration of Semliki Forest virus expressing interleukin-12 combined with attenuated *Salmonella* eradicates breast cancer metastasis and achieves long-term survival in immunocompetent mice. *BMC Cancer* 15:620. doi: 10.1186/s12885-015-1618-x

Leschner, S., and Weiss, S. (2010). *Salmonella*-allies in the fight against cancer. *J. Mol. Med.* 88, 763–773. doi: 10.1007/s00109-010-0636-z

Liu, F., Zhang, L., Hoffman, R. M., and Zhao, M. (2010). Vessel destruction by tumor-targeting *Salmonella typhimurium* A1-R is enhanced by high tumor vascularity. *Cell Cycle* 9, 4518–4524. doi: 10.4161/cc.9.22.13744

Low, K. B., Ittensohn, M., Le, T., Platt, J., Sodi, S., Amoss, M., et al. (1999). Lipid A mutant *Salmonella* with suppressed virulence and TNFalpha induction retain tumor-targeting in vivo. *Nat. Biotechnol.* 17, 37–41. doi: 10.1038/5205

Matsumoto, Y., Miwa, S., Zhang, Y., Zhao, M., Yano, S., Uehara, F., et al. (2015). Intraperitoneal administration of tumor-targeting *Salmonella typhimurium* A1-R inhibits disseminated human ovarian cancer and extends survival in nude mice. *Oncotarget* 6, 11369–11377. doi: 10.18632/oncotarget.3607

McCarthy, E. F. (2006). The toxins of William B. Coley and the treatment of bone and soft-tissue sarcomas. *Iowa Orthop. J.* 26, 154–158.

Min, J. J., Kim, H. J., Park, J. H., Moon, S. M., Jeong, J. H., Hong, Y. J., et al. (2008a). Noninvasive real-time imaging of tumors and metastases using tumor-targeting light-emitting *Escherichia coli*. *Mol. Imaging Biol.* 10, 54–61.

Min, J. J., Nguyen, V. H., Kim, H. J., Hong, Y., and Choy, H. E. (2008b). Quantitative bioluminescence imaging of tumor-targeting bacteria in living animals. *Nat. Protoc.* 3, 629–636. doi: 10.1038/nprot.2008.32

Moreno, M., Kramer, M. G., Yim, L., and Chabalgoity, J. A. (2010). Salmonella as live trojan horse for vaccine development and cancer gene therapy. *Curr. Gene Ther.* 10, 56–76. doi: 10.2174/156652310790945566

Murakami, T., DeLong, J., Eilber, F. C., Zhao, M., Zhang, Y., Zhang, N., et al. (2016). Tumor-targeting *Salmonella typhimurium* A1-R in combination with doxorubicin eradicate soft tissue sarcoma in a patient-derived orthotopic xenograft (PDOX) model. *Oncotarget* 7, 12783–12790. doi: 10.18632/oncotarget.7226

Murakami, T., Igarashi, K., Kawaguchi, K., Kiyuna, T., Zhang, Y., Zhao, M., et al. (2017). Tumor-targeting *Salmonella typhimurium* A1-R regresses an osteosarcoma in a patient-derived xenograft model resistant to a molecular-targeting drug. *Oncotarget* 8, 8035–8042. doi: 10.18632/oncotarget.14040

Na, H. S., Kim, H. J., Lee, H. C., Hong, Y., Rhee, J. H., and Choy, H. E. (2006). Immune response induced by *Salmonella typhimurium* defective in ppGpp synthesis. *Vaccine* 24, 2027–2034. doi: 10.1016/j.vaccine.2005.11.031

Nagakura, C., Hayashi, K., Zhao, M., Yamauchi, K., Yamamoto, N., Tsuchiya, H., et al. (2009). Efficacy of a genetically-modified *Salmonella typhimurium* in an orthotopic human pancreatic cancer in nude mice. *Anticancer. Res.* 29, 1873–1878.

Nauts, H. C. (2004). *Coley's Toxins- the First Century*. Available at: https://www.highbeam.com/doc/1G1-117854757.html

Nauts, H. C., Swift, W. E., and Coley, B. L. (1946). The treatment of malignant tumors by bacterial toxins as developed by the late William B. Coley, M.D., reviewed in the light of modern research. *Cancer Res.* 6, 205–216.

Nguyen, V. H., Kim, H. S., Ha, J. M., Hong, Y., Choy, H. E., and Min, J. J. (2010). Genetically engineered *Salmonella typhimurium* as an imageable therapeutic probe for cancer. *Cancer Res.* 70, 18–23. doi: 10.1158/0008-5472.CAN-09-3453

Nguyen, V. H., and Min, J. J. (2017). *Salmonella*-mediated cancer therapy: roles and potential. *Nucl. Med. Mol. Imaging* 51, 118–126. doi: 10.1007/s13139-016-0415-z

Noe, G. (2016). A hybrid fever therapy for increased tumor selectivity. *Med. Hypotheses* 89, 63–64. doi: 10.1016/j.mehy.2016.02.005

Oelschlaeger, T. A. (2010). Bacteria as tumor therapeutics? *Bioeng. Bugs* 1, 146–147. doi: 10.4161/bbug.1.2.11248

Othman, S., Roe, A. J., Parton, R., and Coote, J. G. (2013). Use of a dual reporter plasmid to demonstrate Bactofection with an attenuated $AroA^-$ derivative of *Pasteurella multocida* B:2. *PLOS ONE* 8:e71524. doi: 10.1371/journal.pone.0071524

Patyar, S., Joshi, R., Byrav, D. S. P., Prakash, A., Medhi, B., and Das, B. K. (2010). Bacteria in cancer therapy: a novel experimental strategy. *J. Biomed. Sci.* 17, 21. doi: 10.1186/1423-0127-17-21

Pawelek, J. M., Low, K. B., and Bermudes, D. (1997). Tumor-targeted *Salmonella* as a novel anticancer vector. *Cancer Res.* 57, 4537–4544.

Payette, P. J., and Davis, H. L. (2001). History of vaccines and positioning of current trends. *Curr. Drug Targets Infect. Disord.* 1, 241–247. doi: 10.2174/1568005014606017

Phan, T. X., Nguyen, V. H., Duong, M. T., Hong, Y., Choy, H. E., and Min, J. J. (2015). Activation of inflammasome by attenuated *Salmonella typhimurium* in bacteria-mediated cancer therapy. *Microbiol. Immunol.* 59, 664–675. doi: 10.1111/1348-0421.12333

Roberts, N. J., Zhang, L., Janku, F., Collins, A., Bai, R.-Y., Staedtke, V., et al. (2014). Intratumoral injection of *Clostridium novyi*-NT spores induces antitumor responses. *Sci. Transl. Med* 6, 249ra111. doi: 10.1126/scitranslmed.3008982

Saccheri, F., Pozzi, C., Avogadri, F., Barozzi, S., Faretta, M., Fusi, P., et al. (2010). Bacteria-induced gap junctions in tumors favor antigen cross-presentation and antitumor immunity. *Sci. Transl. Med.* 2:44ra57. doi: 10.1126/scitranslmed.3000739

Silva-Valenzuela, C. A., Desai, P. T., Molina-Quiroz, R. C., Pezoa, D., Zhang, Y., Porwollik, S., et al. (2014). Solid tumors provide niche-specific conditions that lead to preferential growth of *Salmonella*. *Oncotarget* 7, 35169–35180. doi: 10.18632/oncotarget.9071

Toso, J. F., Gill, V. J., Hwu, P., Marincola, F. M., Restifo, N. P., Schwartzentruber, D. J., et al. (2002). Phase I study of the intravenous administration of attenuated *Salmonella typhimurium* to patients with metastatic melanoma. *J. Clin. Oncol.* 20, 142–152. doi: 10.1200/JCO.2002.20.1.142

Uchugonova, A., Zhao, M., Zhang, Y., Weinigel, M., König, K., and Hoffman, R. M. (2012). Cancer-cell killing by engineered *Salmonella* imaged by multiphoton tomography in live mice. *Anticancer. Res.* 32, 4331–4337.

van Riet, E., Ainai, A., Suzuki, T., Kersten, G., and Hasegawa, H. (2014). Combatting infectious diseases; nanotechnology as a platform for rational vaccine design. *Adv. Drug Deliv. Rev.* 74, 28–34. doi: 10.1016/j.addr.2014.05.011

Wang, Q., Imamura, R., Motani, K., Kushiyama, H., Nagata, S., and Suda, T. (2013). Pyroptotic cells externalize eat-me and release find-me signals and are efficiently engulfed by macrophages. *Int. Immunol.* 25, 363–372. doi: 10.1093/intimm/dxs161

Weiss, S., and Chakraborty, T. (2001). Transfer of eukaryotic expression plasmids to mammalian host cells by bacterial carriers. *Curr. Opin. Biotechnol.* 12, 467–472. doi: 10.1016/S0958-1669(00)00247-0

Westphal, K., Leschner, S., Jablonska, J., Loessner, H., and Weiss, S. (2008). Containment of tumor-colonizing bacteria by host neutrophils. *Cancer Res.* 68, 2952–2960. doi: 10.1158/0008-5472.CAN-07-2984

Wishar, D. S. (2016). Emerging applications of metabolomics in drug discovery and precision medicine. *Nat. Rev. Drug Discov.* 15, 473–484. doi: 10.1038/nrd.2016.32

Wouters, B. G., Koritzinsky, M., Chiu, R. K., Theys, J., Buijsen, J., and Lambin, P. (2003). Modulation of cell death in the tumor microenvironment. *Semin. Radiat. Oncol.* 13, 31–41. doi: 10.1053/srao.2003.50004

Yamamoto, M., Zhao, M., Hiroshima, Y., Zhang, Y., Shurell, E., Eilber, F. C., et al. (2016). Efficacy of tumor-targeting *Salmonella* A1-R on a melanoma patient-derived orthotopic xenograft (PDOX) nude-mouse model. *PLOS ONE* 11:e0160882. doi: 10.1371/journal.pone.0160882

Yano, S., Takehara, K., Zhao, M., Tan, Y., Han, Q., Li, S., et al. (2016). Tumor-specific cell-cycle decoy by *Salmonella typhimurium* A1-R combined with tumor-selective cell-cycle trap by methioninase overcome tumor intrinsic chemoresistance as visualized by FUCCI imaging. *Cell Cycle* 15, 1715–1723. doi: 10.1080/15384101.2016.1181240

Yazawa, K., Fujimori, M., Amano, J., Kano, Y., and Taniguchi, S. (2000). Bifidobacterium longum as a delivery system for cancer gene therapy: selective localization and growth in hypoxic tumors. *Cancer Gene Ther.* 7, 269–274. doi: 10.1038/sj.cgt.7700122

Yoon, W. S., Choi, W. C., Sin, J.-I., and Park, Y. K. (2007). Anti-tumor therapeutic effects of *Salmonella typhimurium* containing Flt_3 Ligand expression plasmids in melanoma-bearing mouse. *Biotechnol. Lett.* 29, 511–516. doi: 10.1007/s10529-006-9270-9

Yu, Y. A., Shabahang, S., Timiryasova, T. M., Zhang, Q., Beltz, R., Gentschev, I., et al. (2004). Visualization of tumors and metastases in live animals with bacteria and vaccinia virus encoding light-emitting proteins. *Nat. Biotechnol.* 22, 313–320. doi: 10.1038/nbt937

Zhang, Y., Tome, Y., Suetsugu, A., Zhang, L., Zhang, N., Hoffman, R. M., et al. (2012). Determination of the optimal route of administration of *Salmonella typhimurium* A1-R to target breast cancer in nude mice. *Anticancer.Res.* 32, 2501–2508.

Zhang, Y., Zhang, N., Su, S., Hoffman, R. M., and Zhao, M. (2013). *Salmonella typhimurium* A1-R tumor targeting in immunocompetent mice is enhanced by a traditional Chinese medicine herbal mixture. *Anticancer.Res.* 33, 1837–1843.

Zhang, Y., Zhang, N., Zhao, M., and Hoffman, R. M. (2015). Comparison of the selective targeting efficacy of *Salmonella typhimurium* A1-R and VNP20009 on the Lewis lung carcinoma in nude mice. *Oncotarget* 6, 14625–14631. doi: 10.18632/oncotarget.3342

Zhao, M., Suetsugu, A., Ma, H., Zhang, L., Liu, F., Zhang, Y., et al. (2012). Efficacy against lung metastasis with a tumor-targeting mutant of *Salmonella typhimurium* in immunocompetent mice. *Cell Cycle* 11, 187–193. doi: 10.4161/cc.11.1.18667

Zhao, M., Yang, M., Li, X.-M., Jiang, P., Baranov, E., Li, S., et al. (2005). Tumor-targeting bacterial therapy with amino acid auxotrophs of GFP-expressing *Salmonella typhimurium*. *Proc. Natl. Acad. Sci. U.S.A.* 102, 755–760. doi: 10.1073/pnas.0408422102

Zhao, M., Yang, M., Ma, H., Li, X., Tan, X., Li, S., et al. (2006). Targeted therapy with a *Salmonella typhimurium* leucine-arginine auxotroph cures orthotopic human breast tumors in nude mice. *Cancer Res.* 66, 7647–7652. doi: 10.1158/0008-5472.CAN-06-0716

Zheng, J. H., Nguyen, V. H., Jiang, S.-N., Park, S.-H., Tan, W., Hong, S. H., et al. (2017). Two-step enhanced cancer immunotherapy with engineered *Salmonella typhimurium* secreting heterologous flagellin. *Sci. Transl. Med.* 9:eaak9537. doi: 10.1126/scitranslmed.aak9537

Zitvogel, L., Pitt, J. M., Daillère, R., Smyth, M. J., and Kroemer, G. (2016). Mouse models in oncoimmunology. *Nat. Rev. Cancer* 16, 759–773. doi: 10.1038/nrc.2016.91

Zlotta, A. R., Fleshner, N. E., and Jewett, M. A. (2009). The management of BCG failure in non-muscle-invasive bladder cancer: an update. *Can. Urol. Assoc. J.* 3, S199–S205.

2

Biofilm-Forming Clinical *Staphylococcus* Isolates Harbor Horizontal Transfer and Antibiotic Resistance Genes

Sandra Águila-Arcos[1], *Itxaso Álvarez-Rodríguez*[1], *Olatz Garaiyurrebaso*[1], *Carlos Garbisu*[2], *Elisabeth Grohmann*[3] *and Itziar Alkorta*[1*]

[1] *Instituto Biofisika (UPV/EHU, CSIC), Department of Biochemistry and Molecular Biology, University of the Basque Country, Bilbao, Spain,* [2] *Department of Conservation of Natural Resources, Soil Microbial Ecology Group, NEIKER-Tecnalia, Derio, Spain,* [3] *Life Sciences and Technology, Beuth University of Applied Sciences, Berlin, Germany*

Edited by:
Manuel Espinosa,
Centro de Investigaciones Biológicas (CSIC), Spain

Reviewed by:
Guenther Muth,
Universität Tübingen, Germany
Fabián Lorenzo,
Universidad de La Laguna, Spain
Gloria Del Solar,
Consejo Superior de Investigaciones Científicas (CSIC), Spain

***Correspondence:**
Itziar Alkorta
itzi.alkorta@ehu.eus

Infections caused by staphylococci represent a medical concern, especially when related to biofilms located in implanted medical devices, such as prostheses and catheters. Unfortunately, their frequent resistance to high doses of antibiotics makes the treatment of these infections a difficult task. Moreover, biofilms represent a hot spot for horizontal gene transfer (HGT) by bacterial conjugation. In this work, 25 biofilm-forming clinical staphylococcal isolates were studied. We found that *Staphylococcus epidermidis* isolates showed a higher biofilm-forming capacity than *Staphylococcus aureus* isolates. Additionally, horizontal transfer and relaxase genes of two common staphylococcal plasmids, pSK41 and pT181, were detected in all isolates. In terms of antibiotic resistance genes, *aac6-aph2a*, *ermC,* and *tetK* genes, which confer resistance to gentamicin, erythromycin, and tetracycline, respectively, were the most prevalent. The horizontal transfer and antibiotic resistance genes harbored on these staphylococcal clinical strains isolated from biofilms located in implanted medical devices points to the potential risk of the development and dissemination of multiresistant bacteria.

Keywords: Staphylococci, biofilm, relaxases, antibiotic resistance, nosocomial infections

INTRODUCTION

Staphylococci, mainly *Staphylococcus aureus* and *Staphylococcus epidermidis*, are well-known causative agents of a large number of human infectious diseases, including skin, soft tissue, respiratory tract, bone, joint and endovascular infections, as well as infections related to implanted medical devices (Otto, 2012; Le et al., 2014). Their pathogenicity is due not only to the virulence factors that they express, but also to the ability of these bacteria to form biofilms (i.e., deeply seated microbial communities attached to inert or living surfaces; Costerton et al., 1999; Otto, 2008). The treatment of biofilm-associated infections is considered a challenging task owing to their inherent resistance to (i) antimicrobial agents and (ii) the host immune system (Hoiby et al., 2010). Moreover, nowadays, the incidence of antibiotic resistant pathogenic bacteria in clinical settings is dramatically increasing, making treatment of bacterial infections one of our most serious health threats (Guridi et al., 2015). This problem arises from the resistance phenotype of bacteria that harbor resistance genes in their chromosomal and/or plasmid DNA.

Bacteria can acquire resistance genes by horizontal gene transfer (HGT). Actually, conjugative plasmid-mediated HGT is considered the most important process in the emergence of new resistant pathogens (Schiwon et al., 2013). It is well-documented that bacterial conjugation can occur within biofilms since they provide an ideal situation for the exchange of genetic material of various origins (Christensen et al., 1998; Hausner and Wuertz, 1999). On the other hand, bacterial conjugation can induce biofilm formation since the cell-to-cell contact established for gene exchange favors the close proximity of bacteria required for biofilm formation (Ghigo, 2001; Molin and Tolker-Nielsen, 2003; Reisner et al., 2006; Yang et al., 2008; D'Alvise et al., 2010). This link between biofilms and bacterial conjugation increases both the risk of biofilm-related infections and the conjugative spread of virulence factors.

In this work, we studied 25 staphylococcal biofilm-forming clinical isolates belonging to the following species: *S. aureus*, *S. epidermidis*, *S. hominis*, and *S. capitis*. These species are commonly found on human skin and can cause biofilm-forming healthcare-associated infections. Both horizontal transfer and antibiotic resistance genes were detected in these staphylococcal clinical isolates. This work adds valuable information on the risk of development and dissemination of antibiotic resistance in *Staphylococcus* biofilm-forming clinical isolates.

MATERIALS AND METHODS

Bacterial Strains

A total of 25 staphylococcal biofilm-forming clinical isolates were kindly provided by Hospital Universitario Donostia, Spain. In addition, they provided data on their antibiotic resistance phenotype, determined by diffusion discs on agar. The origin and antibiotic resistance phenotype of each isolate are shown in **Table 1**.

Growth Conditions

Swabs from the clinical isolates were plated on tryptic soy agar (TSA) and incubated at 37°C. Subsequently, a single colony of each isolate was grown in 10 ml of tryptic soy broth (TSB) supplemented with at least two antibiotics to which the strain was phenotypically resistant (see **Table 1**), at 37°C overnight. The culture was centrifuged at 8,000 × g for 10 min. Then, the pellet was resuspended in 2 ml of TSB medium containing 40% (v/v) glycerol and stored at −80°C.

For this study, strains were grown in TSB medium at 37°C and 200 rpm. TSB medium and TSA plates were supplemented, when required, with amoxicillin (8 μg/ml), cloxacillin [2 μg/ml for *S. aureus* and 0.5 μg/ml for coagulase negative staphylococci (CoNS)], erythromycin (4 μg/ml), mupirocin (520 μg/ml), tetracycline (8 μg/ml), gentamicin (20 μg/ml), rifampicin (2 μg/ml), or levofloxacin (2 μg/ml).

DNA Extraction

Plasmid DNA was extracted from the 25 clinical isolates with the ATP™ Plasmid Midi kit (ATP biotech Inc., Taiwan), according to the manufacturer's instructions.

Detection of Small Plasmids by Agarose Gel Electrophoresis

To detect small plasmids (molecular size < 20 kb), 1 μg of total extracted plasmid DNA was linearized by incubation with 30 U of *Aspergillus oryzae* nuclease S1 (Sigma, Spain) at 37°C for 45 min. Nuclease S1 cuts one strand of the DNA at the nick site and its activity results in linearized plasmids (Germond et al., 1974). Different enzyme concentrations were studied to optimize nuclease S1 digestion (data not shown). Linearized plasmids were visualized on 1% (w/v) agarose gels in 1 × TAE buffer.

Detection of Large Plasmids by Pulsed Field Gel Electrophoresis

Detection of large plasmids (molecular size > 20 kb) was carried out by Pulsed Field Gel Electrophoresis (PFGE) as described by Barton et al. (1995) with modifications. Bacteria were grown in 2 ml of TSB medium overnight at 37°C and 200 rpm. Cultures were diluted in PIV buffer [10 mM Tris-HCl (pH 8), 1 M NaCl] until $OD_{600} = 1$. Then, 600 μl of diluted culture were centrifuged at 11,000 × g for 2 min. Subsequently, the pellet was washed with 500 μl of PIV buffer and centrifuged again. The pellet was resuspended in 300 μl of PIV buffer and incubated at 42°C for 10 min. Next, 150 μl of the sample were mixed with 150 μl of 2% (w/v) low-melting agarose (BioRad) which had been preincubated at 42°C. The mixture was transferred into the plugs, incubated at room temperature for 10 min and, subsequently, for 15 min at 4°C. Once solidified, gel plugs were incubated at 37°C for 5–6 h with shaking (600 rpm) in 1 ml of lysis buffer EC [6 mM Tris-HCl (pH 8), 1 M NaCl, 100 mM EDTA (pH 8), 0.2% (w/v) sodium deoxycholate, 0.5% (w/v) n-lauroylsarcosine, 100 μg/ml lysozyme, 50 μg/ml lysostaphin]. After cell lysis, gel plugs were transferred to new tubes containing 1 ml of EPS solution [1% (w/v) n-lauroylsarcosine, 0.5 M EDTA (pH 8), 100 μg/ml proteinase K] and then incubated at 56°C for 16–20 h. Next, five washes with 1 ml of TE buffer [10 mM Tris-HCl (pH 8), 1 mM EDTA (pH 8)] at 50°C for 30 min each were carried out. For nuclease S1 digestion, each gel plug was cut into two slices. Each slice was incubated twice in 100 μl of digestion solution [50 mM NaCl, 30 mM sodium acetate (pH 4.5), 5 mM $ZnSO_4$] at room temperature for 15 min. Then, slices were incubated at 37°C for 45 min with 1 U of *A. oryzae* nuclease S1 (Sigma) in 100 μl of digestion solution. The reaction was stopped by transferring the slices to 1 ml of TE buffer for 1 h. Digested slices were applied to wells in 1% (w/v) Pulsed Field Certified Agarose (BioRad) prepared in 0.5 × TBE buffer [45 mM Tris (pH 8), 45 mM boric acid, 1 mM EDTA] and run in CHEF-DR® III System (BioRad) at 6 V/cm, a field angle of 120°, and switch times of 5 to 35 s for 22 h. Lambda Ladder PFGE (New England Biolabs, Ispwich, U.S) was used as molecular size marker and pSK41 plasmid (46.4 kb) was used as positive control. Gels were stained with GelRed Nucleic Acid Stain (Biogen Científica,

TABLE 1 | Origin and antibiotic resistance phenotype of the Staphylococcal biofilm-forming clinical isolates used in this work.

No.	Isolate	Origin	Antibiotic resistance[a,b]
1	*S. aureus* 312042	Prosthesis	AMX, AMC, CFZ, CLI, CLOX, ERY, LVX
2	*S. aureus* 410099	Prosthesis	AMX, AMC, CFZ, CLOX, GEN, LVX, MUP
3	*S. aureus* 218154	Prosthesis	AMX, AMC, CFZ, CLI, CLOX, ERY, LVX, MUP, RIF
4	*S. aureus* 339031	Catheter	AMX, AMC, CFZ, CLI, CLOX, ERY, LVZ
5	*S. aureus* 215642	Prosthesis	AMX
6	*S. epidermidis* 213303	Prosthesis	AMX, AMC, CFZ, CLI, CLOX, CTX, ERY, GEN, LVX, MUP, RIF
7	*S. hominis* 313732	Prosthesis	AMX, AMC, CFZ, CLI, CLOX, CTX, ERY, GEN, LVX, MUP
8	*S. capitis* 316479	Prosthesis	AMX, AMC, CFZ, CLOX, GEN, LVX, MUP
9	*S. epidermidis* 319622	Prosthesis	AMX, ERY, TET
10	*S. epidermidis* 219691	Prosthesis	AMX, ERY, LVX, MUP, RIF
11	*S. aureus* 214967	Ulcer	AMX, MUP
12	*S. epidermidis* 239879	Catheter	AMX, AMC, CFZ, CLI, CLOX, ERY, GEN, MUP, TET
13	*S. epidermidis* 239891	Catheter	AMX, AMC, CFZ, CLOX, GEN, LVX, MUP, RIF
14	*S. aureus* 337423-1	Catheter	AMX, AMC, CFZ, CLOX, LVX
15	*S. aureus* 338550-1	Catheter	AMX, AMC, CFZ, CLOX, ERY, LVX
16	*S. aureus* 339031-2	Catheter	AMX, AMC, CFZ, CLOX, ERY, LVX
17	*S. aureus* 339056-2	Catheter	AMX, AMC, CFZ, CLOX, ERY, LVX
18	*S. aureus* 339300	Catheter	AMX, AMC, CFZ, CLOX, LVX
19	*S. aureus* 338503	Catheter	AMX
20	*S. epidermidis* 214627-A	Articular fluid from patient with prosthesis	AMX, AMC, CFZ, CLI, CLOX, CTX, ERY, GEN, LVX, MUP, RIF
21	*S. epidermidis* 310301-1	Articular fluid from patient with prosthesis	AMX, AMC, CFZ, CLI, CLOX, CTX, ERY, GEN, LVX, MUP, RIF
22	*S. epidermidis* 338400-1	Catheter	AMX, AMC, CFZ, CLOX, MUP
23	*S. epidermidis* 338515-1	Catheter	AMX, AMC, CFZ, CLI, CLOX, CTX, ERY, GEN, MUP, RIF
24	*S. epidermidis* 338684	Catheter	AMX, AMC, CFZ, CLI, CLOX, CTX, GEN, LVX, MUP, RIF
25	*S. epidermidis* 216663	Articular fluid from patient with prosthesis	AMX, AMC, CFZ, CLI, CLOX, ERY, GEN, RIF

[a] *Resistance to antibiotics was analyzed by diffusion discs on agar by Hospital Universitario Donostia.*
[b] *AMX, amoxicillin; AMC, amoxicillin + clavulanic acid; CFZ, cefazolin; CLI, clindamycin; CLOX, cloxacillin; CTX, cotrimoxazol; ERY, erythromycin; GEN, gentamicin; LVX, levofloxacin; MUP, mupirocin; RIF, rifampicin; TET, tetracycline; VAN, vancomycin.*

Madrid, Spain). Bands were visualized by ChemiDoc XRS System (BioRad). Images were analyzed by Quantity One 1-D Analysis Software (BioRad).

Polymerase Chain Reaction (PCR) and Southern Blotting

PCR and Southern blotting assays, specific for horizontal transfer and antibiotic resistance genes, were performed using the strains and plasmids indicated in **Table 2** as reference DNA. Oligonucleotides used for gene detection are listed in **Table 3**. Each 25 μl PCR reaction mixture contained 1.25 U Taq polymerase (New England Biolabs, Ipswich, U.S.), 1 × PCR buffer, 0.5 μM of each primer, 0.2 mM deoxynucleoside triphosphates and 20 ng of template DNA (plasmid DNA). Amplifications were carried out in a C1000™ Thermal Cycler (BioRad). PCR temperature profiles are shown in **Table 4**. PCR products were separated by agarose gel electrophoresis, transferred to a membrane (Sambrook and Russel, 2001), and then hybridized with the corresponding specific DIG-labeled probe using the PCR DIG Probe Synthesis Kit (Roche, Mannheim, Germany). Detection of DNA sequences was performed with the DIG Luminescent Detection Kit (Roche) according to the manufacturer's instructions.

TABLE 2 | Bacterial strains and plasmids used as reference for PCR and Southern blotting.

Strain	Characteristics	References
Bacillus subtilis		
BD662	pBD90, *ermD*	Gryczan et al., 1984
BD1156	pBD370, *ermG*	Monod et al., 1987
Enterococcus faecalis		
RE25	pRE25, *ermB*, *tetM*	Schwarz et al., 2001
V583	pTEF1, pTEF2, pTEF3, *vanB*	Paulsen et al., 2003
S. aureus		
RN3259	pT181, *tetK*, pre_{pT181}	Khan et al., 1981
SK5428	pSK41, *acc(6′)-Ie-aph(2′)-Ia*, tra$^+$, pre_{pSK41}, nes_{pSK41}	Firth et al., 1993
S. haemolyticus		
VPS617	*tetK*, *ermC*	Perreten et al., 2005

Biofilm Formation

To test the 25 clinical isolates for biofilm formation, a quantitative adherence assay (Christensen et al., 1985) with some modifications was used. Briefly, 200 μl of TSB medium

TABLE 3 | Oligonucleotides used for the detection of antibiotic resistance and transfer genes.

Gene	Oligonucleotide	Sequence (5′ →3′)	Acc. No[a]	Nucleotide position	Reference
Antibiotic resistance genes	*aac6-aph2a* fw	GCCAGAACATGAATTACACGAG	NC_005024	42,981–43,002	Schiwon et al., 2013
	aac6-aph2a rev	CTGTTGTTGCATTTAGTCTTTCC		43,569–43,591	
	ermB fw	GCATTTAACGACGAAACTGGCT	U00453	6,796–6,817	Böckelmann et al., 2009
	ermB rev	GACAATACTTGCTCATAAGTAATGGT		7,343–7,368	
	ermC fw	CGTAACTGCCATTGAAATAGACC	V01278	2,555–2,577	Schiwon et al., 2013
	ermC rev	TCCTGCATGTTTTAAGGAATTG		2,079–2,100	
	ermD fw	CGGGCAAATATTAGCATAGACG	M29832	544–565	Schiwon et al., 2013
	ermD rev	ATTCTGACCATTGCCGAGTC		988–1,007	
	ermG fw	TGCAGGGAAAGGTCATTTTAC	M15332	785–805	Schiwon et al., 2013
	ermG rev	AACCCATTTCATTACAAAAGTTTC		1,245–1,268	
	tetK fw	TTTGAGCTGTCTTGGTTCATTG	CP000045	1,398–1,419	Schiwon et al., 2013
	tetK rev	AGCCCACCAGAAAACAAACC		1,918–1,937	
	tetM fw	GAACTCGAACAAGAGGAAAGC	M85225	1,114–1,134	Tenover and Rasheed, 2004
	tetM rev	ATGGAAGCCCAGAAAGGAT		1,835–1,853	
	vanB fw	CCCGAATTTCAAATGATTGAAAA	L06138	113–135	Miele et al., 1995
	vanB rev	CGCCATCCTCCTGCAAAA		552–569	
Transfer genes	pSK41 *pre* fw	CTGGACTAAAAGGCATGCAA	AF051917	20,674–20,693	Schiwon et al., 2013
	pSK41 *pre* rev	GCAGTTTTCCATCACGCATA		20,298–20,317	
	pSK41 *nes* fw	AGCGCTAGTAGGATTAAAG	AF051917	10,016–10,034	Schiwon et al., 2013
	pSK41 *nes* rev	CATAATAAATGTGCGTGAGG		9,706–9,725	
	pT181 *pre* fw	TCGAACAGAATTATACAGGCAA	CP000045	2,708–2,729	Schiwon et al., 2013
	pT181 *pre* rev	CTGACTTATTTGCTCATGTTTAGC		3,082–3,105	
	traE pSK41 fw	TATCATTGATCC(T/C)GAA(A/G)ATGAAT	AF051917	27,456–27,478	Schiwon et al., 2013
	traE pSK41 rev	TCTTTTGT(T/G)ATTTCGTCCCATAA		28,060–28,082	
	traG pSK41 fw	GTGTTGACGGTTCGGGTATC	AF051917	30,132–30,151	Schiwon et al., 2013
	traG pSK41 rev	TTTTCCGTCTGAACCTCCAC		30,570–30,589	
	traK pSK41 fw	TATCTAAAGACCACCCAGCTAGAG	AF051917	34,636–34,660	Schiwon et al., 2013
	traK pSK41 rev	TACTTGTTTCAAACTCTACAGTAGC		35,185–35209	
	traL pSK41 fw	ATGGGGACTATGGCAGGTAG	AF051917	36,279–36,298	Schiwon et al., 2013
	traL pSK41 rev	AAGTTTTGCACCACTTCCAG		36,667–36686	
	traM pSK41 fw	TGTTGTATGGGGAAAACAAGC	AF051917	36,870–36,890	Schiwon et al., 2013
	traM pSK41 rev	GCTGGGCTTATAGC(A/G)TCATC		37,051–37,070	

[a]*Accession Number from Gene Bank.*

TABLE 4 | PCR conditions.

Genes amplified	Denaturation	Primer annealing	Elongation
aac6-aph2a, ermD, ermG,	95°C, 30 s	55°C, 30 s	72°C, 30 s
ermC, tetK, tetM	95°C, 30 s	58°C, 30 s	72°C, 30 s
ermB, vanB	95°C, 30 s	60°C, 45 s	72°C, 60 s
traE, traG, traK, traL, traM, pre_{pT181}	95°C, 30 s	55°C, 45 s	72°C, 60 s
pre_{pSK41}, nes_{pSK41}	95°C, 60 s	50°C, 60 s	72°C, 120 s

An initial denaturation step was performed, consisting of 2 min at 95°C, except for pre_{pSK41}, nes_{pSK41}, *that were denaturated for 4 min. Then, 30 cycles consisting of denaturation, primer annealing and elongation steps were performed at the conditions (temperature and time) specified. A final elongation step at 72°C was performed during 5 min, except for* pre_{pSK41}, nes_{pSK41} *in which it lasted 10 min.*

in 96-well flat-bottom polystyrene plates were inoculated with 10 μl overnight bacterial cultures and grown without shaking at 37°C for 24 h. Planktonic bacteria were removed from each well. Then, three washes with distilled water per well were carried out. Next, 125 μl of 0.1% (w/v) crystal violet solution were added to each well and incubated for 10 min at room temperature. Subsequently, three washes with distilled water were again performed. To solubilize the dye, 200 μl of 33% (v/v) glacial acetic acid solution were added to each stained well and incubated for 10 min at room temperature. TSB medium was used as negative control. The optical density of the attached bacteria was measured in a microplate reader at 570 nm (in triplicate for each strain). The ability to form biofilm was attributed as: $OD_{570} < 0.120$, no biofilm-forming; $0.120 < OD_{570} < 0.240$, weak biofilm-forming; $OD_{570} > 0.240$, strong biofilm-forming (Christensen et al., 1985; Di Rosa et al., 2006); and $OD_{570} > 1.5$, very strong biofilm-forming. Dilutions were performed when absorbance values were higher than the limit of accurate detection. To classify the isolates into significant groups, statistical analysis was performed using SigmaPlot program and Student's *t*-test or Mann–Whitney *U*-test.

RESULTS

All Clinical Isolates, Except One, Harbored Plasmids

Plasmid DNA was extracted from the 25 clinical isolates and then analyzed by agarose gel electrophoresis and PFGE (**Figures 1, 2**). Since plasmid DNA samples are sometimes contaminated with chromosomal DNA, as suggested in **Figure 1** for some of our isolates (i.e., 1, 3, 17, 21), after the extraction of plasmid DNA, we decided to test for such contamination. To this purpose, 16S rRNA from isolates 1, 3, 17, and 21 was amplified by PCR as explained in Broszat et al. (2014). The obtained amplicons were analyzed by 1% (w/v) agarose gel electrophoresis in 1 × TAE buffer. As observed in Supplementary Figure 1, some of our plasmid DNA samples appear to be contaminated with chromosomal DNA. Nonetheless, as reflected in **Figures 1, 2**, the majority of the extracted DNA corresponds to plasmid DNA.

As shown in **Figures 1, 2** and **Table 5**, a total of 54 plasmids of sizes ranging from 2 to 200 kb were detected using both methods: 15 small plasmids (size < 20 kb) and 39 large plasmids (size > 20 kb; Shearer et al., 2011). All clinical isolates contained at least one plasmid, except isolate 3. The combination of agarose gel electrophoresis and PFGE is unable to detect plasmids between 13 and 45 kb. Then, *a priori*, our clinical isolates could harbor more plasmids than observed here. In particular, isolate 3 could harbor a plasmid between 13 and 45 kb, which could explain the apparent lack of plasmid observed for this isolate.

When agarose gel electrophoresis was used, it was observed that 44% of the clinical isolates contained at least one plasmid with a size <20 kb (**Figure 1** and **Table 5**). In particular, nine of the isolates contained only one plasmid smaller than 20 kb.

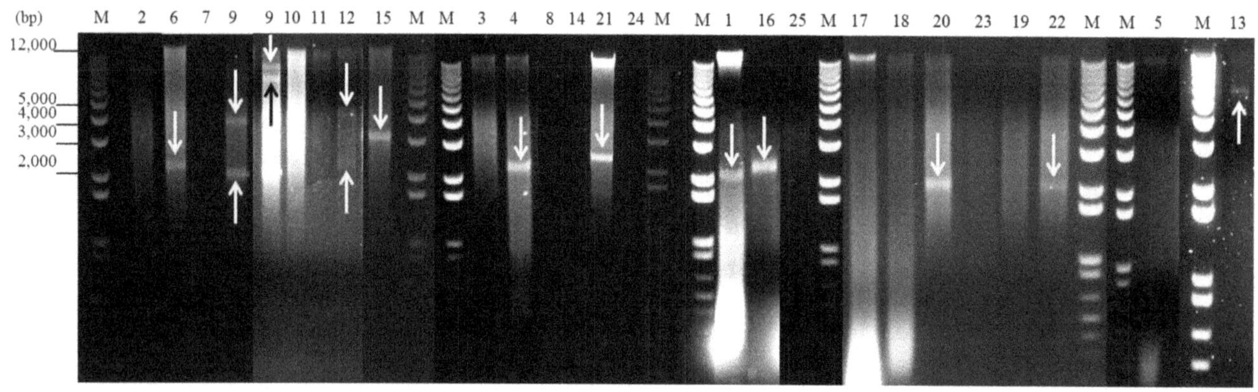

FIGURE 1 | Detection of plasmids from 25 staphylococcal clinical isolates by agarose gel electrophoresis after digestion with nuclease S1. One microgram of plasmid DNA from each isolate was digested with 30 U of nuclease S1 at 37°C for 45 min. After digestion, the plasmids were analyzed by 1% (w/v) agarose gel electrophoresis in 1 × TAE buffer. Lanes 1–25: digested plasmid DNA from each strain (lane numbers correspond to the number of the isolate). Lanes M: DNA molecular weight marker 1 kb Plus DNA Ladder. Bands corresponding to plasmids are indicated with arrows.

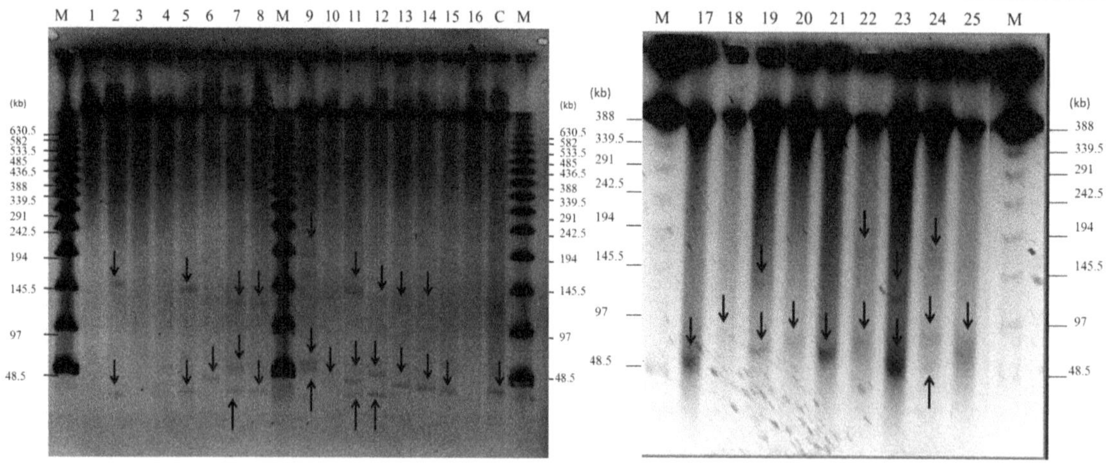

FIGURE 2 | Detection of plasmids in 25 staphylococcal clinical isolates by PFGE. Large plasmids (>30 kb) were analyzed by PFGE after digestion with nuclease S1 at 37°C for 45 min. Lanes 1–25: clinical isolates; lane number corresponds to the number of the isolate. Lane C: positive control, plasmid pSK41 (46.4 kb) extracted from SK5428 strain. Lane M: Lambda Ladder PFGE molecular size marker. Arrows point to detected plasmids.

TABLE 5 | Antibiotic resistance profiles, transfer genes, plasmid content, and biofilm-forming capacity of staphylococcal clinical isolates.

Isolate	Antibiotic resistance phenotype[a]	Antibiotic resistance genotype[b]	Transfer genes	Plasmids[c]		Biofilm-forming capacity 24 h[d]
				<20 kb	>20 kb	
1	ERY	ermB, ermC, tetK, aac6-aph2a	pre_{pSK41}, nes_{pSK41}, pre_{pT181}, traE, traG, traK, traL, traM*	1	0	0
2	GEN	ermB, ermC, ermG, tetK, aac6-aph2a	pre_{pSK41}, pre_{pT181}, traG, traL	0	2	1
3	ERY	ermC, tetK, aac6-aph2a	pre_{pSK41}, pre_{pT181}, traL*	0	0	1
4	ERY	ermB*, ermC, tetK, aac6-aph2a	pre_{pSK41}, nes_{pSK41}, pre_{pT181}, traL	1	0	0
5	–	ermB, ermC, tetK*, aac6-aph2a*	pre_{pSK41}, pre_{pT181}, traK, traL	0	2	3
6	ERY, GEN	ermB, ermC, tetK, aac6-aph2a	pre_{pSK41}, pre_{pT181}, traG, traK, traL	1	1	2
7	ERY, GEN	ermB, ermC, tetK, aac6-aph2a	pre_{pSK41}, nes_{pSK41}, pre_{pT181}, traE, traK, traL	0	3	2
8	GEN	ermC, tetK*, aac6-aph2a	pre_{pSK41}, nes_{pSK41}, pre_{pT181}, traG, traK	0	2	3
9	ERY, TET	ermC, tetK*	pre_{pSK41}*, pre_{pT181}, traL*	4	3	2
10	ERY	ermB, ermC, tetK, aac6-aph2a, vanB	pre_{pSK41}, pre_{pT181}, traG, traL	0	1	2
11	–	ermB, ermC, tetK	pre_{pSK41}, nes_{pSK41}, pre_{pT181}, traE, traG, traL, traM	0	3	1
12	ERY, GEN, TET	ermC, tetK, aac6-aph2a	pre_{pSK41}*, pre_{pT181}, traG*, traL	2	3	3
13	GEN	ermB, ermC, tetK, aac6-aph2a	pre_{pSK41}, pre_{pT181}, traE*, traG*, traK, traL*, traM*	1	2	2
14	-	ermC, tetK*, aac6-aph2a	pre_{pT181}, traE, traG*, traK, traL	0	2	0
15	ERY	ermB, ermC, ermG*, tetK, aac6-aph2a	pre_{pSK41}, pre_{pT181}, traL*	1	1	1
16	ERY	ermB, ermC, tetK, aac6-aph2a	pre_{pSK41}, nes_{pSK41}, pre_{pT181}, traE*, traG*, traK, traL	1	0	0
17	ERY	ermB, ermC, tetK, aac6-aph2a	pre_{pSK41}, nes_{pSK41}, pre_{pT181}, traE, traG, traK, traL, traM	0	1	2
18	–	ermB*, ermC, tetK, aac6-aph2a	pre_{pSK41}, nes_{pSK41}, pre_{pT181}, traE, traG, traK, traL	0	1	0
19	–	ermC, tetK, aac6-aph2a	pre_{pSK41}, pre_{pT181}, traK*, traM	0	2	2
20	ERY, GEN	ermB*, ermC, tetK, aac6-aph2a	pre_{pSK41}, pre_{pT181}, traE*, traG*, traK*, traL, traM	1	1	2
21	ERY, GEN	ermB*, ermC, tetK	pre_{pSK41}, pre_{pT181}, traE, traG, traK, traL, traM	1	1	2
22	–	ermB, ermC, tetK, aac6-aph2a	pre_{pSK41}, pre_{pT181}, traE*, traG*, traL, traM	1	2	3
23	ERY, GEN	ermC, aac6-aph2a	pre_{pSK41}, pre_{pT181}, traE*, traG, traL	0	2	2
24	GEN	ermB*, ermC, aac6-aph2a	pre_{pSK41}, pre_{pT181},	0	3	2
25	ERY, GEN	ermB, ermC, tetK, tetM, aac6-aph2a	pre_{pSK41}, pre_{pT181}, traG*, traK*, traL, traM	0	1	3

[a]GEN, gentamicin; ERY, erythromycin; TET, tetracycline.
[b]aac6-aph2a, gentamicin; ermB/ermC/ermG, erythromycin; tetK/tetM, tetracycline; vanB, vancomycin resistance genes.
[c]Numbers indicate the number of plasmid bands observed in the 1% agarose gel or in the PFGE.
[d]0, no biofilm-forming capacity; 1, weak biofilm-forming capacity; 2, strong biofilm-forming capacity, 3, very strong biofilm-forming capacity.
*Weak signal intensity in the Southern blot.

Isolate 12 harbored two plasmids smaller than 20 kb, while 4 plasmids of this size were identified in isolate 9.

According to our PFGE data, 84% of the clinical isolates (all except isolates 1, 3, 4, and 16) contained at least one large plasmid (**Figure 2** and **Table 5**): 32% of the isolates (6, 10, 15, 17, 18, 20, 21, and 25) harbored one large plasmid; 32% of the isolates (2, 5, 8, 13, 14, 19, 22, and 23) contained two large plasmids; and 20% of the isolates (7, 9, 11, 12, and 24) harbored three large plasmids.

All Clinical Isolates Contained Antibiotic Resistance Genes

Eight resistance genes commonly found in staphylococci were investigated by PCR and Southern blotting: genes encoding resistance to erythromycin (*ermB*, *ermC*, *ermD*, *ermG*), tetracycline (*tetK*, *tetM*), gentamicin (*aac6-aph2a*), and vancomycin (*vanB*). The presence of these genes was tested in our extracted DNA (i.e., putative plasmid DNA) because, initially, we were only interested in the risk of dissemination of antibiotic resistance from these clinical strains through bacterial conjugation.

Concerning erythromycin resistance, 15 of the strains had an erythromycin resistance phenotype (**Table 1**). Data at the genotype level for the different clinical isolates are shown in **Table 5**. *ermC* gene was observed in all the isolates (**Figure 3**), while *ermD* was not detected in any of the isolates (Supplementary Figure 2). Likewise, 72% of the isolates were *ermB*-positive (Supplementary Figure 3), whereas only 8% of the isolates (2 and 15) harbored the *ermG* gene (Supplementary Figure 4).

With respect to tetracycline, only two isolates (9 and 12) were observed to be tetracycline resistant at the phenotype level (**Table 1**). Regarding this antibiotic, 23 out of 25 isolates contained the *tetK* gene (Supplementary Figure 5), while only isolate 25 harbored the *tetM* gene (Supplementary Figure 6).

Similarly, the 25 isolates were analyzed for the occurrence of the gentamicin resistance *aac6-aph2a* gene. As shown in Supplementary Figure 7, this gene was detected in 88% of the isolates (all the isolates except 9, 11, and 21 showed a positive result for the *aac6-aph2a* gene). However, according to the phenotype (**Table 1**), only 44% of the isolates showed gentamicin resistance.

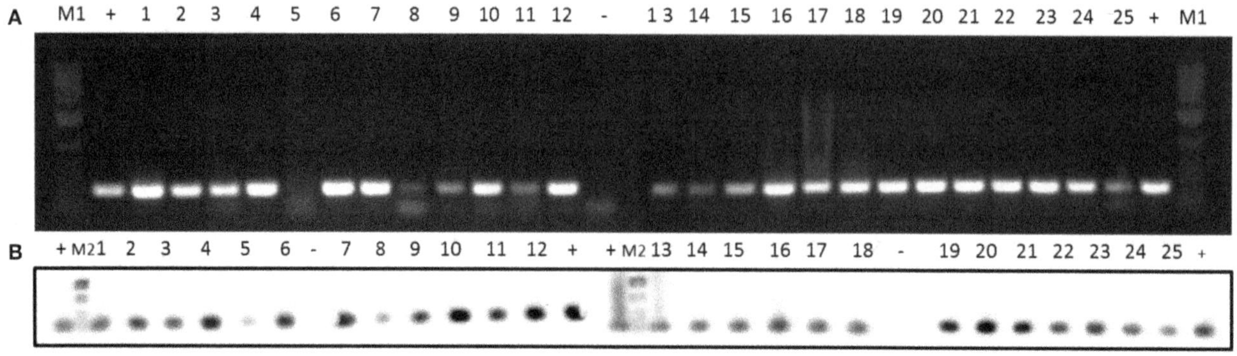

FIGURE 3 | Detection of erythromycin resistance gene *ermC* in 25 clinical isolates by PCR **(A)** and Southern blotting **(B)**. Amplicons of *ermC* (477 bp) were visualized on 1% (w/v) agarose gels. Lanes 1–25: clinical isolates. Lanes +: positive control. Lanes –: negative control. Lanes M1: DNA molecular weight marker 1 kb Plus DNA Ladder. Lanes M2: DNA molecular weight marker VI DIG-labeled.

Finally, regarding vancomycin resistance, all the isolates were phenotypically sensitive to this antibiotic (**Table 1**). At the genotype level, only isolate 10 proved to be *vanB*-positive (**Figure 4**).

All Clinical Isolates Encoded Relaxase and/or Horizontal Transfer Genes Commonly Found in *Staphylococcus* Conjugative/Mobilizable Plasmids

In order to find out whether the abovementioned antibiotic resistance genes were likely to be disseminated via conjugative transfer, we searched for horizontal transfer genes from two common staphylococcal plasmids: (i) conjugative pSK41 and (ii) mobilizable pT181 (Novick, 1989; Berg et al., 1998).

In relation to pSK41, the *pre* relaxase gene was found in all the isolates except isolate 14 (Supplementary Figure 8). In addition, isolates 1, 4, 7, 8, 11, 16, 17, and 18 contained the *nes* relaxase gene of pSK41 (Supplementary Figure 9). Five genes (*traE, traG, traK, traL,* and *traM*) from the transfer region of pSK41 were also analyzed: *traE* gene was present in 48% of the isolates (Supplementary Figure 10), *traG* gene was detected in 68% of the isolates (Supplementary Figure 11), *traK* gene was found in 56% of the isolates (**Figure 5**), and *traL* gene was detected in 88% of the isolates (Supplementary Figure 12). Finally, *traM* gene was found in only 36% of the isolates (Supplementary Figure 13).

In addition, we tested for the presence of the *pre* relaxase gene of the staphylococcal mobilizable plasmid pT181. As shown in **Figure 6**, this gene was detected in all the clinical isolates.

Clinical Isolates Differed in their Biofilm-Forming Capacity

All the clinical strains were isolated from biofilms formed on medical devices such as catheters and prostheses, as well as from ulcer and articular fluids from patients with prostheses (**Table 1**). In order to confirm their biofilm forming capacity, we used the *in vitro* assay described above (Christensen et al., 1985; Di Rosa et al., 2006).

As shown in **Figure 7**, isolates were divided into four groups: (i) no biofilm-forming isolates: 1, 4, 14, 16, 18; (ii) weak biofilm-forming isolates: 2, 3, 11, 15; (iii) strong biofilm-forming isolates: 6, 7, 9, 10, 13, 17, 19, 20, 21, 23, 24; and (iv) very strong biofilm-forming isolates: 5, 8, 12, 22, 25. Additionally, our statistical analysis showed that the "strong biofilm-forming" group could be further divided into three different sub-groups with increasing biofilm forming capacity from "strong biofilm-forming (1)" to "strong biofilm-forming (3)." The distribution of isolates in these three sub-groups was as follows: "strong biofilm-forming (1)": 7, 10, 17, 23; "strong biofilm-forming (2)": 9, 19; and "strong biofilm-forming (3)": 6, 13, 20, 21, 24 (**Figure 7**). Furthermore, the relationship between biofilm-forming capacity and Staphylococcus species was studied. *S. epidermidis* isolates had a significantly higher ($p < 0.001$) biofilm-forming capacity than *S. aureus* isolates.

DISCUSSION

Staphylococci nosocomial pathogens are frequently involved in biomaterial-associated infections (Pfaller and Herwaldt, 1988; Kloos and Bannerman, 1994; Huebner and Goldmann, 1999; Otto, 2008). The eradication of these biofilm-associated infections with antibiotic treatment is usually impossible without the removal of the medical device (Stewart and Costerton, 2001; Mack et al., 2004; Otto, 2012; Tong et al., 2012). Furthermore, conjugative plasmid-mediated dissemination of antibiotic resistance is favored in bacterial biofilms (Ghigo, 2001; Molin and Tolker-Nielsen, 2003; Reisner et al., 2006; Yang et al., 2008; D'Alvise et al., 2010).

In this work, 25 staphylococcal biofilm-forming clinical isolates were studied. First, plasmids of different sizes were detected. Secondly, antibiotic resistance and transfer genes were detected by PCR and Southern blotting. Finally, the capacity of these isolates to form biofilms *in vitro* was studied.

Fifteen plasmids smaller than 20 kb and 39 plasmids larger than 20 kb were found in 11 (44%) and 21 (84%) isolates, respectively. This higher percentage of isolates with large

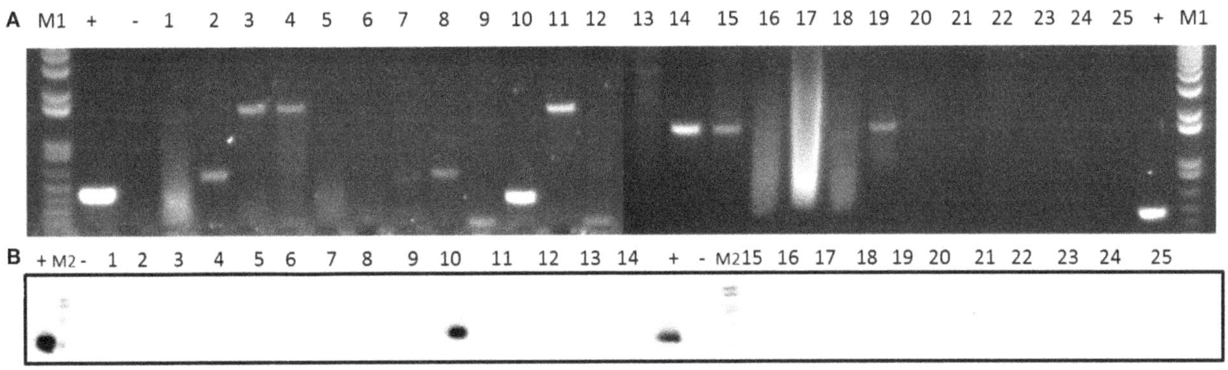

FIGURE 4 | Detection of vancomycin resistance gene *vanB* in 25 clinical isolates by PCR **(A)** and Southern blotting **(B)**. Amplicons of *vanB* (539 bp) were visualized on 1% (w/v) agarose gels. Lanes 1–25: clinical isolates. Lanes +: positive control. Lanes −: negative control. Lanes M1: DNA molecular weight marker 1 kb Plus DNA Ladder. Lanes M2: DNA molecular weight marker VI DIG-labeled.

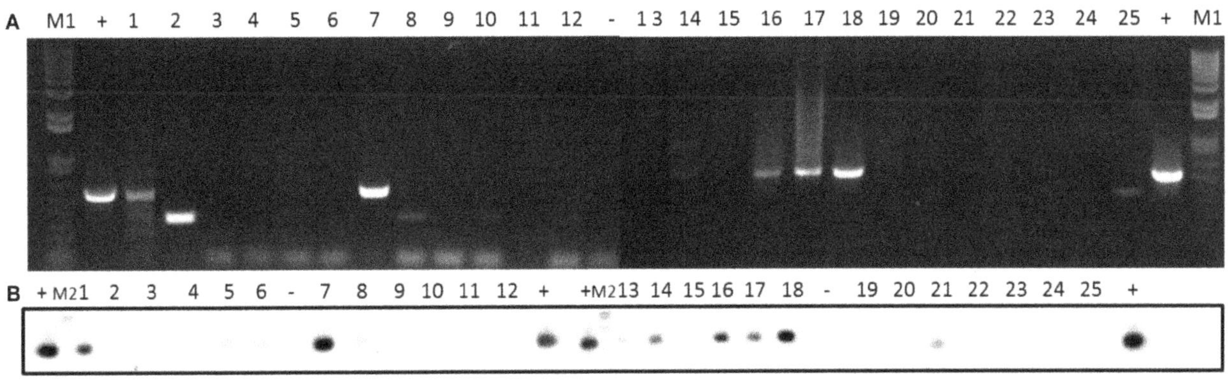

FIGURE 5 | Detection of *traK* gene in 25 clinical isolates by PCR **(A)** and Southern blotting **(B)**. Amplicons of *traK* gene (573 bp) were visualized on 1% (w/v) agarose gels. Lanes 1–25: clinical isolates. Lanes +: positive control. Lanes −: negative control. Lanes M1: DNA molecular weight marker 1 kb Plus DNA Ladder. Lanes M2: DNA molecular weight marker VI DIG-labeled.

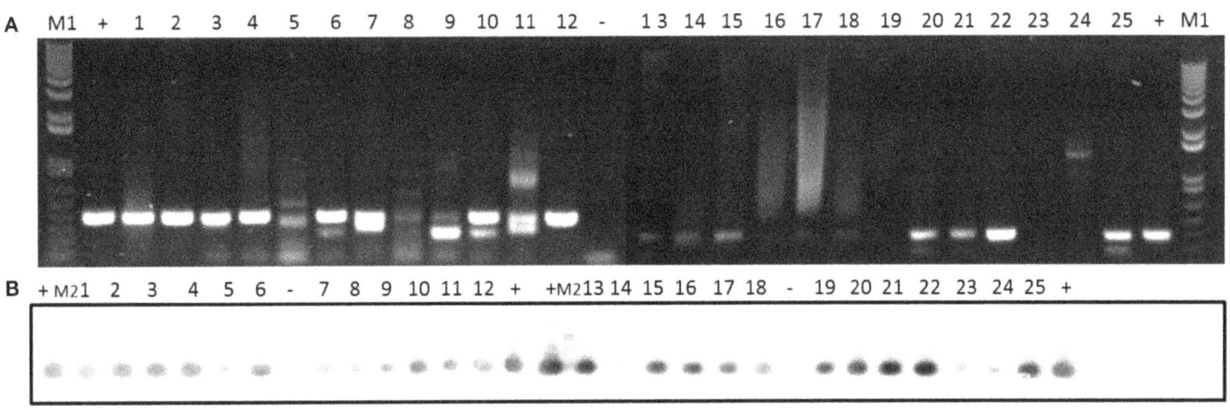

FIGURE 6 | Detection of pre_{pT181} gene in 25 clinical isolates by PCR **(A)** and Southern blotting **(B)**. Amplicons of pre_{pT181} gene (397 bp) were visualized on 1% (w/v) agarose gels. Lanes 1–25: clinical isolates. Lanes +: positive control. Lanes −: negative control. Lanes M1: DNA molecular weight marker 1 kb Plus DNA Ladder. Lanes M2: DNA molecular weight marker VI DIG-labeled.

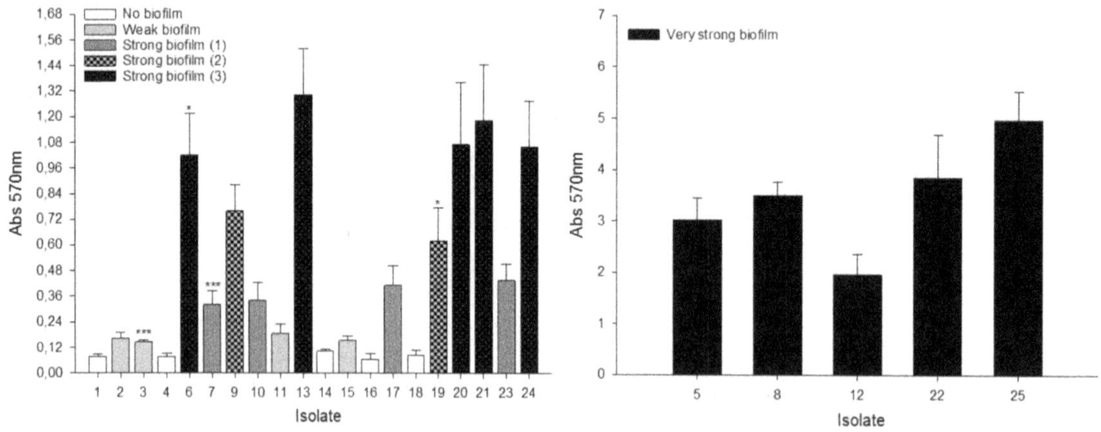

FIGURE 7 | Biofilm-forming capacity of staphylococcal clinical isolates. 96-well flat-bottom polystyrene plates were incubated for 24 h at 37°C without shaking. Cells attached to the wells were stained with 0.1% (w/v) crystal violet. Absorption at 570 nm was measured to quantify biofilm formation. Results are the mean ± SEM of at least 4 independent biological experiments performed in triplicate. To classify the isolates into significant groups, statistical analysis was performed using Student's *t*-test or Mann–Whitney *U*-test (*$p < 0.05$, ***$p < 0.001$).

plasmids, compared to isolates with small plasmids, is in agreement with results obtained by Shearer et al. (2011) who found that 79% of their isolates harbored at least one large (>20 kb) plasmid. According to Smillie et al. (2010), in proteobacteria, 58% of the plasmids larger than 20 kb are mobilizable. Therefore, our results suggest that almost all our *Staphylococcus* clinical isolates could harbor conjugative and/or mobilizable plasmids.

On the other hand, the presence of antibiotic resistance and horizontal transfer genes commonly found staphylococci was investigated. Antibiotic sensitivity tests (to obtain the well-known antibiograms) are the most common method to determine antibiotic resistance of pathogenic bacteria. Nonetheless, the study of antibiotic resistance at the genotype level is crucial to get information on the potential of those bacteria to develop and disseminate resistance against antibiotics (Palmer and Kishony, 2013). Erythromycin, tetracyclines, gentamicin, and vancomycin are the most used antibiotics for the treatment of staphylococcal infections, but resistance genes against these antibiotics have been described in staphylococcal clinical isolates (Duran et al., 2012; Emaneini et al., 2013). Therefore, we searched in our 25 clinical isolates for the presence of genes involved in the resistance to these antibiotics.

Macrolide antibiotics, such as erythromycin, are broad-spectrum antibiotics; relevantly, anti-biofilm activities have been assigned to them (Parra-Ruiz et al., 2012; Zhao et al., 2015). In this work, 60% of the isolates were phenotypically resistant to erythromycin. At the genotype level, different studies have reported the prevalence of the *ermC* gene in staphylococci (Duran et al., 2012; Schiwon et al., 2013); in our study, the *ermC* gene was identified in all the clinical isolates. Although a low prevalence of the *ermB* gene has been described in Staphylococcus (Zmantar et al., 2011), in our study, 72% of the isolates harbored the *ermB* gene.

After penicillin, tetracyclines are the second most widely used group of antibiotics worldwide (van Hoek et al., 2011). Resistance to tetracycline can be encoded in plasmid-located *tet* genes such as *tetK* and *tetL*, or, alternatively, in genes located in the chromosome or transposons such as *tetM* and *tetO* (Emaneini et al., 2013). A high incidence of the *tetK* gene (92%) was observed here, whereas only one isolate contained the *tetM* gene. Several studies have reported the coexistence of both *tetM* and *tetK* genes in staphylococci strains (Duran et al., 2012; Camoez et al., 2013; Emaneini et al., 2013; Schiwon et al., 2013). Here, a disagreement between phenotypic and genotypic data was observed, since only 8% of the isolates were phenotypically resistant to tetracycline.

Aminoglycosides, such as gentamicin, are broad-spectrum antibiotics used against *S. aureus* infections. Aminoglycoside modifying enzymes (AME) are used by bacteria to abolish the effect of these antibiotics. In *S. aureus* strains, one of the most common genes encoding AME is the *aac6-aph2a* gene (Emaneini et al., 2013). In our study, 88% of the isolates contained this gene, in agreement with other studies on staphylococcal isolates (Duran et al., 2012; Emaneini et al., 2013). In terms of the phenotype, 44% of the isolates showed resistance to gentamicin (**Table 1**). Similar discrepancies between phenotypic and genotypic results have been observed by other authors (Duran et al., 2012; Emaneini et al., 2013).

The lack of correlation between resistance phenotypic and genotypic data could be due to mutations in genes resulting in non-functional proteins, as well as to the lack of gene expression (Martineau et al., 2000). Also, methods to detect antibiotic resistance phenotype are influenced by technical variables such as temperature, incubation time, inoculum density and so on (Baddour et al., 2007). Likewise, the pattern of negative resistance phenotype together with a positive resistance genotype can be due to the presence of pseudogenes (Davis et al., 2011). As a consequence, it is essential to take this fact into account because it indicates that bacteria have the potential to be resistant to more antibiotics than those shown phenotypically.

Vancomycin has been used to treat staphylococcal infections, mainly methicillin resistant *S. aureus* (Huebner and Goldmann, 1999). In the late 1980s, the emergence of vancomycin resistance was reported for the first time (van Hoek et al., 2011). One of the genes responsible for vancomycin resistance is the *vanB* gene (van Hoek et al., 2011), which was only found in isolate 10. This low incidence of the *vanB* gene, together with the fact that all isolates were phenotypically sensitive to vancomycin (**Table 1**), suggest that (i) vancomycin is still one of the best options for the treatment of staphylococcal infections and (ii) it should be then used judiciously.

Concerning the presence of horizontal transfer genes, plasmid pSK41 is a prototypical multiresistance plasmid of 46 kb from *S. aureus* (Berg et al., 1998). Therefore, we searched for pre_{pSK41} and nes_{pSK41} genes, as well as for five different *tra* genes involved in the conjugative transfer of plasmid pSK41, in our clinical isolates. Although all the isolates, except for one, contained the pre_{pSK41} gene, only 32% of the isolates harbored the nes_{pSK41} gene. In addition, 20% of the isolates contained the five *tra* genes tested here. According to these results, and taking into account that isolates 11 and 17 harbored plasmids of around 46 kb, we speculate that these strains could contain pSK41-type plasmids. Other studies have identified plasmids of the pSK41 family in geographically diverse isolates of both *S. aureus* and CoNS (Berg et al., 1998). The coexistence of pre_{pSK41} and nes_{pSK41} genes, together with *tra* genes, in some of our isolates, points to a risk of dissemination of resistance traits.

pT181 plasmid is also common among staphylococci (Khan and Novick, 1983; Novick, 1989). Then, we tested for the presence of the relaxase pre_{pT181} gene responsible for pT181 mobilization. All isolates harbored the pre_{pT181} relaxase gene, suggesting that pT181-type plasmids could be present in all of the samples. pT181 is a low copy number plasmid and then, not surprisingly, we could not detect it in the electrophoretic gels; however, it may be present at undetectable amounts in some strains. This is of great concern especially in those strains where potentially conjugative plasmids that could mobilize these small plasmids are present.

Finally, in general, *S. epidermidis* isolates have shown a higher biofilm-forming capacity than *S. aureus* isolates Although, the 25 isolates studied here were obtained from biofilms present in the clinical environment, some of the *S. aureus* isolates were unable to form biofilms under our experimental conditions. This is probably because biofilms in clinical conditions take longer times to form, in comparison to the standardized *in vitro* method used here, in which 24 h was the biofilm-forming time.

The fact that our clinical isolates contained both antibiotic resistance and horizontal transfer genes, as well as conjugative and/or mobilizable plasmids, suggest the possibility of their disseminating antibiotic resistance to other bacteria. Here, it must be stated that, due to the abovementioned presence of chromosomal DNA in our extracted DNA samples, we cannot rule out the possibility that the antibiotic resistance genes identified here were encoded in the chromosomal DNA. However, as reflected in **Figures 1**, **2**, the majority of the extracted DNA corresponds to plasmid DNA. For example, the high incidence of the plasmid-encoded *tetK* gene in our isolates could support this fact. In any case, genes encoded in the chromosome can also be mobilized between bacterial cells. For instance, transposons can mobilize chromosomal genes by jumping into plasmids or phages which can then be transferred into other cells (Frost et al., 2014). On the other hand, the conjugation process can also occur via chromosomally integrated conjugative elements, such as conjugative transposons. Integrated conjugative elements are known to encode proteins that facilitate their own transfer and sometimes the transfer of other cellular DNA from the donor (Frost et al., 2014). Indeed, as reported by Wilkins and Frost (2001), many plasmids and integrated conjugative elements can effect the transfer of chromosomal DNA. Then, if some of the antibiotic resistance genes identified here were encoded in the chromosomal DNA present in some of our samples, the risk of transfer to other bacterial cells would still exist, although *a priori* lower than if they were encoded in the observed plasmids.

Recent studies underline the importance of collecting more epidemiological data on antibiotic resistance, in order to design novel control strategies for this growing global health problem (Frieri et al., 2016). The isolation and molecular characterization of plasmids from nosocomial pathogens will provide valuable information in the search for new strategies to control the dissemination of antibiotic resistance among clinical pathogens.

AUTHOR CONTRIBUTIONS

SÁ and IÁR: Acquisition of the data, writing and revision of the content, approval of the last version of the work. OG: Revision of the content, approval of the last version of the work. CG: Writing and revision of the content, approval of the last version and ensuring accuracy and integrity of the work. EG: Design of the work, revision of the content, approval of the last version, and ensuring accuracy and integrity of the work. IA: Design of the work and the acquisition of the data, writing, and revision of the content, approval of the last version and ensuring accuracy and integrity of the work.

ACKNOWLEDGMENTS

This work was financially supported by the Spanish Ministry of Economy (Grant No BFU2012-36241) and MICINN (Grant No. BFU2010-22103). SÁ and OG were graduate students supported by the Basque Government and Fundación Biofísica Bizkaia. At the moment, IÁR is a graduate student supported by the Basque Government. We thank Eneritz Bilbao for excellent technical assistance. We thank Prof. Pérez-Trallero and Dr. Alonso from the Hospital Universitario Donostia for providing the clinical isolates, and Prof. Quindós and Aketza Varona from the University of the Basque Country, and Dr. Rodríguez-Lázaro and Dr. Hernández from the Instituto Tecnológico Agrario de Castilla y León, Spain for assistance with the PFGE.

REFERENCES

Baddour, M. M., AbuElKheir, M. M., and Fatani, A. J. (2007). Comparison of mecA polymerase chain reaction with phenotypic methods for the detection of methicillin-resistant *Staphylococcus aureus*. *Curr. Microbiol.* 55, 473–479. doi: 10.1007/s00284-007-9015-6

Barton, B. M., Harding, G. P., and Zuccarelli, A. J. (1995). A general method for detecting and sizing large plasmids. *Anal. Biochem.* 226, 235–240. doi: 10.1006/abio.1995.1220

Berg, T., Firth, N., Apisiridej, S., Hettiaratchi, A., Leelaporn, A., and Skurray, R. A. (1998). Complete nucleotide sequence of pSK41: evolution of staphylococcal conjugative multiresistance plasmids. *J. Bacteriol.* 180, 4350–4359.

Böckelmann, U., Dorries, H., Ayuso-Gabella, M. N., de Salgot Marcay, M., Tandoi, V., Levantesi, C., et al. (2009). Quantitative PCR monitoring of antibiotic resistance genes and bacterial pathogens in three European artificial groundwater recharge systems. *Appl. Environ. Microbiol.* 75, 154–163. doi: 10.1128/AEM.01649-08

Broszat, M., Nacke, H., Blasi, R., Siebe, C., Huebner, J., Daniel, R., et al. (2014). Wastewater irrigation increases the abundance of potentially harmful gammaproteobacteria in soil in Mezquital Valley, Mexico. *Appl. Environ. Microbiol.* 80, 5282–5291. doi: 10.1128/AEM.01295-14

Camoez, M., Sierra, J. M., Pujol, M., Hornero, A., Martin, R., and Domínguez, M. A. (2013). Prevalence and molecular characterization of methicillin-resistant *Staphylococcus aureus* ST398 resistant to tetracycline at a Spanish Hospital over 12 years. *PLoS ONE* 8:e72828. doi: 10.1371/journal.pone.0072828

Christensen, B. B., Sternberg, C., Andersen, J. B., Eberl, L., Moller, S., Givskov, M., et al. (1998). Establishment of new genetic traits in a microbial biofilm community. *Appl. Environ. Microbiol.* 64, 2247–2255.

Christensen, G. D., Simpson, W. A., Younger, J. J., Baddour, L. M., Barrett, F. F., Melton, D. M., et al. (1985). Adherence of coagulase-negative staphylococci to plastic tissue culture plates: a quantitative model for the adherence of staphylococci to medical devices. *J. Clin. Microbiol.* 22, 996–1006

Costerton, J. W., Stewart, P. S., and Greenberg, E. P. (1999). Bacterial biofilms: a common cause of persistent infections. *Science* 284, 1318–1322. doi: 10.1126/science.284.5418.1318

D'Alvise, P. W., Sjoholm, O. R., Yankelevich, T., Jin, Y., Wuertz, S., and Smets, B. F. (2010). TOL plasmid carriage enhances biofilm formation and increases extracellular DNA content in *Pseudomonas putida* KT2440. *FEMS Microbiol. Lett.* 312, 84–92. doi: 10.1111/j.1574-6968.2010.02105.x

Davis, M. A., Besser, T. E., Orfe, L. H., Baker, K. N., Lanier, A. S., Broschat, S. L., et al. (2011). Genotypic-phenotypic discrepancies between antibiotic resistance characteristics of *Escherichia coli* isolates from calves in management settings with high and low antibiotic use. *Appl. Environ. Microbiol.* 77, 3293–3299. doi: 10.1128/AEM.02588-10

Di Rosa, R., Creti, R., Venditti, W., D'Amelio, R., Arciola, C. R., Montanaro, L., et al. (2006). Relationship between biofilm formation, the enterococcal surface protein and gelatinase in clinical isolates of *Enterococcus faecalis* and *Enterococcus faecium*. *FEMS Micobiol. Lett.* 256, 145–150. doi: 10.1111/j.1574-6968.2006.00112.x

Duran, N., Ozer, B., Duran, G. G., Onlen, Y., and Demir, C. (2012). Antibiotic resistance genes and susceptibility patterns in staphylococci. *Indian J. Med. Res.* 135, 389–396.

Emaneini, M., Bigverdi, R., Kalantar, D., Soroush, S., Jabalameli, F., Noorazar Khoshgnab, B., et al. (2013). Distribution of genes encoding tetracycline resistance and aminoglycoside modifying enzymes in *Staphylococcus aureus* strains isolated from a burn center. *Ann. Burns Fire Disasters* 26, 76–80.

Firth, N., Ridgway, K. P., Byrne, M. E., Fink, P. D., Johnson, L., Paulsen, I. T., et al. (1993). Analysis of a transfer region from the staphylococcal conjugative plasmid pSK41. *Gene* 136, 13–25. doi: 10.1016/0378-1119(93)90442-6

Frieri, M., Kumar, K., and Boutin, A. (2016). Antibiotic resistance. *J. Infect. Public Health* 10, 369–378. doi: 10.1016/j.jiph.2016.08.007

Frost, L. S., Leplae, R., Summers, A. O., and Toussaint, A. (2014). Mobile genetic elements: the agents of open source evolution. *Nat. Rev. Microbiol.* 3, 722–732. doi: 10.1038/nrmicro1235

Germond, J. E., Vogt, V. M., and Hirt, B. (1974). Characterization of the single-strand-specific nuclease S1 activity on double-stranded supercoiled polyoma DNA. *Eur. J. Biochem.* 43, 591–600. doi: 10.1111/j.1432-1033.1974.tb03446.x

Ghigo, J. M. (2001). Natural conjugative plasmids induce bacterial biofilm development. *Nature* 412, 442–445. doi: 10.1038/35086581

Gryczan, T., Israeli-Reches, M., Del Bue, M., and Dubnau, D. (1984). DNA sequence and regulation of ermD, a macrolide-lincosamidestreptogramin B resistance element from *Bacillus licheniformis*. *Mol. Gen. Genet.* 194, 349–356. doi: 10.1007/BF00425543

Guridi, A., Diederich, A. K., Aguila-Arcos, S., Garcia-Moreno, M., Blasi, R., Broszat, M., et al. (2015). New antimicrobial contact catalyst killing antibiotic resistant clinical and waterborne pathogens. *Mater. Sci. Eng. C Mater. Biol. Appl.* 50, 1–11. doi: 10.1016/j.msec.2015.01.080

Hausner, M., and Wuertz, S. (1999). High rates of conjugation in bacterial biofilms as determined by quantitative *in situ* analysis. *Appl. Environ. Microbiol.* 65, 3710–3713

Hoiby, N., Bjarnsholt, T., Givskov, M., Molin, S., and Ciofu, O. (2010). Antibiotic resistance of bacterial biofilms. *Int. J. Antimicrob. Agents* 35, 322–332. doi: 10.1016/j.ijantimicag.2009.12.011

Huebner, J., and Goldmann, D. A. (1999). Coagulase-negative staphylococci: role as pathogens. *Annu. Rev. Med.* 50, 223–236. doi: 10.1146/annurev.med.50.1.223

Khan, S. A., and Novick, R. P. (1983). Complete nucleotide sequence of pT181, a tetracycline resistance plasmid from *Staphylococcus aureus*. *Plasmid* 10, 251–259. doi: 10.1016/0147-619X(83)90039-2

Khan, S. A., Carleton, S. M., and Novick, R. P. (1981). Replication of plasmid pT181 DNA *in vitro*: requirement for a plasmid-encoded product. *Proc. Natl. Acad. Sci. U.S.A.* 78, 4902–4906. doi: 10.1073/pnas.78.8.4902

Kloos, W. E., and Bannerman, T. L. (1994). Update on clinical significance of coagulase-negative staphylococci. *Clin. Microbiol. Rev.* 7, 117–140. doi: 10.1128/CMR.7.1.117

Le, K. Y., Dastgheyb, S., Ho, T. V., and Otto, M. (2014). Molecular determinants of staphylococcal biofilm dispersal and structuring. *Front. Cell. Infect. Microbiol.* 4:167. doi: 10.3389/fcimb.2014.00167

Mack, D., Becker, P., Chatterjee, I., Dobinsky, S., Knobloch, J. K., Peters, G., et al. (2004). Mechanisms of biofilm formation in *Staphylococcus epidermidis* and *Staphylococcus aureus*: functional molecules, regulatory circuits, and adaptive responses. *Int. J. Med. Microbiol.* 294, 203–212. doi: 10.1016/j.ijmm.2004.06.015

Martineau, F., Picard, F. J., Lansac, N., Menard, C., Roy, P. H., Ouellette, M., et al. (2000). Correlation between the resistance genotype determined by multiplex PCR assays and the antibiotic susceptibility patterns of *Staphylococcus aureus* and *Staphylococcus epidermidis*. *Antimicrob. Agents Chemother.*. 44, 231–238. doi: 10.1128/AAC.44.2.231-238.2000

Miele, A., Bandera, M., and Goldstein, B. P. (1995). Use of primers selective for vancomycin resistance genes to determine van genotype in enterococci and to study gene organization in VanA isolates. *Antimicrob. Agents Chemother.* 39, 1772–1778. doi: 10.1128/AAC.39.8.1772

Molin, S., and Tolker-Nielsen, T. (2003). Gene transfer occurs with enhanced efficiency in biofilms and induces enhanced stabilisation of the biofilm structure. *Curr. Opin. Biotechnol.* 14, 255–261. doi: 10.1016/S0958-1669(03)00036-3

Monod, M., Mohan, S., and Dubnau, D. (1987). Cloning and analysis of ermG, a new macrolide-lincosamide-streptogramin B resistance element from *Bacillus sphaericus*. *J. Bacteriol.* 169, 340–350. doi: 10.1128/jb.169.1.340-350.1987

Novick, R. P. (1989). Staphylococcal plasmids and their replication. *Annu. Rev. Microbiol.* 43, 537–565. doi: 10.1146/annurev.mi.43.100189.002541

Otto, M. (2008). Staphylococcal biofilms. *Curr. Top. Microbiol. Immunol.* 322, 207–228. doi: 10.1007/978-3-540-75418-3_10

Otto, M. (2012). Molecular basis of *Staphylococcus epidermidis* infections. *Semin. Immunopathol.* 34, 201–214. doi: 10.1007/s00281-011-0296-2

Palmer, A. C., and Kishony, R. (2013). Understanding, predicting and manipulating the genotypic evolution of antibiotic resistance. *Nat. Rev. Genet.* 14, 243–248. doi: 10.1038/nrg3351

Parra-Ruiz, J., Vidaillac, C., and Rybak, M. J. (2012). Macrolides and staphylococcal biofilms. *Rev. Esp. Quimioter.* 25, 10–16.

Paulsen, I. T., Banerjei, L., Myers, G. S., Nelson, K. E., Seshadri, R., Read, T. D., et al. (2003). Role of mobile DNA in the evolution of vancomycin resistant *Enterococcus faecalis*. *Science* 299, 2071–2074. doi: 10.1126/science.1080613

Perreten, V., Vorlet-Fawer, L., Slickers, P., Ehricht, R., Kuhnert, P., and Frey, J. (2005). Microarray-based detection of 90 antibiotic resistance genes of gram-positive bacteria. *J. Clin. Microbiol.* 43, 2291–2302. doi: 10.1128/JCM.43.5.2291-2302.2005

Pfaller, M. A., and Herwaldt, L. A. (1988). Laboratory, clinical and epidemiological aspects of coagulase-negative staphylococci. *Clin. Microbiol. Rev.* 1, 281–299. doi: 10.1128/CMR.1.3.281

Reisner, A., Höller, B. M., Molin, S., and Zechner, E. L. (2006). Synergistic effects in mixed *Escherichia coli* biofilms: conjugative plasmid transfer drives biofilm expansion. *J. Bacteriol.* 188, 3582–3588. doi: 10.1128/JB.188.10.3582-3588.2006

Sambrook, J., and Russel, D. W. (2001). *Molecular Cloning. A Laboratory Manual.* New York, NY: Cold Spring Harbor Laboratory Press.

Schiwon, K., Arends, K., Rogowski, K. M., Fürch, S., Prescha, K., Sakinc, T., et al. (2013). Comparison of antibiotic resistance, biofilm formation and conjugative transfer of Staphylococcus and Enterococcus isolates from International Space Station and Antarctic Research Station Concordia. *Microb. Ecol.* 65, 638–651. doi: 10.1007/s00248-013-0193-4

Schwarz, F. V., Perreten, V., and Teuber, M. (2001). Sequence of the 50-kb conjugative multiresistance plasmid pRE25 from *Enterococcus faecalis* RE25. *Plasmid* 46, 170–187. doi: 10.1006/plas.2001.1544

Shearer, J. E., Wireman, J., Hostetler, J., Forberger, H., Borman, J., Gill, J., et al. (2011). Major families of multiresistant plasmids from geographically and epidemiologically diverse staphylococci. *G3* 1, 581–591. doi: 10.1534/g3.111.000760

Smillie, C., Garcillán-Barcia, M. P., Francia, M. V., Rocha, E. P., and de la Cruz, F. (2010). Mobility of plasmids. *Microbiol. Mol. Biol. Rev.* 74, 3434–3452. doi: 10.1128/MMBR.00020-10

Stewart, P. S., and Costerton, J. W. (2001). Antibiotic resistance of bacteria in biofilms. *Lancet* 358, 135–138. doi: 10.1016/S0140-6736(01)05321-1

Tenover, F. C., and Rasheed, J. K. (2004). "Detection of antimicrobial resistance genes and mutations associated with antimicrobial resistance in microorganisms," in *Molecular Microbiology: Diagnostics Principles and Practice, Vol. 1*, eds D. H. Persing, F. G. Tenover, J. Versalovic, T. Yuer, W. T. J, Relman (Washington, DC: ASM Press), 391–406.

Tong, S. Y., Chen, L. F., and Fowler, V. G. Jr. (2012). Colonization, pathogenicity, host susceptibility, and therapeutics for *Staphylococcus aureus*: what is the clinical relevance? *Semin. Immunopathol.* 34, 185–200. doi: 10.1007/s00281-011-0300-x

van Hoek, A. H., Mevius, D., Guerra, B., Mullany, P., Roberts, A. P., and Aarts, H. J. (2011). Acquired antibiotic resistance: an overview. *Front. Microbiol.* 2:203. doi: 10.3389/fmicb.2011.00203

Wilkins, B. M., and Frost, L. S. (2001). "Mechanisms of gene exchange between bacteria," in *Molecular Medical Microbiology*, ed M. Sussman (London: Academic Press), 355–400.

Yang, X., Ma, Q., and Wood, T. K. (2008). The R1 conjugative plasmid increases *Escherichia coli* biofilm formation through an envelope stress response. *Appl. Environ. Microbiol.* 74, 2690–2699. doi: 10.1128/AEM.02809-07

Zhao, Y. L., Zhou, Y. H., Chen, J. Q., Huang, Q. Y., Han, Q., Liu, B., et al. (2015). Quantitative proteomic analysis of sub-MIC erythromycin inhibiting biofilm formation of *S. suis in vitro*. *J. Proteomics* 116, 1–14. doi: 10.1016/j.jprot.2014.12.019

Zmantar, T., Kouidhi, B., Miladi, H., and Bakhrouf, A. (2011). Detection of macrolide and disinfectant resistance genes in clinical *Staphylococcus aureus* and coagulase-negative staphylococci. *BMC Res. Notes* 4:453. doi: 10.1186/1756-0500-4-453

3

Prediction of Phenotypic Antimicrobial Resistance Profiles From Whole Genome Sequences of Non-typhoidal *Salmonella enterica*

Saskia Neuert[1,2], Satheesh Nair[2], Martin R. Day[2], Michel Doumith[2], Philip M. Ashton[2], Kate C. Mellor[3,4], Claire Jenkins[1,2], Katie L. Hopkins[2], Neil Woodford[2], Elizabeth de Pinna[2], Gauri Godbole[1,2] and Timothy J. Dallman[1,2]*

[1] National Institute for Health Research Health Protection Research Unit in Gastrointestinal Infections, University of Liverpool, Liverpool, United Kingdom, [2] Bacteriology Reference Department, National Infection Service, Public Health England, London, United Kindom, [3] Department of Pathobiology and Population Sciences, Royal Veterinary College, London, United Kingdom, [4] London School of Hygiene & Tropical Medicine, London, United Kingdom

Edited by:
Chew Chieng Yeo,
Sultan Zainal Abidin University,
Malaysia

Reviewed by:
Jeanette Teo,
National University Hospital,
Singapore
Debashree Basu,
University of Minnesota Twin Cities,
United States

***Correspondence:**
Timothy J. Dallman
tim.dallman@phe.gov.uk

Surveillance of antimicrobial resistance (AMR) in non-typhoidal *Salmonella enterica* (NTS), is essential for monitoring transmission of resistance from the food chain to humans, and for establishing effective treatment protocols. We evaluated the prediction of phenotypic resistance in NTS from genotypic profiles derived from whole genome sequencing (WGS). Genes and chromosomal mutations responsible for phenotypic resistance were sought in WGS data from 3,491 NTS isolates received by Public Health England's Gastrointestinal Bacteria Reference Unit between April 2014 and March 2015. Inferred genotypic AMR profiles were compared with phenotypic susceptibilities determined for fifteen antimicrobials using EUCAST guidelines. Discrepancies between phenotypic and genotypic profiles for one or more antimicrobials were detected for 76 isolates (2.18%) although only 88/52,365 (0.17%) isolate/antimicrobial combinations were discordant. Of the discrepant results, the largest number were associated with streptomycin (67.05%, n = 59). Pan-susceptibility was observed in 2,190 isolates (62.73%). Overall, resistance to tetracyclines was most common (26.27% of isolates, n = 917) followed by sulphonamides (23.72%, n = 828) and ampicillin (21.43%, n = 748). Multidrug resistance (MDR), i.e., resistance to three or more antimicrobial classes, was detected in 848 isolates (24.29%) with resistance to ampicillin, streptomycin, sulphonamides and tetracyclines being the most common MDR profile (n = 231; 27.24%). For isolates with this profile, all but one were *S*. Typhimurium and 94.81% (n = 219) had the resistance determinants bla_{TEM-1}, *strA-strB*, *sul2* and *tet*(A). Extended-spectrum β-lactamase genes were identified in 41 isolates (1.17%) and multiple mutations in chromosomal genes associated with ciprofloxacin resistance in 82 isolates (2.35%). This study showed that WGS is suitable as a rapid means of determining AMR patterns of NTS for public health surveillance.

Keywords: antimicrobial resistance, multidrug resistance, non-typhoidal *Salmonella enterica*, whole genome sequencing, public health surveillance, One Health

INTRODUCTION

Salmonella enterica subspecies *enterica* is responsible for 99% of salmonellosis cases in humans and animals, and can be further subdivided into the host-restricted typhoidal salmonellae and the more generalist non-typhoidal salmonellae (NTS) (Langridge et al., 2015; Wain et al., 2015). As host-adapted or generalist organisms, NTS can be transferred from animals to humans causing zoonotic infections and therefore fall under the World Health Organization's One Health approach. Globally, NTS were estimated to cause 93.8 million enteric infections resulting in 155,000 deaths annually (Majowicz et al., 2010), and in the United Kingdom they are the third most common cause of bacterial gastroenteritis (Tam et al., 2012). Although NTS symptoms are often limited to the gastrointestinal tract, invasive disease can occur, especially in high-risk groups such as immunocompromised patients and the elderly (Parry et al., 2013). Invasive disease has also been observed in several low-income settings (Kingsley et al., 2009; Feasey et al., 2016; Ashton et al., 2017) and was estimated to result in 3.4 million cases and 681,000 deaths worldwide in 2010, with the heaviest burden on the African continent (Ao et al., 2015).

While the use of antimicrobial agents to treat invasive and severe gastrointestinal cases has decreased mortality rates for NTS infections, and veterinary antimicrobial therapy has lowered the risk of zoonoses, these interventions have come with a price. Increased use of the traditional first-line drugs ampicillin, chloramphenicol, streptomycin, sulphonamides and tetracycline quickly led to the emergence of ACSSuT-type *S. enterica* serovar Typhimurium strains in the 1980s, resistant to exactly these drugs (Threlfall et al., 1996; Boyd et al., 2002). Resistance to fluoroquinolones, introduced to circumvent this problem, developed as a consequence of the veterinary use of enrofloxacin (Threlfall et al., 1997). NTS strains resistant to extended-spectrum cephalosporins, an alternative to fluoroquinolones for the treatment of invasive disease, have been detected throughout Europe since the 1990s (Tassios et al., 1999; Villa et al., 2002; Burke et al., 2014). By 2015, 29.3% of the NTS isolates in the European Union were categorized as multidrug-resistant (MDR) (EFSA, 2017). More recently, the spread of an extensively drug-resistant strain of *S.* Kentucky, non-susceptible to ciprofloxacin, extended-spectrum cephalosporins, carbapenems, most aminoglycosides, trimethoprim-sulfamethoxazole, and azithromycin, has sparked concern (Le Hello et al., 2013). Resistance to azithromycin has been reported in other NTS serovars (Villa et al., 2015; Nair et al., 2016). Acquired resistance to colistin, considered the antimicrobial of last resort for the treatment of many MDR Gram-negative pathogens, has also been detected in NTS (Doumith et al., 2016).

Due to the association of MDR NTS infection with increased mortality and higher costs to the healthcare system, determination of antimicrobial resistance (AMR) profiles is an essential part of NTS surveillance in reference laboratories. Phenotypic serotyping and phage typing at Public Health England's (PHE) Gastrointestinal Bacteria Reference Unit (GBRU) has been replaced by the routine implementation of whole genome sequencing (WGS) for identification and surveillance of *Salmonella* since April 2014 (Ashton et al., 2016). As well as providing information about phylogenetic relationships between isolates, the sequencing data can be used to identify resistance determinants and therefore constitutes a rapid alternative to monitor emerging trends in AMR patterns of NTS. With this study, we sought to evaluate the suitability of inferring AMR profiles from genotype in NTS in comparison with traditional phenotypic susceptibility testing.

MATERIALS AND METHODS

Bacterial Isolates

Between April 2014 and March 2015, PHE received 7,009 NTS *S. enterica* subspecies *enterica* isolates for surveillance purposes. After deduplication of outbreak cases and exclusion of isolates with WGS of insufficient quality, results of phenotypic susceptibility testing and genotypic profiling were available for 3,491 isolates (49.81%). These comprised 227 different serovars plus 66 isolates that could not successfully be subtyped to serovar level. GBRU's routine phenotypic testing strategy for surveillance of NTS attempts to maximize the detection of AMR by focussing on serovars known to have high resistance rates. This leads to an under-representation of some serovars, such as *S.* Enteritidis, and an over-representation of others, such as *S.* Infantis and *S.* Kentucky, in this dataset. Amongst the isolates included in the analysis, the ten most common serovars were *S.* Typhimurium (23.69%, n = 827), *S.* Enteritidis (8.42%, n = 294), *S.* Virchow (4.01%, n = 140), *S.* Stanley (3.98%, n = 139), *S.* Newport (3.75%, n = 131), *S.* Infantis (3.47%, n = 121), *S.* Kentucky (3.12%, n = 109), *S.* Oranienburg (2.06%, n = 72), *S.* Java (2.03%, n = 71) and *S.* Saint-Paul (1.78%, n = 62). The majority (n = 3487) of isolates were of human origin, three were derived from food and one from an unknown source.

Whole Genome Sequencing

Sequencing libraries were prepared from extracted genomic DNA using the Nextera XT DNA Sample Preparation kit (Illumina, Cambridge, United Kingdom). Short-read sequence fragments of 100 bp were produced by paired-end sequencing on an Illumina HiSeq platform (Illumina, Cambridge, United Kingdom). FASTQ sequences were deposited in the NCBI Short Read Archive under the BioProject PRJNA315192. Short read archive accession numbers are available in **Supplementary Table S1**.

Serovar Prediction

Serovars were inferred from the sequencing data using the seven-gene MLST and eBurst Group approach (Achtman et al., 2012; Ashton et al., 2016). Traditional serotyping was not performed.

Detection of Antimicrobial Resistance Determinants

For the identification of AMR determinants, the 'Genefinder' algorithm was employed, which maps the sequencing reads to a set of reference sequences using Bowtie 2 followed by generation of an mpileup file using Samtools (Langmead and Salzberg, 2012). To establish the presence of the reference sequence or nucleotide variations within the read set, a positive match had to meet the following criteria: query coverage 100%, base-call variation > 85% and nucleotide identity > 90%.

The reference database used included acquired genes and mutations known to confer resistance to β-lactams (including penicillins, 2nd-, 3rd- and 4th-generation cephalosporins and carbapenems), fluoroquinolones, aminoglycosides, sulphonamides, tetracyclines, trimethoprim and phenicols (Day et al., 2017b; Sadouki et al., 2017). Variants of β-lactamase genes were identified with 100% identity based on reference sequences downloaded from the Lahey[1] or NCBI β-lactamase data resources[2]. Further reference sequences for acquired resistance genes were obtained from the Comprehensive Antimicrobial Resistance Database[3] and the Resfinder datasets[4]. Chromosomal mutations were limited to previously published variations within the quinolone resistance-determining regions (QRDRs) of *gyrA* and *parC*.

Antimicrobial Susceptibility Testing

Isolates were recovered from the PHE archive and retrospective susceptibility testing was performed and interpreted using EUCAST breakpoints and screening concentrations[5]. For the purpose of epidemiologically screening the large numbers of *S. enterica* isolates received by the reference laboratory, agar dilution with Mueller–Hinton agar was used to determine breakpoint values of ampicillin, cefotaxime, ceftazidime, cefpirome, ertapenem, chloramphenicol, gentamicin, streptomycin, tobramycin, sulphonamides, tetracycline, trimethoprim and ciprofloxacin. Decreased susceptibility (MIC 0.06–0.25 mg/L) and resistance (MIC > 0.5 mg/L) were distinguished for ciprofloxacin. If required, MICs were confirmed by Etest® (bioMérieux, Marcy-l'Étoile, France) or by agar dilution. To aid detection of OXA-48-like carbapenemases and acquired AmpC genes, breakpoint testing of temocillin and cefoxitin, respectively, was included in the panel.

Statistical Analysis

Comparisons were made between the prevalence of resistance determinants associated with isolates, for which a travel history was available, and those for which there was no information about recent travel. Travel destinations were grouped according to the United Nations geoscheme. Statistical significance was assessed using the chi-square test. A p-value $\leq$ 0.05 was considered statistically significant. Statistical analysis was performed using R's chisq.test function.

[1] www.lahey.org

[2] https://www.ncbi.nlm.nih.gov/pathogens/beta-lactamase-data-resources

[3] http://arpcard.mcmaster.ca

[4] https://cge.cbs.dtu.dk/services/data.php

[5] http://www.eucast.org/fileadmin/src/media/PDFs/EUCAST_files/Breakpoint_tables/v_8.0_Breakpoint_Tables.pdf

RESULTS

Comparison Between Phenotypic and Genotypic AMR Profiles

Phenotypic and genotypic AMR profiling was highly correlated, with the profiles of 3,415 isolates (97.82%) being entirely in agreement for both approaches for all 15 antimicrobials from nine different classes. For the 76 isolates with discordant results, the genotype wrongly predicted pan-susceptibility for one isolate (1.32%). This isolate was phenotypically resistant to one antimicrobial. For a further 64 discrepant isolates (84.21%), the mismatch was based on false or missing prediction of resistance to a single antimicrobial, and for 11 (14.47%) based on two antimicrobials.

Overall, 88 (0.17%) out of a possible 52,365 isolate/antimicrobial combinations did not match (**Table 1**). Of these discrepant results, 69/88 (78.41%) constituted major errors (MEs), i.e., isolates were genotypically predicted to be resistant but showed phenotypic susceptibility, rather than very major errors (VMEs), which were genotypically susceptible but phenotypically resistant. The largest fraction of the 88 mismatches could be attributed to streptomycin (n = 59, 67.05%), 51 of these were MEs. Sensitivity of resistance prediction from genotype was ≥95% for all antimicrobials except temocillin. However, only a single isolate was found to be phenotypically resistant to temocillin. Specificity of prediction exceeded 98% for all fifteen antimicrobials tested.

Resistance to β-lactams

Of the 3,491 isolates in this study, 749 (21.46%) carried genes conferring resistance to β-lactam antibiotics (**Supplementary Table S2**). The most common genes were the penicillinase-encoding $bla_{\text{TEM}-1}$ (n = 603) and $bla_{\text{PSE}-1}/bla_{\text{CARB}-2}$ (n = 75). Additionally, other TEM-type β-lactamase genes were detected in 36 isolates, including $bla_{\text{TEM}-117}$ (n = 12) and $bla_{\text{TEM}-135}$ (n = 7). The single ME associated with predicted ampicillin resistance was due to the presence of $bla_{\text{TEM}-1}$ without phenotypic consequences. Seven isolates (0.20%) carried OXA-type class D β-lactamases. Of these, four were found in S. Typhimurium and two in S. Kentucky.

Genes for CTX-M-type extended-spectrum β-lactamases (ESBLs) were present in 41 isolates (1.17%), most commonly $bla_{\text{CTX}-\text{M}-9}$ (n = 10) and $bla_{\text{CTX}-\text{M}-55}$ (n = 9). Twenty of these were S. Typhimurium and five S. Kentucky. Additionally, four isolates carried the $bla_{\text{SHV}-12}$ ESBL gene. No ESBL genes were detected in S. Enteritidis. Combinations of penicillinase and ESBL genes occurred in 16 isolates, most frequently $bla_{\text{TEM}-1}$ with $bla_{\text{CTX}-\text{M}-55}$ (n = 9).

Sixteen isolates (0.46%), seven of these S. Typhimurium and one S. Kentucky, had the acquired AmpC resistance gene $bla_{\text{CMY}-2}$. Carbapenemase genes were not detected.

Resistance to Quinolones

Multiple mutations in the QRDR of the DNA gyrase subunit gene *gyrA* in combination with multiple mutations in the DNA topoisomerase gene *parC* are expected to confer full ciprofloxacin

TABLE 1 | Comparison of phenotypic antimicrobial susceptibility testing and genome-derived resistance prediction for non-typhoidal *Salmonella enterica* (*n* = 3491).

Antimicrobial	Phenotype: susceptible		Phenotype: resistant		Sensitivity (%)	Specificity (%)
	Genotype: resistant	Genotype: susceptible	Genotype: resistant	Genotype: susceptible		
Ampicillin	1	2742	747	1	99.87	99.96
Temocillin	0	3490	0	1	0	100
Cefoxitin	0	3471	19	1	95.0	100
Cefotaxime	0	3434	57	0	100	100
Ceftazidime	0	3444	47	0	100	100
Cefpirome	0	3444	47	0	100	100
Ertapenem	0	3481	10	0	100	100
Chloramphenicol	4	3284	201	2	99.01	99.88
Gentamicin	1	3351	138	1	99.28	99.97
Streptomycin	51	2821	613	8	98.71	98.22
Tobramycin	2	3392	97	0	100	99.94
Sulphonamides	2	2661	828	0	100	99.92
Tetracycline	6	2568	917	0	100	99.77
Trimethoprim	1	3185	301	4	98.69	99.97
Ciprofloxacin	1	3352	137	1	99.28	99.97

Values shown designate the number of isolates. For ciprofloxacin only isolates with a MIC > 0.5 mg/L are shown.

resistance (MIC > 0.5 mg/L) and were observed in 82 isolates (2.35%) (**Table 2**). The most common combinations were either *gyrA*[83:S-F;87:D-Y] (*n* = 41) or *gyrA*[83:S-F;87:D-N] (*n* = 25) in conjunction with *parC*[57:T-S;80:S-I]. For *S.* Kentucky, multiple QRDR mutations were identified in 77 isolates. Neither *S.* Typhimurium nor *S.* Enteritidis had any of these mutations (**Supplementary Table S2**).

A further 599 isolates (17.16%) harbored determinants responsible for reduced susceptibility to ciprofloxacin (MIC 0.06–0.25 mg/L) with or without *parC* mutations. These included a single *gyrA* mutation in the QRDR (*n* = 430), most commonly *gyrA*[87:D-Y] (*n* = 155) or *gyrA*[83:S-Y] (*n* = 112) and/or one or multiple plasmid-mediated quinolone resistance (PMQR) genes (*n* = 195). The most frequent PMQR genes detected were *qnrS1* (*n* = 95) and *qnrB19* (*n* = 49). PMQR genes were rare in *S.* Kentucky (*n* = 1 compared with *n* = 188 for chromosomal mutations). One or more PMQR determinants in combination with a single *gyrA* mutation were found in twenty isolates. Of the isolates carrying both multiple *parC* and *gyrA* mutations, only one *S.* Indiana had additional PMQR genes, namely the efflux pump-encoding *oqxA* and *oqxB*.

Seven isolates carried the fluoroquinolone- and aminoglycoside-modifying *N*-acetyltransferase gene variant *aac(6′)-Ib-cr*, six of these in combination with other quinolone resistance determinants. Of the 138 isolates showing full ciprofloxacin resistance, nineteen carried a single *gyrA* mutation only, 17 a single *gyrA* mutation together with a PMQR gene, 20 had one or more PMQR genes and a single isolate carried *parC*[57:T-S] only (**Table 2**). The single ME associated with predicted ciprofloxacin resistance was based on the presence of *gyrA*[83:S-F;87:D-N] and *parC*[57:T-S;80:S-I] resulting in reduced susceptibility instead of full resistance.

Resistance to Aminoglycosides

Genes predicted to confer resistance to streptomycin were detected in 728 isolates (20.85%): 436 had *strA-strB* only and 292 carried genes encoding aminoglycoside adenylyltransferases, most commonly *aadA2* (*n* = 189) and *aadA17* (*n* = 107) (**Supplementary Table S2**). Both *strA-strB* and an *aadA* variant were observed in 101 isolates. Of the 51 MEs associated with streptomycin resistance, 27 were due to the presence of *strA-strB* and twelve had *aadA2* and *aadA17* without phenotypic consequences.

All but eight of the total 3,491 isolates carried an aminoglycoside acetyltransferase *aac(6′)*-type gene. However, the majority either had the *aac(6′)-Iy* (*n* = 1997), more common in *S.* Enteritidis (*n* = 297), or *aac(6′)-Iaa* variant (*n* = 1486), more common in *S.* Typhimurium (*n* = 869) and *S.* Kentucky (*n* = 81). Of the 2,726 isolates carrying either of these two genes as the only aminoglycoside resistance determinant, only eleven showed phenotypic resistance to an aminoglycoside antimicrobial.

Aminoglycoside acetyltransferase *aac(3)* variants associated with resistance to gentamicin and tobramycin were detected in 130 isolates (3.72%), most notably *aac(3)-Id* (*n* = 50) and *aac(3)-IIa* (*n* = 36). *aac(3)-IVa*, which confers resistance to the veterinary aminoglycoside apramycin, was present in 24 isolates. No *aac(3)* variants were found in *S.* Enteritidis. Furthermore, the aminoglycoside adenylyltransferase gene *ant(2″)-Ia* (*n* = 12) and the aminoglycoside phosphotransferase genes *aph(4)-Ia* (*n* = 23) and *aph(3′)-IIa* (*n* = 10) were identified. None of these were present in *S.* Enteritidis or *S.* Kentucky. No 16S rRNA methyltransferase genes were detected. In the single isolate predicted to be resistant to gentamicin but showing phenotypic susceptibility, *aac(3)-IId* was observed. For prediction of tobramycin resistance, one ME was associated with the presence of *ant(2″)-Ia* and the second one with *aac(3)-IIa*.

TABLE 2 | Relationship between decreased ciprofloxacin susceptibility (<CIP, MIC 0.06–0.25 mg/L), full ciprofloxacin resistance (>CIP, MIC > 0.5 mg/L) and the most common genotypic quinolone resistance determinants in non-typhoidal *Salmonella enterica*.

Number of isolates	Phenotype				Genotype
	<CIP		>CIP		
	S	R	S	R	
116	6	110	114	2	*gyrA*[87:D-Y]
51	1	50	42	9	*gyrA*[83:S-Y];*parC*[57:T-S]
51	4	47	51	0	*gyrA*[87:D-N]
47	0	47	43	4	*gyrA*[83:S-Y]
44	0	44	39	5	*parC*[57:T-S];*qnrS1*
41	0	41	0	41	*gyrA*[83:S-F;87:D-Y];*parC*[57:T-S;80:S-I]
37	1	36	28	9	*qnrS1*
34	2	32	34	0	*gyrA*[87:D-Y];*parC*[57:T-S]
32	0	32	31	1	*parC*[57:T-S];*qnrB19*
29	1	28	28	1	*gyrA*[83:S-F];*parC*[57:T-S]
25	0	25	1	24	*gyrA*[83:S-F;87:D-N];*parC*[57:T-S;80:S-I]
23	0	23	20	3	*gyrA*[83:S-F]
19	2	17	19	0	*gyrA*[87:D-G]
17	2	15	17	0	*gyrA*[87:D-G];*parC*[57:T-S]
12	6	6	12	0	*parC*[57:T-S];*qnrD*
12	0	12	9	3	*qnrB19*
11	0	11	0	11	*gyrA*[83:S-F;87:D-G];*parC*[57:T-S;80:S-I]
11	0	11	11	0	*gyrA*[87:D-N];*parC*[57:T-S]
8	0	8	8	0	*qnrA1*
7	1	6	3	4	*aac(6')-Ib-cr*
4	0	4	0	4	*gyrA*[83:S-Y];*parC*[57:T-S];*qnrS1*
4	0	4	0	4	*gyrA*[87:D-Y];*qnrS1*
4	0	4	4	0	*parC*[57:T-S];*qnrB9*
4	0	4	4	0	*parC*[57:T-S];*qnrS2*
4	0	4	2	2	*qnrB6*
3	0	3	0	3	*gyrA*[83:S-F;87:D-G];*parC*[57:T-S;80:S-R]
3	0	3	1	2	*gyrA*[83:S-Y];*parC*[57:T-S];*qnrD*
3	0	3	0	3	*gyrA*[83:S-Y];*qnrS1*
2	0	2	2	0	*parC*[57:T-S];*qnrA1*
2	0	2	2	0	*parC*[57:T-S];*qnrB1*

Values shown are the number of isolates. S, susceptible; R, Resistant.

Resistance to Sulphonamides, Tetracyclines and Trimethoprim

Sulphonamide resistance genes were found in 830 isolates (23.78%): 490 carried *sul2*, 350 *sul1* and 75 *sul3* (**Supplementary Table S2**). Seventy-seven isolates had a combination of two different *sul* genes, most notably *sul1* and *sul2* (n = 37), and four isolates carried all three variants. Of the two MEs that occurred for the prediction of sulphonamide resistance, one was based on the presence of *sul2* and one on the presence of *sul1* without phenotypic consequences.

Tetracycline resistance genes occurred in 927 isolates (26.55%), mostly *tet*(A) (n = 843). Additional, less frequently encountered genes were the efflux pump-encoding *tet*(G) (n = 68), *tet*(C) (n = 10) and *tet*(D) (n = 5), and the ribosomal protection protein-producing *tet*(M) (n = 57). Fifty-six isolates carried a combination of two different genes, mainly *tet*(A) and *tet*(M) (n = 51). Five of the six isolates with predicted but not phenotypic tetracycline resistance harbored *tet*(M).

Trimethoprim resistance-conferring *dfrA* gene variants were identified in 302 isolates (8.65%), most commonly *dfrA12* (n = 84), *dfrA1* (n = 81) and *dfrA14* (n = 65). The remaining isolates carried eight additional variants of *dfrA*. Only one isolate harbored a combination of two different genes (*dfrA1;dfrA12*). The single ME associated with prediction of trimethoprim resistance was due to the presence of *dfrA14* without phenotypic consequences.

Resistance to Phenicols

Genes linked to chloramphenicol resistance were identified in 215 isolates (6.16%) (**Supplementary Table S2**). Efflux pump genes were found in 194 isolates: *floR* (n = 147) and/or *cmlA1* (n = 67). All four MEs were associated with the presence of *cmlA1*. Chloramphenicol acetyltransferase genes of the *catA*- or *catB*-type were detected in 32 isolates. Eleven isolates harbored genes encoding both an efflux pump and an acetyltransferase.

Multidrug Resistance

Out of a total 3,491 isolates, 1,301 (37.27%) were phenotypically resistant to at least one antimicrobial of the testing panel (**Table 3**). For the two most common serovars, S. Typhimurium and S. Enteritidis, this applied to 568/827 (68.68%) and 130/294 (44.22%) isolates, respectively, and for S. Kentucky to 82/109 isolates (75.23%).

MDR, i.e., resistance to three or more antimicrobial classes, was observed in 848 of all the NTS isolates (24.29%), 467 S. Typhimurium (56.47%), 70 S. Kentucky (64.22%) and only 13 S. Enteritidis (4.42%). One S. Typhimurium isolate exhibited resistance to all nine antimicrobial classes tested.

Detected in 231 isolates overall (6.62%), resistance to ampicillin, streptomycin, sulphonamides and tetracyclines was the most commonly occurring MDR profile, all but one isolate were S. Typhimurium. In 219 isolates with this profile, the underlying genotype was bla_{TEM-1}, *strA-strB sul2*, *tet*(A). For S. Enteritidis, decreased susceptibility to ciprofloxacin was observed most frequently (n = 87) with *gyrA*[87:D-Y] being the

TABLE 3 | Most common combinations of antimicrobial resistance phenotypes and genotypes in non-typhoidal *Salmonella enterica* for all serovars, S. Typhimurium and S. Enteritidis.

Serovar	Antimicrobial classes	Number of isolates (%)	Most common phenotypic combination (number of isolates)	Most common genotypic combination (number of isolates)
Total (n = 3491)	0	2190 (62.73)	–	–
	1/2	453 (12.98)	<CIP (196)	*gyrA*[87:D-Y] (52)
	3/4	514 (14.72)	AMP/STR/SUL/TET (231)	*sul2;strA-strB;tet*(A);bla_{TEM-1} (219)
	5/6/7	315 (9.02)	AMP/CHL/STR/SUL/TET (37)	*sul1;tet*(G)*;aadA17;aadA2;floR;* bla_{PSE-1}/bla_{CARB-2}(30)
	8/9	19 (0.54)	AMP/CAZ/CHL/< CIP/CPR/CTX/ETP/ FOX/GEN/STR/SUL/TET/ TMP/TOB (1)	*qnrS1;sul1;sul2;sul3;strA-strB;tet*(M)*;tet*(A)*;dfrA12;aac(3)-Iva;aadA2;aadA12;aph(4)-Ia;cml1;floR;*bla_{CMY-2};bla_{TEM-1} (1)
S. Typhimurium (n = 827)	0	259 (31.32)	–	–
	1/2	101 (12.21)	TET (40)	*tet*(A) (38)
	3/4	317 (38.33)	AMP/STR/SUL/TET (230)	*sul2;strA-strB;tet*(A);bla_{TEM-1} (220)
	5/6/7	138 (16.69)	AMP/CHL/STR/SUL/TET (33)	*sul1;tet*(G)*;aadA17;aadA2;floR;* bla_{PSE-1}/bla_{CARB-2}(28)
	8/9	12 (1.45)	AMP/CAZ/CHL/< CIP/CPR/CTX/ETP/ FOX/GEN/STR/SUL/TET/ TMP/TOB (1)	*qnrS1;sul1;sul2;sul3;strA-strB;tet*(M)*;tet*(A)*;dfrA12;aac(3)-Iva;aadA2;aadA12;aph(4)-Ia;cml1;floR;*bla_{CMY-2};bla_{TEM-1} (1)
S. Enteritidis (n = 294)	0	164 (55.78)	–	–
	1/2	117 (39.80)	<CIP (87)	*gyrA*[87:D-Y] (24)
	≥3	13(4.42)	AMP/ < CIP/STR/SUL/TET (4)	*gyrA*[87:D-N]*;sul2;strA-strB;tet*(A);bla_{TEM-1} (3)
S. Kentucky (n = 109)	0	27 (24.77)	–	-
	1/2	12 (11.01)	AMP/> CIP (5)	*gyrA*[83:S-F;87:D-Y]*;parC*[57:T-S;80:S-I];bla_{TEM-1} (4)
	3/4	19 (17.43)	AMP/> CIP/SUL/TET (11)	*gyrA*[83:S-F;87:D-Y]*;parC*[57:T-S;80:S-I]*;sul1;tet*(A)*;aadA7;*bla_{TEM-1} (3)
	5/6/7	51 (46.79)	AMP/> CIP/GEN /STR/SUL/TET (24)	*gyrA*[83:S-F;87:D-Y]*;parC*[57:T-S;80:S-I]*;sul1;tet*(A)*;aac(3)-Id;aadA7;*bla_{TEM-1} (20)

AMP, ampicillin; FOX, cefoxitin; CTX, cefotaxime; CAZ, ceftazidime; CPR, cefpirome; ETP, ertapenem; CHL, chloramphenicol; GEN, gentamicin; STR, streptomycin; TOB, tobramycin; SUL, sulphonamides; TET, tetracycline; TMP, trimethoprim; <CIP, ciprofloxacin MIC 0.06–0.25 mg/L; >CIP, ciprofloxacin MIC > 0.5 mg/L. Values in the column antimicrobial class denote the numbers of antimicrobial classes the isolates are resistant to.

most common genotypic determinant (n = 24). The majority of resistant S. Kentucky showed phenotypic resistance to ampicillin, ciprofloxacin, gentamicin, streptomycin, sulphonamides and tetracyclines (n = 24). In 20 isolates, this profile was based on the presence of $bla_{\mathrm{TEM-1}}$, *gyrA[83:S-F;87:D-Y]*, *parC[57:T-S;80:S-I]*, *aac(3)-Id*, *aadA7*, *sul1* and *tet*(A) (n = 20). Thirty-three S. Typhimurium isolates (3.99%) exhibited the penta-resistant phenotype with resistance to ampicillin, chloramphenicol, streptomycin, sulphonamides and tetracyclines. Of these, 28 carried a combination of $bla_{\mathrm{PSE-1}}/bla_{\mathrm{CARB-2}}$, *floR*, *aadA17*, *aadA2*, *sul1* and *tet*(G).

AMR and International Travel

Travel history data was available for 1,070 isolates (30.65%) (**Supplementary Table S3**). The proportion of isolates resistant to at least one antimicrobial of the testing panel was significantly higher for isolates known to be travel-associated ($p = 4.5 \times 10^{-5}$) (**Figure 1**). MDR, on the other hand, was correlated with travel to specific regions, namely Eastern Africa (p = 0.04), North Africa (p = 0.005), Western Africa (p = 0.03), Southeast Asia ($p = 3.4 \times 10^{-5}$) and the Caribbean ($p = 1.3 \times 10^{-4}$). ESBL genes were more likely to be found in isolates related to travel to North Africa (p = 0.01) and South America (p = 0.03). Mutations and acquired genes conferring decreased susceptibility or resistance to ciprofloxacin were more likely to occur in travel-associated isolates (resistance-conferring mutations: $p = 8.5 \times 10^{-6}$; single *gyrA* mutations: $p = 4.5 \times 10^{-9}$; PMQRs: $p = 1.3 \times 10^{-7}$). The presence of genes conferring ciprofloxacin resistance was associated with travel to Southern Asia ($p = 5.6 \times 10^{-6}$). Determinants of aminoglycoside resistance were more prevalent in travel-related isolates, particularly for travel destinations in North Africa, Asia and the Caribbean. The presence of sulphonamide and tetracycline resistance genes was linked to travel to Southeast and Western Asia and the Caribbean while *dfrA* genes were commonly found in isolates associated with travel to North Africa ($p = 5.2 \times 10^{-7}$) and South Asia (p = 0.01). Furthermore, travel to North Africa or Southeast Asia was a risk factor for acquisition of isolates carrying chloramphenicol resistance genes (p = 0.009 and $p = 4.7 \times 10^{-10}$, respectively).

DISCUSSION

The implementation of WGS for surveillance of enteric pathogens has revolutionized the work of public health laboratories, as it allows inference of a multitude of pathogen characteristics in a single sequencing run, which would traditionally require a series of independent laboratory tests. A prominent example of the added value provided by WGS is the generation of AMR profiles from the sequences in real-time.

WGS has previously proven successful for prediction of AMR profiles in a variety of gastrointestinal pathogens, including *Shigella sonnei* (Sadouki et al., 2017), *Escherichia coli* (Stoesser et al., 2013; Tyson et al., 2015; Day et al., 2017a), S. Typhi (Day et al., 2017b) and smaller datasets of NTS (Zankari et al., 2013; Nair et al., 2016; McDermott et al., 2016). Our present comparison of phenotypic susceptibility testing and genotypic prediction of AMR profiles based on WGS data for a much larger dataset, comprising 3,491 NTS isolates, identified 88 discordant results (0.17%) out of a possible 52,365 isolate/antimicrobial combinations, with the AMR profiles of 3,415 isolates (97.82%) completely matching for both approaches. Zankari et al. (2013) observed complete agreement of the two approaches for fifty S. Typhimurium isolates but only when excluding ciprofloxacin from the testing panel. Similar to our results, McDermott et al. (2016) found lower sensitivity and specificity for prediction of streptomycin resistance than for other antimicrobials tested.

Despite being an invaluable tool for surveillance purposes, AMR prediction based on WGS data is not yet deemed suitable to guide treatment choices (Ellington et al., 2017). Many MEs, where an isolate is phenotypically susceptible but carries genetic resistance determinants, seem to be associated with the breakpoints used for phenotypic testing. In some cases, the MICs are just below the recommended breakpoints but slight technical variations of the agar dilution method are possible so that the isolate would be falsely classified as susceptible. This seems to be an issue especially when testing for streptomycin resistance (Garcia-Migura et al., 2012), which would explain the relatively large number of mismatches in the present study. Recently, it has been suggested to adapt the breakpoint values to take into account MICs associated with the presence of specific resistance determinants (Tyson et al., 2017). Additionally, many of the resistance genes detected by the algorithm are plasmid-encoded but phenotypic susceptibility testing was carried out retrospectively. During storage and sub-culture of the isolates plasmids may be lost. Thus, genes detected during sequencing after initial cultivation might not be present when retrospective phenotypic testing is performed on a different colony. Furthermore, silent resistance genes, such as $bla_{\mathrm{CMY-2}}$ and *tet* variants, have been observed previously in *Salmonella* (Heider et al., 2009; Adesiji et al., 2014). Other genes, such as the *aac(6′)* variants, are normally silent and only become transcriptionally active in rare cases (Magnet et al., 1999).

The other mismatch category, the VMEs, where an isolate is genotypically predicted to be susceptible but exhibits phenotypic resistance, highlight the importance of active curation of the resistance gene database used for genotypic prediction. Mismatches are likely based on the presence of resistance determinants not included in the reference database used for prediction or on novel, unknown resistance mechanisms, the genetic determinants of which have not yet been described. Our pipeline, for instance, does not detect impermeability or efflux pump genes potentially contributing to ciprofloxacin resistance (Hopkins et al., 2005). Continuous scanning for new research findings should be carried out to enable identification of novel resistance mechanisms. These novel mechanisms will then be incorporated into the reference databases to maintain a high level of prediction sensitivity. Only recently, for example, computational methods identified previously unknown *qnr*-type fluoroquinolone resistance genes (Boulund et al., 2017). Despite these issues, the overall ME and VME rates of 0.13 and 0.04%, respectively, obtained in this study fall below the cut-offs of 3 and 1.5% from the US Food and Drug Administration for authorizing new susceptibility testing devices (FDA, 2009).

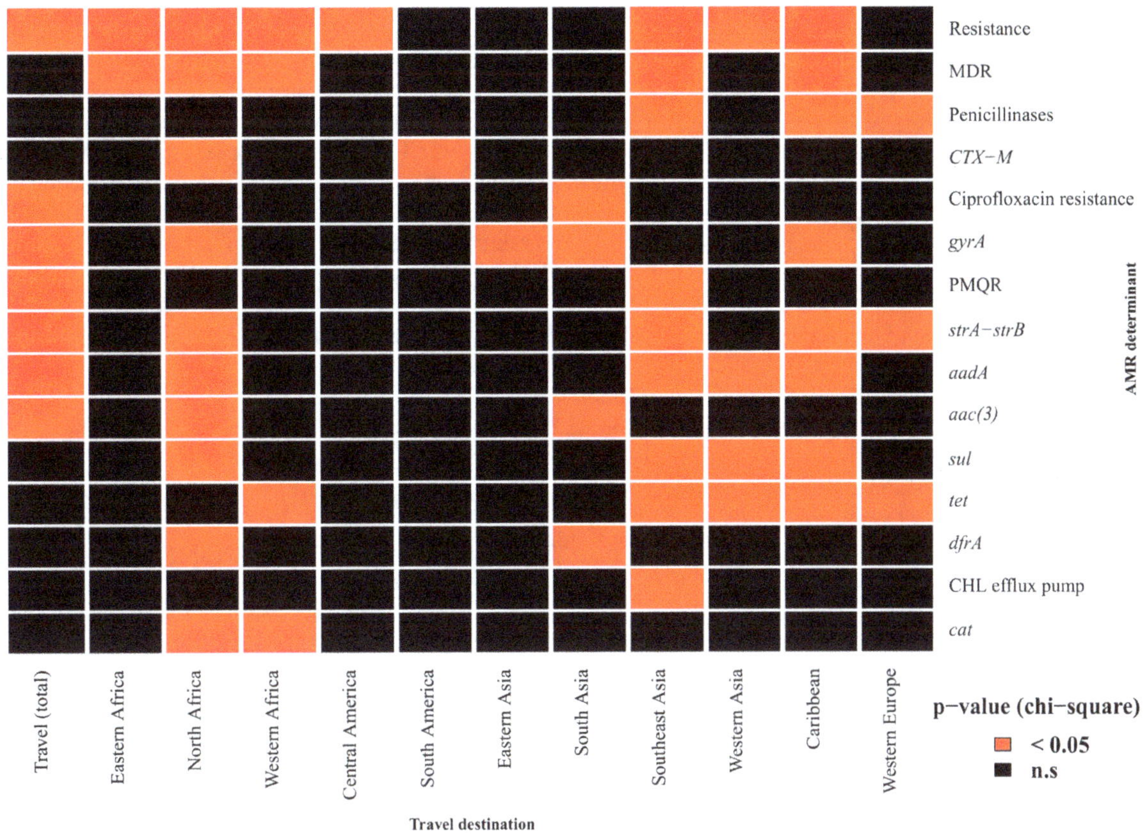

FIGURE 1 | Association of resistance determinants and travel history. Red cells indicate a significant association (p-value < 0.05) between patient travel and the occurrence of phenotypic resistance to at least one antimicrobial, multidrug resistance (MDR) or the presence of specific resistance determinants. *gyrA* denotes isolates with single mutations in the gene responsible for reduced ciprofloxacin susceptibility. Only travel destinations for which there was an association with at least one resistance determinant are shown. PMQR, plasmid-mediated quinolone resistance; CHL, chloramphenicol.

Specificity and sensitivity of ciprofloxacin resistance prediction exceeded 99% but we only considered isolates with an MIC > 0.5 mg/L for this evaluation. Traditionally, *gyrA* mutations in combination with *parC* mutations were thought to be required for ciprofloxacin resistance (Ruiz et al., 1997) and PMQRs on their own were not considered sufficient. Indeed, in our study, the majority of isolates showing resistance carried at least two mutations in both *gyrA* and *parC*. Thirty-seven had a PMQR gene, alone or in conjunction with a single *gyrA* mutation, which would normally be expected to result in reduced susceptibility instead of full resistance. Ciprofloxacin MICs for isolates carrying PMQR genes alone were found to range between 0.25 and 1 mg/L (Garcia-Fernandez et al., 2009) so that some isolates with this profile would be classed as resistant and some as having reduced susceptibility during phenotypic testing. Although the QRDR of *gyrA* is located between amino acids 67 and 106, mutations at positions 83 and 87 are most common (Yoshida et al., 1990). In our study, none of the isolates with mutations at other positions of the QRDR alone exhibited reduced ciprofloxacin susceptibility.

It has been suggested previously that an increased use of alternative antimicrobials, such as ciprofloxacin and extended-spectrum β-lactams, favored the re-emergence of susceptibility to classical first-line drugs (Sood et al., 1999; Rahman et al., 2002). The limitation of our study was that it was biased toward serovars selected for their known high resistance rates, and therefore not a true representation of the expected serovar distribution in England and Wales over this time frame. We were therefore unable to assess changes in incidence of resistance to specific antimicrobials over the years. However, moving forward, genome-derived AMR profiling will provide a robust framework to explore longitudinal trends.

A worrying trend is the increase in resistance to extended-spectrum cephalosporins (Su et al., 2005; Mataseje et al., 2009). Since these antimicrobials are used as an alternative for treatment of invasive disease in case of resistance to ciprofloxacin, the emergence of co-resistance to both antimicrobial classes is of great concern. Co-resistance is especially prevalent in Asia (Lee et al., 2009) and was identified in thirteen isolates (0.37%) in this study, a slight increase from the 0.25% prevalence observed in the UK between 2010 and 2012 (Burke et al., 2014). Seven of the eighteen isolates carrying CTX-M-type ESBLs were associated with travel to Asia and a further seven with travel to North Africa. Similarly, as observed previously (Hopkins et al., 2007), PMQR genes were more likely to be found in isolates from patients who had traveled to Asia. Extensively drug-resistant S. Typhimurium, like the one isolate in this study resistant to all eleven antimicrobial classes tested, have been found in Southeast Asia before (Benacer et al., 2010; Vo et al., 2010). Unfortunately, no travel history data was available for this isolate.

In addition to providing information on AMR for the entire NTS population, WGS-based prediction was able to highlight some interesting genotypic differences between the most common serovars S. Typhimurium and S. Enteritidis and the extensively drug-resistant S. Kentucky: ESBL genes and *aac(3)* variants, while found in S. Typhimurium and S. Kentucky, were absent in S. Enteritidis. Only S. Kentucky carried multiple mutations in the QRDRs of *gyrA* and *parC* but PMQR genes were less common than in other serovars. A more detailed investigation of these differences might lead to a better understanding of the varying outcomes associated with infections caused by different serovars of NTS (Jones et al., 2008).

CONCLUSION

This large-scale study supports the suitability of WGS-based prediction to reliably replace phenotypic susceptibility testing for rapid monitoring of emerging trends in AMR patterns in NTS and for studying the spread of AMR genes in this pathogen population. Since sequencing is routinely used in public health laboratories already, it constitutes a time-saving alternative to traditional approaches that can further our understanding of resistance mechanisms as long as constant curation of the resistance gene database used is warranted. Prediction for further antimicrobials such as macrolides, fosfomycin and colistin will be validated in the near future to increase the robustness of the pipeline. Information derived from WGS-based studies can then be used to inform public health interventions aimed at limiting further dissemination of AMR genes and thus aid in the fight against the global AMR threat.

AUTHOR CONTRIBUTIONS

SNa, EP, TD, PA, and GG conceived the study. SNe, MRD, MD, CJ, PA, KM, KH, NW, and TD contributed to the data analysis. SNe, CJ, and TD wrote the manuscript. All authors contributed to, read, and approved the final manuscript.

REFERENCES

Achtman, M., Wain, J., Weill, F. X., Nair, S., Zhou, Z., Sangal, V., et al. (2012). Multilocus sequence typing as a replacement for serotyping in *Salmonella enterica*. *PLoS Pathog.* 8:e1002776. doi: 10.1371/journal.ppat.1002776

Adesiji, Y. O., Deekshit, V. K., and Karunasagar, I. (2014). Antimicrobial-resistant genes associated with *Salmonella* spp. isolated from human, poultry, and seafood sources. *Food Sci. Nutr.* 2, 436–442. doi: 10.1002/fsn3.119

Ao, T. T., Feasey, N. A., Gordon, M. A., Keddy, K. H., Angulo, F. J., and Crump, J. A. (2015). Global burden of invasive nontyphoidal *Salmonella* disease, 2010[1]. *Emerg. Infect. Dis.* 21, 941–949. doi: 10.3201/eid2106.140999

Ashton, P. M., Nair, S., Peters, T. M., Bale, J. A., Powell, D. G., Painset, A., et al. (2016). Identification of *Salmonella* for public health surveillance using whole genome sequencing. *PeerJ* 4:e1752. doi: 10.7717/peerj.1752

Ashton, P. M., Owen, S. V., Kaindama, L., Rowe, W. P. M., Lane, C. R., Larkin, L., et al. (2017). Public health surveillance in the UK revolutionises our understanding of the invasive *Salmonella* Typhimurium epidemic in Africa. *Genome Med.* 9:92. doi: 10.1186/s13073-017-0480-7

Benacer, D., Thong, K.-L., Watanabe, H., and Puthucheary, S. D. (2010). Characterization of drug resistant *Salmonella* enterica serotype Typhimurium by antibiograms, plasmids, integrons, resistance genes and PFGE. *J. Microbiol. Biotechnol.* 20, 1042–1052. doi: 10.4014/jmb.0910.10028

Boulund, F., Berglund, F., Flach, C.-F., Bengtsson-Palme, J., Marathe, N. P., Larsson, D. G. J., et al. (2017). Computational discovery and functional validation of novel fluoroquinolone resistance genes in public metagenomic data sets. *BMC Genomics* 18:682. doi: 10.1186/s12864-017-4064-0

Boyd, D., Cloeckaert, A., Chaslus-Dancla, E., and Mulvey, M. R. (2002). Characterization of variant *Salmonella* genomic island 1 multidrug resistance regions from serovars Typhimurium DT104 and Agona. *Antimicrob. Agents Chemother.* 46, 1714–1722. doi: 10.1128/AAC.46.6.1714-1722.2002

Burke, L., Hopkins, K. L., Meunier, D., Pinna, E., Fitzgerald-Hughes, D., Humphreys, H., et al. (2014). Resistance to third-generation cephalosporins in human non-typhoidal *Salmonella* enterica isolates from England and Wales, 2010-12. *J. Antimicrob. Chemother.* 69, 977–981. doi: 10.1093/jac/dkt469

Day, M., Doumith, M., Jenkins, C., Dallman, T. J., Hopkins, K. L., Elson, R., et al. (2017b). Antimicrobial resistance in Shiga toxin-producing *Escherichia coli* serogroups O157 and O26 isolated from human cases of diarrhoeal disease in England, 2015. *J. Antimicrob. Chemother.* 72, 145–152. doi: 10.1093/jac/dkw371

Day, M. R., Doumith, M., Do Nascimento, V., Nair, S., Ashton, P. M., Jenkins, C., et al. (2017a). Comparison of phenotypic and whole genome sequencing-derived antimicrobial resistance profiles of *Salmonella enterica* serovars Typhi and Paratyphi. *J. Antimicrob. Chemother.* 73, 365–372. doi: 10.1093/jac/dkx379

Doumith, M., Godbole, G., Ashton, P., Larkin, L., Dallman, T., et al. (2016). Detection of the plasmid-mediated *mcr-1* gene conferring colistin resistance in human and food isolates of *Salmonella enterica* and *Escherichia coli* in England and Wales. *J. Antimicrob. Chemother.* 71, 2300–2305. doi: 10.1093/jac/dkw093

EFSA (2017). The European Union summary report on antimicrobial resistance in zoonotic and indicator bacteria from humans, animals and food in 2015. *EFSA J.* 15:212.

Ellington, M. J., Ekelund, O., Aarestrup, F. M., Canton, R., Doumith, M., Giske, C., et al. (2017). The role of whole genome sequencing in antimicrobial susceptibility testing of bacteria: report from the EUCAST Subcommittee. *Clin. Microbiol. Infect.* 23, 2–22. doi: 10.1016/j.cmi.2016.11.012

FDA (2009). *Class II Special Controls Guidance Document: Antimicrobial Susceptibility Test (AST) Systems; Guidance for Industry and FDA.* Rockville, MD: US Food and Drug Administration, 42.

Feasey, N. A., Hadfield, J., Keddy, K. H., Dallman, T. J., Jacobs, J., Deng, X., et al. (2016). Distinct *Salmonella* Enteritidis lineages associated with enterocolitis in high-income settings and invasive disease in low-income settings. *Nat. Genet.* 48, 1211–1217. doi: 10.1038/ng.3644

Garcia-Fernandez, A., Fortini, D., Veldman, K., Mevius, D., and Carattoli, A. (2009). Characterization of plasmids harbouring *qnrS1*, *qnrB2* and *qnrB19* genes in *Salmonella*. *J. Antimicrob. Chemother.* 63, 274–281. doi: 10.1093/jac/dkn470

Garcia-Migura, L., Sunde, M., Karlsmose, S., Veldman, K., Schroeter, A., Guerra, B., et al. (2012). Establishing streptomycin epidemiological cut-off values for *Salmonella* and *Escherichia coli*. *Microb. Drug Resist.* 18, 88–93. doi: 10.1089/mdr.2011.0064

Heider, L. C., Hoet, A. E., Wittum, T. E., Khaitsa, M. L., Love, B. C., Huston, C. L., et al. (2009). Genetic and phenotypic characterization of the bla_{Cmy} gene from *Escherichia coli* and *Salmonella enterica* isolated from food-producing animals, humans, the environment, and retail meat. *Foodborne Pathog. Dis.* 6, 1235–1240. doi: 10.1089/fpd.2009.0294

Hopkins, K. L., Davies, R. H., and Threlfall, E. J. (2005). Mechanisms of quinolone resistance in *Escherichia coli* and *Salmonella*: recent developments. *Int. J. Antimicrob. Agents* 25, 358–373. doi: 10.1016/j.ijantimicag.2005.02.006

Hopkins, K. L., Wootton, L., Day, M. R., and Threlfall, E. J. (2007). Plasmid-mediated quinolone resistance determinant *qnrS1* found in *Salmonella enterica* strains isolated in the UK. *J. Antimicrob. Chemother.* 59, 1071–1075. doi: 10.1093/jac/dkm081

Jones, T. F., Ingram, L. A., Cieslak, P. R., Vugia, D. J., Tobin-D'angelo, M., Hurd, S., et al. (2008). Salmonellosis outcomes differ substantially by serotype. *J. Infect. Dis.* 198, 109–114. doi: 10.1086/588823

Kingsley, R. A., Msefula, C. L., Thomson, N. R., Kariuki, S., Holt, K. E., Gordon, M. A., et al. (2009). Epidemic multiple drug resistant *Salmonella* Typhimurium causing invasive disease in sub-Saharan Africa have a distinct genotype. *Genome Res.* 19, 2279–2287. doi: 10.1101/gr.091017.109

Langmead, B., and Salzberg, S. L. (2012). Fast gapped-read alignment with Bowtie 2. *Nat. Methods* 9, 357–359. doi: 10.1038/nmeth.1923

Langridge, G. C., Fookes, M., Connor, T. R., Feltwell, T., Feasey, N., Parsons, B. N., et al. (2015). Patterns of genome evolution that have accompanied host adaptation in *Salmonella*. *Proc. Natl. Acad. Sci. U.S.A.* 112, 863–868. doi: 10.1073/pnas.1416707112

Lee, H.-Y., Su, L.-H., Tsai, M.-H., Kim, S.-W., Chang, H.-H., Jung, S.-I., et al. (2009). High rate of reduced susceptibility to ciprofloxacin and ceftriaxone among nontyphoid *Salmonella* clinical isolates in Asia. *Antimicrob. Agents Chemother.* 53, 2696–2699. doi: 10.1128/AAC.01297-08

Le Hello, S., Harrois, D., Bouchrif, B., Sontag, L., Elhani, D., Guibert, V., et al. (2013). Highly drug-resistant *Salmonella enterica* serotype Kentucky St198-X1: a microbiological study. *Lancet Infect. Dis.* 13, 672–679. doi: 10.1016/S1473-3099(13)70124-5

Magnet, S., Courvalin, P., and Lambert, T. (1999). Activation of the cryptic *aac(6′)-Iy* aminoglycoside resistance gene of *Salmonella* by a chromosomal deletion generating a transcriptional fusion. *J. Bacteriol.* 181, 6650–6655.

Majowicz, S. E., Musto, J., Scallan, E., Angulo, F. J., Kirk, M., O'brien, S. J., et al. (2010). The global burden of nontyphoidal *Salmonella* gastroenteritis. *Clin. Infect. Dis.* 50, 882–889. doi: 10.1086/650733

Mataseje, L. F., Xiao, J., Kost, S., Ng, L. K., Dore, K., Mulvey, M. R., et al. (2009). Characterization of Canadian cefoxitin-resistant non-typhoidal *Salmonella* isolates, 2005-06. *J. Antimicrob. Chemother.* 64, 723–730. doi: 10.1093/jac/dkp249

McDermott, P. F., Tyson, G. H., Kabera, C., Chen, Y., Li, C., Folster, J. P., et al. (2016). Whole-genome sequencing for detecting antimicrobial resistance in nontyphoidal *Salmonella*. *Antimicrob. Agents Chemother.* 60, 5515–5520. doi: 10.1128/AAC.01030-16

Nair, S., Ashton, P., Doumith, M., Connell, S., Painset, A., Mwaigwisya, S., et al. (2016). WGS for surveillance of antimicrobial resistance: a pilot study to detect the prevalence and mechanism of resistance to azithromycin in a UK population of non-typhoidal *Salmonella*. *J. Antimicrob. Chemother.* 71, 3400–3408. doi: 10.1093/jac/dkw318

Parry, C. M., Thomas, S., Aspinall, E. J., Cooke, R. P., Rogerson, S. J., Harries, A. D., et al. (2013). A retrospective study of secondary bacteraemia in hospitalised adults with community acquired non-typhoidal *Salmonella* gastroenteritis. *BMC Infect. Dis.* 13:107. doi: 10.1186/1471-2334-13-107

Rahman, M., Ahmad, A., and Shoma, S. (2002). Decline in epidemic of multidrug resistant *Salmonella* Typhi is not associated with increased incidence of antibiotic-susceptible strain in Bangladesh. *Epidemiol. Infect.* 129, 29–34. doi: 10.1017/S0950268802007203

Ruiz, J., Castro, D., Goni, P., Santamaria, J. A., Borrego, J. J., and Vila, J. (1997). Analysis of the mechanism of quinolone resistance in nalidixic acid-resistant clinical isolates of *Salmonella* serotype Typhimurium. *J. Med. Microbiol.* 46, 623–628. doi: 10.1099/00222615-46-7-623

Sadouki, Z., Day, M. R., Doumith, M., Chattaway, M. A., Dallman, T. J., Hopkins, K. L., et al. (2017). Comparison of phenotypic and WGS-derived antimicrobial resistance profiles of *Shigella sonnei* isolated from cases of diarrhoeal disease in England and Wales, 2015. *J. Antimicrob. Chemother.* 72, 2496–2502. doi: 10.1093/jac/dkx170

Sood, S., Kapil, A., Das, B., Jain, Y., and Kabra, S. K. (1999). Re-emergence of chloramphenicolsensitive *Salmonella typhi*. *Lancet*, 353, 1241–1242. doi: 10.1016/S0140-6736(99)00637-6

Stoesser, N., Batty, E. M., Eyre, D. W., Morgan, M., Wyllie, D. H., Del Ojo Elias, C., et al. (2013). Predicting antimicrobial susceptibilities for *Escherichia coli* and *Klebsiella pneumoniae* isolates using whole genomic sequence data. *J. Antimicrob. Chemother.* 68, 2234–2244. doi: 10.1093/jac/dkt180

Su, L. H., Wu, T. L., Chia, J. H., Chu, C., Kuo, A. J., and Chiu, C. H. (2005). Increasing ceftriaxone resistance in *Salmonella* isolates from a university hospital in Taiwan. *J. Antimicrob. Chemother.* 55, 846–852. doi: 10.1093/jac/dki116

Tam, C. C., O'brien, S. J., Tompkins, D. S., Bolton, F. J., Berry, L., Dodds, J., et al. (2012). Changes in causes of acute gastroenteritis in the United Kingdom over 15 years: microbiologic findings from 2 prospective, population-based studies of infectious intestinal disease. *Clin. Infect. Dis.* 54, 1275–1286. doi: 10.1093/cid/cis028

Tassios, P. T., Gazouli, M., Tzelepi, E., Milch, H., Kozlova, N., Sidorenko, S., et al. (1999). Spread of a *Salmonella typhimurium* clone resistant to expanded-spectrum cephalosporins in three European countries. *J. Clin. Microbiol.* 37, 3774–3777.

Threlfall, E. J., Frost, J. A., Ward, L. R., and Rowe, B. (1996). Increasing spectrum of resistance in multiresistant Salmonella typhimurium. *Lancet* 347, 1053–1054. doi: 10.1016/S0140-6736(96)90199-3

Threlfall, E. J., Ward, L. R., and Rowe, B. (1997). Increasing incidence of resistance to trimethoprim and ciprofloxacin in epidemic *Salmonella* typhimurium DT104 in England and Wales. *Euro Surveill.* 2, 81–84. doi: 10.2807/esm.02.11.00187-en

Tyson, G. H., Mcdermott, P. F., Li, C., Chen, Y., Tadesse, D. A., Mukherjee, S., et al. (2015). WGS accurately predicts antimicrobial resistance in *Escherichia coli*. *J. Antimicrob. Chemother.* 70, 2763–2769. doi: 10.1093/jac/dkv186

Tyson, G. H., Zhao, S., Li, C., Ayers, S., Sabo, J. L., Lam, C., et al. (2017). Establishing genotypic cutoff values to measure antimicrobial resistance in *Salmonella*. *Antimicrob. Agents Chemother.* 61:e02140-16. doi: 10.1128/AAC.02140-16

Villa, L., Guerra, B., Schmoger, S., Fischer, J., Helmuth, R., Zong, Z., et al. (2015). IncA/C plasmid carrying bla_{NDM-1}, bla_{CMY-16}, and fosA3 in a *Salmonella* enterica serovar corvallis strain isolated from a migratory wild bird in Germany. *Antimicrob. Agents Chemother.* 59, 6597–6600. doi: 10.1128/AAC.00944-15

Villa, L., Mammina, C., Miriagou, V., Tzouvelekis, L. S., Tassios, P. T., Nastasi, A., et al. (2002). Multidrug and broad-spectrum cephalosporin resistance among *Salmonella enterica* serotype enteritidis clinical isolates in Southern Italy. *J. Clin. Microbiol.* 40, 2662–2665. doi: 10.1128/JCM.40.7.2662-2665.2002

Vo, A. T. T., Van Duijkeren, E., Gaastra, W., and Fluit, A. C. (2010). Antimicrobial resistance, class 1 integrons, and genomic island 1 in *Salmonella* isolates from Vietnam. *PLoS One* 5:e9440. doi: 10.1371/journal.pone.0009440

Wain, J., Hendriksen, R. S., Mikoleit, M. L., Keddy, K. H., and Ochiai, R. L. (2015). Typhoid fever. *Lancet* 385, 1136–1145. doi: 10.1016/S0140-6736(13)62708-7

Yoshida, H., Bogaki, M., Nakamura, M., and Nakamura, S. (1990). Quinolone resistance-determining region in the DNA gyrase *gyrA* gene of *Escherichia coli*. *Antimicrob. Agents Chemother.* 34, 1271–1272. doi: 10.1128/AAC.34.6.1271

Zankari, E., Hasman, H., Kaas, R. S., Seyfarth, A. M., Agersø, Y., Lund, O., et al. (2013). Genotyping using whole-genome sequencing is a realistic alternative to surveillance based on phenotypic antimicrobial susceptibility testing. *J. Antimicrob. Chemother.* 68, 771–777. doi: 10.1093/jac/dks496

4

Mycobacterium tuberculosis Acquires Limited Genetic Diversity in Prolonged Infections, Reactivations and Transmissions Involving Multiple Hosts

Marta Herranz [1,2,3], Ilva Pole [4,5], Iveta Ozere [4,6], Álvaro Chiner-Oms [7], Miguel Martínez-Lirola [8], Felipe Pérez-García [1,2], Paloma Gijón [1,2,3], María Jesús Ruiz Serrano [1,2,3], Laura Clotet Romero [9], Oscar Cuevas [10], Iñaki Comas [11,12], Emilio Bouza [1,2,3,13], Laura Pérez-Lago [1,2,3†] and Darío García-de-Viedma [1,2,3*†]*

[1] Servicio Microbiología Clínica y Enfermedades Infecciosas, Hospital General Universitario Gregorio Marañón, Madrid, Spain, [2] Instituto de Investigación Sanitaria Gregorio Marañón, Madrid, Spain, [3] CIBER Enfermedades Respiratorias (CIBERES), Madrid, Spain, [4] Childhood Tuberculosis Department, Centre of Tuberculosis and Lung Diseases, Riga East University Hospital, Riga, Latvia, [5] Latvian Biomedical Research and Study Centre, Riga, Latvia, [6] Department of Infectology and Dermatology, Riga Stradinš University, Riga, Latvia, [7] Unidad Mixta Genómica y Salud, Centro Superior de Investigación en Salud Pública (FISABIO)-Universitat de València, Valencia, Spain, [8] Servicio de Microbiología, Complejo Hospitalario Torrecárdenas, Almería, Spain, [9] Servei de Vigilància Epidemiològica i Resposta a Emergències de Salut Pública al Vallès Occidental i Vallès Oriental, Subdirecció General de Vigilància i Resposta a Emergències de Salut Pública, Agència de Salut Pública de Catalunya, Barcelona, Spain, [10] Servicio de Laboratorio, Institut d'Investigació i Innovació Parc Taulí, I3PT Parc Taulí Hospital Universitari, Universitat Autònoma de Barcelona, Barcelona, Spain, [11] Instituto de Biomedicina de Valencia, Consejo Superior de Investigaciones Científicas, Valencia, Spain, [12] CIBER en Epidemiología y Salud Pública, Madrid, Spain, [13] Departamento de Medicina, Facultad de Medicina, Universidad Complutense de Madrid, Madrid, Spain

Edited by:
Manuel Espinosa,
Centro de Investigaciones Biológicas (CSIC), Spain

Reviewed by:
Marco Rinaldo Oggioni,
University of Leicester, United Kingdom
Giovanni Delogu,
Università Cattolica del Sacro Cuore, Italy
Andrej Trauner,
Swiss Tropical and Public Health Institute, Switzerland

***Correspondence:**
Laura Pérez-Lago
lperezg00@gmail.com
Darío García-de-Viedma
dgviedma2@gmail.com

† These authors have contributed equally to this work.

Background: *Mycobacterium tuberculosis* (MTB) has limited ability to acquire variability. Analysis of its microevolution might help us to evaluate the pathways followed to acquire greater infective success. Whole-genome sequencing (WGS) in the analysis of the transmission of MTB has elucidated the magnitude of variability in MTB. Analysis of transmission currently depends on the identification of clusters, according to the threshold of variability (<5 SNPs) between isolates.

Objective: We evaluated whether the acquisition of variability in MTB, was more frequent in situations which could favor it, namely intrapatient, prolonged infections or reactivations and interpatient transmissions involving multiple sequential hosts.

Methods: We used WGS to analyze the accumulation of variability in sequential isolates from prolonged infections or translations from latency to reactivation. We then measured microevolution in transmission clusters with prolonged transmission time, high number of involved cases, simultaneous involvement of latency and active transmission.

Results: Intrapatient and interpatient acquisition of variability was limited, within the ranges expected according to the thresholds of variability proposed, even though bursts of variability were observed.

Conclusions: The thresholds of variability proposed for MTB seem to be valid in most circumstances, including those theoretically favoring acquisition of variability. Our data point to multifactorial modulation of microevolution, although further studies are necessary to elucidate the factors underlying this modulation.

Keywords: tuberculosis, variability, whole genome sequencing, SNPs, microevolution

INTRODUCTION

Whole-genome sequencing (WGS) has transformed the way we analyze transmission of tuberculosis (TB) (Nikolayevskyy et al., 2016; Comas, 2017; Satta et al., 2017). Identification of transmission clusters is currently based on determination of the magnitude of genomic diversity among the isolates in a population. An exhaustive analysis of the magnitude of this variability acquired in different clinical/epidemiological situations led to the definition of thresholds to determine whether 2 isolates were part of the same transmission chain, i.e., clustered (<5 SNPs) or unrelated (>12 SNPs) (Walker et al., 2013). Since then, these thresholds have been used as a consensus reference, and it has been accepted that microevolution in *Mycobacterium tuberculosis* (MTB) infection occurs within those ranges and with no wider deviations expected (Casali et al., 2016; Seto et al., 2017; Witney et al., 2017).

Acquisition of variability in MTB has been studied not only to determine the SNP thresholds to be applied in genomic epidemiology, but also to help us to understand one of the microevolution paths in MTB infection which may lead to more successful variants, in addition to other molecular events such as insertion/deletions or intragenomic recombination.

Different studies have addressed the emergence of clonal variants in MTB infections according to different molecular markers (Warren et al., 2002; Shamputa et al., 2006; Al-Hajoj et al., 2010; Navarro et al., 2011; Black et al., 2015). The magnitude of intrapatient diversity which can be measured for a strain within the same patient has been found to be comparable to the one observed after sequential rounds of multiple hosts sequentially along transmission chains (Pérez-Lago et al., 2014). Sometimes, these variants have been found associated to prolonged infections or to a history of previous TB (Navarro et al., 2011), suggesting that infection time longer than the average and reactivation events could be associated with the accumulation of genetic diversity. The characteristics of the healthcare system, the efficiency of the drug treatment and the socioeconomic status also have a role on the genetic diversity expected. The analysis of diversity in the population of MTB has been found to be associated in certain cases with substandard life conditions, delayed diagnosis and improper therapeutical management and poor adherence to treatment, factors which can be responsible for prolonged infections (Shamputa et al., 2006; Navarro et al., 2011).

Post-mortem genetic analysis of the MTB population within individuals (Cohen et al., 2011) has demonstrated the accumulation of diversity by MTB after a prolonged infection period. It has been shown with special exhaustivity in a recent article by Lieberman et al. (2016), based on post-mortem multiple sampling and WGS analysis in patients who had received only minimal anti-TB treatment.

The identification of genetic diversity is not a meaningless finding and different authors have found an effect of certain subtle genetic modifications and the expression of neighboring genes and even in the infectivity between closely related variants and the potential functional role of microevolution has been explored (Soto et al., 2004; Tantivitayakul et al., 2010; Pérez-Lago et al., 2011, 2013).

Therefore, although the similarity thresholds defined in the genomic epidemiology of TB (Walker et al., 2013) are based on a wide and solid analysis of patients, additional analysis of microevolution in MTB in specific clinical/epidemiological situations that might represent greater opportunities to acquire variability, which were not extensively represented in the studies defining these cutoffs, could help us to evaluate the robustness of the similarity thresholds proposed and to understand the influence of various infection scenarios on microevolution dynamics.

MATERIALS AND METHODS

Genotyping by MIRU-VNTR Analysis

24-Locus MIRU-VNTR multiplex analysis was performed from cultured isolates. DNA was purified using the Qiagen DNA MiniKit (Qiagen, Hilden, Germany).

The final reaction mixture (50 μl) included 25 μl of PCR Master Mix (QIAGEN multiplex PCR kit), 5 μl of Q solution, and 0.25 μM of each labeled and unlabeled oligonucleotide (3.9 μM for loci 4156 and 2059). The primers used for PCR amplification and PCR conditions have been reported elsewhere (Supply et al., 2006; Oelemann et al., 2007). PCR products were sized using capillary electrophoresis in an ABI Prism 3100 genetic analyzer (Applied Biosystems, NLLab Centraal B.V., Haarlem, The Netherlands). MIRU-VNTR types were compared using Bionumerics (4.6 Applied Maths, Sint-Martens-Latem, Belgium).

Whole-Genome Analysis

DNA was extracted, after recovering all the bacteria present in the liquid cultures by centrifugation, using the standard cetyl trimethyl ammonium bromide (CTAB) method, and DNA libraries were generated following the Nextera XT Illumina protocol (Nextera XT Library Prep kit [FC-131-1024]). Library quality and size distribution were checked by running 2 μl on a 2200 TapeStation Bioanalyzer (Agilent Technologies, USA). The libraries were then normalized based on the average fragment size observed, the library concentration was measured using Qubit 2.0 Fluorometer (Life Technologies, US), and the libraries were pooled. Paired-end sequences were obtained using a MiSeq platform, with an average per base coverage of 87x (range 62x–113).

SNP calling was performed as indicated elsewhere (Pérez-Lago et al., 2014). In summary, after mapping to a hypothetical MTB ancestral genome (identical to H37Rv according to structure but including the maximum-likelihood-inferred ancestral nucleotide positions from a virtual ancestor; Comas et al., 2010), we extracted all variable positions in the strain of interest. In order to avoid false-positive calls, a series of quality filters were applied to data associated with the SNP. First of all, a minimum coverage >20x and mapping quality 20 were required. From all the variants detected, we divide them into homozygous (present at least in 90% of the reads) and heterozygous calls (present in less than 90% of the reads). Only those SNPs in heterozygosis that appear in homozygosis in other member of the cluster, were selected to be included in the analyses. False positive variants could appear due to mapping errors in genome repetitive regions or near indels. So, we filter out these potential errors by omitting from our analysis variants detected in repetitive regions, phages and PE/PPE regions. Also, those variants found near indel areas and in regions with an anomalous accumulation of SNPs (3 or more SNPs in a 10 bp window) were omitted.

Alignments and SNP variants were visualized and checked using the IGV program. Multiple comparisons between the SNPs from different isolates were performed using an in-house script written in R.

Fastq files with the raw data for isolates are deposited (http://www.ebi.ac.uk) under accession numbers ERS2016357-ERS2016427 and ERP002297.

The dN/dS ratio was calculated using the total number of synonymous and non-synonymous variants for the intra- and inter-patient groups. The potential synonymous and non-synonymous sites for the *M. tuberculosis* genome were obtained using the SNAP tool (Ota and Nei, 1994). All the regions omitted in the variant calling pipeline (repeats, phages and PE/PPE) were also omitted when calculating the dN/dS ratio. The reference dataset used to compare the dN/dS ratio was obtained from a previously published work (12808 synonymous SNPs and 21118 non-synonymous SNPs; Comas et al., 2013). Fisher's exact test was applied in the comparison of nonsynonimous/synonymous ratio.

The median-joining networks were constructed from the SNP matrix generated for each case using NETWORK 5.0.0.1.

RESULTS

We aimed to select representative examples to illustrate deviations in the standard infective circumstances faced by MTB that could provide MTB with better opportunities to acquire variability. We examined both intrapatient scenarios (single infections in selected cases) and interpatient scenarios (transmission chains involving serial hosts). We selected 2 settings in which we could expect a more marked intrapatient microevolution: prolonged infections due to poor adherence to treatment and cases in which reactivation occurred sometime after the microbiological cure of the first episode. For the interpatient evaluation, extensive active transmission clusters involving a high number of cases and clusters involving both latency and active transmission events were selected to attempt to obtain a selective pressure that was higher than average.

Intrapatient Analysis: Prolonged Infections

We selected 4 cases involving poor adherence to treatment in which 2 sequential MTB isolates (**Supplementary Table 1**, cases 1–4; all pansusceptible) had been obtained for the same episode over a longer period than usual (20–29 months). Reinfection was excluded because the paired isolates shared identical MIRU-VNTR patterns (data not shown). The comparative WGS analysis of the paired isolates revealed the occurrence of variability, with a range of 1-3 SNPs between isolates (**Figure 1A**; **Supplementary Table 1**).

Intrapatient Analysis: Reactivations

Two cases with complete adherence to treatment and a second isolate with an identical MIRU-VNTR pattern obtained 27 and 56 months after the resolution of the first episode were selected as representatives of reactivation. We evaluated whether the stress arising from the transition from latency to reactivation could lead to greater accumulation of variability. Microevolution was identified in both cases, and 1 and 5 SNPs were found between the sequential isolates (**Supplementary Table 1**, cases 5 and 6) (**Figure 1B**).

Interpatient Analysis: Extensive Active Transmission Clusters

We selected 3 clusters (cluster F, 6 and B) that fulfilled the double criteria of prolonged infection (6, 11, and 12 years, respectively) and a high number of cases (9–17 cases). In all cases, the strains involved were pansusceptible. The microevolution observed in each cluster led to accumulation of, i.e., 11, 20, and 15 SNPs (one of which was in heterozygosis), respectively (**Supplementary Tables 2–4**). However, the accumulation dynamics of SNPs between each 2 linked cases in the transmission networks was moderate (**Figure 2**), with the maximum distances between any of 2 clustered isolates in clusters 6, B, and F being 9, 7, and 6.

Interpatient Analysis: Clusters Including Both Latency and Active Transmission

We now aimed to analyze events with coincidental involvement of latency and active transmission through sequential hosts and added the requirements of prolonged observation periods and/or the involvement of a high number of cases. We identified 3 clusters in 3 different geographic settings (Cluster 1 in Madrid, Spain; Cluster 2 in Sabadell, Spain; and Cluster 3 in Latgale, Latvia) fulfilling as many coincidental factors as possible (see above).

Cluster 1 corresponded to an 11-case outbreak in a school in Madrid. The event consisted of a double outbreak involving the same setting but 3 years apart (**Figure 3**). In 2012, 2 MTB isolates with identical MIRU-VNTR patterns were obtained from

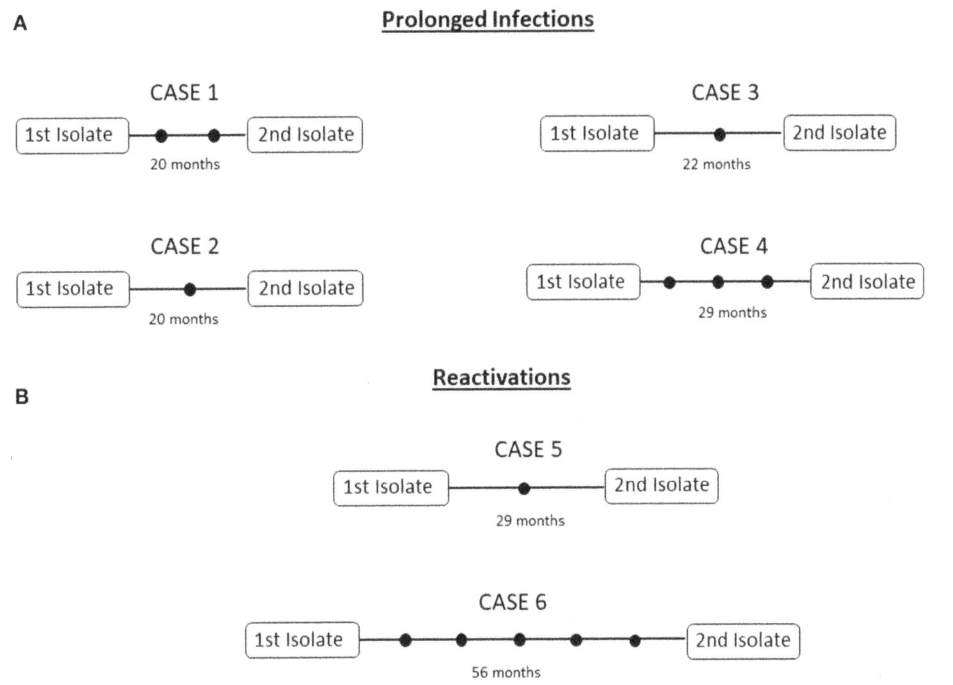

FIGURE 1 | Networks of relationships for the intrapatient **(A)** prolonged infections, **(B)** reactivations in analysis. Each black dot corresponds to a SNP.

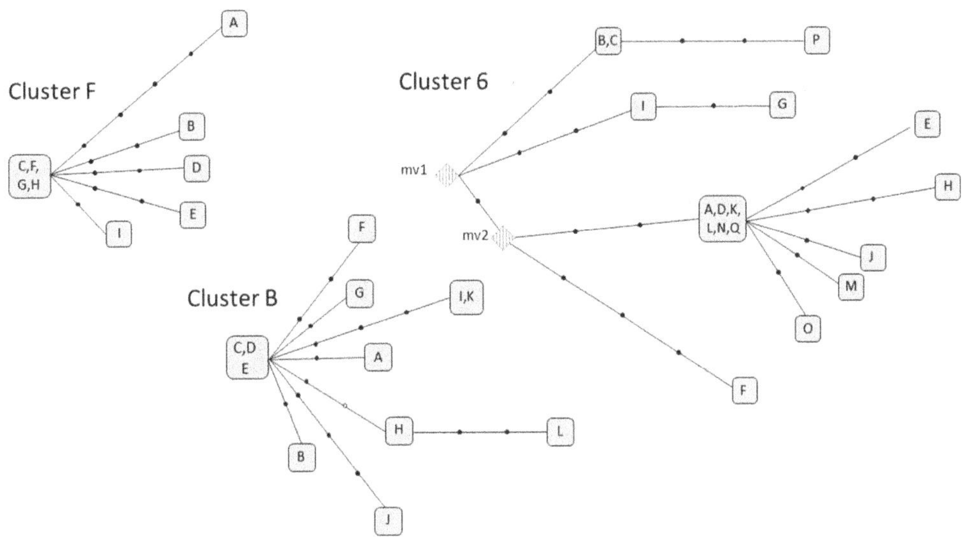

FIGURE 2 | Networks of relationships for the extensive active transmission clusters in analysis. Each black dot corresponds to a SNP. The letters correspond to the isolates. Those sharing the same squared box shared the same polymorphisms. The white dot in Cluster B indicates a heterozygous SNP. mv: (median vector) non-sampled isolates inferred from the distribution of SNPs in the network.

2 epidemiologically linked cases (a child from the school and the father of a child from another class). At the time, contact tracing was incomplete. One of the teachers (case 3)—likely exposed in the 2012 event—developed active TB 3 years later (2015), and 8 new cases of culture-positive TB were diagnosed among the children in the school. All 11 isolates were identical according to the MIRU-VNTR analysis. Unfortunately, 3 isolates were not available for WGS, although we included 4 isolates from the teacher to study intrapatient variability.

We expected that the combination of these coincidental factors (latency, active transmission, and multi-host infection) in this complex event would reveal marked variability between

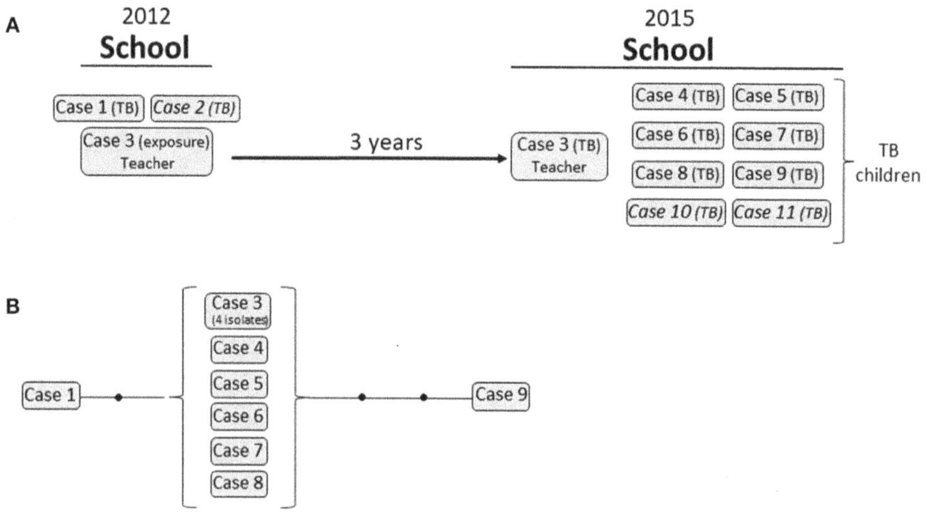

FIGURE 3 | Cluster 1. **(A)** Graphical scheme of the cluster including chronology, transmission settings, and cases involved. **(B)** Network of relationships for the isolates analyzed by WGS. Each black dot corresponds to a SNP. Isolates not available for the WGS analysis are in italic.

the strains involved. However, global variability in SNPs was limited (**Figure 3**; **Supplementary Table 5**). All 4 isolates from the teacher and 5 out 6 isolates from 2015 were identical (0 SNPs) and showed only 1 SNP of difference compared with the 2012 isolate (case 1). One of the isolates from 2015 (case 9) showed 2 additional SNPs.

Cluster 2 corresponded to a complex family outbreak in Sabadell involving 9 cases and spanning 6 years owing to the overlap of active transmission events and latency periods between the exposure of specific cases where the disease developed some years later and eventually coincided with 3 reactivations (**Figure 4**). Seven isolates were available for study, and all of them were identical according to MIRU-VNTR analysis. Despite the complexity of the cluster, WGS revealed global variability for only 5 SNPs (**Figure 4**; **Supplementary Table 5**).

Cluster 3 corresponded to a school and household outbreak in Latvia (**Figure 4**) that involved 5 cases with active TB [1 child (index case), her 3 sisters, and 1 adult (the teacher)] in 2012. Three cases exposed in 2012 developed the disease 1 year later (case 6) and 2 years later (cases 8 and 9). The scenario was further complicated when one of the 2012 cases (case 4) reactivated 2 years later and transmitted TB to her husband (case 7). All 10 isolates were available and shared an identical MIRU-VNTR pattern. WGS analysis revealed that all but 1 of the cases was identical (0 SNPs); however, in the remaining case (case 2), we observed a burst of variability that led to the accumulation of 5 SNPs (**Figure 5**; **Supplementary Table 5**).

DISCUSSION

Microevolution is a phenomenon which can be modulated by various factors. We have to consider it at least a bifactorial event in which both bacterial and host factors intervene. The role of bacterial factors is illustrated by the finding that certain strains from the Beijing lineage show a hypermutator phenotype (Ebrahimi-Rad et al., 2003)- Other non-Beijing strains have been also described to be prone to acquire diversity (Navarro et al., 2017). Microevolution events leading to higher than expected accumulation of variability have been partially addressed in a previous study from our group (Pérez-Lago et al., 2014). Some of these events involved sequential host-to-host transmissions, but an equivalent degree of diversity was observed also within individuals. This intrapatient diversity was detected between isolates infecting different organs, in patients with respiratory and extrarespiratory infection, but also in isolates recovered strictly from the respiratory site. These data mean that the intrapatient diversity might impact on the inference of recent transmission clusters (Pérez-Lago et al., 2014).

As our previous study was based on a convenience sample of isolates selected once they had been shown to have microevolved, we now aimed to perform a more systematic evaluation (i.e., with no preselection of microevolved isolates) of the impact of several circumstances on the acquisition of variability that MTB infection could meet in various clinical or epidemiological scenarios. As with any evolutionary process, microevolution in MTB is the combination of the opportunities the microorganism has to acquire variability and the occurrence of selective pressure bottlenecks that force the selection of clonal variants from among those which emerge. Following this rationale, we selected clinical/epidemiological situations that could lead to increased acquisition of variability (both intrapatient and interpatient).

We began with the intrapatient scenario by selecting single case infections, looking for situations with longer periods of active infection (prolonged infections) in order to increase the likelihood of the emergence of variants and/or looking for factors involved in selective pressure (such as intermittent poor adherence to treatment or transitions from latency to active infection in reactivations). In general, our findings highlight the limited variability acquired, which was within

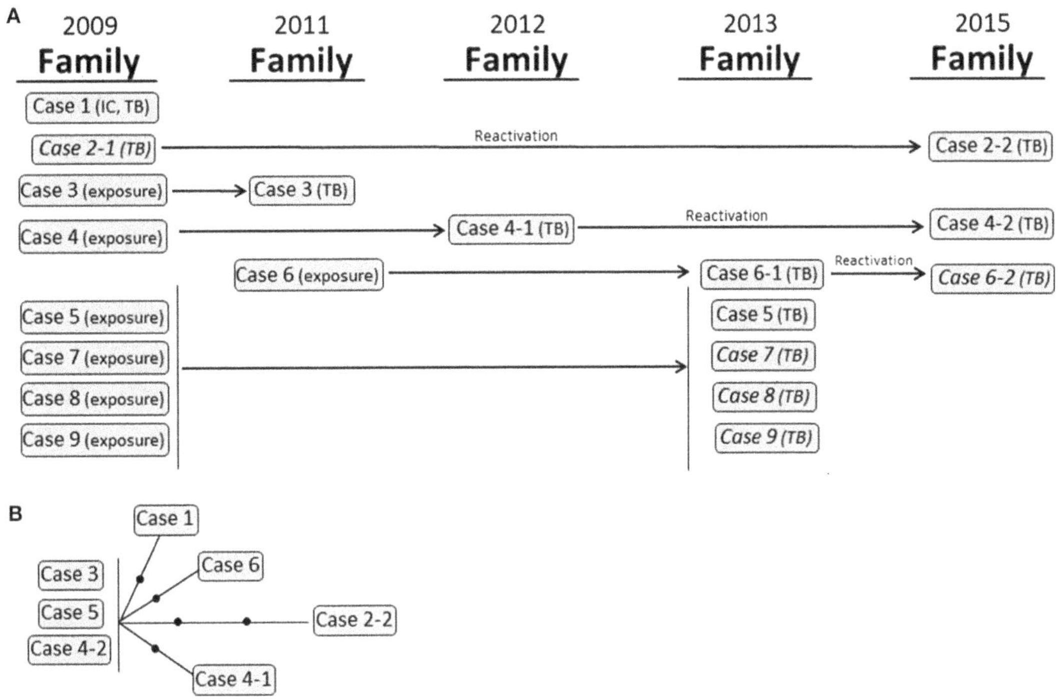

FIGURE 4 | Cluster 2. **(A)** Graphical scheme of the cluster including chronology, transmission settings, and cases involved. **(B)** Network of relationships for the isolates analyzed by WGS. Each black dot corresponds to a SNP. Isolates not available for the WGS analysis are in italic. IC: index case.

the ranges expected. However, in 1 reactivation (case 6, **Supplementary Table 1**), we observed greater accumulation of variability (5 SNPs), thus reaching the lower threshold for considering 2 cases as not epidemiologically linked (Walker et al., 2013). Previously, we had also observed intrapatient situations leading to a higher accumulation of SNPs (Pérez-Lago et al., 2014; Navarro et al., 2017).

The non-systematic association between a greater accumulation of variability and clinical situations that could theoretically favor it underlines the multifactorial nature of this phenomenon. Observations from other studies also lead to contradictory findings. Whereas no variability is found in circumstances where it is highly likely, such as long-term persistent infection by a Beijing strain with constant intermittent treatment (Pérez-Lago et al., 2015a), marked variability is detected in less extreme situations (Pérez-Lago et al., 2014; Navarro et al., 2017). It is also true that greater diversity is described in studies involving MDR strains in which a more dynamic process of acquisition of variability is expected owing to constant competition between resistant and compensatory mutations (Merker et al., 2013), although competition between clonal variants is also expected in heterogeneous populations involving susceptible strains.

We then moved on to the interpatient analysis, which focused first on extensive clusters to evaluate the acquisition of variability when 2 key factors coincided, namely, prolonged observation time and adaptation to a high number of sequentially infected hosts. However, the maximum genetic distance accumulated never surpassed 9 SNPs, and the distance between sequential isolates ranged from 1 to 6 SNPs, i.e., within the previously established thresholds although approaching in some cases to their limits (Walker et al., 2013).

We then looked for the coincidence of several of the factors that had been evaluated independently, making it possible to observe amplification of effects that could only have a subtle impact when evaluated one by one. It was not easy to find transmission events with the intervention of as many as possible of these factors, namely, extended time to allow microevolution to occur, involvement of a high number of hosts to ensure sequential selective pressures due to the establishment of each infection in different hosts, and overlap of transitions from latency to active infections. These kind of clusters with such a complex epidemiological peculiarities were insufficiently represented in the studies defining the SNP thresholds to be applied as a reference in genomic epidemiology (Walker et al., 2013). We found 3 clusters fulfilling all or most of our requirements, each from a different population and involving 2 countries. However, despite the greater opportunities to acquire diversity that the clinical/epidemiological situations in these complex clusters offered, the global variability identified generally lay within the thresholds established, which limits the impact of this diversity on the inference of recent transmission in genomic epidemiology studies.

Although our main aim was to perform a quantitative analysis of the number of SNPs acquired, we had the opportunity to record interesting qualitative observations. The first was the finding that some SNPs (1 each in case 4 and case 6, **Supplementary Table 1**; and one each in case 1 and 4-1 in

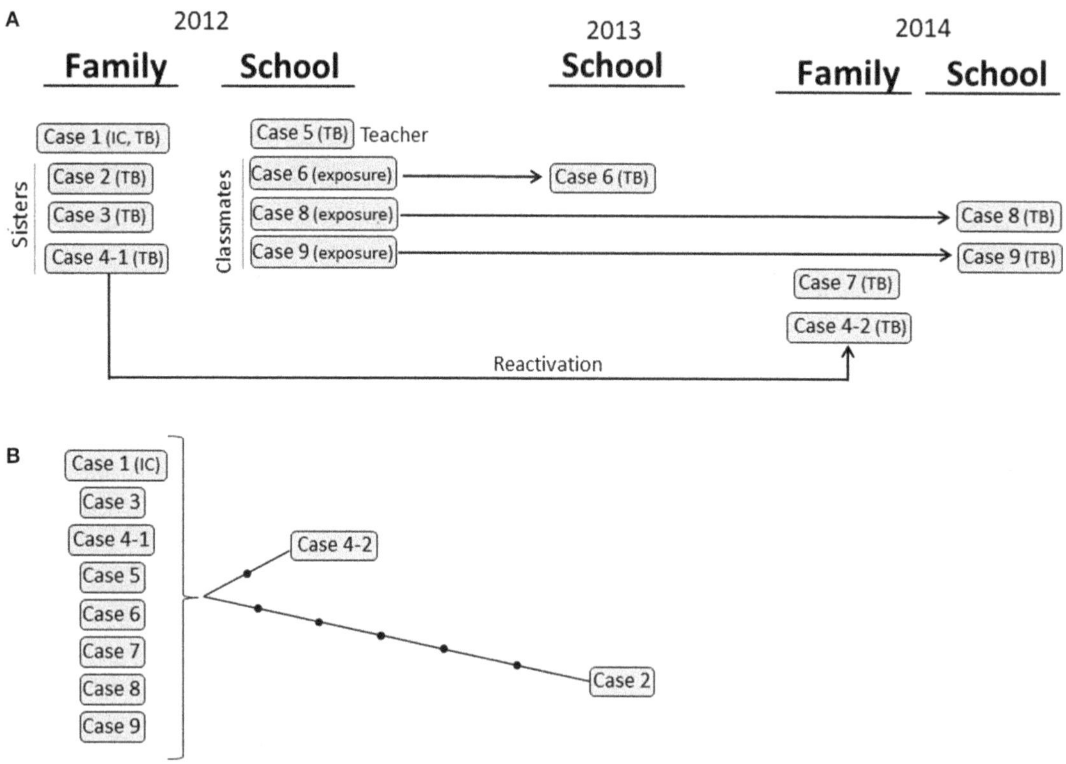

FIGURE 5 | Cluster 3. **(A)** Graphical scheme of the cluster including chronology, transmission settings, and cases involved. **(B)** Network of relationships for the isolates analyzed by WGS. Each black dot corresponds to a SNP. Isolates not available for the WGS analysis are in italic. IC: index case.

cluster 2, **Supplementary Table 5**) were only observed in the first isolate of a chronological series. While this observation may at first sight seem incongruent, owing to the negligible degree of homoplasy described for *Mycobacterium tuberculosis complex* (Hershberg et al., 2008; Comas et al., 2009) and the highly unlikely reversion of mutations once acquired, we believe that it might be explained by undetected coexistence of the variant which acquired the SNP together with the parental variant, which did not have the SNP. Given the unavoidable limitations of sputum sampling with respect to the whole bacterial population in the lung, we likely only detected 1 of the coexisting variants at each sampling point. The observation that isolates recovered from 1 sputum sample do not represent true clonal complexity in the lung has been addressed elsewhere (Black et al., 2015; Pérez-Lago et al., 2015b).

About the functional significance of SNPs (13 synonymous and 40 non-synonymous SNPs), it must be highlighted a lower nonsynonymous/synonimous ratio (0.325) compared with a global database including 220 strains (0.606, $p = 0.04$ Fisher's exact test). The difference is mainly driven by the accumulation of a lower number of nonsynonymous changes in intrapatient cases compared to inter-patient transmission cluster (dN/dS 0.353 vs. 1.714). This observation is in contrast with the expectation that shorter timespan should be associated to higher dN/dS ratios. The lower number of nonysnonymous changes intrapatient may indicate the action of purifying selection. Purifying selection is expected during treatment course (Trauner et al., 2017) and can remove diversity faster, given the clonal nature of MTBC populations. However, the heterogeneity of the dataset involving different patients and transmission clusters as well as the low number of mutations prevents us to support any general conclusion.

Our global analysis did not enable us to identify a clinical/epidemiological situation that could predict greater accumulation of diversity in MTB. We must admit certain limitations resulting of not having considered the diversity based on structural variants and indels or not having performed a high-depth analysis or minority variants. We have focused on diversity based on SNPs mainly due to the fact that this is the criterion in which genomic epidemiology lies, and thus to evaluate the impact of this variability on the inference of recent transmission. The number of SNPs identified under circumstances that could theoretically favor microevolution was limited, within the ranges expected according to the proposed similarity thresholds for considering 2 isolates as related or unrelated (Walker et al., 2013). These findings support the robustness of the cutoffs. However, we also found asymmetric bursts of variability, as reported elsewhere (Pérez-Lago et al., 2014; Navarro et al., 2017). The coincidence of several such cases in the same transmission event could eventually lead us to overlook some transmission clusters.

We must also accept that despite trying to force the coincidence of factors theoretically favoring microevolution, some of these factors, such as length of the active infection period, serial involvement of sequentially infected hosts due to active recent transmission, and lack of adherence to treatment, could be more frequent and of a greater magnitude in settings with weaker diagnostic and control programs. Studies similar to ours should be replicated in these settings to enrich our knowledge of the circumstances which could favor diversity in TB and to determine more precisely its impact on the tracking of transmission.

AUTHOR CONTRIBUTIONS

Experimental load: all authors; Writing: LP-L, DG-d-V; Design conceptualization: LP-L, DG-d-V; Bioinformatic Analysis and design: ÁC-O, IC.

ACKNOWLEDGMENTS

We thank Thomas O'Boyle for proofreading the manuscript. This study was funded by a grant from the Ministry of Economy and Competitiveness ISCIII FIS (grant 15/01554) and cofunded by ERDF (FEDER) Funds from the European Commission: "A way of making Europe." IP and IO were supported by the Latvian National Research program VPP 5.7 "BIOMEDICINE." LP-L was supported by M. Servet contract MS15/00075 and CP15/00075. IC was supported by Ministerio de Educación y Ciencia (Ref SAF2016-77346-R) and the European Research Council (ERC) (638553-TB-ACCELERATE). AC-O is recipient of a FPU fellowship (FPU13/00913) from Ministerio de Educación y Ciencia (Spanish Government).

REFERENCES

Al-Hajoj, S. A., Akkerman, O., Parwati, I., al-Gamdi, S., Rahim, Z., van Soolingen, D., et al. (2010). Microevolution of *Mycobacterium tuberculosis* in a tuberculosis patient. *J. Clin. Microbiol.* 48, 3813–3816. doi: 10.1128/JCM.00556-10

Black, P. A., de Vos, M., Louw, G. E., van der Merwe, R. G., Dippenaar, A., Streicher, E. M., et al. (2015). Whole genome sequencing reveals genomic heterogeneity and antibiotic purification in *Mycobacterium tuberculosis* isolates. *BMC Genomics* 16, 857. doi: 10.1186/s12864-015-2067-2

Casali, N., Broda, A., Harris, S. R., Parkhill, J., Brown, T., and Drobniewski, F. (2016). Whole genome sequence analysis of a large isoniazid-resistant tuberculosis outbreak in London: a retrospective observational study. *PLoS Med.* 13:e1002137. doi: 10.1371/journal.pmed.1002137

Cohen, T., Wilson, D., Wallengren, K., Samuel, E. Y., and Murray, M. (2011). Mixed-strain *Mycobacterium tuberculosis* infections among patients dying in a hospital in KwaZulu-Natal, South Africa. *J. Clin. Microbiol.* 49, 385–388. doi: 10.1128/JCM.01378-10

Comas, I. (2017). Genomic epidemiology of tuberculosis. *Adv. Exp. Med. Biol.* 1019, 79–93. doi: 10.1007/978-3-319-64371-7_4

Comas, I., Chakravartti, J., Small, P. M., Galagan, J., Niemann, S., Kremer, K., et al. (2010). Human T cell epitopes of *Mycobacterium tuberculosis* are evolutionarily hyperconserved. *Nat. Genet.* 42, 498–503. doi: 10.1038/ng.590

Comas, I., Coscolla, M., Luo, T., Borrell, S., Holt, K. E., Kato-Maeda, M., et al. (2013). Out-of-Africa migration and neolithic coexpansion of *Mycobacterium tuberculosis* with modern humans. *Nat. Genet.* 45, 1176–1182. doi: 10.1038/ng.2744

Comas, I., Homolka, S., Niemann, S., and Gagneux, S. (2009). Genotyping of genetically monomorphic bacteria: DNA sequencing in *Mycobacterium tuberculosis* highlights the limitations of current methodologies. *PLoS ONE* 4:e7815. doi: 10.1371/journal.pone.0007815

Ebrahimi-Rad, M., Bifani, P., Martin, C., Kremer, K., Samper, S., Rauzier, J., et al. (2003). Mutations in putative mutator genes of *Mycobacterium tuberculosis* strains of the W-Beijing family. *Emerging Infect. Dis.* 9, 838–845. doi: 10.3201/eid0907.020803

Hershberg, R., Lipatov, M., Small, P. M., Sheffer, H., Niemann, S., Homolka, S., et al. (2008). High functional diversity in *Mycobacterium tuberculosis* driven by genetic drift and human demography. *PLoS Biol* 6:e311. doi: 10.1371/journal.pbio.0060311

Lieberman, T. D., Wilson, D., Misra, R., Xiong, L. L., Moodley, P., Cohen, T., et al. (2016). Genomic diversity in autopsy samples reveals within-host dissemination of HIV-associated *Mycobacterium tuberculosis*. *Nat. Med.* 22, 1470–1474. doi: 10.1038/nm.4205

Merker, M., Kohl, T. A., Roetzer, A., Truebe, L., Richter, E., Rüsch-Gerdes, S., et al. (2013). Whole genome sequencing reveals complex evolution patterns of multidrug-resistant *Mycobacterium tuberculosis* Beijing strains in patients. *PLoS ONE* 8:e82551. doi: 10.1371/journal.pone.0082551

Navarro, Y., Herranz, M., Pérez-Lago, L., Martínez Lirola, M., Indal, T. B., Ruiz-Serrano, M. J., et al. (2011). Systematic survey of clonal complexity in tuberculosis at a populational level and detailed characterization of the isolates involved. *J. Clin. Microbiol.* 49, 4131–4137. doi: 10.1128/JCM.05203-11

Navarro, Y., Pérez-Lago, L., Herranz, M., Sierra, O., Comas, I., Sicilia, J., et al. (2017). In-depth characterization and functional analysis of clonal variants in a *Mycobacterium tuberculosis* strain prone to microevolution. *Front. Microbiol.* 8:694. doi: 10.3389/fmicb.2017.00694

Nikolayevskyy, V., Kranzer, K., Niemann, S., and Drobniewski, F. (2016). Whole genome sequencing of *Mycobacterium tuberculosis* for detection of recent transmission and tracing outbreaks: a systematic review. *Tuberculosis* 98, 77–85. doi: 10.1016/j.tube.2016.02.009

Oelemann, M. C., Diel, R., Vatin, V., Haas, W., Rüsch-Gerdes, S., Locht, C., et al. (2007). Assessment of an optimized mycobacterial interspersed repetitive- unit-variable-number tandem-repeat typing system combined with spoligotyping for population-based molecular epidemiology studies of tuberculosis. *J. Clin. Microbiol.* 45, 691–697. doi: 10.1128/JCM.01393-06

Ota, T., and Nei, M. (1994). Variance and covariances of the numbers of synonymous and nonsynonymous substitutions per site. *Mol. Biol. Evol.* 11, 613–619.

Pérez-Lago, L., Comas, I., Navarro, Y., Gonzalez-Candelas, F., Herranz, M., Bouza, E., et al. (2014). Whole genome sequencing analysis of intrapatient microevolution in *Mycobacterium tuberculosis*: potential impact on the inference of tuberculosis transmission. *J. Infect. Dis.* 209, 98–108. doi: 10.1093/infdis/jit439

Pérez-Lago, L., Herranz, M., Lirola, M. M., Group, I.-T., Bouza, E., and García de Viedma, D. (2011). Characterization of microevolution events in *Mycobacterium tuberculosis* strains involved in recent transmission clusters. *J. Clin. Microbiol.* 49, 3771–3776. doi: 10.1128/JCM.01285-11

Pérez-Lago, L., Navarro, Y., Herranz, M., Bouza, E., and García-de-Viedma, D. (2013). Differences in gene expression between clonal variants of *Mycobacterium tuberculosis* emerging as a result of microevolution. *Int. J. Med. Microbiol.* 303, 674–677. doi: 10.1016/j.ijmm.2013.09.010

Pérez-Lago, L., Navarro, Y., Montilla, P., Comas, I., Herranz, M., Rodríguez-Gallego, C., et al. (2015a). Persistent infection by a *Mycobacterium tuberculosis* strain that was theorized to have advantageous properties, as it was responsible for a massive outbreak. *J. Clin. Microbiol.* 53, 3423–3429. doi: 10.1128/JCM.01405-15

Pérez-Lago, L., Palacios, J. J., Herranz, M., Ruiz Serrano, M. J., Bouza, E., and García-de-Viedma, D. (2015b). Revealing hidden clonal complexity in *Mycobacterium tuberculosis* infection by qualitative and quantitative improvement of sampling. *Clin. Microbiol. Infect.* 21, 147.e1–7. doi: 10.1016/j.cmi.2014.09.015

Satta, G., Lipman, M., Smith, G. P., Arnold, C., Kon, O. M., and McHugh, T. D. (2017). *Mycobacterium tuberculosis*and Whole-genome sequencing: how close are we to unleash its full potential? *Clin. Microbiol. Infect.* doi: 10.1016/j.cmi.2017.10.030. [Epub ahead of print].

Seto, J., Wada, T., Suzuki, Y., Ikeda, T., Mizuta, K., Yamamoto, T., et al. (2017). *Mycobacterium tuberculosis* transmission among elderly persons, yamagata prefecture, Japan, 2009-2015. *Emerging Infect. Dis.* 23, 448–455. doi: 10.3201/eid2303.161571

Shamputa, I. C., Jugheli, L., Sadradze, N., Willery, E., Portaels, F., Supply, P., et al. (2006). Mixed infection and clonal representativeness of a single sputum sample in tuberculosis patients from a penitentiary hospital in Georgia. *Respir. Res.* 7:99. doi: 10.1186/1465-9921-7-99

Soto, C. Y., Menéndez, M. C., Pérez, E., Samper, S., Gómez, A. B., García, M. J., et al. (2004). IS6110 mediates increased transcription of the phoP virulence gene in a multidrug-resistant clinical isolate responsible for tuberculosis outbreaks. *J. Clin. Microbiol.* 42, 212–219. doi: 10.1128/JCM.42.1.212-219.2004

Supply, P., Allix, C., Lesjean, S., Cardoso-Oelemann, M., Rüsch-Gerdes, S., Willery, E., et al. (2006). Proposal for standardization of optimized mycobacterial interspersed repetitive unit-variable-number tandem repeat typing of *Mycobacterium tuberculosis*. *J. Clin. Microbiol.* 44, 4498–4510. doi: 10.1128/JCM.01392-06

Tantivitayakul, P., Panapruksachat, S., Billamas, P., and Palittapongarnpim, P. (2010). Variable number of tandem repeat sequences act as regulatory elements in *Mycobacterium tuberculosis*. *Tuberculosis* 90, 311–318. doi: 10.1016/j.tube.2010.08.003

Trauner, A., Liu, Q., Via, L. E., Liu, X., Ruan, X., Liang, L., et al. (2017). The within-host population dynamics of *Mycobacterium tuberculosis* vary with treatment efficacy. *Genome Biol.* 18:71. doi: 10.1186/s13059-017-1196-0

Walker, T. M., Ip, C. L., Harrell, R. H., Evans, J. T., Kapatai, G., Dedicoat, M. J., et al. (2013). Whole-genome sequencing to delineate *Mycobacterium tuberculosis* outbreaks: a retrospective observational study. *Lancet Infect. Dis.* 13, 137–146. doi: 10.1016/S1473-3099(12)70277-3

Warren, R. M., van der Spuy, G. D., Richardson, M., Beyers, N., Booysen, C., Behr, M. A., et al. (2002). Evolution of the IS6110-based restriction fragment length polymorphism pattern during the transmission of *Mycobacterium tuberculosis*. *J. Clin. Microbiol.* 40, 1277–1282. doi: 10.1128/JCM.40.4.1277-1282.2002

Witney, A. A., Bateson, A. L., Jindani, A., Phillips, P. P., Coleman, D., Stoker, N. G., et al. (2017). Use of whole-genome sequencing to distinguish relapse from reinfection in a completed tuberculosis clinical trial. *BMC Med.* 15:71. doi: 10.1186/s12916-017-0834-4

5

Replication of Staphylococcal Resistance Plasmids

Stephen M. Kwong[1]*, Joshua P. Ramsay[2], Slade O. Jensen[3] and Neville Firth[1]

[1] School of Life and Environmental Sciences, University of Sydney, Sydney, NSW, Australia, [2] School of Biomedical Sciences, Curtin University, Perth, WA, Australia, [3] Antimicrobial Resistance and Mobile Elements Group, Ingham Institute for Applied Medical Research, Sydney, NSW, Australia

Edited by:
Chew Chieng Yeo,
Sultan Zainal Abidin University,
Malaysia

Reviewed by:
Elisabeth Grohmann,
Beuth University of Applied Sciences
Berlin, Germany
Gloria Del Solar,
Consejo Superior de Investigaciones
Científicas (CSIC), Spain

*Correspondence:
Stephen M. Kwong
stephen.kwong@sydney.edu.au

The currently widespread and increasing prevalence of resistant bacterial pathogens is a significant medical problem. In clinical strains of staphylococci, the genetic determinants that confer resistance to antimicrobial agents are often located on mobile elements, such as plasmids. Many of these resistance plasmids are capable of horizontal transmission to other bacteria in their surroundings, allowing extraordinarily rapid adaptation of bacterial populations. Once the resistance plasmids have been spread, they are often perpetually maintained in the new host, even in the absence of selective pressure. Plasmid persistence is accomplished by plasmid-encoded genetic systems that ensure efficient replication and segregational stability during cell division. Staphylococcal plasmids utilize proteins of evolutionarily diverse families to initiate replication from the plasmid origin of replication. Several distinctive plasmid copy number control mechanisms have been studied in detail and these appear conserved within plasmid classes. The initiators utilize various strategies and serve a multifunctional role in (i) recognition and processing of the cognate replication origin to an initiation active form and (ii) recruitment of host-encoded replication proteins that facilitate replisome assembly. Understanding the detailed molecular mechanisms that underpin plasmid replication may lead to novel approaches that could be used to reverse or slow the development of resistance.

Keywords: staphylococci, multiresistance plasmid, plasmid replication, replication initiation protein, plasmid copy number control, antisense RNA

INTRODUCTION

Plasmids are accessory extra-chromosomal genetic elements that provide bacteria with various adaptive qualities that have contributed to their success in diverse environmental niches. Over the last seven decades, the use of antimicrobial compounds in medical, veterinary, and agricultural practices has provided strong evolutionary selection for the acquisition of pre-existing and newly evolved antimicrobial resistance genes for bacterial survival. Plasmids have been instrumental in the dissemination of these resistance genes, and the rapid evolution of multiply resistant strains of *Staphylococcus aureus* in hospitals throughout the world provides an exemplar of this process.

Staphylococcus aureus is commonly normal flora of healthy individuals but is capable of causing serious life-threatening conditions, predominantly in debilitated individuals or patients undergoing surgical procedures. They have long been a primary cause of nosocomial infections and are notorious for the propensity to develop resistance to multiple antimicrobial agents.

Particularly concerning has been the emergence of multidrug-resistant community-associated strains of *S. aureus* capable of causing highly virulent infections in healthy populations (Nimmo, 2012; Planet, 2017). In staphylococci, resistance genes are primarily associated with mobile genetic elements such as plasmids, genomic islands, and transposons (Lyon and Skurray, 1987; Schwarz et al., 2014). Many of these elements are capable of horizontal transfer between bacterial cells by conjugation, mobilization, and/or phage-mediated mechanisms, thus promoting the spread of resistance genes. In *S. aureus*, conjugative plasmids play a central role in enabling intercellular DNA transfer of both conjugative and mobilizable plasmids, which are each capable of accruing multiple resistance and/or virulence genes. It has recently been demonstrated that many of the large staphylococcal multiresistance plasmids, that were previously thought to be non-mobilizable, can in fact be mobilized through the carriage of *oriT* mimic sequences (O'Brien et al., 2015; Ramsay et al., 2016; Ramsay and Firth, 2017). The transfer of these types of plasmids can facilitate multidrug resistance evolution in a single step. Thus, both conjugative and mobilizable plasmid classes are important adaptive tools that have had a major impact on the evolution of antimicrobial resistance.

In the event of plasmid horizontal transfer, a single-stranded copy of the plasmid is transmitted from the donor cell to a recipient cell where it is re-circularized and replicated into a double-stranded form. If the transferred plasmid is capable of efficient autonomous replication in the recipient and becomes established, the plasmid and its resistance genes are rarely lost. Most small plasmids counteract loss during cell division by replicating at high copy numbers. In this situation, there are many plasmid copies that are randomly distributed to daughter cells as the cytoplasmic contents of the parent cell are shared. It is uneconomical for larger plasmids to use this strategy due to the metabolic and genetic loads imposed upon the host cell that would encumber fitness. Large plasmids instead encode various segregational stability systems, often including active partitioning, post-segregational killing, and multimer resolution, that work together to maintain extremely high inherent stability, enabling them to replicate with low copy numbers, presumably to minimize the burden on the host. Crucial to plasmid survival, in terms of both segregational stability and fitness cost, is replication, which can be initiated when needed during cell division and turned off once the ideal plasmid copy number is established after division. This essential capacity to sense and adjust plasmid quantity leads to a defined average copy number in any given host. Plasmids display considerable diversity in their replication systems, the components used to initiate replication and the mechanisms by which replication is controlled (del Solar et al., 1998; Brantl, 2014). Comprehensive reviews on the replication of circular bacterial plasmids (del Solar et al., 1998) and of rolling-circle plasmids (Khan, 1997; Ruiz-Masó et al., 2015) are available in the literature. This article reviews our current understanding of the replication initiation mechanisms of staphylococcal plasmids (last reviewed by Novick, 1989), the systems they use to control plasmid copy number, and includes an updated view of the distribution of *S. aureus* plasmid types.

CLASSIFICATION OF STAPHYLOCOCCAL PLASMIDS

Staphylococcal plasmids range from just over 1 kb to greater than 60 kb in size. In general, the smaller plasmids (between 1 and 8 kb) are cryptic or encode a single resistance determinant and replicate via a rolling-circle replication (RCR) mechanism that is hallmarked by the production of single-stranded intermediates during replication. Many similarities can be drawn between the replication of RCR plasmids and some classes of bacteriophages (e.g., ΦX174), with the major difference being the control of replication frequency. Bacteriophage often propagates their genomes in bursts without consideration to the host's survival, whereas plasmids are host-dependent and thus control their replication to be synergetic with that of the host. Staphylococcal RCR plasmids (also known as class I staphylococcal plasmids) can be further sub-grouped into plasmid families based on the evolutionary relationships of their essential replication initiation proteins. Four main subclasses of RCR plasmids have been found in staphylococci and are represented by the prototypes pT181, pC194, pE194, and pSN2, which are distinguished by the type of replication initiation gene that they carry (Novick, 1989; Firth and Skurray, 2006). Some RCR plasmids have been extensively studied including closely related tetracycline resistance plasmid pT181 and chloramphenicol resistance plasmid pC221 of the pT181 family (Thomas et al., 1995; Khan, 1997). Characterized members of the pC194 family include pC194 itself, conferring resistance to chloramphenicol (Gruss et al., 1987), and the aminoglycoside resistance plasmid pUB110 (Maciag et al., 1988). The best studied plasmid of the pE194 family is tetracycline resistance plasmid pMV158, originally isolated from streptococci but shown to be capable of stable replication in a wide range of bacterial species (Meijer et al., 1995). pSN2 family members have been studied to a lesser extent in regard to their replication and control mechanisms.

Staphylococcal plasmids greater than 8 kb in size typically utilize a theta (θ)-type replication mechanism and have historically been divided into two main classes depending upon their conjugative ability (Novick, 1989; Firth and Skurray, 2006). Non-conjugative theta-replicating plasmids include the well-known β-lactamase/heavy-metal resistance plasmids, pSK1-family multiresistance plasmids, and pSK639-family plasmids (Firth and Skurray, 2006). The conjugative plasmids have traditionally included closely related multiresistance plasmids exemplified by pSK41 (Liu et al., 2013) and pGO1 (Caryl and O'Neill, 2009); however, two new distinct families have recently been described (Ramsay et al., 2016). In general, the conjugative plasmids are larger than the non-conjugative plasmids due to the carriage of extensive gene arrays that encode a type IV secretion system (T4SS), a large multiprotein pore complex through which single-stranded DNA can be transferred to the recipient cell, a nicking relaxase enzyme and its DNA substrate (*oriT*), relaxase accessory proteins, and a coupling protein that provides the basis of recognition between the relaxase and the mating pore. RCR plasmids of the pT181, pC194, and pE194 families and some members of the theta-replicating pSK639-family are known to carry only the relaxase unit (*pre* or *mob* genes and *oriT*), which enables horizontal transfer via mobilization. Plasmid mobilization can occur if these plasmids are able to exploit the mating pore provided by a suitable co-resident conjugative plasmid (or other conjugative element).

PLASMID INCOMPATIBILITY

Non-identical plasmids that share nearly identical replication/maintenance components (DNA, RNA, and/or proteins) display plasmid incompatibility. That is, they are unable to be maintained efficiently through continued rounds of cell division in the absence of plasmid selection. Incompatibility is caused by the inability of the *trans*-acting replication or maintenance components to distinguish "self" from "non-self," and has historically been used as an indication of plasmid relatedness. Staphylococcal plasmids have been placed into at least 15 incompatibility groups (Ruby and Novick, 1975; Iordanescu and Surdeanu, 1980; Novick, 1989; Udo and Grubb, 1991). Ten groups corresponded to RCR plasmids with the remainder being larger, and hence, probably theta-replicating. It was noted earlier that very closely related RCR plasmids were in different incompatibility groups and quite dissimilar theta-replicating plasmids were often in the same incompatibility group (Novick, 1989). This indicated that while incompatibility tests can yield biologically relevant information regarding the ability of plasmids to coexist stably, they do not necessarily indicate relatedness on a whole, particularly in regard to their resistance or other phenotypes. The recombinatory systems (described below) that have shaped the evolution of both RCR and theta-replicating plasmids provide an explanation for this phenomenon.

EVOLUTION OF RESISTANCE PLASMIDS

As the nucleotide sequences of RCR plasmids became available, it was apparent that they were composed of interchangeable modules or gene cassettes (Gruss and Ehrlich, 1989; Novick, 1989). Genetic features, such as resistance genes, mobilization systems, and lagging-strand replication origins were not conserved among plasmids of the same incompatibility group or replicon type. The gene cassette junctions were noted to be abrupt with the level of sequence identity dropping from near perfect to no homology across a single pair of nucleotides (Gruss and Ehrlich, 1989; Novick, 1989). The production of ssDNA in the RCR mechanism is critical to this type of cassette exchange due to the greatly increased capacity for homologous and illegitimate recombination events (Niaudet et al., 1984; Jannière and Ehrlich, 1987). Thus, a cassette can insert or replace another cassette if the appropriate flanking sequences are present in the target plasmid. An example of cassette dissemination by this mechanism is evident with the multidrug resistance locus *qacC* in members of the pC194 family where the gene is located between conserved elements required for leading and lagging strand replication (Leelaporn et al., 1995; Wassenaar et al., 2016). The composition and arrangement of gene cassettes and the associated flanking regions are shuffled by occurrences such as plasmid co-integration and aberrant replication events to generate new combinations (Ballester et al., 1989; Gruss and Ehrlich, 1989).

pT181 was discovered to have defined recombination sites, termed RS_A and RS_B, that promote the formation of plasmid co-integrates by site-specific recombination (Novick et al., 1984). RS_B is a conserved sequence found in the pT181 lagging strand replication origin *palA* (aka *ssoA*), which is broadly carried by diverse staphylococcal RCR plasmids (Novick et al., 1989). Plasmids bearing lagging stand replication origin types *ssoU*, *ssoT*, and *ssoW* also possess the conserved RS_B sequence (Kramer et al., 1999). RS_A is found in the region upstream of *pre*, the protein product (Pre) of which is essential for recombination at this site (Gennaro et al., 1987). Later it was discovered that Pre is a member of the pMV158 Mob protein family, which was essential for the mobilization of this plasmid (Priebe and Lacks, 1989). Therefore, it is now understood that the primary role of pT181 Pre is in mobilization and its role in recombination is a by-product of its nicking activity at RS_A, which functions as the pT181 *oriT*. Pre-mediated nicking of RS_A generates an efficient substrate for co-integrate formation between two plasmids (Priebe and Lacks, 1989).

Theta-replicating plasmids are known to carry transposons but often also contain one or more assemblies of resistance gene clusters inserted into a conserved plasmid backbone (Firth and Skurray, 2006). These regions often correspond to IS*257*-flanked co-integrated copies of small plasmids. The genetic arrangement of clusters resembles that of composite transposons but probably have not inserted as such. Rather the clusters are likely the result of IS*257* transposition by a non-resolved replicative mechanism leading to co-integration of two plasmids and resulting in directly repeated IS*257* copies at each junction (Needham et al., 1995; Leelaporn et al., 1996; Firth and Skurray, 1998). Often the resulting co-integrant is then fine tuned by intra-molecular transposition of IS*257*, which promotes sequence deletions of flanking DNA (e.g., the replication region of co-integrated RCR plasmids; Berg et al., 1998), and sequences within the terminal inverted repeats of IS*257* can modulate the expression of adjoining genes by generating hybrid promoters (Leelaporn et al., 1994; Simpson et al., 2000; Pérez-Roth et al., 2010). Thus, theta-replicating plasmids accrue resistance genes through the incorporation of smaller resistance plasmids and subsequent deletion of problematic or unnecessary sequences (Firth and Skurray, 1998). In addition to influencing the evolution of theta-replicating multiresistance plasmids, IS*257*-mediated RCR plasmid integration events also appear to have played a role in the evolution of staphylococcal genomic islands, such as SCC*mec* (Stewart et al., 1994; Firth and Skurray, 2006).

PLASMID REPLICATION REGIONS

Staphylococcal plasmids carry a 1- to 2-kb region that contains genetic information for autonomous replication and its control. The essential components include (i) an origin of replication (*dso* in RCR plasmids or *ori* in theta-replicating plasmids), (ii) a replication control element (antisense RNA and/or protein), and (iii) a gene encoding the replication initiation protein, Rep. In RCR plasmids, the double-stranded origin, *dso*, contains a sequence-specific binding site for the Rep protein and a short, partially palindromic sequence. In pT181, Rep binding to the *dso* alters the conformation of this palindromic sequence to a cruciform structure that is efficiently nicked by Rep (Koepsel and Khan, 1986; Noirot et al., 1990). In most RCR plasmids, the Rep binding site and the nick site are adjacent, and in other plasmids (e.g., pMV158), the two sites are separated by a short distance of up to 100 bp. The *dso* is often positioned upstream of the *rep* coding region, except in pT181 family plasmids, where the *dso* is found within the *rep* coding region (**Figure 1**). In RCR plasmids, an additional element, the single-stranded origin, *sso*, is required for efficient lagging strand synthesis using the leading strand as template. Therefore, the *sso* is only functional on the leading strand and its orientation cannot be reversed (Gruss et al., 1987). The *sso* displays more variety in its position relative to the *rep* gene (**Figure 1**) and is often separated from it by various gene cassettes indicating that its position is flexible. *sso* sequences are known to limit the host range of RCR plasmids and currently five classes are known to exist: *ssoA*, *ssoU*, *ssoT*, *ssoW*, and *ssoL* (Ruiz-Masó et al., 2015). The ~160 nt *ssoA* carried by pT181 family plasmids restricts their stable replication to staphylococcal species (Gruss et al., 1987) and *ssoA* sequences carried by other plasmid families also appear host-restricted (Kramer et al., 1998). In a single-stranded state, the *ssoA* is capable of forming a large secondary structure that is recognized and primed by RNA polymerase (RNAP) (Kramer et al., 1997). In addition to DNA polymerase III (PolC), DNA polymerase I (PolA) is required during elongation and termination of lagging strand synthesis (Diaz et al., 1994; Kramer et al., 1997). Members of the pC194 and pE194 families have been shown to carry a widely recognized RNAP-dependent *ssoU* permitting them to replicate in all *firmicutes* tested (Boe et al., 1989; Kramer et al., 1995; Lorenzo-Díaz and Espinosa, 2009). *ssoW* of lactococcal plasmid pWVO1 appears to be capable of both RNAP-dependent priming and primosome-dependent priming via a *ssoW*-located primosome assembly site; however, lagging strand replication of this plasmid is restricted to lactococcal species (Seegers et al., 1995).

In theta-replicating plasmids, the position of the origin of replication (*ori*) is variable. In pSK41, the *ori* has been shown to be contained centrally within *rep*, whereas in pSK639 and pCH91 family members, the predicted location of *ori* is upstream and downstream of *rep*, respectively (**Figure 1**). Although *dso* and *ori* both contain one or more binding sites for the replication initiator, they are functionally quite different. Cleavage of the parental DNA strands does not occur during theta-replication and instead a short region of the *ori* (or immediately next to it) is melted upon Rep binding to form an open single-stranded initiation complex.

REPLICATION INITIATORS

To date, staphylococcal plasmids have been found to encode one or more of seven distinct types of replication initiation protein, defined by the conserved domains Rep_trans (pfam02486; pT181), Rep_1 (pfam01446; pC194), Rep_2 (pfam01719; pE194), and RepL (pfam05732; pSN2) for the RCR plasmids, and Rep_3 (pfam01051; pSK639), RepA_N (pfam06970; pSK41), and PriCT_1 (pfam08708; pCH91) for the theta-replicating plasmids. The domain organizations for representatives of each protein family are illustrated in **Figure 2**. All of the initiators contain DNA-binding domains (DBD) for specific binding to their cognate *dso* or *ori*. In some cases, oligomerization domains (OD) have been identified. The RCR initiators are all expected to have topoisomerase activity essential for *dso* cleavage at initiation of replication and for cleavage and ligation during the termination of leading strand synthesis. Conserved catalytic tyrosine residues needed for these activities are also indicated where known (**Figure 2**).

The large number of completely sequenced staphylococcal plasmids now available has allowed us to present an updated view of the distribution of replication initiator genes that are typical of each plasmid family. Due to its clinical significance, plasmids from *S. aureus* clearly dominate the databases. It is currently not known whether distribution of the plasmid classes presented here would vary markedly in other species of the genus. Our analysis of the replication genes of 278 completely sequenced non-identical *S. aureus* plasmids is presented in **Figure 3** (Supplementary Data Sheet 1). Slightly more than half of the plasmids (57%) are predicted to replicate via a theta-type mechanism with the remainder using an RCR mechanism (**Figure 3**). As noted previously (Novick, 1989), the RCR plasmids appear restricted in size with > 90% less than 5 kb. Of the RCR plasmids, 45% were expected to utilize an initiator containing the Rep_1 conserved domain, 28% Rep_trans, 22% RepL, and only 3% Rep_2, with two novel plasmids that could not be grouped. Interestingly, there also appears to be a trend between RCR initiator type and plasmid size where RepL < Rep_1 < Rep_trans although there are also numerous exceptions to this trend (**Figure 3**).

Of 160 *S. aureus* theta-replicating plasmids, those using RepA_N initiators are clearly the most common, being encoded by approximately 80% (**Figure 3**). Rep_3 domain initiators are encoded by approximately 20% and PriCT_1 by about 6% of sequenced plasmids. A total of 11% of theta-replicating plasmids contained two different potential replication initiation genes of these types. In some cases, there is evidence that one of the replicons has been inactivated through genetic alterations (mutations/truncations). In plasmids with both *repA_N*- and *rep_3*-type genes, there are examples where either type appears inactivated. In other cases, there is no obvious defect in either replication region. In regard to the conjugative plasmids, use of RepA_N-type initiators is even more pronounced being used by 90% (18 out of 20), with only two cases of a PriCT_1-type

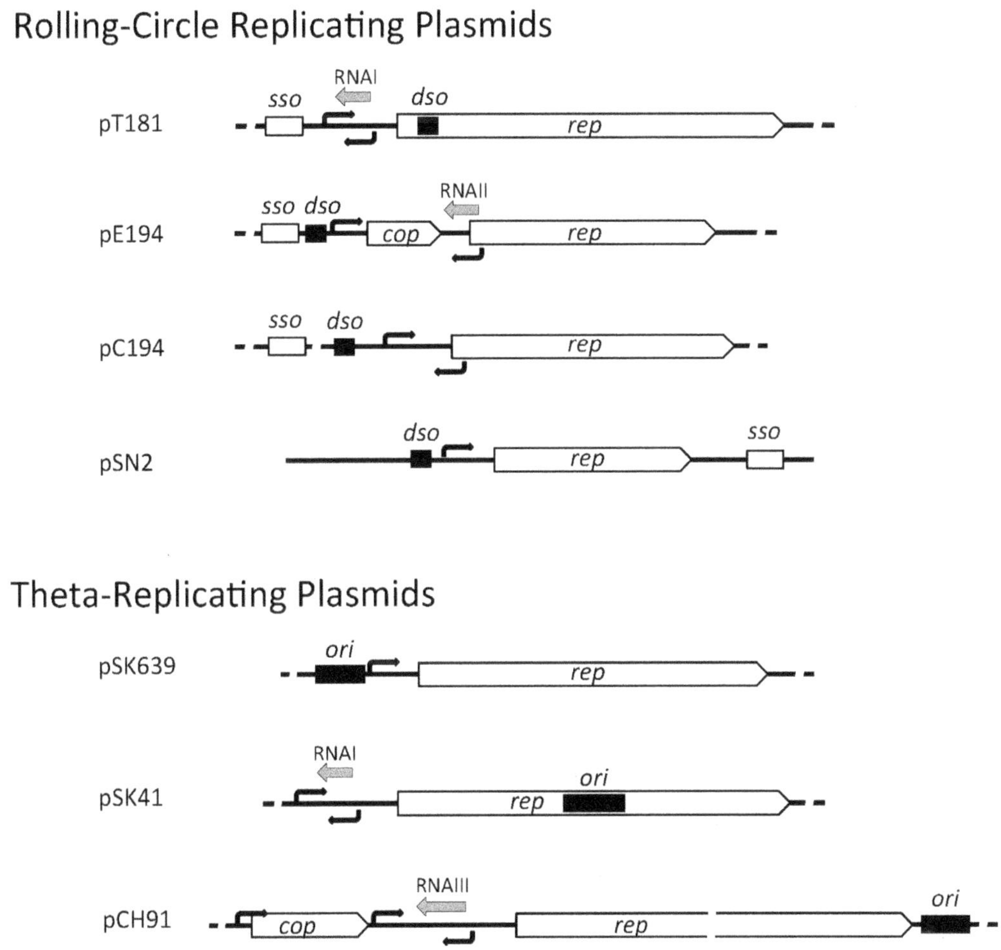

FIGURE 1 | Genetic organization of staphylococcal plasmid replication regions. A representative of each of the seven plasmid initiator types are illustrated and include the *rep* gene for the replication initiation protein, *cop* gene encoding a *rep* repressor, origin of replication (*dso* or *ori*, black rectangles), predicted or determined promoters (black arrows), antisense RNAs (gray arrows), and single-stranded origin of replication (*sso*).

replicon. Interestingly, there are no Rep_3-type plasmids that are predicted to be conjugative.

ROLLING-CIRCLE REPLICATION

Rep_trans Plasmids – the pT181 Family

pT181 family plasmids are widespread in the staphylococci and frequently carry tetracycline or chloramphenicol resistance genes. The two best studied members include pT181 and pC221, and their respective initiators, RepC and RepD, share 82% overall sequence identity. Sequence identity (90%) increases across the common Rep_trans domains, which contain the essential catalytically active tyrosine residues (RepC Y188; RepD Y191). The major region of sequence divergence between the initiators lies in the C-terminal regions, which, for both proteins, has been shown to bind to their cognate *dso*s (Koepsel and Khan, 1986; Thomas et al., 1990). These variable regions have been shown to contain a six amino acid determinant that governs *dso*-binding specificity and limits cross reactivity between initiators and origins of different Rep_trans plasmids (Dempsey et al., 1992; Wang et al., 1992). As such, multiple incompatibility groups exist in the pT181 plasmid family potentially allowing them to stably coexist in the same host cell.

In pT181 family plasmids, the Rep protein binds as a dimer to the *dso* bending the DNA into a cruciform structure that exposes the nick site in its single-stranded form (Noirot et al., 1990). The topoisomerase activity of the initiator results in nicking at the *dso* and formation of a phosphodiester bond between the reactive tyrosine hydroxyl group and the newly created 5′ phosphate of the nicked strand (Thomas et al., 1990). The Rep protein also aids recruitment of PcrA helicase, which is essential to plasmid replication and unwinds and separates the leading parental strand from the lagging template strand (Slatter et al., 2009). Synthesis is initiated from the Rep-generated 3′ hydroxyl group and is known to require DNA polymerase III (Majumder and Novick, 1988). Synthesis of the leading strand continues full circle and 10–12 nt past the original nick site (Rasooly and Novick, 1993). The covalently attached Rep dimer then catalyses a second series of strand-transfer

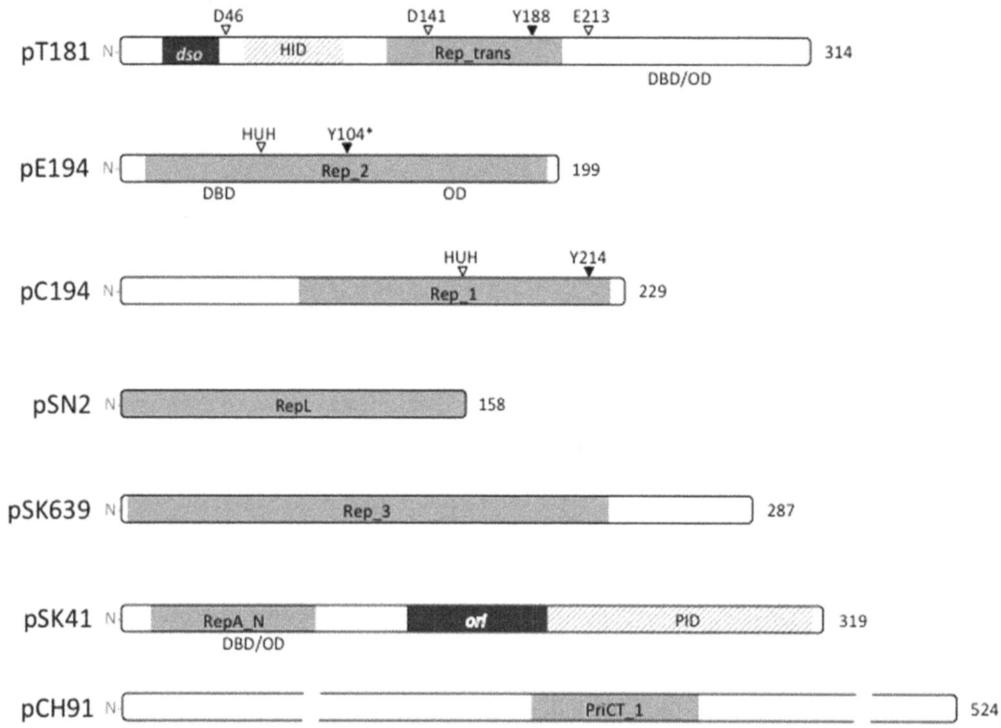

FIGURE 2 | Replication initiation proteins of staphylococcal plasmids. The initiation proteins of prototypical plasmids are represented by bars that indicate the relative sizes of the proteins. Conserved domains within each plasmid replication initiator type are shaded gray. The HUH and DDE metal ion-binding motifs and catalytic tyrosine residues (Y) are labeled where known. *The Y104 of pE194 was predicted by generating an amino acid sequence alignment with the pMV158 RepB protein. In pT181 and pSK41 plasmid families, the origins of replication (*dso* and *ori*) are located within the *rep* coding regions and derived protein sequences in these regions are dispensable for replication. The pT181 RepC–PcrA interaction domain (HID) and the pSK41 Rep–DnaG primase interaction domain (PID) are indicated. Where known, the positions of DNA-binding domains (DBD) and oligomerization domains (OD) are also shown.

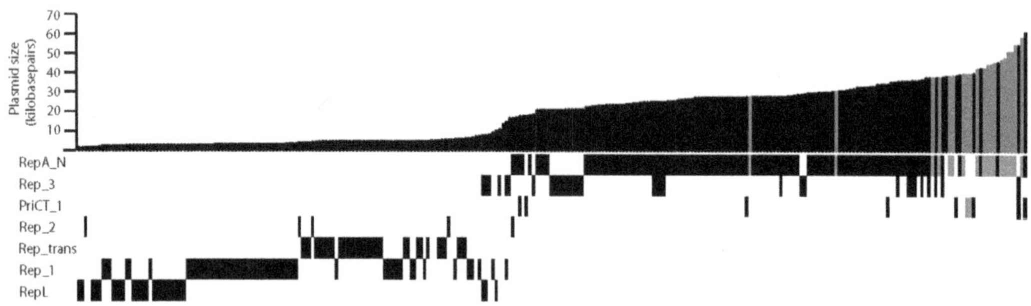

FIGURE 3 | Distribution of replicon types among sequenced *Staphylococcus aureus* plasmids. The graph at the top indicates plasmid sizes for 278 plasmids in the databases. The conserved domains for each replication initiation protein are indicated by a black/gray bar below the graph. Some plasmids encode more than one Rep type. Plasmids predicted to be non-conjugative are represented by black bars and plasmids predicted to be conjugative are represented by gray bars.

reactions that terminates leading strand replication, resulting in a double-stranded plasmid containing a newly synthesized leading strand, a re-circularized single-stranded displaced strand and a Rep dimer (RepC/RepC*) that remains attached to the 10–12 nt adduct rendering the protein inactive for replication (Rasooly et al., 1994). This inactivation mechanism prevents reuse of the initiator for further rounds of replication and is a fundamental prerequisite for pT181 plasmid copy number control.

Crystallography and 3D modeling of Rep_trans initiators suggests that the dimeric structure resembles a horseshoe with a basic channel sitting atop the structure (containing the six amino acid specificity determinant) that could accommodate dsDNA (Carr et al., 2016). The catalytic residues, required for DNA nicking and ligation, including the Tyr active site, are positioned on the inner face. The divalent metal ion-binding site that is essential for topoisomerase activity is coordinated by three distally located residues DDE (**Figure 2**) brought together within antiparallel β-sheets (Carr et al., 2016). Residues affecting PcrA helicase interaction were mapped to the open end of the horseshoe (Carr et al., 2016).

Rep_1 and Rep_2 Plasmids – the pC194 and pE194 Families

Although possessing distinguishable conserved domains, the Rep_1 and Rep_2 families of plasmid replication initiators are united by the HUH metal ion-binding motif and a catalytic Tyr residue which are both necessary for topoisomerase activity (Chandler et al., 2013). The HUH superfamily also includes bacteriophage replication proteins, conjugative DNA relaxases and transposases. Members of the HUH superfamily may have either one or two active tyrosine residues (Chandler et al., 2013). Plasmids in both families exhibit a broad host range with some members having been shown to replicate efficiently in a wide range of Gram-positive and Gram-negative bacteria and even eukaryotic cells (Weisblum et al., 1979; Goursot et al., 1982; del Solar et al., 1987; Coffey et al., 1994; Aleshin et al., 1999).

The pMV158 initiator RepB (210 residues) has been studied in detail and is one of only a small number of Rep proteins with structure determined by crystallography (Boer et al., 2009). Unlike the dimeric state of pT181 initiators, pMV158 RepB is instead shown to form a hexameric ring structure that may encircle the DNA and increase processivity of leading strand synthesis (Ruiz-Masó et al., 2004; Boer et al., 2009). The N-terminal region of RepB contains both the DNA binding and topoisomerase activities of the protein, while the C-terminal domain is required for hexameric oligomerization (Boer et al., 2016). The pMV158 *dso* contains a set of three directly repeated sequences (*bind*) located 84 bp downstream of the nick site, which is found in the loop of a palindromic secondary structure (Puyet et al., 1988). It has been shown that, at least *in vitro*, RepB can bind to DNA fragments containing either the nick site or the *bind* locus, although the affinity of the protein for the latter is much higher (Ruiz-Masó et al., 2007). On the other hand, RepB catalytic activity requires a single-stranded DNA substrate, and the requirement for plasmid supercoiling indicates that the nick site is presented in a single-stranded conformation only when in this topological state (del Solar et al., 1987; Moscoso et al., 1995). In contrast to the Rep_trans type initiators, the active Tyr residue in pMV158 RepB does not appear to form a stable covalent phosphodiester bond with the 5′-end of the cleaved strand (Moscoso et al., 1995). However, a more labile RepB-DNA covalent adduct was observed after rapid treatment of cleavage reactions with SDS and proteinase K (Ruiz-Masó et al., 2016), indicating that RepB inactivation after one round of leading strand synthesis might still occur through a similar mechanism.

The pC194 and pUB110 replication initiation proteins, RepA and RepU, are monomers in solution but bind to their *dso*s cooperatively in pairs (Noirot-Gros et al., 1994; Müller et al., 1995). At high concentrations, approximately six RepU monomers coat the *dso* region and this complex is able to extend upon the adjacent *repU* promoter leading to *repU* transcriptional silencing (Müller et al., 1995). The pC194 RepA protein has three catalytic residues, Tyr214, Glu142, and Glu210 that are essential for nicking and closing activities of the protein but that do not affect *dso* binding and these residues are universally conserved in all plasmids of the pC194 family (Noirot-Gros et al., 1994). The proposed roles of the essential residues are as follows: (i) nucleophilic attack at the nick site during initiation by Tyr214; (ii) hydrolysis of the regenerated nick site at the leading strand termination step promoted by Glu210 (which would act as a general base catalyst); and (iii) metal ion coordination involving Glu142 (Noirot-Gros et al., 1994). The equivalent residues in pUB110 RepU are Tyr241, Glu163, and Glu237 (Noirot-Gros et al., 1994; **Figure 3**). The 37 kDa RepU protein has two forms that show slightly different molecular weights in polyacrylamide gels and it has been suggested that the larger form could be attached to a short oligo (like pT181) or modified in another way that renders it inactive for replication (Müller et al., 1995). Although the precise mechanism used to prevent Rep recycling has not yet been elucidated, such a mechanism would be a prerequisite for effective copy number control (Gros et al., 1989). Interestingly, it has been reported that the Rep proteins of several RCR plasmids, including pC194, can also act as mobilization relaxases in the presence of the integrative conjugative element ICE*Bs1* (Lee et al., 2012).

RepL Plasmids – the pSN2 Family

pSN2 family members are among the smallest RCR plasmids in staphylococci (**Figure 3**). Many pSN2 family plasmids are cryptic or carry a single resistance determinant, with erythromycin resistance being the most common. No members of this family are known to carry mobilization functions. The pSN2 *dso* and *ssoA* can be readily identified based on similarity to other RCR plasmids (Khan and Novick, 1982; Novick, 1989; Khan, 1997) and the *ssoA* was shown to be required for efficient lagging strand synthesis by an RNAP-dependent priming mechanism (Dempsey et al., 1995). Notably the pSN2 family RepL proteins contain a helix-turn-helix (HTH) DNA domain and are predicted to be about 18 kDa, whereas other RCR initiators are considerably larger (Novick, 1989; Catchpole and Dyke, 1992). At least two incompatibility groups have been shown to exist in the pSN2 family (Oliveira et al., 1993) with sequence variation most often occurring in the central part of the protein (residues 85–95), indicating the location of a possible DNA-binding specificity determinant. At this time, no copy number control mechanism has been identified although many pSN2 family members carry a short open reading frame (that is only sometimes annotated) encoding a small, basic protein (45–60 aa) that could potentially play a regulatory role.

THETA-MODE REPLICATION

RepA_N Plasmids

Plasmid replication initiators containing the conserved RepA_N domain are frequently used by theta-replicating *S. aureus* multiresistance plasmids, both conjugative and non-conjugative, and are widespread in many other coagulase-negative staphylococcal species, including plasmids from *S. epidermidis*, *S. haemolyticus*, *S. saprophyticus*, *S. xylosis*, and *S. warneri*. The RepA_N-type initiators are also broadly distributed among other large plasmids of the low G+C Gram-positive firmicutes such as enterococci, lactobacilli, lactococci, and bacilli (Firth et al., 2000; Weaver et al., 2009).

The best studied RepA_N plasmid is the conjugative multiresistance plasmid pSK41, which confers resistance to the aminoglycosides kanamycin, tobramycin, gentamycin, and neomycin, as well as to bleomycin and antiseptics and disinfectants (Berg et al., 1998; Liu et al., 2013). The pSK41 Rep protein (319 aa) has been divided into three functional domains. The N-terminal 120 residues (NTD) contains the conserved RepA_N domain and mediates *ori* specific binding at four Rep boxes found centrally within the *rep* gene (Kwong et al., 2004; Liu et al., 2012). The NTD also contains sequences necessary for oligomerization of the protein (Schumacher et al., 2014). The central domain (121–199 aa), encoded by DNA sequences corresponding to the *ori*, functions as a linker and proteins carrying an in-frame deletion of the central domain can rescue replication of Rep-defective plasmids containing a functional *ori* (Liu et al., 2012). The Rep C-terminal domain (CTD; 200–319) is essential for replication and displays considerably high sequence conservation, but only in plasmids from the same genera, suggesting that it may perform a host-specific function (Weaver et al., 2009). Recently, it was shown that the pSK41 Rep CTD interacts directly with the *S. aureus* DnaG primase (Schumacher et al., 2014).

Crystallographic analysis of RepA_N proteins indicated that the Rep NTD readily forms tetramers and contains a winged HTH that allows interaction with both the major and the minor grooves of Rep box DNA, inducing a bend of approximately 30° (Schumacher et al., 2014). This was consistent with the intrinsic bend associated with A+T-rich tracts and substitution of this Rep box sequence to a G+C-rich tract reduced pSK41 Rep binding by 10-fold (Schumacher et al., 2014). The Rep CTD was shown to form a compact structure composed of five helices and on its own was monomeric. Interestingly, the pSK41 Rep structures of the NTD and CTD were found to display structural similarity to the primosomal protein DnaD, suggesting a common evolutionary origin (Schumacher et al., 2014). Structural similarity between the *B. subtilis* DnaD and DnaB proteins that was not evident in pairwise sequence alignments has also been described (Marston et al., 2010). In *B. subtilis*, DnaD and DnaB play a central role in both DnaA-mediated initiation of replication at *oriC* and also in restart of stalled replication forks by primosome assembly (Bruand et al., 1995; Smits et al., 2011).

Multiple incompatibility groups are likely to exist within the staphylococcal RepA_N plasmids since pSK41 and pSK1, which encode divergent RepA_N proteins, are known to be compatible (Firth et al., 2000). A novel chimeric replication initiation gene was identified in the high-level mupirocin resistance plasmid pPR9 from Spain (Pérez-Roth et al., 2010). pPR9 displays high-level nucleotide sequence homology to pSK41 throughout the backbone including replication, maintenance, and transfer regions, except for a ~500-bp region corresponding to the pSK41 RepA_N domain and replication origin (Pérez-Roth et al., 2010). The pPR9 Rep NTD was instead found to share homology to a putative phage replication protein while retaining > 97% amino acid sequence identity to the pSK41 Rep CTD and 99% nucleotide sequence identity in the upstream *rep* control region. Instead of the winged-HTH present in the RepA_N domain, the pPR9 Rep NTD contains a putative HTH belonging to pfam13730. Construction of a pPR9 mini-replicon showed that the pPR9 *rep* region supported autonomous replication and that it was compatible with a pSK41 mini-replicon (Pérez-Roth et al., 2010). This intriguing modular pPR9 initiator represents a new initiator type that could have evolved to overcome incompatibility barriers. This type of hybrid initiation gene is found in a small number of other plasmids such as pUSA03 from caMRSA strain USA300 (Diep et al., 2006).

Rep_3 Plasmids

One fifth of theta-replicating *S. aureus* plasmids were found to carry a replication initiation gene that gives rise to a product containing the conserved Rep_3 domain (pfam01051). In staphylococci, this type of initiator was first observed on the small (8 kb) *S. epidermidis* trimethoprim resistance plasmid pSK639 (Apisiridej et al., 1997). The pSK639 *rep* gene encodes a protein of 287 residues in length with the conserved Rep_3 domain spanning the first ~220 residues and a ~50 residue CTD. Like the RepA_N plasmids, Rep_3 domain initiators are widely distributed in plasmids from low G+C Gram-positive bacteria including coagulase-positive and -negative staphylococci, enterococci, and lactococci and are distantly related to a large number of iteron-regulated plasmid initiators of Gram-negative bacteria including *Pseudomonas syringae* plasmid pPS10, *Escherichia coli* plasmids F, pSC101, R6K, P1, and broad host range plasmid RK2. The Gram-negative Rep_3 initiators are variously dependent upon DnaA for their replication and often contain dimerization motifs critical for copy number control (Ingmer et al., 1995; Matsunaga et al., 1997; del Solar et al., 1998; Toukdarian and Helinski, 1998; Das and Chattoraj, 2004; Giraldo and Fernández-Tresguerres, 2004; Swan et al., 2006; Konieczny et al., 2014).

Upstream of the pSK639 *rep* coding region, in the vicinity of the promoter, is a series of five, 22 bp tandemly repeated sequences that most likely represent Rep binding sites and constitute the origin of replication (Apisiridej et al., 1997). The position of these potential Rep binding sites suggests that pSK639 Rep may autoregulate its own transcription (Apisiridej et al., 1997) although this has yet to be demonstrated. Iteron-mediated regulation has been shown to be the primary copy number control mechanism in many of the Rep_3 domain plasmid replicons from Gram-negative bacteria and could also play a role in pSK639 copy number regulation. This form of regulation relies upon dimerization domains that allow plasmids to pair (handcuff). In these plasmids, Rep proteins are only active for initiation as monomers and at higher Rep concentrations dimerization promotes plasmid pairing and the inhibition of replication initiation.

PriCT_1 Plasmids

A smaller number of staphylococcal theta-replicating plasmids (~6%) encode a replication initiator belonging to the broad host range Inc18 family, which includes the enterococcal conjugative plasmid pAMβ1 and streptococcal conjugative resistance plasmids pSM19035 and pIP501. The three plasmids share a high degree of sequence identity (Lioy et al., 2010) and utilize closely related Rep proteins containing the conserved PriCT_1 domain (**Figure 2**), which are considerably larger than most plasmid replication initiators (~60 kDa). Staphylococcal plasmids that were detected to contain a PriCT_1 type replication initiator include pCH91 (17 kb), which encodes a type II toxin-antitoxin system *pemIK* (Bukowski et al., 2013), the exfoliative toxin B plasmid pETB (38 kb; Yamaguchi et al., 2001), pWBG707 (Udo et al., 1992), and *cfr*-carrying conjugative plasmid pSA737 (39 kb; Mendes et al., 2013). The other PriCT_1 plasmids detected in *S. aureus* were also found to encode a RepA_N or Rep_3 initiator gene.

Replication of Inc18 plasmids has been characterized in considerable detail using pAMβ1 to study the replication initiation mechanism and pIP501 the copy number control mechanism. These plasmids exhibit a broad host range. In pAMβ1 replication, the RepE monomer binds to a single 25-bp sequence in the origin, which is located immediately downstream of the *repE* gene and induces localized melting of a short DNA region (15 nt) found next to the binding site (Le Chatelier et al., 2001). The RepE protein has higher affinity for non-specific single-stranded DNA than for its double-stranded binding site and this activity is believed to play a role in extending *ori* strand opening. Transcription through the *ori* is essential for the replication process, which is independent of DnaA but requires DNA polymerase I (Bruand et al., 1993; Ceglowski et al., 1993; Bruand and Ehrlich, 1998). It has been proposed that the Rep transcript synthesized by RNAP stalls at *ori* when Rep is bound, the transcript is cleaved (by RNAP or Rep) leaving a ~20-nt RNA that acts as the replication primer, which is extended by DNA polymerase I (Le Chatelier et al., 2001). The D-loop structure generated by DNA polymerase I is then an efficient substrate for PriA-mediated primosome assembly that requires the host-encoded replication proteins DnaB, DnaD, and DnaI (Polard et al., 2002). Sequences central to the pAMβ1 *ori* (5′-TGCCATTACATTTAT-3′) that constitute the RepE binding site (Le Chatelier et al., 2001) and are also found in the minimal *ori* of pIP501 and pSM19035 (Brantl and Behnke, 1992a; Lioy et al., 2010) can be detected in an analogous position downstream of the *rep* genes in staphylococcal plasmids pCH91, pETB, pWBG707, and pSA737, suggesting that they utilize a similar replication initiation mechanism. Furthermore, *cop* and antisense RNA genes similarly positioned to copy number control elements in pIP501 (see below; Brantl and Behnke, 1992b) can also be detected in each of the staphylococcal plasmids.

COPY NUMBER CONTROL MECHANISMS

Antisense RNA-mediated copy number control is broadly utilized by both RCR and theta-replicating staphylococcal plasmids where regulation of replication has been investigated, including members of the pT181, pE194, pC194, and pSK41 families. Copy number control in the pSN2, pSK639, and pCH91 families have not yet been studied in detail, however, as described above, members of the latter family appear likely to utilize copy number control systems similar to the Inc18 broad host range conjugative pIP501 employing both a small protein repressor (Cop) and an antisense RNA-mediated attenuation system to regulate Rep expression (reviewed in Brantl, 2014 and Grohmann et al., 2016). Dual-regulation of copy number by both Cop repressor and antisense RNA has also been well established in members of the pE194 family via pMV158 (del Solar and Espinosa, 2000), although the mechanistic details of the systems are distinctive for each family.

Antisense RNA-Mediated Copy Number Control of Rep_trans Plasmids

pT181 family plasmids use small, untranslated antisense RNAs to regulate expression of the Rep protein and thereby control replication initiation. In pT181, the 87-nt antisense RNA (RNAI) is counter-transcribed to *repC* and is complementary to the *repC* mRNA untranslated leader region. RNA–RNA interaction between RNAI and the *repC* mRNA causes the formation of a thermodynamically stable secondary structure (stem-loop IV) immediately upstream of the *repC* start codon (Novick et al., 1989). Stem-loop IV is predicted to contain the ribosome binding site, however, the main effect of antisense RNA binding is transcriptional termination at stem-loop IV (which resembles a σ-independent terminator) and the attenuated transcripts are incapable of producing RepC (Novick et al., 1989). In the absence of RNAI, sequences in the 5′-proximal arm of stem-loop IV preferentially pair with another complementary sequence in the *repC* leader, termed the pre-emptor, preventing formation of stem-loop IV and allowing full-length *repC* mRNA to be transcribed (Novick et al., 1989).

Copy Number Control in Rep_2 Family Plasmids

pE194 family plasmids are predicted to utilize a copy number control system that has been studied extensively in pMV158 and its deletion derivative pLS1. pMV158 encodes two *trans*-acting negative regulators of the replication initiation gene, *repB*, an antisense RNA (RNAII) and a small repressor protein (CopG; del Solar and Espinosa, 1992, 2000). *copG* is found upstream of *repB* and the two genes form an operon. RNAII is a 48-nt transcript that is counter-transcribed from a promoter within the 5′-end of the *repB* coding sequence and is complementary to a region found immediately upstream of an atypical ribosome binding site essential for RepB translation (López-Aguilar et al., 2013). It has been proposed that mRNA-RNAII duplex formation hinders binding of the ribosome to the translation initiation region (López-Aguilar et al., 2013). RNA–RNA interactions initiate through base contacts in the RNAII 5′ single-stranded tail, while the RNAII stem-loop appears to only play an auxiliary role in RepB translational repression (López-Aguilar et al., 2015). CopG is dimeric in solution and has a ribbon-helix-helix structure (RHH_1, pfam01402; Gomis-Rüth et al., 1998). It has been shown that four dimers bind cooperatively to the *copG* promoter leading to transcriptional repression of the *copG-repB* transcript by competitively inhibiting the RNAP-promoter interaction (Hernández-Arriaga et al., 2009). A recent study has indicated 'crosstalk' between the pMV158 mobilization and replication systems. MobM was found to bind and repress the RNAII promoter, leading to elevated levels of RepB and an increase in plasmid copy number (Lorenzo-Díaz et al., 2017). Staphylococcal pE194 (Rep_2) family plasmids including pCPS49, pDLK3, SAP085B each have analogously positioned elements that are predicted to give rise to a CopG-like repressor and a RNAII-like antisense RNA. Thus, the staphylococcal plasmids carrying a Rep_2 initiator are all expected to utilize similar copy number control system to pMV158. pE194, pCPS49 and SAP085B also carry a *mob/pre* gene distantly related to pMV158 *mobM*. In Rep_1 family plasmids pC194 and pUB110, antisense RNAs are also predicted to be counter-transcribed in the respective Rep translation initiation regions and have been proposed to directly block translation of the initiator (Alonso and Tailor, 1987; Maciag et al., 1988), although the intricacies of these antisense RNA-mediated control systems have yet to be described in the same detail.

Antisense RNA-Mediated Copy Number Control of RepA_N Plasmids

Staphylococcal RepA_N plasmids that have so far been examined are found to employ closely related antisense RNA-mediated copy number control systems, comparable to the prototype pSK41 (Kwong et al., 2008). It is noteworthy that similar RNA-mediated control systems do not appear to be present in RepA_N plasmids from other genera. Expression of pSK41 Rep is negatively regulated by a ~83-nt antisense RNA (RNAI) that is counter-transcribed to the *rep* mRNA and is complementary to its leader region in a position ~100 nt upstream of the translation initiation region (Kwong et al., 2004). It has been proposed that binding of RNAI to the Rep mRNA leader induces formation of a stem-loop secondary structure in the *rep* translation initiation region. However, unlike pT181 family plasmids, the antisense RNA induced stem-loop does not attenuate transcription but rather sequesters the ribosome binding site in the stem-loop rendering it inaccessible to the translation machinery (Kwong et al., 2006). Secondary structure probing of pSK41 RNAI revealed the presence of two stem-loops separated by an 8-nt single-stranded spacer and an unstructured 18-nt 5′-tail (Kwong and Firth, 2015). Mutations in either stem-loop significantly reduced RNAI repressor activity but the single-stranded regions could be deleted without affecting RNAI function (Kwong and Firth, 2015), indicating that complete base pairing between the antisense RNA and its target was not required.

Copy Number Control in PriCT_1 Plasmids

pIP501 copy number is controlled by two *trans*-acting negative regulators, a small repressor protein, CopR, and an antisense RNA, designated RNAIII (see **Figure 1**; pCH91). *copR* is found upstream of the initiator gene, *repR*, but the genes are independently transcribed. CopR consists of 92 amino acid residues and contains a conserved HTH domain (HTH_XRE; pfam01381) that facilitates operator DNA binding as a dimer (Steinmetzer et al., 1998). CopR does not autoregulate but binds and represses transcription from the *repR* promoter (Brantl, 1994). It has also been demonstrated that CopR-mediated repression of the *rep* promoter effectively increases expression of RNAIII by preventing convergent transcription (Brantl and Wagner, 1997). RNAIII interacts with the leader of the RepR mRNA to induce the formation of a terminator-like structure that results in attenuation of the RepR transcript in a similar manner to that observed in pT181 replication control (Brantl et al., 1993). In *S. aureus* plasmids pCH91, pWBG707, and pSA737 all of the copy number control elements present in pIP501 can be detected even though the predicted replication initiators of the staphylococcal plasmids only share ~30% amino acid sequence identity to pIP501 RepR. These include a small HTH domain protein of the XRE family (Cop), a *rep* promoter that gives rise to a long (~320 nt) leader, an antisense RNA promoter positioned midway through the leader that could give rise to an antisense RNA (**Figure 2**), and inverted repeats followed by a poly[T] tract just 5′ of the *rep* translation initiation region that appears capable of forming a σ-independent terminator-like structure. The presence of these elements indicates that the staphylococcal PriCT_1 family plasmids are likely to use an analogous copy number control system to plasmid pIP501.

HOST-ENCODED PROTEINS IN PLASMID REPLICATION

As we have described above, plasmids encode their own replication components for the initiation of replication, including a replication initiation protein and an origin of replication, and a mechanism that controls the expression/activity of the initiation protein. The interaction between initiator and origin prepares the DNA for replication, either by strand-specific cleavage at *dso* (generating a 3′-OH) or melting of strands at *ori*. Both of these replication mechanisms then depend on helicase enzymes to facilitate further duplex melting. In contrast to RCR plasmids, and as part of the initiation process, theta-replicating plasmids require synthesis of a leading strand replication primer to generate 3′-OH. Once initiated, plasmids then rely on replisomes consisting of host-encoded replication proteins that are normally used for chromosomal replication and repair. In theta-replication of plasmids and the chromosome, the replisome is composed of DNA polymerase III holoenzyme, primase, sliding clamps, helicase, and other accessory factors (Kornberg and Baker, 1992). Fundamental differences between the RC replication mechanism and theta-type replication mechanism (assymetric vs. semi-conservative) suggest that the replisome components could be quite different. In this section, we discuss some of the host-encoded proteins that are known to play a role in the replication of staphylococcal plasmids.

In most bacteria, DnaA is the essential replication initiator of the chromosomal origin, *oriC*. In many theta-replicating plasmids of *E. coli*, the Rep proteins have been shown to recruit DnaA to their *ori* to assist in the initiation step and often the *oris* possess DnaA boxes homologous to DnaA-binding sites in *oriC* (del Solar et al., 1998). DnaA has not yet been directly implicated in the replication mechanism of any staphylococcal plasmids and plasmid DnaA boxes have not so far been detected.

Polymerases

Escherichia coli possesses five different DNA polymerases, Pols I, II, III, IV, and V. Pols II, IV, and V are translesion polymerases, Pol III is the core processive polymerase involved in the replisome and PolI is required in lagging strand theta-replication to remove RNA primers and fill in the gaps of Okazaki fragments (Kornberg and Baker, 1992). Low G+C, Gram-positive bacteria usually possess three DNA polymerase enzymes, PolC, DnaE, and PolA, which are thought to be functionally equivalent to PolIII, PolII, and PolI, respectively. DnaE has been shown to be essential for viability in both *S. aureus* and *B. subtilis* (Dervyn et al., 2001; Inoue et al., 2001). In *B. subtilis*, DnaE was not involved in leading strand synthesis but was essential in lagging strand synthesis for initial extension of RNA primers (Sanders et al., 2010). The role of PolA would likely be in removing RNA primers and joining Okazaki fragments as in *E. coli*. PolA has also been shown to be involved in specific stages of replication of some staphylococcal plasmids including lagging strand synthesis of RCR plasmids (Diaz et al., 1994; Kramer et al., 1997) and initial extension of the leading strand RNA primer in PriCT_1-family plasmids (Bruand et al., 1993). In both of these stages of plasmid replication, RNAP is involved in generating the replication primer at the *sso* or by transcription through *ori* as described above.

Helicases

Bacteria have multiple helicases that have specialized roles (Hall and Matson, 1999). In *E. coli*, DnaB helicase is the primary replicative helicase and is required for replication of the chromosome and theta-replicating plasmids (Kornberg and Baker, 1992). Specialized helicases include the misleadingly named "Rep" helicase protein involved in the replication of some phages (Lane and Denhardt, 1975; Takahashi et al., 1979), and UvrD that engages in DNA repair and replication of some viruses and RCR plasmids (Bierne et al., 1997; Bruand and Ehrlich, 2000). In low G+C Gram-positive bacteria, the DnaB homolog, DnaC, is expected to fulfill the main replicative helicase role.

Rolling-circle replication plasmids of the pT181 family have been shown to require the host-encoded helicase PcrA for replication. Mutations in *S. aureus pcrA* led to the accumulation of pT181 initiation complexes indicating that a transition to elongation phase of replication had stalled (Iordanescu and Basheer, 1991). Suppressor mutations that restored replication were mapped to the pT181 Rep protein, suggesting a direct interaction between the two proteins (Iordanescu, 1993). Rep loads the helicase onto the lagging strand of the nicked *dso* and remains engaged with PcrA increasing its processivity and enabling it to displace DNA from a nicked substrate (Soultanas et al., 1999; Chang et al., 2002; Anand and Khan, 2004; Zhang et al., 2007). pT181 Rep displays an interaction with the PcrA helicases of *S. aureus*, *Bacillus anthracis*, *Bacillus cereus*, and *Streptococcus pneumoniae* but fails to stimulate full unwinding activity in the latter (Anand et al., 2004; Ruiz-Masó et al., 2006). Together with previous observations that pT181 can replicate in bacilli (albeit unstably) but not streptococci, these results indicate that Rep-mediated activation of PcrA is a requirement for efficient replication and lack of interaction is likely to limit plasmid host range. In *B. subtilis* carrying the *pcrA3* mutation, pT181 was incapable of replication; however, pC194 and pE194 plasmids could still replicate at normal copy number. This suggested that either another helicase may be required for leading strand synthesis in pC194 and pE194 or that the *pcrA3* mutation does not effect PcrA interaction with their respective Rep proteins (Petit et al., 1998). Interestingly, in *E. coli* host cells, the PcrA homolog UvrD was found to be essential for pC194 and pE194 replication (Bruand and Ehrlich, 2000).

Primases

The bacterial primosome is a multiprotein complex containing helicase, primase, and accessory proteins that assist in helicase loading and is required for generating RNA primers on single-stranded DNA. Primosomes are assembled during the initiation of chromosome replication at *oriC* (DnaA-dependent) and in the restart of stalled or collapsed replication forks (PriA-dependent). In Gram-positive bacteria, the replicative helicase (DnaC) is loaded through DnaI with the assistance of DnaB and DnaD. Once DnaC has been loaded it recruits DnaG primase and this primosome complex may then associate with the PolC holoenzyme and other factors to constitute the replisome. pAMβ1 carries a primosome assembly site (*ssiA*) downstream of *ori* that requires PriA, DnaB, DnaD, and DnaI suggesting that the activated *ori* (containing a D-loop) is recognized and targeted by PriA in a process that resembles re-combinational DNA repair (Polard et al., 2002). pSK41 Rep was shown to share structural similarity to DnaD primosomal helicase loader and interact directly with DnaG primase (Schumacher et al., 2014). Thus, it is possible that in pSK41, Rep assists loading of the DnaC helicase, perhaps in combination with DnaB and DnaI, or it may recruit DnaC indirectly through helicase interaction domains conserved in DnaG.

CONCLUDING REMARKS

The different types of plasmid replication systems described here encompass the diversity of plasmids recognized in staphylococci. Plasmids using each of these systems have been shown to act as vehicles for the carriage of antimicrobial resistance genes. Our view of plasmid diversity in staphylococci is heavily skewed by a historical focus on clinical isolates, and the consequential bias toward *S. aureus* and hence under-representation of coagulase negative species. The extent to which current understanding represents a comprehensive or distorted description of the staphylococcal plasmidome is an open question that awaits far broader sampling of the disparate environments occupied by staphylococci. Advances in sequencing capacity provide an opportunity to address this knowledge gap, while increasing evidence of transmission pathways linking bacteria that impact human health with those in the broader biosphere should provide motivation.

Just as our understanding of plasmid diversity is likely incomplete, the level to which the differing replication systems used by staphylococcal plasmids have been studied varies tremendously. While RCR plasmid replication has been analyzed in considerable detail, the replication systems of theta-replicating plasmids have been largely ignored in comparison, despite the established significance of these plasmids in the expression of resistance and virulence properties. There are several areas where information about these larger staphylococcal plasmids is particularly lacking. This includes how their replication systems interface with the chromosomally encoded replication machinery, and how they integrate and cooperate with other plasmid modules associated with plasmid propagation, such as partitioning, conjugation, and mobilization systems. There seems to be a general view that plasmid biology is well understood, but this is really not the case excepting a handful of model systems. This point has been highlighted by the recent characterization of new types of staphylococcal conjugative plasmids and identification of previously unrecognized mobilization determinants that are widespread on staphylococcal plasmids (O'Brien et al., 2015; Ramsay et al., 2016; Ramsay and Firth, 2017). Given the pivotal role plasmids play in bacterial adaptation, not least in the emergence of staphylococcal resistance, a renewed emphasis on studies elucidating the properties and mechanisms of plasmids is required if we are to meaningfully appreciate their roles in bacterial evolution and its consequences.

AUTHOR CONTRIBUTIONS

All authors listed have made a substantial, direct and intellectual contribution to the work, and approved it for publication.

REFERENCES

Aleshin, V. V., Semenova, E. V., Doroshenko, V. G., Jomantas, Y. V., Tarakanov, B. V., and Livshits, V. A. (1999). The broad host range plasmid pLF1311 from *Lactobacillus fermentum* VKM1311. *FEMS Microbiol. Lett.* 178, 47–53. doi: 10.1111/j.1574-6968.1999.tb13758.x

Alonso, J. C., and Tailor, R. H. (1987). Initiation of plasmid pC194 replication and its control in *Bacillus subtilis*. *Mol. Gen. Genet.* 210, 476–484. doi: 10.1007/BF00327200

Anand, S. P., and Khan, S. A. (2004). Structure-specific DNA binding and bipolar helicase activities of PcrA. *Nucleic Acids Res.* 32, 3190–3197. doi: 10.1093/nar/gkh641

Anand, S. P., Mitra, P., Naqvi, A., and Khan, S. A. (2004). *Bacillus anthracis* and *Bacillus cereus* PcrA helicases can support DNA unwinding and *in vitro* rolling-circle replication of plasmid pT181 of *Staphylococcus aureus*. *J. Bacteriol.* 186, 2195–2199. doi: 10.1128/JB.186.7.2195-2199.2004

Apisiridej, S., Leelaporn, A., Scaramuzzi, C. D., Skurray, R. A., and Firth, N. (1997). Molecular analysis of a mobilizable theta-mode trimethoprim resistance plasmid from coagulase-negative staphylococci. *Plasmid* 38, 13–24. doi: 10.1006/plas.1997.1292

Ballester, S., Lopez, P., Espinosa, M., Alonso, J. C., and Lacks, S. A. (1989). Plasmid structural instability associated with pC194 replication functions. *J. Bacteriol.* 171, 2271–2277. doi: 10.1128/jb.171.5.2271-2277.1989

Berg, T., Firth, N., Apisiridej, S., Hettiaratchi, A., Leelaporn, A., and Skurray, R. A. (1998). Complete nucleotide sequence of pSK41: evolution of staphylococcal conjugative multiresistance plasmids. *J. Bacteriol.* 180, 4350–4359.

Bierne, H., Seigneur, M., Ehrlich, S. D., and Michel, B. (1997). *uvrD* mutations enhance tandem repeat deletion in the *Escherichia coli* chromosome via SOS induction of the RecF recombination pathway. *Mol. Microbiol.* 26, 557–567. doi: 10.1046/j.1365-2958.1997.6011973.x

Boe, L., Gros, M. F., te Riele, H., Ehrlich, S. D., and Gruss, A. (1989). Replication origins of single-stranded-DNA plasmid pUB110. *J. Bacteriol.* 171, 3366–3372. doi: 10.1128/jb.171.6.3366-3372.1989

Boer, D. R., Ruiz-Masó, J. A., López-Blanco, J. R., Blanco, A. G., Vives-Llàcer, M., Chacón, P., et al. (2009). Plasmid replication initiator RepB forms a hexamer reminiscent of ring helicases and has mobile nuclease domains. *EMBO J.* 28, 1666–1678. doi: 10.1038/emboj.2009.125

Boer, D. R., Ruiz-Masó, J. A., Rueda, M., Petoukhov, M. V., Machón, C., Svergun, D. I., et al. (2016). Conformational plasticity of RepB, the replication initiator protein of promiscuous streptococcal plasmid pMV158. *Sci. Rep.* 6:20915. doi: 10.1038/srep20915

Brantl, S. (1994). The *copR* gene product of plasmid pIP501 acts as a transcriptional repressor at the essential *repR* promoter. *Mol. Microbiol.* 14, 473–483. doi: 10.1111/j.1365-2958.1994.tb02182.x

Brantl, S. (2014). Plasmid replication control by antisense RNAs. *Microbiol. Spectr.* 2:PLAS-0001-2013. doi: 10.1128/microbiolspec.PLAS-0001-2013

Brantl, S., and Behnke, D. (1992a). Characterization of the minimal origin required for replication of the streptococcal plasmid pIP501 in *Bacillus subtilis*. *Mol. Microbiol.* 6, 3501–3510.

Brantl, S., and Behnke, D. (1992b). Copy number control of the streptococcal plasmid pIP501 occurs at three levels. *Nucleic Acids Res.* 20, 395–400.

Brantl, S., Birch-Hirschfeld, E., and Behnke, D. (1993). RepR protein expression on plasmid pIP501 is controlled by an antisense RNA-mediated transcription attenuation mechanism. *J. Bacteriol.* 175, 4052–4061. doi: 10.1128/jb.175.13.4052-4061.1993

Brantl, S., and Wagner, E. G. (1997). Dual function of the *copR* gene product of plasmid pIP501. *J. Bacteriol.* 179, 7016–7024. doi: 10.1128/jb.179.22.7016-7024.1997

Bruand, C., and Ehrlich, S. D. (1998). Transcription-driven DNA replication of pAMß1 in *Bacillus subtilis*. *Mol. Microbiol.* 30, 135–145. doi: 10.1046/j.1365-2958.1998.01044.x

Bruand, C., and Ehrlich, S. D. (2000). UvrD-dependent replication of rolling-circle plasmids in *Escherichia coli*. *Mol. Microbiol.* 35, 204–210. doi: 10.1046/j.1365-2958.2000.01700.x

Bruand, C., Ehrlich, S. D., and Jannière, L. (1995). Primosome assembly site in *Bacillus subtilis*. *EMBO J.* 14, 2642–2650.

Bruand, C., Le Chatelier, E., Ehrlich, S. D., and Jannière, L. (1993). A fourth class of theta-replicaing plasmids: the pAMß1family from gram-positive bacteria. *Proc. Natl. Acad. Sci. U.S.A.* 90, 11668–11672. doi: 10.1073/pnas.90.24.11668

Bukowski, M., Lyzen, R., Helbin, W. M., Bonar, E., Szalewska-Palasz, A., Wegrzyn, G., et al. (2013). A regulatory role for the *Staphylococcus aureus* toxin-antitoxin system PemIKSa. *Nat. Commun.* 4, 2012. doi: 10.1038/ncomms3012

Carr, S. B., Phillips, S. E., and Thomas, C. D. (2016). Structures of replication initiation proteins from staphylococcal antibiotic resistance plasmids reveal protein asymmetry and flexibility are necessary for replication. *Nucleic Acids Res.* 44, 2417–2428. doi: 10.1093/nar/gkv1539

Caryl, J. A., and O'Neill, A. J. (2009). Complete nucleotide sequence of pGO1, the prototype conjugative plasmid from the staphylococci. *Plasmid* 62, 35–38. doi: 10.1016/j.plasmid.2009.03.001

Catchpole, I. R., and Dyke, K. G. (1992). Replication of staphylococcal plasmid pT48. *FEMS Microbiol. Lett.* 70, 55–62. doi: 10.1111/j.1574-6968.1992.tb05183.x

Ceglowski, P., Lurz, R., and Alonso, J. C. (1993). Functional analysis of pSM19035-derived replicons in *Bacillus subtilis*. *FEMS Microbiol. Lett.* 109, 145–150. doi: 10.1111/j.1574-6968.1993.tb06159.x

Chandler, M., de la Cruz, F., Dyda, F., Hickman, A. B., Moncalian, G., and Ton-Hoang, B. (2013). Breaking and joining single stranded DNA: the HUH endonuclease superfamily. *Nat. Rev. Microbiol.* 11, 525–538. doi: 10.1038/nrmicro3067

Chang, T. L., Naqvi, A., Anand, S. P., Kramer, M. G., Munshi, R., and Khan, S. A. (2002). Biochemical characterization of the *Staphylococcus aureus* PcrA helicase and its role in plasmid rolling circle replication. *J. Biol. Chem.* 277, 45880–45886. doi: 10.1074/jbc.M207383200

Coffey, A., Harrington, A., Kearney, K., Daly, C., and Fitzgerald, G. (1994). Nucleotide sequence and structural organization of the small, broad-host-range plasmid pCI411 from *Leuconostoc lactis* 533. *Microbiology* 140, 2263–2269. doi: 10.1099/13500872-140-9-2263

Das, N., and Chattoraj, D. K. (2004). Origin pairing ('hand-cuffing') and unpairing in the control of P1 plasmid replication. *Mol. Microbiol.* 54, 836–849. doi: 10.1111/j.1365-2958.2004.04322.x

del Solar, G., Diaz, R., and Espinosa, M. (1987). Replication of the streptococcal plasmid pMV158 and derivatives in cell-free extracts of *Escherichia coli*. *Mol. Gen. Genet.* 206, 428–435. doi: 10.1007/BF00428882

del Solar, G., and Espinosa, M. (1992). The copy number of plasmid pLS1 is regulated by two *trans*-acting plasmid products: the antisense RNAII and the repressor protein, RepA. *Mol. Microbiol.* 6, 83–94. doi: 10.1111/j.1365-2958.1992.tb00840.x

del Solar, G., and Espinosa, M. (2000). Plasmid copy number control: an ever-growing story. *Mol. Microbiol.* 37, 492–500. doi: 10.1046/j.1365-2958.2000.02005.x

del Solar, G., Giraldo, R., Ruiz-Echevarria, M. J., Espinosa, M., and Diaz-Orejas, R. (1998). Replication and control of circular bacterial plasmids. *Microbiol. Mol. Biol. Rev.* 62, 434–464.

Dempsey, L. A., Birch, P., and Khan, S. A. (1992). Six amino acids determine the sequence-specific DNA binding and replication specificity of the initiator proteins of the pT181 family. *J. Biol. Chem.* 267, 24538–24543.

Dempsey, L. A., Zhao, A. C., and Khan, S. A. (1995). Localization of the start sites of lagging-strand replication of rolling-circle plasmids from gram-positive bacteria. *Mol. Microbiol.* 15, 679–687. doi: 10.1111/j.1365-2958.1995.tb02377.x

Dervyn, E., Suski, C., Daniel, R., Bruand, C., Chapuis, J., Errington, J., et al. (2001). Two essential DNA polymerases at the bacterial replication fork. *Science* 294, 1716–1719. doi: 10.1126/science.1066351

Diaz, A., Lacks, S. A., and López, P. (1994). Multiple roles for DNA polymerase I in establishment and replication of the promiscuous plasmid pLS1. *Mol. Microbiol.* 14, 773–783. doi: 10.1111/j.1365-2958.1994.tb01314.x

Diep, B. A., Gill, S. R., Chang, R. F., Phan, T. H., Chen, J. H., Davidson, M. G., et al. (2006). Complete genome sequence of USA300, an epidemic clone of community-acquired methicillin-resistant *Staphylococcus aureus*. *Lancet* 367, 731–739. doi: 10.1016/S0140-6736(06)68231-7

Firth, N., Apisiridej, S., Berg, T., O'Rourke, B. A., Curnock, S., Dyke, K. G., et al. (2000). Replication of staphylococcal multiresistance plasmids. *J. Bacteriol.* 182, 2170–2178. doi: 10.1128/JB.182.8.2170-2178.2000

Firth, N., and Skurray, R. A. (1998). Mobile elements in the evolution and spread of multiple-drug resistance in staphylococci. *Drug Resist. Updat.* 1, 49–58. doi: 10.1016/S1368-7646(98)80214-8

Firth, N., and Skurray, R. A. (2006). "Genetics: accessory elements and genetic exchange," in *Gram-positive Pathogens*, 2nd Edn, eds V. A. Fischetti, R. P. Novick, J. J. Ferretti, D. A. Portnoy, and J. I. Rood (Washington, DC: ASM Press), 413-426.

Gennaro, M. L., Kornblum, J., and Novick, R. P. (1987). A site-specific recombination function in *Staphylococcus aureus* plasmids. *J. Bacteriol.* 169, 2601–2610. doi: 10.1128/jb.169.6.2601-2610.1987

Giraldo, R., and Fernández-Tresguerres, M. E. (2004). Twenty years of the pPS10 replicon: insights on the molecular mechanism for the activation of DNA replication in iteron-containing bacterial plasmids. *Plasmid* 52, 69–83. doi: 10.1016/j.plasmid.2004.06.002

Gomis-Rüth, F. X., Solá, M., Acebo, P., Párraga, A., Guasch, A., Eritja, R., et al. (1998). The structure of plasmid-encoded transcriptional repressor CopG unliganded and bound to its operator. *EMBO J.* 17, 7404–7415. doi: 10.1093/emboj/17.24.7404

Goursot, R., Goze, A., Niaudet, B., and Ehrlich, S. D. (1982). Plasmids from *Staphylococcus aureus* replicate in yeast *Saccharomyces cerevisiae*. *Nature* 298, 488–490. doi: 10.1038/298488a0

Grohmann, E., Goessweiner-Mohr, N., and Brantl, S. (2016). DNA-binding proteins regulating pIP501 transfer and replication. *Front. Mol. Biosci.* 3:42. doi: 10.3389/fmolb.2016.00042

Gros, M. F., te Riele, H., and Ehrlich, S. D. (1989). Replication origin of a single-stranded DNA plasmid pC194. *EMBO J.* 8, 2711–2716.

Gruss, A., and Ehrlich, S. D. (1989). The family of highly interrelated single-stranded deoxyribonucleic acid plasmids. *Microbiol. Rev.* 53, 231–241.

Gruss, A. D., Ross, H. F., and Novick, R. P. (1987). Functional analysis of a palindromic sequence required for normal replication of several staphylococcal plasmids. *Proc. Natl. Acad. Sci. U.S.A.* 84, 2165–2169. doi: 10.1073/pnas.84.8.2165

Hall, M. C., and Matson, S. W. (1999). Helicase motifs: the engine that powers DNA unwinding. *Mol. Microbiol.* 34, 867–877. doi: 10.1046/j.1365-2958.1999.01659.x

Hernández-Arriaga, A. M., Rubio-Lepe, T. S., Espinosa, M., and del Solar, G. (2009). Repressor CopG prevents access of RNA polymerase to promoter and actively dissociates open complexes. *Nucleic Acids Res.* 37, 4799–4811. doi: 10.1093/nar/gkp503

Ingmer, H., Fong, E. L., and Cohen, S. N. (1995). Monomer-dimer equilibrium of the pSC101 RepA protein. *J. Mol. Biol.* 250, 309–314. doi: 10.1006/jmbi.1995.0378

Inoue, R., Kaito, C., Tanabe, M., Kamura, K., Akimitsu, N., and Sekimizu, K. (2001). Genetic identification of two distinct DNA polymerases, DnaE and PolC, that are essential for chromosomal DNA replication in *Staphylococcus aureus*. *Mol. Genet. Genomics* 266, 564–571. doi: 10.1007/s004380100564

Iordanescu, S. (1993). Plasmid pT181-linked suppressors of the *Staphylococcus aureus pcrA3* chromosomal mutation. *J. Bacteriol.* 175, 3916–3917. doi: 10.1128/jb.175.12.3916-3917.1993

Iordanescu, S., and Basheer, R. (1991). The *Staphylococcus aureus* mutation *pcrA3* leads to the accumulation of pT181 replication initiation complexes. *J. Mol. Biol.* 221, 1183–1189. doi: 10.1016/0022-2836(91)90927-X

Iordanescu, S., and Surdeanu, M. (1980). New incompatibility groups of *Staphylococcus aureus* plasmids. *Plasmid* 4, 256–260. doi: 10.1016/0147-619X(80)90064-5

Jannière, L., and Ehrlich, S. D. (1987). Recombination between short repeated sequences is more frequent in plasmids than in the chromosome of *Bacillus subtilis*. *Mol. Gen. Genet.* 210, 116–121. doi: 10.1007/BF00337766

Khan, S. A. (1997). Rolling-circle replication of bacterial plasmids. *Microbiol. Mol. Biol. Rev.* 61, 442–455.

Khan, S. A., and Novick, R. P. (1982). Structural analysis of plasmid pSN2 in *Staphylococcus aureus*: no involvement in enterotoxin B production. *J. Bacteriol.* 149, 642–649.

Koepsel, R. R., and Khan, S. A. (1986). Static and initiator protein-enhanced bending of DNA at a replication origin. *Science* 233, 1316–1318. doi: 10.1126/science.3749879

Konieczny, I., Bury, K., Wawrzycka, A., and Wegrzyn, K. (2014). Iteron Plasmids. *Microbiol. Spectr.* 2:PLAS-0026-2014. doi: 10.1128/microbiolspec.PLAS-0026-2014

Kornberg, A., and Baker, T. A. (1992). *DNA Replication*, 2nd Edn. New York, NY: W.H. Freeman.

Kramer, M. G., del Solar, G., and Espinosa, M. (1995). Lagging-strand origins of the promiscuous plasmid pMV158: physical and functional characterization. *Microbiology* 141, 655–662. doi: 10.1099/13500872-141-3-655

Kramer, M. G., Espinosa, M., Misra, T. K., and Khan, S. A. (1998). Lagging strand replication of rolling-circle plasmids: specific recognition of the *ssoA*-type origins in different gram-positive bacteria. *Proc. Natl. Acad. Sci. U.S.A.* 95, 10505–10510. doi: 10.1073/pnas.95.18.10505

Kramer, M. G., Espinosa, M., Misra, T. K., and Khan, S. A. (1999). Characterization of a single-strand origin, *ssoU*, required for broad host range replication of rolling-circle plasmids. *Mol. Microbiol.* 33, 466–475. doi: 10.1046/j.1365-2958.1999.01471.x

Kramer, M. G., Khan, S. A., and Espinosa, M. (1997). Plasmid rolling circle replication: identification of the RNA polymerase-directed primer RNA and requirement for DNA polymerase I for lagging strand synthesis. *EMBO J.* 16, 5784–5795. doi: 10.1093/emboj/16.18.5784

Kwong, S. M., and Firth, N. (2015). Structural and sequence requirements for the antisense RNA regulating replication of staphylococcal multiresistance plasmid pSK41. *Plasmid* 78, 17–25. doi: 10.1016/j.plasmid.2015.01.002

Kwong, S. M., Lim, R., Lebard, R. J., Skurray, R. A., and Firth, N. (2008). Analysis of the pSK1 replicon, a prototype from the staphylococcal multiresistance plasmid family. *Microbiology* 154, 3084–3094. doi: 10.1099/mic.0.2008/017418-0

Kwong, S. M., Skurray, R. A., and Firth, N. (2004). *Staphylococcus aureus* multiresistance plasmid pSK41: analysis of the replication region, initiator protein binding and antisense RNA regulation. *Mol. Microbiol.* 51, 497–509. doi: 10.1046/j.1365-2958.2003.03843.x

Kwong, S. M., Skurray, R. A., and Firth, N. (2006). Replication control of staphylococcal multiresistance plasmid pSK41: an antisense RNA mediates dual-level regulation of Rep expression. *J. Bacteriol.* 188, 4404–4412. doi: 10.1128/JB.00030-06

Lane, H. E., and Denhardt, D. T. (1975). The *rep* mutation. IV. Slower movement of replication forks in *Escherichia coli rep* strains. *J. Mol. Biol.* 97, 99–112.

Le Chatelier, E., Jannière, L., Ehrlich, S. D., and Canceille, D. (2001). The RepE initiator is a double-stranded and single-stranded DNA binding protein that forms an atypical open complex at the onset of replication of plasmid pAMß1 from Gram-positive bacteria. *J. Biol. Chem.* 276, 10234–10246. doi: 10.1074/jbc.M010118200

Lee, C. A., Thomas, J., and Grossman, A. D. (2012). The *Bacillus subtilis* conjugative transposon ICE*Bs1* mobilizes plasmids lacking dedicated mobilization functions. *J. Bacteriol.* 194, 3165–3172. doi: 10.1128/JB.00301-12

Leelaporn, A., Firth, N., Byrne, M. E., Roper, E., and Skurray, R. A. (1994). Possible role of insertion sequence IS*257* in dissemination and expression of high- and low-level trimethoprim resistance in staphylococci. *Antimicrob. Agents Chemother.* 38, 2238–2244. doi: 10.1128/AAC.38.10.2238

Leelaporn, A., Firth, N., Paulsen, I. T., Hettiaratchi, A., and Skurray, R. A. (1995). Multidrug resistance plasmid pSK108 from coagulase-negative staphylococci; relationships to *Staphylococcus aureus qacC* plasmids. *Plasmid* 34, 62–67. doi: 10.1006/plas.1995.1034

Leelaporn, A., Firth, N., Paulsen, I. T., and Skurray, R. A. (1996). IS*257*-mediated cointegration in the evolution of a family of staphylococcal trimethoprim resistance plasmids. *J. Bacteriol.* 178, 6070–6073. doi: 10.1128/jb.178.20.6070-6073.1996

Lioy, V. S., Pratto, F., de la Hoz, A. B., Ayora, S., and Alonso, J. C. (2010). Plasmid pSM19035, a model to study stable maintenance in Firmicutes. *Plasmid* 64, 1–17. doi: 10.1016/j.plasmid.2010.04.002

Liu, M. A., Kwong, S. M., Jensen, S. O., Brzoska, A. J., and Firth, N. (2013). Biology of the staphylococcal conjugative multirestance plasmid pSK41. *Plasmid* 70, 42–51. doi: 10.1016/j.plasmid.2013.02.001

Liu, M. A., Kwong, S. M., Pon, C. K., Skurray, R. A., and Firth, N. (2012). Genetic requirements for replication initiation of the staphylococcal multiresistance plasmid pSK41. *Microbiology* 158, 1456–1467. doi: 10.1099/mic.0.057620-0

López-Aguilar, C., Romero-López, C., Espinosa, M., Berzal-Herranz, A., and del Solar, G. (2015). The 5'-tail of antisense RNAII of pMV158 plays a critical role in binding to the target mRNA and in translation inhibition of *repB*. *Front. Genet.* 6:255. doi: 10.3389/fgene.2015.00225

López-Aguilar, C., Ruiz-Masó, J. A., Rubio-Lepe, T. S., Sanz, M., and del Solar, G. (2013). Translation initiation of the replication initiator *repB* gene of promiscuous plasmid pMV158 is led by an extended non-SD sequence. *Plasmid* 70, 69–77. doi: 10.1016/j.plasmid.2013.01.011

Lorenzo-Díaz, F., and Espinosa, M. (2009). Lagging-strand DNA replication origins are required for conjugal transfer of the promiscuous plasmid pMV158. *J. Bacteriol.* 191, 720–727. doi: 10.1128/JB.01257-08

Lorenzo-Díaz, F., Fernández-López, C., Lurz, R., Bravo, A., and Espinosa, M. (2017). Crosstalk between vertical and horizontal gene transfer: plasmid replication control by a conjugative relaxase. *Nucleic Acids Res.* 45, 7774–7785. doi: 10.1093/nar/gkx450

Lyon, B. R., and Skurray, R. A. (1987). Antimicrobial resistance of *Staphylococcus aureus*: genetic basis. *Microbiol. Rev.* 51, 88–134.

Maciag, I. E., Viret, J. F., and Alonso, J. C. (1988). Replication and incompatibility properties of plasmid pUB110 in *Bacillus subtilis*. *Mol. Gen. Genet.* 212, 232–240. doi: 10.1007/BF00334690

Majumder, S., and Novick, R. P. (1988). Intermediates in plasmid pT181 DNA replication. *Nucleic Acids Res.* 16, 2897–2912. doi: 10.1093/nar/16.7.2897

Marston, F. Y., Grainger, W. H., Smits, W. K., Hopcroft, N. H., Green, M., Hounslow, A. M., et al. (2010). When simple sequence comparison fails: the cryptic case of the shared domains of the bacterial replication initiation proteins DnaB and DnaD. *Nucleic Acids Res.* 38, 6930–6942. doi: 10.1093/nar/gkq465

Matsunaga, F., Ishiai, M., Kobayashi, G., Uga, H., Yura, T., and Wada, C. (1997). The central region of RepE initiator protein of mini-F plasmid plays a crucial role in dimerization required for negative replication control. *J. Mol. Biol.* 274, 27–38. doi: 10.1006/jmbi.1997.1373

Meijer, W. J., van der Lelie, D., Venema, G., and Bron, S. (1995). Effects of the generation of single-stranded DNA on the maintenance of plasmid pMV158 and derivatives in *Lactococcus lactis*. *Plasmid* 33, 91–99. doi: 10.1006/plas.1995.1011

Mendes, R. E., Deshpande, L. M., Bonilla, H. F., Schwarz, S., Huband, M. D., Jones, R. N., et al. (2013). Dissemination of a pSCFS3-like *cfr*-carrying plasmid in *Staphylococcus aureus* and *Staphylococcus epidermidis* clinical isolates recovered from hospitals in Ohio. *Antimicrob. Agents Chemother.* 57, 2923–2928. doi: 10.1128/AAC.00071-13

Moscoso, M., del Solar, G., and Espinosa, M. (1995). Specific nicking-closing activity of the initiator of replication protein RepB of plasmid pMV158 on supercoiled or single-stranded DNA. *J. Biol. Chem.* 270, 3772–3779. doi: 10.1074/jbc.270.8.3772

Müller, A. K., Rojo, F., and Alonso, J. C. (1995). The level of the pUB110 replication initiator protein is autoregulated, which provides an additional control for plasmid copy number. *Nucleic Acids Res.* 23, 1894–1900. doi: 10.1093/nar/23.11.1894

Needham, C., Noble, W. C., and Dyke, K. G. (1995). The staphylococcal insertion sequence IS*257* is active. *Plasmid* 34, 198–205. doi: 10.1006/plas.1995.0005

Niaudet, B., Jannière, L., and Ehrlich, S. D. (1984). Recombination between repeated DNA sequences occurs more often in plasmids than in the chromosome of *Bacillus subtilis*. *Mol. Gen. Genet.* 197, 46–54. doi: 10.1007/BF00327921

Nimmo, G. R. (2012). USA300 abroad: global spread of a virulent strain of community-associated methicillin-resistant *Staphylococcus aureus*. *Clin. Microbiol. Infect.* 18, 725–734. doi: 10.1111/j.1469-0691.2012.03822.x

Noirot, P., Bargonetti, J., and Novick, R. P. (1990). Initiation of rolling-circle replication in pT181 plasmid: initiator protein enhances cruciform extrusion at the origin. *Proc. Natl. Acad. Sci. U.S.A.* 87, 8560–8564. doi: 10.1073/pnas.87.21.8560

Noirot-Gros, M. F., Bidnenko, V., and Ehrlich, S. D. (1994). Active site of the replication protein of the rolling circle plasmid pC194. *EMBO J.* 13, 4412–4420.

Novick, R. P. (1989). Staphylococcal plasmids and their replication. *Annu. Rev. Microbiol.* 43, 537–563. doi: 10.1146/annurev.mi.43.100189.002541

Novick, R. P., Iordanescu, S., Projan, S. J., Kornblum, J., and Edelman, I. (1989). pT181 plasmid replication is regulated by a countertranscript-driven transcriptional attenuator. *Cell* 59, 395–404. doi: 10.1016/0092-8674(89)90300-0

Novick, R. P., Projan, S. J., Rosenblum, W., and Edelman, I. (1984). Staphylococcal plasmid cointegrates are formed by host- and phage-mediated general *rec* systems that act on short regions of homology. *Mol. Gen. Genet.* 195, 374–377. doi: 10.1007/BF00332777

O'Brien, F. G., Yui Eto, K., Murphy, R. J., Fairhurst, H. M., Coombs, G. W., Grubb, W. B., et al. (2015). Origin-of-transfer sequences facilitate mobilisation of non-conjugative antimicrobial-resistance plasmids in *Staphylococcus aureus*. *Nucleic Acids Res.* 43, 7971–7983. doi: 10.1093/nar/gkv755

Oliveira, S. S., Murphy, E., Gamon, M. R., and Bastos, M. C. (1993). pRJ5: a naturally occurring *Staphylococcus aureus* plasmid expressing constitutive macrolide-lincosamide-streptogramin B resistance contains a tandem duplication in the leader region of the *ermC* gene. *J. Gen. Microbiol.* 139, 1461–1467. doi: 10.1099/00221287-139-7-1461

Pérez-Roth, E., Kwong, S. M., Alcoba-Florez, J., Firth, N., and Méndez-Alvarez, S. (2010). Complete nucleotide sequence and comparative analysis of pPR9, a 41.7-kilobase conjugative staphylococcal multiresistance plasmid conferring high-level mupirocin resistance. *Antimicrob. Agents Chemother.* 54, 2252–2257. doi: 10.1128/AAC.01074-09

Petit, M. A., Dervyn, E., Rose, M., Entian, K. D., McGovern, S., Ehrlich, S. D., et al. (1998). PcrA is an essential DNA helicase of *Bacillus subtilis* fulfilling functions both in repair and rolling-circle replication. *Mol. Microbiol.* 29, 261–273. doi: 10.1046/j.1365-2958.1998.00927.x

Planet, P. J. (2017). Life after USA300: The rise and fall of a superbug. *J. Infect. Dis.* 215(Suppl. 1), S71–S77. doi: 10.1093/infdis/jiw444

Polard, P., Marsin, S., McGovern, S., Velton, M., Wigley, D. B., Ehrlich, S. D., et al. (2002). Restart of DNA replication in Gram-positive bacteria: functional characterisation of the *Bacillus subtilis* PriA initiator. *Nucleic Acids Res.* 30, 1593–1605. doi: 10.1093/nar/30.7.1593

Priebe, S. D., and Lacks, S. A. (1989). Region of the streptococcal plasmid pMV158 required for conjugative mobilisation. *J. Bacteriol.* 171, 4778–4784. doi: 10.1128/jb.171.9.4778-4784.1989

Puyet, A., del Solar, G. H., and Espinosa, M. (1988). Identification of the origin and direction of replication of the broad-host-range plasmid pLS1. *Nucleic Acids Res.* 16, 115–133. doi: 10.1093/nar/16.1.115

Ramsay, J. P., and Firth, N. (2017). Diverse mobilization strategies facilitate transfer of non-conjugative mobile genetic elements. *Curr. Opin. Microbiol.* 38, 1–9. doi: 10.1016/j.mib.2017.03.003

Ramsay, J. P., Kwong, S. M., Murphy, R. J., Yui Eto, K., Price, K. J., Nguyen, Q. T., et al. (2016). An updated view of plasmid conjugation and mobilization in Staphylococcus. *Mob. Genet. Elements* 6:e1208317. doi: 10.1080/2159256X.2016.1208317

Rasooly, A., and Novick, R. P. (1993). Replication-specific inactivation of the pT181 plasmid initiator protein. *Science* 262, 1048–1050. doi: 10.1126/science.8235621

Rasooly, A., Wang, P. Z., and Novick, R. P. (1994). Replication-specific conversion of the *Staphylococcus aureus* pT181 initiator protein from an active homodimer to an inactive heterodimer. *EMBO J.* 13, 5245–5251.

Ruby, C., and Novick, R. P. (1975). Plasmid interactions in *Staphylococcus aureus*: nonadditivity of compatible plasmid DNA pools. *Proc. Natl, Acad. Sci. U.S.A.* 72, 5031–5035. doi: 10.1073/pnas.72.12.5031

Ruiz-Masó, J. A., Anand, S. P., Espinosa, M., Khan, S. A., and del Solar, G. (2006). Genetic and biochemical characterization of the *Streptococcus pneumoniae* PcrA helicase and its role in plasmid rolling circle replication. *J. Bacteriol.* 188, 7416–7425. doi: 10.1128/JB.01010-06

Ruiz-Masó, J. A., Bordanaba-Ruiseco, L., Sanz, M., Menéndez, M., and del Solar, G. (2016). Metal-induced stabilization and activation of plasmid replication initiator RepB. *Front. Mol. Biosci.* 3:56. doi: 10.3389/fmolb.2016.00056

Ruiz-Masó, J. A., López-Zumel, C., Menéndez, M., Espinosa, M., and del Solar, G. (2004). Structural features of the initiator of replication protein RepB encoded by the promiscuous plasmid pMV158. *Biochim. Biophys. Acta* 1696, 113–119. doi: 10.1016/j.bbapap.2003.09.010

Ruiz-Masó, J. A., Lurz, R., Espinosa, M., and del Solar, G. (2007). Interactions between the RepB initiator protein of plasmid pMV158 and two distant DNA regions within the origin of replication. *Nucleic Acids Res.* 35, 1230–1244. doi: 10.1093/nar/gkl1099

Ruiz-Masó, J. A., Machón, C., Bordanaba-Ruiseco, L., Espinosa, M., Coll, M., and del Solar, G. (2015). Plasmid rolling-circle replication. *Microbiol. Spectr.* 3:PLAS-0035-2014. doi: 10.1128/microbiolspec.PLAS-0035-2014

Sanders, G. M., Dallmann, H. G., and McHenry, C. S. (2010). Reconstitution of the *B. subtilis* replisome with 13 proteins including two distinct replicases. *Mol. Cell* 37, 273–281. doi: 10.1016/j.molcel.2009.12.025

Schumacher, M. A., Tonthat, N. K., Kwong, S. M., Chinnam, N. B., Liu, M. A., Skurray, R. A., et al. (2014). Mechanism of staphylococcal multiresistance plasmid replication origin assembly by the RepA protein. *Proc. Natl. Acad. Sci. U.S.A.* 111, 9121–9126. doi: 10.1073/pnas.1406065111

Schwarz, S., Shen, J., Wendlandt, S., Fessler, A. T., Wang, Y., Kadlec, K., et al. (2014). Plasmid-mediated antimicrobial resistance in staphylococci and other firmicutes. *Microbiol. Spectr.* 2:PLAS-0020-2014. doi: 10.1128/microbiolspec.PLAS-0020-2014

Seegers, J. F., Zhao, A. C., Meijer, W. J., Khan, S. A., Venema, G., and Bron, S. (1995). Structural and functional analysis of the single-strand origin of replication from the lactococcal plasmid pWV01. *Mol. Gen. Genet.* 249, 43–50. doi: 10.1007/BF00290234

Simpson, A. E., Skurray, R. A., and Firth, N. (2000). An IS*257*-derived hybrid promoter directs transcription of a *tetA*(K) tetracycline resistance gene in the *Staphylococcus aureus* chromosomal *mec* region. *J. Bacteriol.* 182, 3345–3352. doi: 10.1128/JB.182.12.3345-3352.2000

Slatter, A. F., Thomas, C. D., and Webb, M. R. (2009). PcrA helicase tightly couples ATP hydrolysis to unwinding double-stranded DNA, modulated by the initiator protein for plasmid replication, RepD. *Biochemistry* 48, 6326–6334. doi: 10.1021/bi900101h

Smits, W. K., Merrikh, H., Bonilla, C. Y., and Grossman, A. D. (2011). Primosomal proteins DnaD and DnaB are recruited to chromosomal regions bound by DnaA in *Bacillus subtilis*. *J. Bacteriol.* 193, 640–648. doi: 10.1128/JB.01253-10

Soultanas, P., Dillingham, M. S., Papadopoulos, F., Phillips, S. E., Thomas, C. D., and Wigley, D. B. (1999). Plasmid replication initiator protein RepD increases the processivity of PcrA DNA helicase. *Nucleic Acids Res.* 27, 1421–1428. doi: 10.1093/nar/27.6.1421

Steinmetzer, K., Behlke, J., and Brantl, S. (1998). Plasmid pIP501 encoded transcriptional repressor CopR binds to its target DNA as a dimer. *J. Mol. Biol.* 283, 595–603. doi: 10.1006/jmbi.1998.2122

Stewart, P. R., Dubin, D. T., Chikramane, S. G., Inglis, B., Matthews, P. R., and Poston, S. M. (1994). IS*257* and small plasmid insertions in the *mec* region of the chromosome of *Staphylococcus aureus*. *Plasmid* 31, 12–20. doi: 10.1006/plas.1994.1002

Swan, M. K., Bastia, D., and Davies, C. (2006). Crystal structure of pi initiator protein-iteron complex of plasmid R6K: implications for initiation of plasmid DNA replication. *Proc. Natl. Acad. Sci. U.S.A.* 103, 18481–18486. doi: 10.1073/pnas.0609046103

Takahashi, S., Hours, C., Chu, A., and Denhardt, D. T. (1979). The *rep* mutation. VI. Purification and properties of the *Escherichia coli* rep protein, DNA helicase III. *Can. J. Biochem.* 57, 855–866. doi: 10.1139/o79-105

Thomas, C. D., Balson, D. F., and Shaw, W. V. (1990). *In vitro* studies of the initiation of staphylococcal plasmid replication. Specificity of RepD for its origin (*oriD*) and characterization of the Rep-*ori* tyrosyl ester intermediate. *J. Biol. Chem.* 265, 5519–5530.

Thomas, C. D., Nikiforov, T. T., Connolly, B. A., and Shaw, W. V. (1995). Determination of sequence specificity between a plasmid replication initiator protein and the origin of replication. *J. Mol. Biol.* 254, 381–391. doi: 10.1006/jmbi.1995.0625

Toukdarian, A. E., and Helinski, D. R. (1998). TrfA dimers play a role in copy-number control of RK2 replication. *Gene* 223, 205–211. doi: 10.1016/S0378-1119(98)00370-9

Udo, E. E., and Grubb, W. B. (1991). A new incompatibility group plasmid in *Staphylococcus aureus*. *FEMS Microbiol. Lett.* 62, 33–36. doi: 10.1111/j.1574-6968.1991.tb04412.x

Udo, E. E., Wei, M. Q., and Grubb, W. B. (1992). Conjugative trimethoprim resistance in *Staphylococcus aureus*. *FEMS Microbiol. Lett.* 76, 243–248. doi: 10.1111/j.1574-6968.1992.tb05470.x

Wang, P. Z., Projan, S. J., Henriquez, V., and Novick, R. P. (1992). Specificity of origin recognition by replication initiator protein in plasmids of the pT181 family is determined by a six amino acid residue element. *J. Mol. Biol.* 223, 145–158. doi: 10.1016/0022-2836(92)90722-V

Wassenaar, T. M., Ussery, D. W., and Ingmer, H. (2016). The *qacC* gene has recently spread between rolling circle plasmids of *Staphylococcus*, indicative of a novel gene transfer mechanism. *Front. Microbiol.* 7:1528. doi: 10.3389/fmicb.2016.01528

Weaver, K. E., Kwong, S. M., Firth, N., and Francia, M. V. (2009). The RepA_N replicons of Gram-positive bacteria: a family of broadly distributed but narrow host range plasmids. *Plasmid* 61, 94–109. doi: 10.1016/j.plasmid.2008.11.004

Weisblum, B., Graham, M. Y., Gryczan, T., and Dubnau, D. (1979). Plasmid copy number control: isolation and characterization of high-copy-number mutants of plasmid pE194. *J. Bacteriol.* 137, 635–643.

Yamaguchi, T., Hayashi, T., Takami, H., Ohnishi, M., Murata, T., Nakayama, K., et al. (2001). Complete nucleotide sequence of a *Staphylococcus aureus* exfoliative toxin B plasmid and identification of a novel ADP-ribosyltransferase, EDIN-C. *Infect. Immun.* 69, 7760–7771. doi: 10.1128/IAI.69.12.7760-7771.2001

Zhang, W., Dillingham, M. S., Thomas, C. D., Allen, S., Roberts, C. J., and Soultanas, P. (2007). Directional loading and stimulation of PcrA helicase by the replication initiator protein RepD. *J. Mol. Biol.* 371, 336–348. doi: 10.1016/j.jmb.2007.05.050

6

Antibiotic-Induced Alterations in Gut Microbiota are Associated with Changes in Glucose Metabolism in Healthy Mice

Richard R. Rodrigues[1], Renee L. Greer[2], Xiaoxi Dong[1], Karen N. DSouza[1], Manoj Gurung[2], Jia Y. Wu[2], Andrey Morgun[1] and Natalia Shulzhenko[2]**

[1] Department of Pharmaceutical Sciences, Oregon State University, Corvallis, OR, United States, [2] Department of Biomedical Sciences, Oregon State University, Corvallis, OR, United States

Edited by:
Tatiana Venkova,
Fox Chase Cancer Center,
United States

Reviewed by:
Amanda Ellen Ramer-Tait,
University of Nebraska–Lincoln,
United States
Jonathan Badger,
National Cancer Institute (NIH),
United States

***Correspondence:**
Natalia Shulzhenko
natalia.shulzhenko@oregonstate.edu
Andrey Morgun
andriy.morgun@oregonstate.edu

The gut microbiome plays an important role in health and disease. Antibiotics are known to alter gut microbiota, yet their effects on glucose tolerance in lean, normoglycemic mice have not been widely investigated. In this study, we aimed to explore mechanisms by which treatment of lean mice with antibiotics (ampicillin, metronidazole, neomycin, vancomycin, or their cocktail) influences the microbiome and glucose metabolism. Specifically, we sought to: (i) study the effects on body weight, fasting glucose, glucose tolerance, and fasting insulin, (ii) examine the changes in expression of key genes of the bile acid and glucose metabolic pathways in the liver and ileum, (iii) identify the shifts in the cecal microbiota, and (iv) infer interactions between gene expression, microbiome, and the metabolic parameters. Treatment with individual or a cocktail of antibiotics reduced fasting glucose but did not affect body weight. Glucose tolerance changed upon treatment with cocktail, ampicillin, or vancomycin as indicated by reduced area under the curve of the glucose tolerance test. Antibiotic treatment changed gene expression in the ileum and liver, and shifted the alpha and beta diversities of gut microbiota. Network analyses revealed associations between *Akkermansia muciniphila* with fasting glucose and liver farsenoid X receptor (Fxr) in the top ranked host-microbial interactions, suggesting possible mechanisms by which this bacterium can mediate systemic changes in glucose metabolism. We observed *Bacteroides uniformis* to be positively and negatively correlated with hepatic Fxr and Glucose 6-phosphatase, respectively. Overall, our transkingdom network approach is a useful hypothesis generating strategy that offers insights into mechanisms by which antibiotics can regulate glucose tolerance in non-obese healthy animals. Experimental validation of our predicted microbe-phenotype interactions can help identify mechanisms by which antibiotics affect host phenotypes and gut microbiota.

Keywords: antibiotics, gut microbiota, glucose tolerance, lean, non-obese, transkingdom networks

Abbreviations: AUC, area under the curve; Fgf15, fibroblast growth factor 15; Fxr, farsenoid x receptor; G6pase, glucose 6-phosphatase; Glut1, glucose transporter 1; GTT, glucose tolerance test; Hk1, hexokinase 1; Hk2, hexokinase 2; Insr, insulin receptor; Pck1, phosphoenolpyruvate carboxykinase 1; Shp, small heterodimer partner; Tgr5, G protein-coupled bile acid receptor 1.

INTRODUCTION

The human gastrointestinal tract contains a multitude of microbiota, including bacteria, viruses, and fungi (Utzschneider et al., 2016). Their genome, although variable between individuals (Human Microbiome Project Consortium, 2012), is capable of a diverse set of functions that may influence the host's metabolic and immune systems (Tremaroli and Backhed, 2012; Greer et al., 2013; Sanz et al., 2015), including normal homeostasis (Utzschneider et al., 2016). Changes in the gut microbes have recently been associated with various diseases (Qin et al., 2012; Karlsson et al., 2013; Wu et al., 2015). For example, changes in Lactobacillus, Clostridium, Ruminococcus sp., *E. coli*, Bacteroides, *Akkermansia muciniphila* are observed in diabetic and obese patients (Qin et al., 2012; Karlsson et al., 2013; Murri et al., 2013; Chakraborti, 2015; Kasai et al., 2015; Sanz et al., 2015). These diverse results indicate a need for a better understanding of the mechanistic roles specific taxa play in the regulation of host metabolic functions.

Antibiotics add an interesting dynamic to the host-microbiome relationship. Although, antibiotics are well-known to cause short (Perez-Cobas et al., 2013; Pallav et al., 2014; Panda et al., 2014) and long-term (De La Cochetiere et al., 2005; Jernberg et al., 2007; Dethlefsen et al., 2008; Jakobsson et al., 2010; Dethlefsen and Relman, 2011; Raymond et al., 2016) alterations in the gut microbiome, there is a lack of consensus on their effects on glucose tolerance, body weight and other metabolic parameters (Francino, 2015; Mikkelsen et al., 2016). Moreover, effects of antibiotics in lean, normoglycemic mice as compared to mouse obesity models have not been widely investigated. An intervention study in healthy, glucose tolerant young human males treated with 4-days broad-spectrum antibiotics cocktail showed shifts in the cultivable gut microbiota but no changes in postprandial plasma glucose and serum insulin (Mikkelsen et al., 2016). Due to the use of a broad-acting antibiotic cocktail in a short course as well as the use of fecal samples for culture-based bacterial assessment, this study provides limited insight on a comprehensive picture of changes in intestinal microbes and on associations between individual antibiotics and specific intestinal microbes. Understanding antibiotic-microbiome interactions and their effects on glucose metabolism in healthy mammals is critical for identifying initial changes in microbiota that eventually may lead to diseases such as obesity and diabetes.

In this study, we aimed to understand the regulatory mechanisms by which individual antibiotics and their cocktail influence the cecal microbiome and host phenotypes in lean mice, namely, gene expression and metabolic parameters. By treating lean mice with different antibiotics we sought to: (i) study the effects on body weight, fasting glucose, glucose tolerance, and fasting insulin, (ii) examine the changes in expression of key genes of the bile acid and glucose metabolic pathways in the liver and ileum, (iii) identify the shifts in the cecal microbiota, and (iv) infer interactions between gene expression, microbiome, and the metabolic parameters. We repeated the entire experiment twice and performed meta-analyses to increase the confidence of our results.

MATERIALS AND METHODS

Mice and Antibiotics Treatment

Eight weeks old adult male Swiss Webster mice were initially purchased from Taconic Biosciences (Germantown, MD, United States). Mice were housed at the Laboratory Animal Resource Center at Oregon State University for 3–5 days for acclimation under standard 12-h light cycle with free access to food (5001, Research Diets) and water. Experimental procedures were carried out in accordance with protocols approved by the Oregon State University Institutional Animal Care and Use Committee. Mice were given single, cocktail, or no antibiotics for 4 weeks to create a stable altered microbiome. Antibiotics were administered in autoclaved drinking water individually, or in a cocktail for 4 weeks in the following concentrations: ampicillin (1 gl^{-1}), metronidazole (1 gl^{-1}), neomycin trisulfate (1 gl^{-1}), and vancomycin (0.5 gl^{-1}). This time course is consistent with standard antibiotic administration used in multiple studies for altering microbiota (Rakoff-Nahoum et al., 2004; Morgun et al., 2015; Greer R.L. et al., 2016). Each group consisted of five mice per experiment, total 30 mice per experiment, except for four mice in the cocktail group from the second experiment. Water consumption was monitored over the 4 weeks treatment period and all groups showed consumption equivalent to control water.

Glucose Tolerance Testing

Mice were fasted for 6 h during the light phase with free access to water. A concentration of 2 mg kg^{-1} glucose (Sigma–Aldrich) was injected intraperitoneally. Blood glucose was measured at 0 (immediately before glucose injection), 15, 30, 60, and 120 min with a Freestyle Lite glucometer (Abbot Diabetes Care).

Serum Collection and Hormone Measures

Mice were fasted for 6 h during the light phase with free access to water. Serum was collected via submandibular bleed using BD microtainer serum separator tubes. Fasting insulin was measured by ultrasensitive ELISA (Crystal Chem) according to manufacturer's protocols.

Bacterial DNA Extraction, 16S rRNA Gene Library Preparation and PCR

Unflushed cecal tissue and content was suspended in 1.4 ml ASL buffer (Qiagen) and homogenized with 2.8 mm ceramic beads followed by 0.5 mm glass beads using an OMNI Bead Ruptor (OMNI International). DNA was extracted from the entire resulting suspension using QIAamp DNA Stool Mini Kit (Qiagen) according to manufacturer's protocol. DNA was quantified using Qubit broad range DNA assay (Life Technologies). The V4 region of 16s rRNA gene was amplified using universal primers (515f and 806r) (Caporaso et al., 2012). Individual samples were barcoded, pooled to construct the sequencing library, and then sequenced using an Illumina Miseq (Illumina, San Diego, CA, United States) to generate pair-ended 250 nt reads. Quantitative PCR was performed for *A. muciniphila* as described in Schneeberger et al. (2015) with DNA for standard curve isolated from the cultivated microbe.

RNA Preparation and Gene Expression Analysis

Liver and ileum (flushed out of content) were collected and snap frozen prior to RNA extraction. Liver was homogenized using OMNI Rotor-Stator Homogenizer in Trizol and RNA was extracted using Trizol/chloroform extraction followed by the RNeasy Mini kit (Qiagen). Ileum RNA was extracted using OMNI Bead Ruptor and 2.8 mm ceramic beads (OMNI International) in RLT buffer followed by Qiashredder and RNeasy kit using Qiacube (Qiagen) automated extraction according to manufacturer's specifications. Total RNA was quantified using Nanodrop (Thermo Scientific). Complementary DNA was prepared using iScript reverse transcription kit (Bio-Rad) and qPCR was performed using QuantiFast SYBR mix (Qiagen) and StepOne Plus Real Time PCR system and software (Applied Biosystems). Primers used for qPCR are listed in **Supplementary Table S1**.

Statistical Analysis of Phenotypic Data

An outlier value per group per experiment was removed (if p-value $< 5\%$) for each phenotype (metabolic parameters and genes) using the default Grubb's test from R package *outliers* v0.14 (Komsta, 2011). The data was $\log_2$ transformed and differential phenotypes (antibiotics vs. control) were detected using *limma* (Ritchie et al., 2015) (Bioconductor 3.4, BiocInstaller 1.24.0, R 3.3.2) per experiment. A combined Fisher's p-value was calculated for each phenotype from the p-values for the limma t-statistic from each experiment. A false discovery rate (FDR) was calculated on the combined p-values. Change in phenotype was considered statistically significant if the phenotype had same direction of (abx/control) fold change in both experiments, individual p-value $< 20\%$ in each experiment, Fisher's combined p-value (Fisher, 1932) $<5\%$ and FDR $< 10\%$. The dot plots for the phenotypes were generated using R package *ggplot2* (Wickham, 2009) and the GTT curves were generated using GraphPad Prism software v7.03.

Analyses of 16S rRNA Gene Sequencing Data

The samples were demultiplexed and forward-end fastq files were analyzed using QIIME v. 1.9.1 (Caporaso et al., 2010). The default quality filter parameters from QIIME's *split_libraries_fastq.py* were applied to retain high quality reads (Phred quality score $> = 20$ and minimum read length $= 75\%$ of 250 nucleotides). A closed reference OTU picking with 97% sequence similarity was performed using UCLUST (Edgar, 2010) and Greengenes reference database v13.8 (DeSantis et al., 2006; McDonald et al., 2012) to cluster 16S rRNA gene sequence reads into OTUs and assign taxonomy. The reference sequence of an OTU from the Greengenes database was used to obtain species level taxonomic assignment using Megablast (Altschul et al., 1997; Morgulis et al., 2008) (top hit using default parameters). A threshold of 99% cumulative abundance across all samples in an experiment was used to retain abundant microbes, thus removing OTUs with approximately $<0.01\%$ abundance across all samples in that experiment. The read counts were normalized using cumulative sum scaling (Paulson et al., 2013) followed by quantile normalization.

The normalized OTU tables were used for diversity and statistical analysis. Briefly, a sampling depth of 200,000 sequences per sample was used for rarefaction. The alpha diversity metrics were calculated on unrarefied and rarefied OTU tables (**Supplementary Table S2**). The Shannon diversity index (from rarefied data) for samples with and without antibiotics treatment was compared with a non-parametric t-test. The difference was considered to be statistically significant if the direction of (abx/control) fold change in both experiments is the same, individual p-value $< 2\%$ in each experiment, Fisher's combined p-value $< 0.1\%$ and FDR $< 0.1\%$. Beta diversity was calculated using weighted UniFrac (Lozupone and Knight, 2005) and the distances were used for PCoA (Gower, 1998) and visualized using EMPeror (Vazquez-Baeza et al., 2013). The taxonomic summary bar plots were used to visualize abundance at the phylum and order levels.

The $\log_2$ transformed OTU tables were used for limma analysis. Meta-analysis was performed using the same criteria as applied for phenotypes to identify differentially abundant OTUs. A heatmap with row scaling was generated for each experiment using R packages *ggfortify* v0.2 (Horikoshi and Tang, 2016) and *gplots* v3.0.1 (Warnes et al., 2016). Hierarchical clustering was used to group OTUs (rows) based on similar abundance patterns across the groups in the first experiment and the same row order was used for the second experiment without row-wise clustering.

Network Reconstruction and Prioritizing Microbe-Phenotype Edges

Spearman rank correlations were calculated between all pairs of genes, microbes, and metabolic parameters across all samples or per-group in an experiment. A combined Fisher's p-value was calculated for each pair from the p-values for the correlation from each experiment. A FDR was calculated on the combined p-values separately for the following correlations: (i) within genes, (ii) within metabolic parameters, (iii) between genes and metabolic parameters, and (iv) between OTUs and phenotypes (genes or metabolic parameters).

We retained edges that satisfy the following criteria: the sign of correlation coefficients in the two experiments should be consistent, individual p-value of correlation within each experiment is $<20\%$, combined Fisher's p-value of all experiments $<5\%$ and FDR cutoff of 10% for edges without a microbial node (i, ii, and iii), whereas 1% for edges containing at least one microbial node (iv).

Next, the transkingdom network was generated (Dong et al., 2015; Morgun et al., 2015; Greer R.L. et al., 2016; Rodrigues et al., 2017) by keeping the criteria-satisfying phenotypic (i, ii, and iii) and OTU-phenotype (iv) edges, where the OTU has >0.5% median abundance across the two experiments in at least one group.

Finally, an OTU-phenotype edge was retained if it showed consistent sign of per-group Spearman correlation coefficient between the two experiments, principles of causality compliancy (Yambartsev et al., 2016) [i.e., satisfied fold change relationship between the two partners in the appropriate (abx vs. control) comparison] in at least one group, and the same sign of correlation coefficient across different groups. To put this bipartite network in perspective of the phenotypic connections a phenotypic edge was included (only during visualization) if its strength of correlation was stronger than at least one OTU-phenotype edges connecting the phenotypes. Network topology statistics, namely degree and betweenness centrality (BC), were calculated using NetworkAnalyzer (Assenov et al., 2008) in Cytoscape v3.5 (Shannon et al., 2003). These edges were ranked using a score of maximum (per-group OTU abundance) × absolute [median (per-group correlation)] to prioritize OTUs and the phenotypes they potentially affect, where the per-group OTU abundance and correlation are medians across the two experiments. The top hit of BLAST for the Greengenes representative sequence for an OTU was used to obtain species level identification.

Data Availability

Raw reads of 16S rRNA gene sequencing have been deposited at NCBI under BioProject PRJNA394608, Biosamples of SAMN07356206 - SAMN07356264, Sequence Read Archive SRP112596.

RESULTS

Lean, normoglycemic male mice were left untreated, or were treated with ampicillin, metronidazole, neomycin or vancomycin, or a cocktail containing all four antibiotics for 4 weeks to study the effects of antibiotic treatment on glucose tolerance, genes involved in glucose and bile acid metabolism, and the gut microbiota. Antibiotics resulted in different patterns of changes in the metabolic parameters, gene expression, and intestinal microbiome.

Antibiotics Improved Glucose Tolerance in Lean Mice

No metabolic parameter worsened following antibiotics treatment (**Figure 1** and **Supplementary Figure S1**). We observed that treatment with individual or cocktail of antibiotics reduced fasting glucose, but did not change body weight. Glucose tolerance improved upon treatment with cocktail, ampicillin, or vancomycin as indicated by reduced AUC of the GTT. Treatment with all antibiotics, including metronidazole or neomycin reduced fasting glucose levels, however, the latter two did not cause changes in systemic glucose tolerance. Fasting insulin was reduced only when the mice were treated with vancomycin. Overall, glucose metabolism was regulated by antibiotic treatment.

Antibiotics Changed Expression of Genes Involved in Glucose and Bile Acid Metabolism

Tissue specific host gene expression is important in many metabolic processes (Thomas et al., 2008; Chiang, 2013) and regulated by gut microbiota (Larsson et al., 2012). These, along with the knowledge that intestinal glucose metabolism can control systemic glucose levels (Saeidi et al., 2013), led us to examine the expression of key glucose and bile acid metabolic genes in the liver and the ileum.

The majority of the tested genes in the ileum showed changes in expression due to antibiotic treatment (**Figure 1**). Ileum Hk2 and G6pase transcripts showed decreased expressions after treatment with cocktail, ampicillin, or vancomycin. Ileum Pck1 and Tgr5 mRNA were increased after treatment with ampicillin or vancomycin, but showed no changes after with cocktail. Ileal Hk1 and Glut1 did not change gene expression after antibiotics, whereas, Fgf15, Fxr, and Shp showed antibiotic-specific patterns in expression.

Only three genes showed differential expression in the liver following antibiotic treatment (**Figure 1**). Fxr and G6pase showed increased and decreased expression, respectively, in ampicillin or vancomycin treated mice. Pck1 showed lower expressions in ampicillin treated samples. Hk2 and Insr genes in the liver did not change following antibiotics treatment.

Despite some variability in tissue specific behavior of genes in response to antibiotics, the improved glucose tolerance upon antibiotic treatment suggests that relationships between gene expression and metabolic parameters are mostly preserved across all groups. Hence, we constructed a correlation network consisting of (differentially expressed) genes and (differentially abundant) metabolic parameters using all samples per experiment (**Figure 2**). Genes from the ileum, including G6pase, Hk2, and Fxr were strongly connected with the GTT-AUC. The Fxr gene in the liver was positively correlated with the ileum Tgr5 but negatively correlated with ileum Fxr and with fasting glucose and GTT. Altogether, this network indicates opposite effects of intestinal and liver Fxr on glucose metabolism. Furthermore, it also suggests that increased glycolytic gene expression program in ileum is connected to worsening of systemic glucose metabolism.

Antibiotics Caused Shifts in Microbial Communities

Microbiome composition is known to be affected by antibiotics (De La Cochetiere et al., 2005; Jernberg et al., 2007; Dethlefsen et al., 2008; Jakobsson et al., 2010; Dethlefsen and Relman, 2011; Perez-Cobas et al., 2013; Pallav et al., 2014; Panda et al., 2014; Raymond et al., 2016) and involved in metabolic processes (Larsson et al., 2012; Tremaroli and Backhed, 2012; Sanz et al., 2015; Utzschneider et al., 2016), so we hypothesized that gut microbes might play a mechanistic role in the effect of antibiotics (Morgun et al., 2015; Greer R. et al., 2016; Greer

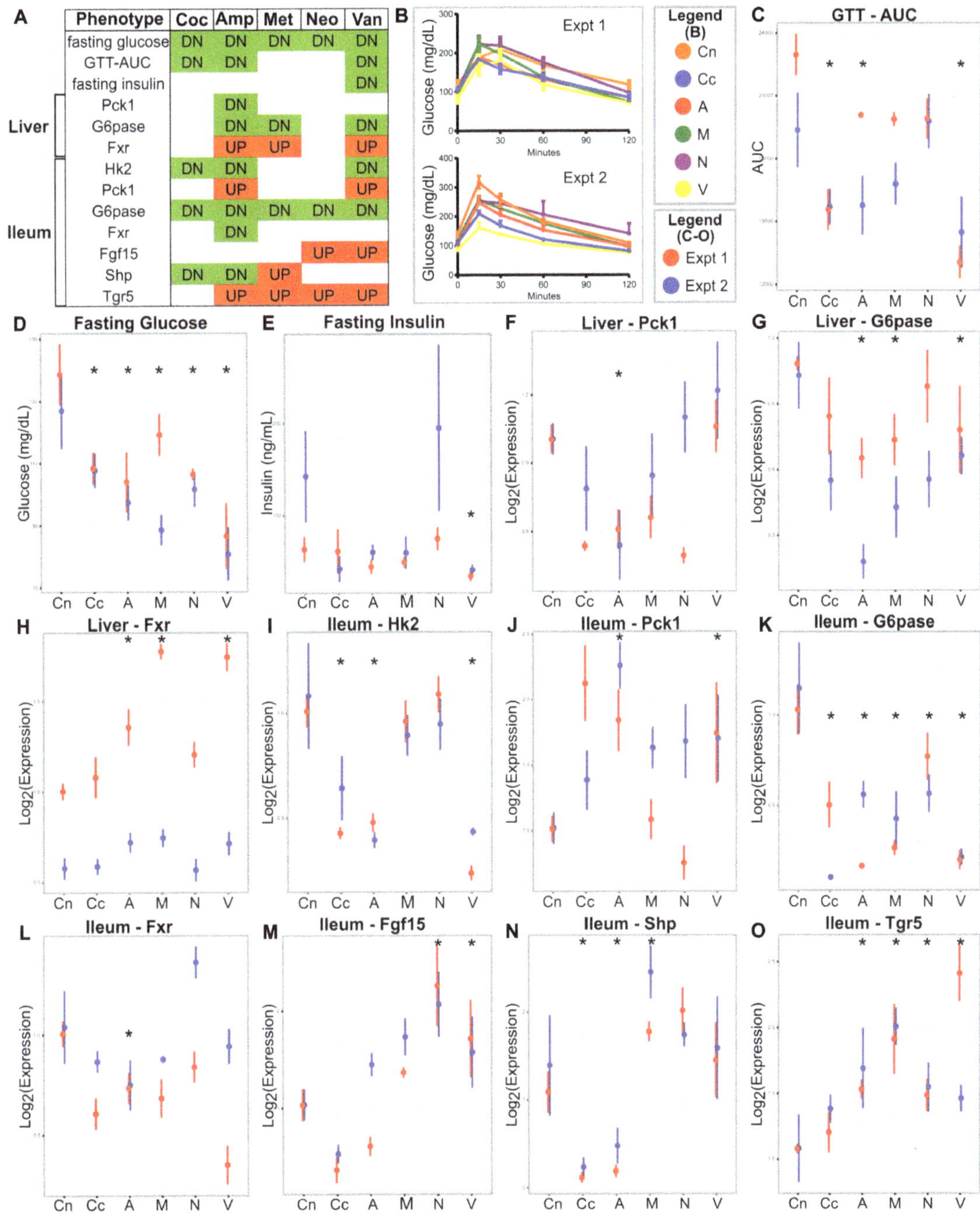

FIGURE 1 | Metabolic parameters and gene expression in antibiotic-treated and control animals. **(A)** Summary table; the red and green colors indicate increase and decrease, respectively, in antibiotic treated group compared to the control. **(B)** GTT curves for the antibiotics treated and control groups in the two experiments. **(C–E)** Metabolic parameters and gene **(F–O)** expression represented as means with standard error bars. The red and blue colors indicate experiments one and two, respectively. Asterisks indicate parameters that show statistically significant differences upon antibiotics treatment compared to untreated control mice [same direction of (abx/control) fold change in both experiments, individual *p*-value < 20% in each experiment, Fisher's combined *p*-value < 5% and FDR < 10%]. Cn, Control; Coc or Cc, cocktail; Amp or A, ampicillin; Met or M, metronidazole; Neo or N, neomycin; Van or V, vancomycin.

R.L. et al., 2016) on host glucose metabolism (Caesar et al., 2012; Greer R.L. et al., 2016). Sequencing the 16S rRNA gene of the cecal microbiome from the two experiments provided a total of 14,321,948 high quality reads with mean length of 248.50 bases and standard deviation of 9.42. A threshold of 99% cumulative abundance across all samples per experiment retained 734 and 677 OTUs in the two experiments (overlap of 561 OTUs) with 5,450,867 and 5,525,927 assigned sequences to the OTUs. The alpha diversity metrics on the normalized and rarefied OTU tables are provided in **Supplementary**

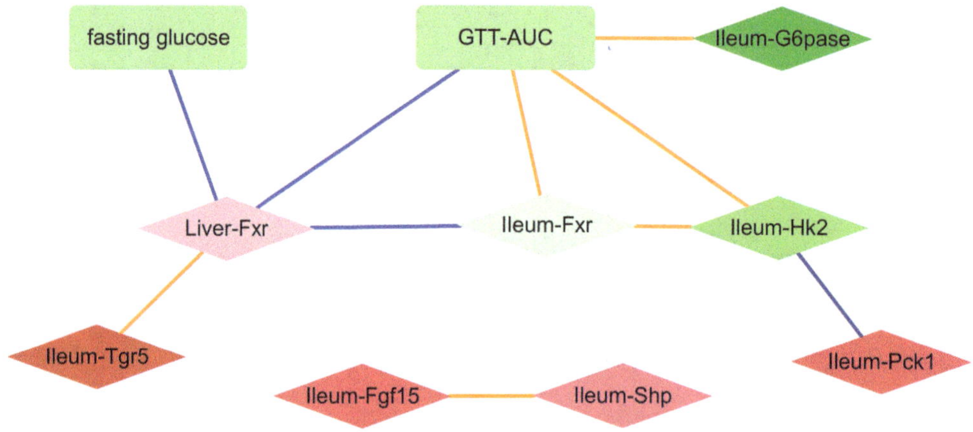

FIGURE 2 | A network consisting of metabolic parameters and gene expression from liver or ileum. An edge indicates the sign of spearman correlation coefficients across all samples in the two experiments are consistent, individual *p*-value of correlation within each experiment is <20%, Fisher's combined *p*-value of all experiments <5% and FDR < 10%. Red and green colors indicate increased and decreased median fold change (abx/control) for nodes, respectively; where the color intensity corresponds to the level of fold change (e.g., dark color indicates fold change ratio is further away from 1); diamond and rectangle shapes indicate genes and metabolic parameters, respectively. Blue and orange colors indicate negative and positive correlated edges, respectively.

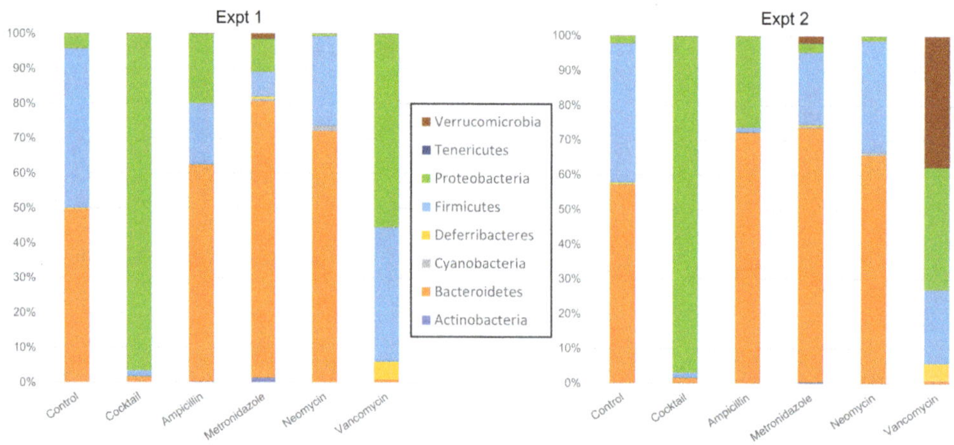

FIGURE 3 | Taxonomic plots showing mean bacterial abundance across the different groups at the phylum level.

Table S2. As expected, the cocktail of antibiotics reduced the diversity of the samples compared to untreated or individual antibiotics (**Supplementary Figures S2, S3**). Shannon diversity comparisons showed that alpha diversity decreased when treated with cocktail, ampicillin, or vancomycin (**Supplementary Figure S2**).

A PCoA analysis using the weighted UniFrac suggested that the overall community composition from vancomycin and ampicillin treatment was closer to that when treated with antibiotics cocktail (**Supplementary Figure S3**). At the phylum level, Firmicutes decreased in cocktail, ampicillin, and metronidazole treated samples. Bacteroidetes decreased upon cocktail and vancomycin treatment but increased when treated with metronidazole (**Figures 3, 4** and **Supplementary Figure S5**; Sheet in **Supplementary Table S3**). The treatment with antibiotics showed similar patterns of change in the abundant bacteria at the order level, while less abundant bacteria showed antibiotic specific changes (**Supplementary Figures S4, S6**). Vancomycin treatment increased Verrucomicrobiales in both experiments compared to control, however, increase in the second experiment was extremely high (fold change = 17,480) compared to the first (fold change = 358). Of note, this order was presented by single member (*A. muciniphila*). Thus, we also analyzed the abundance of this microbe via specific PCR and confirmed differences between two experiments in vancomycin treated groups (0.04 and 9794.4 ng DNA *A. muciniphila*/ g cecal content in the first and second experiments, respectively).

Cocktail, ampicillin, and vancomycin treated samples showed similar patterns of change at the OTU level as compared to the microbiome of control samples (**Figure 4**), which may be related to the fact that only these antibiotic treatments were able to change GTT-AUC (**Figure 1**). *Prevotella* sp. (OTU_189721) was the most abundant OTU in control (median abundance across two groups (24%), neomycin (38%), and metronidazole (17.5%)

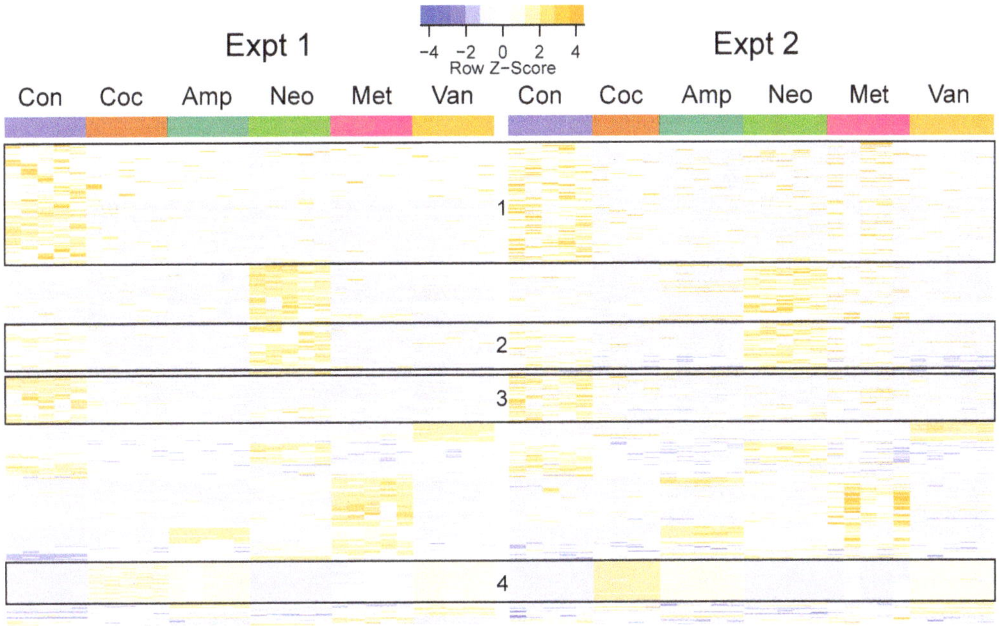

FIGURE 4 | Heatmap of differentially abundant OTUs in the different groups. The represented OTUs passed the same statistical criteria as that used for phenotypes. The rows (OTUs) were clustered based on the abundance in the first experiment and the same row order was used for the second experiment. The rows were scaled, and orange and blue colors indicate increased and decreased abundance, respectively. The ids and taxa names for the rows (OTUs) are provided in **Supplementary Table S3**. The boxes indicate OTU clusters with the same taxonomy: Boxes 1 (113/117), 2 (43/46), and 3 (43/45) primarily contained Clostridiales (phylum: Firmicutes), while Box 4 (38/38) had Enterobacteriales (phylum: Proteobacteria).

treated samples. Enterobacteriaceae family (OTU_1111294) was the most abundant in cocktail (38%) and vancomycin (28%), and the third most abundant in ampicillin (14%) treated samples. Bacteroides uniformis (OTU_589071) was the most abundant upon ampicillin treatment (22.8%), while *A. muciniphila* was the second most abundant in vancomycin treated samples (17.5%) (**Supplementary Table S4**).

Microbes Are Associated with Changed Phenotypes

Gut microbiota can control the expression of many genes in the small intestine (Larsson et al., 2012). Therefore, we asked whether the antibiotic-induced changes in the microbiome were potentially connected to the observed changes in gene expression. We constructed a transkingdom network using all groups, consisting of genes, metabolic parameters, and OTUs, to identify candidate interactions whereby microbes can mediate changes in systemic glucose tolerance and found 131 OTU-phenotype edges.

To focus on microbe-phenotype relationships that are not affected by type of antibiotics, we retained the 40 edges (**Figure 5**) that maintained the same sign of correlation coefficient between the various groups of both experiments and consistent with potential causal relations (Dong et al., 2015; Morgun et al., 2015; Greer R.L. et al., 2016; Rodrigues et al., 2017) in at least one group of both experiments. Overall, this means that while a strength of OTU-phenotype interaction may be weak for a particular antibiotic group, this interaction may still be important in mediating effects of antibiotics on the host in general. The abundance of a microbe and its strength of correlation with a phenotype are expected to be crucial in mediating the effects, hence these 40 edges were ranked using a score that takes into account the maximum per-group OTU abundance and the median per-group correlation strength with a phenotype (**Figure 6**; See formula in section "Materials and Methods").

NCBI BLAST on an OTU's Greengenes reference sequence was used to obtain its (closest) species level identification. Interestingly, associations between *A. muciniphila* with fasting glucose and liver Fxr showed as the top interactions suggesting a possible mechanism through which this bacterium can mediate systemic changes in glucose metabolism. *Proteus mirabilis* was negatively correlated with GTT-AUC. *Bacteroides uniformis* was positively and negatively correlated with hepatic Fxr and G6pase, respectively. Importance of phenotypes in the network was also determined by degrees of connectedness (degree) and BC. GTT-AUC (degree = 14, BC score = 0.65) and liver Fxr (degree = 8, BC score = 0.60) were the highly connected metabolic parameter and gene, respectively, as well as the key nodes in the largest connected component of the network. Overall, it suggests that gut microbiota potentially influences the liver metabolic genes and systemic metabolic parameters and mediates the effects of antibiotics on host phenotypes.

DISCUSSION

Germ-free Swiss Webster mice showed improved glucose metabolism (Caesar et al., 2012), suggesting that microbiota

FIGURE 5 | Transkingdom Network. Red and green colors indicate increased and decreased median fold change (abx/control) for nodes, respectively; diamond, rectangle, and circle shapes indicate genes, metabolic parameters, and OTUs, respectively. Blue and orange colors indicate negative and positive correlated edges, respectively. We indicate a phenotypic edge if its strength of correlation (in the phenotypic network) is stronger than at least one OTU-phenotype edges connecting the phenotypes.

regulate metabolism in this strain. Furthermore, Swiss Webster mice are traditionally outbred, so more similar to human population. Therefore, Swiss Webster mice were selected for this study. While research has been done to study the effects of antibiotics on microbiota and glucose tolerance in diseased models (Francino, 2015), these effects in lean, non-diabetic or normoglycemic mice are not well studied. Such a study can provide meaningful insights into the host-microbial interactions and consequences of antibiotics in healthy population, and may allow the prediction of protective mechanisms and risk factors for development of diabetes.

To the best of our knowledge, our study is the first to show the ability of antibiotics to change glucose metabolism in healthy mice. The reduced fasting glucose and GTT-AUC in two experiments, especially in the ampicillin, vancomycin, and cocktail treated samples suggest that antibiotic treatment cause systemic improvements in glucose tolerance. Although our observations of reduced GTT-AUC contradicts with the absence of change observed in healthy humans (Mikkelsen et al., 2015), the cocktail ingredients and time course of antibiotic treatment (1 week vs. 4 weeks) of the two studies may be more critical factors contributing to this disagreement than differences between two species (i.e., mice and humans). Additionally, the unchanged insulin secretion by cocktail treatment in our study is in agreement with the study performed in humans (Mikkelsen et al., 2015). Noteworthy, one study did not observe any changes in fasting glucose and insulin in chow fed C57BL6 mice when treated with broad-spectrum antibiotics (ampicillin, metronidazole, and neomycin) (Pang et al., 2013). This disagreement might be due to differences in the mice strain, gut bacterial communities in different mouse facilities and the antibiotics used in the cocktail. For example, effect of vancomycin on glucose metabolism can be partially attributed (at least for our second experiment) by increased abundance of *A. muciniphila* which is missing in some mouse colonies. While the two studies (Pang et al., 2013; Mikkelsen et al., 2015) have some discrepancies with our observations, there are numerous supportive studies using germ-free (Caesar et al., 2012) or diet-induced obese (Cani et al., 2008, 2014; Membrez et al., 2008; Carvalho et al., 2012; Hwang et al., 2015; Fujisaka et al., 2016) mice that have shown improved glucose tolerance in the absence of microbiota and with antibiotics usage and the consequently modulated microbiota.

The expression of key genes from the glucose and bile acid metabolism pathways were measured, since bile acid signaling

FIGURE 6 | Ranking the transkingdom edges. The score accounts for the bacterial abundance and its correlation strength with a phenotype. Blue and orange colors indicate negative and positive correlated edges, respectively. A broken axis (indicated by "Z") allows for clear visualization of the other pairs.

plays an important role in glucose homeostasis (Nguyen and Bouscarel, 2008; Trauner et al., 2010; Prawitt et al., 2011; Chiang, 2013; Nie et al., 2015; Trabelsi et al., 2016). We observed well-known (and therefore expected) relationships between the tissue-specific expression patterns of different genes themselves and with the systemic metabolic parameters in vancomycin and ampicillin. Low hepatic Fxr causes increased gluconeogenesis (Ma et al., 2006) and bile acid synthesis (Duran-Sandoval et al., 2004), while increased liver Fxr (Li and Guo, 2015) and intestinal Fgf15 (Holt et al., 2003) suppress hepatic bile acid synthesis (Kong et al., 2012) and regulate hepatic glucose metabolism (Potthoff et al., 2011). Also, increased liver Fxr represses G6pase (Yamagata et al., 2004; Ma et al., 2006; Zhang et al., 2006), Pck1 (De Fabiani et al., 2003; Yamagata et al., 2004; Ma et al., 2006), and like repressed ileum Fxr (Jiang et al., 2015b), improves glucose tolerance (Ma et al., 2006; Zhang et al., 2013), similar to our results. Along the same lines, mice treated with Fgf15 showed improved glucose metabolism (Zhou et al., 2017) and increased intestinal Fgf15 expression represses liver G6pase and Pck1, key enzymes for liver gluconeogenesis (Potthoff et al., 2011). In line with our observations in vancomycin treatment, Fxr agonist obeticholic acid (OCA; Intercept Pharmaceuticals, New York, NY, United States) increased mRNA levels of Fgf15 and Tgr5 in the ileum of C57BL/6J mice without increase in ileum Fxr (Pathak et al., 2017), and an increase in ileum Fgf15 expression decreased plasma glucose levels even with low insulin levels (Potthoff et al., 2011). Supporting our results from ampicillin treatment, a study showed that treating mice on high fat diet with antibiotics cocktail inhibited Fxr signaling in the ileum but not in the liver, and observed decreased expression of Shp in the ileum (Jiang et al., 2015a). Furthermore, Fxr and Shp mRNA in the ileum were also reduced in germ-free Swiss Webster mice on chow diet compared to the conventionally raised (untreated) group (Sayin et al., 2013). Overall, these studies along with ours support the idea that bile acids repress gluconeogenesis (Modica et al., 2010). However, it is interesting to see that while metronidazole, neomycin, and cocktail do not show the above changes in gene expression there is still improvement in fasting glucose and/or glucose tolerance, suggesting that microbiota might play an even bigger role in mediating the effects of antibiotics on phenotypes through additional mechanisms not explored here.

In fact, microbiota can change the expression of many genes in the ileum (Larsson et al., 2012). Their study observed down-regulated Hk2, G6pase, and Shp in the ileum of germ-free mice when compared to conventionally raised mice on chow diet, supporting our results from cocktail (ampicillin, or vancomycin) treatment. Also, some of the changes in gut microbiota to antibiotics that we observe in our data, e.g., the increased Verrucomicrobiales following vancomycin (Hansen et al., 2012) and Enterobacteriales increase upon treatment with ampicillin or vancomycin (Ubeda et al., 2010), are well documented.

While the effects of microbes on systemic glucose tolerance in lean subjects are rarely studied, their ability to influence glucose metabolism is well-recognized (De Vadder et al., 2016). A good example of well-established causal relations between specific bacteria and glucose metabolism is beneficial effect of *A. muciniphila*. For example, it was shown that *A. muciniphila* was able to delay the onset of diabetes in the vancomycin treated mice (Hansen et al., 2012). Furthermore, multiple studies demonstrated that this bacterium can improve glucose metabolism in animal models and in humans (Everard et al., 2013; Zhang et al., 2013; Joyce and Gahan, 2014; Shin et al., 2014; Anhe et al., 2015; Dao et al., 2016; Greer R.L. et al., 2016). It might not be surprising that the negative correlation between *A. muciniphila* and glucose levels was detected as one of the top ranked edges in our unbiased transkingdom network, thus, providing extra confidence for our results about less investigated bacteria inferred in our analyses.

Our predictions provide insights into host-microbial interactions. For instance, our result of *Bacteroides uniformis* being correlated with hepatic G6pase and Fxr might indicate potential mechanisms by which this bacterium improves glucose tolerance (Gauffin Cano et al., 2012). Also, it was shown that colonization with *Bacteroides thetaiotaomicron* makes mice leaner comparing controls despite similar levels of food consumption. The ability and preference of *Bacteroides thetaiotaomicron* and *Bacteroides ovatus* to utilize polysaccharide rich diet (McNulty et al., 2013) may explain these effects. However, our result of the negative correlation between abundance of these two bacteria and Pck1 (simple sugar forming gluconeogenic enzyme) may suggest the effect of these bacteria on liver gluconeogenesis. Similarly, *P. mirabilis* is predicted to have a negative interaction with GTT-AUC in our study, but shown to be positively correlated in rats with and without high fat diet (Lecomte et al., 2015). This disagreement may be explained by different physiological pathways dominating in the same bacterial species in different host that has been clearly shown for other bacteria (Oh et al., 2010). Overall, our study offers testable hypothesis regarding critical microbe-phenotype associations.

CONCLUSION

We show that antibiotics alter systemic glucose metabolism in lean mice. In addition to reporting changes in the microbiota, expression of key genes from the glucose and bile acid metabolism pathways, and concomitant systemic metabolic measures, we delineate potential mechanisms by which microbes mediate these effects. While there is a general understanding of the different players and mechanisms of microbiome-mediated regulation of the glycemic response (Tremaroli and Backhed, 2012; Devaraj et al., 2013; Cani et al., 2014; Hartstra et al., 2015; Janssen and Kersten, 2015; Parekh et al., 2015; Sanz et al., 2015; Boulange et al., 2016; Marchesi et al., 2016; Stenman et al., 2016; Suez et al., 2016), a lot remains to be understood, especially in terms of identifying the precise pathways operating in host-microbiome interactions. Overall, our data strongly suggests that antibiotics affect systemic glucose metabolism via shaping gut microbial communities and consequently regulating gene expression programs in intestine and liver. Yet, treatment of germfree mice with antibiotics as well as colonization of germ-free mice with antibiotic modified microbiota are required to fully support above statement. Also, it is doubtful that different antibiotics use the same mechanisms of gene expression and microbiota changes to affect systemic glucose tolerance, and the limited number of samples per group makes it difficult to obtain antibiotic-specific mechanisms. Furthermore, while the taxonomical assignments of 16S rRNA sequencing of current study present natural challenges, further studies employing shotgun metagenomics sequencing will allow to overcome this limitation. Finally, our experimental design followed by a data-driven, systems biology approach of network analysis offers consistent and statistically significant interactions that may be integral in mediating the host-microbiome communication. Furthermore, this approach is a useful hypothesis generating strategy and future experimentation can help investigate the distinct mechanisms in the different antibiotics and eventually lead to personalized medicine (Zmora et al., 2016).

AUTHOR CONTRIBUTIONS

AM and NS conceived the original idea, designed and supervised the experiments, analyses, and writing. RG conceived the original idea, designed and performed the experiments, and supervised the writing. RR designed and performed the analyses, and drafted the manuscript. XD performed the analyses. JW, MG, and KD performed the experiments and analysis, respectively. All authors wrote the manuscript, read and approved the final draft submitted.

ACKNOWLEDGMENTS

This research was supported by startup funds for AM and NS from Oregon State University (OSU), United States; NIH U01 AI109695 (AM), deLaubenfels Comparative Health Research and Education Fund (NS), and R01 DK103761 (NS).

colors indicate experiments one and two, respectively. These phenotypes did not show statistically significant differences upon antibiotics treatment compared to untreated control mice.

FIGURE S2 | Boxplots showing the Shannon diversity index. Asterisk indicate statistically significant differences upon antibiotics treatment compared to untreated control mice: same direction of (abx/control) fold change in both experiments, individual p-value $< 2\%$ in each experiment, Fisher's combined p-value $< 0.1\%$ and FDR $< 0.1\%$.

FIGURE S3 | PCoA plot showing weighted UniFrac distance for cecal microbiota in the control and antibiotics treated mice. Each circle indicates a sample.

FIGURE S4 | Taxonomic plots showing bacterial abundance across the different groups at the order level.

FIGURE S5 | Taxonomic plots showing bacterial abundance across the different samples at the phylum level.

FIGURE S6 | Taxonomic plots showing bacterial abundance across the different samples at the order level.

TABLE S1 | Primers for the genes tested in this study.

TABLE S2 | Alpha diversity metrics on the unrarefied and rarefied OTU tables in the two experiments.

TABLE S3 | Levels of Firmicutes and Bacteroidetes in the two experiments. Asterisks indicate parameters that show statistically significant differences upon antibiotics treatment compared to untreated control mice: same direction of (abx/control) fold change in both experiments, individual p-value $< 20\%$ in each experiment, Fisher's combined p-value $< 5\%$ and FDR $< 10\%$.

TABLE S4 | The ID, Greengenes taxonomy, and the median frequency per group for the OTUs in the same order as of the heatmap.

REFERENCES

Altschul, S. F., Madden, T. L., Schaffer, A. A., Zhang, J., Zhang, Z., Miller, W., et al. (1997). Gapped BLAST and PSI-BLAST: a new generation of protein database search programs. *Nucleic Acids Res.* 25, 3389–3402. doi: 10.1093/nar/25.17.3389

Anhe, F. F., Roy, D., Pilon, G., Dudonne, S., Matamoros, S., Varin, T. V., et al. (2015). A polyphenol-rich cranberry extract protects from diet-induced obesity, insulin resistance and intestinal inflammation in association with increased Akkermansia spp. population in the gut microbiota of mice. *Gut* 64, 872–883. doi: 10.1136/gutjnl-2014-307142

Assenov, Y., Ramirez, F., Schelhorn, S. E., Lengauer, T., and Albrecht, M. (2008). Computing topological parameters of biological networks. *Bioinformatics* 24, 282–284. doi: 10.1093/bioinformatics/btm554

Boulange, C. L., Neves, A. L., Chilloux, J., Nicholson, J. K., and Dumas, M. E. (2016). Impact of the gut microbiota on inflammation, obesity, and metabolic disease. *Genome Med.* 8:42. doi: 10.1186/s13073-016-0303-2

Caesar, R., Reigstad, C. S., Backhed, H. K., Reinhardt, C., Ketonen, M., Lunden, G. O., et al. (2012). Gut-derived lipopolysaccharide augments adipose macrophage accumulation but is not essential for impaired glucose or insulin tolerance in mice. *Gut* 61, 1701–1707. doi: 10.1136/gutjnl-2011-301689

Cani, P. D., Bibiloni, R., Knauf, C., Waget, A., Neyrinck, A. M., Delzenne, N. M., et al. (2008). Changes in gut microbiota control metabolic endotoxemia-induced inflammation in high-fat diet-induced obesity and diabetes in mice. *Diabetes Metab. Res. Rev.* 57, 1470–1481. doi: 10.2337/db07-1403

Cani, P. D., Geurts, L., Matamoros, S., Plovier, H., and Duparc, T. (2014). Glucose metabolism: focus on gut microbiota, the endocannabinoid system and beyond. *Diabetes Metab.* 40, 246–257. doi: 10.1016/j.diabet.2014.02.004

Caporaso, J. G., Kuczynski, J., Stombaugh, J., Bittinger, K., Bushman, F. D., Costello, E. K., et al. (2010). QIIME allows analysis of high-throughput community sequencing data. *Nat. Methods* 7, 335–336. doi: 10.1038/nmeth.f.303

Caporaso, J. G., Lauber, C. L., Walters, W. A., Berg-Lyons, D., Huntley, J., Fierer, N., et al. (2012). Ultra-high-throughput microbial community analysis on the Illumina HiSeq and MiSeq platforms. *ISME J.* 6, 1621–1624. doi: 10.1038/ismej.2012.8

Carvalho, B. M., Guadagnini, D., Tsukumo, D. M. L., Schenka, A. A., Latuf-Filho, P., Vassallo, J., et al. (2012). Modulation of gut microbiota by antibiotics improves insulin signalling in high-fat fed mice. *Diabetologia* 55, 2823–2834. doi: 10.1007/s00125-012-2648-4

Chakraborti, C. K. (2015). New-found link between microbiota and obesity. *World J. Gastrointest. Pathophysiol.* 6, 110–119. doi: 10.4291/wjgp.v6.i4.110

Chiang, J. Y. L. (2013). "Bile acid metabolism and signaling," in *Comprehensive Physiology*, ed. D. M. Pollock (Hoboken, NJ: John Wiley & Sons, Inc.).

Dao, M. C., Everard, A., Aron-Wisnewsky, J., Sokolovska, N., Prifti, E., Verger, E. O., et al. (2016). *Akkermansia muciniphila* and improved metabolic health during a dietary intervention in obesity: relationship with gut microbiome richness and ecology. *Gut* 65, 426–436. doi: 10.1136/gutjnl-2014-308778

De Fabiani, E., Mitro, N., Gilardi, F., Caruso, D., Galli, G., and Crestani, M. (2003). Coordinated control of cholesterol catabolism to bile acids and of gluconeogenesis via a novel mechanism of transcription regulation linked to the fasted-to-fed cycle. *J. Biol. Chem.* 278, 39124–39132. doi: 10.1074/jbc.M305079200

De La Cochetiere, M. F., Durand, T., Lepage, P., Bourreille, A., Galmiche, J. P., and Dore, J. (2005). Resilience of the dominant human fecal microbiota upon short-course antibiotic challenge. *J. Clin. Microbiol.* 43, 5588–5592. doi: 10.1128/JCM.43.11.5588-5592.2005

De Vadder, F., Kovatcheva-Datchary, P., Zitoun, C., Duchampt, A., Backhed, F., and Mithieux, G. (2016). Microbiota-produced succinate improves glucose homeostasis via intestinal gluconeogenesis. *Cell Metab.* 24, 151–157. doi: 10.1016/j.cmet.2016.06.013

DeSantis, T. Z., Hugenholtz, P., Larsen, N., Rojas, M., Brodie, E. L., Keller, K., et al. (2006). Greengenes, a chimera-checked 16S rRNA gene database and workbench compatible with ARB. *Appl. Environ. Microbiol.* 72, 5069–5072. doi: 10.1128/AEM.03006-05

Dethlefsen, L., Huse, S., Sogin, M. L., and Relman, D. A. (2008). The pervasive effects of an antibiotic on the human gut microbiota, as revealed by deep 16S rRNA sequencing. *PLOS Biol.* 6:e280. doi: 10.1371/journal.pbio.0060280

Dethlefsen, L., and Relman, D. A. (2011). Incomplete recovery and individualized responses of the human distal gut microbiota to repeated antibiotic perturbation. *Proc. Natl. Acad. Sci. U.S.A.* 108(Suppl._1), 4554–4561. doi: 10.1073/pnas.1000087107

Devaraj, S., Hemarajata, P., and Versalovic, J. (2013). The human gut microbiome and body metabolism: implications for obesity and diabetes. *Clin. Chem.* 59, 617–628. doi: 10.1373/clinchem.2012.187617

Dong, X., Yambartsev, A., Ramsey, S. A., Thomas, L. D., Shulzhenko, N., and Morgun, A. (2015). Reverse enGENEering of regulatory networks from big data: a roadmap for biologists. *Bioinform. Biol. Insights* 9, 61–74. doi: 10.4137/BBI.S12467

Duran-Sandoval, D., Mautino, G., Martin, G., Percevault, F., Barbier, O., Fruchart, J. C., et al. (2004). Glucose regulates the expression of the farnesoid X receptor in liver. *Diabetes* 53, 890–898. doi: 10.2337/diabetes.53.4.890

Edgar, R. C. (2010). Search and clustering orders of magnitude faster than BLAST. *Bioinformatics* 26, 2460–2461. doi: 10.1093/bioinformatics/btq461

Everard, A., Belzer, C., Geurts, L., Ouwerkerk, J. P., Druart, C., Bindels, L. B., et al. (2013). Cross-talk between *Akkermansia muciniphila* and intestinal epithelium controls diet-induced obesity. *Proc. Natl. Acad. Sci. U.S.A.* 110, 9066–9071. doi: 10.1073/pnas.1219451110

Fisher, R. A. (1932). "Statistical methods for research workers," in *Biological Monographs and Manuals*, 5 Edn, eds F. A. E. Crew and D. W. Cutler (Edinburgh: Oliver and Boyd).

Francino, M. P. (2015). Antibiotics and the human gut microbiome: dysbioses and accumulation of resistances. *Front. Microbiol.* 6:1543. doi: 10.3389/fmicb.2015.01543

Fujisaka, S., Ussar, S., Clish, C., Devkota, S., Dreyfuss, J. M., Sakaguchi, M., et al. (2016). Antibiotic effects on gut microbiota and metabolism are host dependent. *J. Clin. Invest.* 126, 4430–4443. doi: 10.1172/JCI86674

Gauffin Cano, P., Santacruz, A., Moya, A., and Sanz, Y. (2012). Bacteroides uniformis CECT 7771 ameliorates metabolic and immunological dysfunction in mice with high-fat-diet induced obesity. *PLOS ONE* 7:e41079. doi: 10.1371/journal.pone.0041079

Gower, J. (1998). "Principal coordinate analysis," in *Encyclopedia of Biostatistics*, eds T. Coulton and P. Armitage (Hoboken, NJ: John Wiley and Sons Inc.), 3514–3518.

Greer, R., Dong, X., Morgun, A., and Shulzhenko, N. (2016). Investigating a holobiont: microbiota perturbations and transkingdom networks. *Gut Microbes* 7, 126–135. doi: 10.1080/19490976.2015.1128625

Greer, R. L., Dong, X., Moraes, A. C., Zielke, R. A., Fernandes, G. R., Peremyslova, E., et al. (2016). *Akkermansia muciniphila* mediates negative effects of IFNgamma on glucose metabolism. *Nat. Commun.* 7:13329. doi: 10.1038/ncomms13329

Greer, R. L., Morgun, A., and Shulzhenko, N. (2013). Bridging immunity and lipid metabolism by gut microbiota. *J. Allergy Clin. Immunol.* 132, 253–263. doi: 10.1016/j.jaci.2013.06.025

Hansen, C. H., Krych, L., Nielsen, D. S., Vogensen, F. K., Hansen, L. H., Sorensen, S. J., et al. (2012). Early life treatment with vancomycin propagates *Akkermansia muciniphila* and reduces diabetes incidence in the NOD mouse. *Diabetologia* 55, 2285–2294. doi: 10.1007/s00125-012-2564-7

Hartstra, A. V., Bouter, K. E., Backhed, F., and Nieuwdorp, M. (2015). Insights into the role of the microbiome in obesity and type 2 diabetes. *Diabetes Care* 38, 159–165. doi: 10.2337/dc14-0769

Holt, J. A., Luo, G., Billin, A. N., Bisi, J., McNeill, Y. Y., Kozarsky, K. F., et al. (2003). Definition of a novel growth factor-dependent signal cascade for the suppression of bile acid biosynthesis. *Genes Dev.* 17, 1581–1591. doi: 10.1101/gad.1083503

Horikoshi, M., and Tang, Y. (2016). *ggfortify: Data Visualization Tools for Statistical Analysis Results.* Available at: https://cran.r-project.org/web/packages/ggfortify/index.html

Human Microbiome Project Consortium (2012). A framework for human microbiome research. *Nature* 486, 215–221. doi: 10.1038/nature11209

Hwang, I., Park, Y. J., Kim, Y. R., Kim, Y. N., Ka, S., Lee, H. Y., et al. (2015). Alteration of gut microbiota by vancomycin and bacitracin improves insulin resistance via glucagon-like peptide 1 in diet-induced obesity. *FASEB J.* 29, 2397–2411. doi: 10.1096/fj.14-265983

Jakobsson, H. E., Jernberg, C., Andersson, A. F., Sjolund-Karlsson, M., Jansson, J. K., and Engstrand, L. (2010). Short-term antibiotic treatment has differing long-term impacts on the human throat and gut microbiome. *PLOS ONE* 5:e9836. doi: 10.1371/journal.pone.0009836

Janssen, A. W., and Kersten, S. (2015). The role of the gut microbiota in metabolic health. *FASEB J.* 29, 3111–3123. doi: 10.1096/fj.14-269514

Jernberg, C., Lofmark, S., Edlund, C., and Jansson, J. K. (2007). Long-term ecological impacts of antibiotic administration on the human intestinal microbiota. *ISME J.* 1, 56–66. doi: 10.1038/ismej.2007.3

Jiang, C., Xie, C., Li, F., Zhang, L., Nichols, R. G., Krausz, K. W., et al. (2015a). Intestinal farnesoid X receptor signaling promotes nonalcoholic fatty liver disease. *J. Clin. Invest.* 125, 386–402. doi: 10.1172/JCI76738

Jiang, C., Xie, C., Lv, Y., Li, J., Krausz, K. W., Shi, J., et al. (2015b). Intestine-selective farnesoid X receptor inhibition improves obesity-related metabolic dysfunction. *Nat. Commun.* 6:10166. doi: 10.1038/ncomms10166

Joyce, S. A., and Gahan, C. G. (2014). The gut microbiota and the metabolic health of the host. *Curr. Opin. Gastroenterol.* 30, 120–127. doi: 10.1097/MOG.0000000000000039

Karlsson, F. H., Tremaroli, V., Nookaew, I., Bergstrom, G., Behre, C. J., Fagerberg, B., et al. (2013). Gut metagenome in European women with normal, impaired and diabetic glucose control. *Nature* 498, 99–103. doi: 10.1038/nature12198

Kasai, C., Sugimoto, K., Moritani, I., Tanaka, J., Oya, Y., Inoue, H., et al. (2015). Comparison of the gut microbiota composition between obese and non-obese individuals in a Japanese population, as analyzed by terminal restriction fragment length polymorphism and next-generation sequencing. *BMC Gastroenterol.* 15:100. doi: 10.1186/s12876-015-0330-2

Komsta, L. (2011). *Outliers: Tests for Outliers.* Available at: http://www.r-project.org

Kong, B., Wang, L., Chiang, J. Y., Zhang, Y., Klaassen, C. D., and Guo, G. L. (2012). Mechanism of tissue-specific farnesoid X receptor in suppressing the expression of genes in bile-acid synthesis in mice. *Hepatology* 56, 1034–1043. doi: 10.1002/hep.25740

Larsson, E., Tremaroli, V., Lee, Y. S., Koren, O., Nookaew, I., Fricker, A., et al. (2012). Analysis of gut microbial regulation of host gene expression along the length of the gut and regulation of gut microbial ecology through MyD88. *Gut* 61, 1124–1131. doi: 10.1136/gutjnl-2011-301104

Lecomte, V., Kaakoush, N. O., Maloney, C. A., Raipuria, M., Huinao, K. D., Mitchell, H. M., et al. (2015). Changes in gut microbiota in rats fed a high fat diet correlate with obesity-associated metabolic parameters. *PLOS ONE* 10:e0126931. doi: 10.1371/journal.pone.0126931

Li, G., and Guo, G. L. (2015). Farnesoid X receptor, the bile acid sensing nuclear receptor, in liver regeneration. *Acta Pharm. Sin. B* 5, 93–98. doi: 10.1016/j.apsb.2015.01.005

Lozupone, C., and Knight, R. (2005). UniFrac: a new phylogenetic method for comparing microbial communities. *Appl. Environ. Microbiol.* 71, 8228–8235. doi: 10.1128/AEM.71.12.8228-8235.2005

Ma, K., Saha, P. K., Chan, L., and Moore, D. D. (2006). Farnesoid X receptor is essential for normal glucose homeostasis. *J. Clin. Invest.* 116, 1102–1109. doi: 10.1172/JCI25604

Marchesi, J. R., Adams, D. H., Fava, F., Hermes, G. D., Hirschfield, G. M., Hold, G., et al. (2016). The gut microbiota and host health: a new clinical frontier. *Gut* 65, 330–339. doi: 10.1136/gutjnl-2015-309990

McDonald, D., Price, M. N., Goodrich, J., Nawrocki, E. P., DeSantis, T. Z., Probst, A., et al. (2012). An improved Greengenes taxonomy with explicit ranks for ecological and evolutionary analyses of bacteria and archaea. *ISME J.* 6, 610–618. doi: 10.1038/ismej.2011.139

McNulty, N. P., Wu, M., Erickson, A. R., Pan, C., Erickson, B. K., Martens, E. C., et al. (2013). Effects of diet on resource utilization by a model human gut microbiota containing Bacteroides cellulosilyticus WH2, a symbiont with an extensive glycobiome. *PLOS Biol.* 11:e1001637. doi: 10.1371/journal.pbio.1001637

Membrez, M., Blancher, F., Jaquet, M., Bibiloni, R., Cani, P. D., Burcelin, R. G., et al. (2008). Gut microbiota modulation with norfloxacin and ampicillin enhances glucose tolerance in mice. *FASEB J.* 22, 2416–2426. doi: 10.1096/fj.07-102723

Mikkelsen, K. H., Allin, K. H., and Knop, F. K. (2016). Effect of antibiotics on gut microbiota, glucose metabolism and body weight regulation: a review of the literature. *Diabetes. Obes. Metab.* 18, 444–453. doi: 10.1111/dom.12637

Mikkelsen, K. H., Frost, M., Bahl, M. I., Licht, T. R., Jensen, U. S., Rosenberg, J., et al. (2015). Effect of antibiotics on gut microbiota, gut hormones and glucose metabolism. *PLOS ONE* 10:e0142352. doi: 10.1371/journal.pone.0142352

bile acid receptor FXR paradigm. *Nucl. Recept. Signal.* 8:e005. doi: 10.1621/nrs.08005

Morgulis, A., Coulouris, G., Raytselis, Y., Madden, T. L., Agarwala, R., and Schaffer, A. A. (2008). Database indexing for production MegaBLAST searches. *Bioinformatics* 24, 1757–1764. doi: 10.1093/bioinformatics/btn322

Morgun, A., Dzutsev, A., Dong, X., Greer, R. L., Sexton, D. J., Ravel, J., et al. (2015). Uncovering effects of antibiotics on the host and microbiota using transkingdom gene networks. *Gut* 64, 1732–1743. doi: 10.1136/gutjnl-2014-308820

Murri, M., Leiva, I., Gomez-Zumaquero, J. M., Tinahones, F. J., Cardona, F., Soriguer, F., et al. (2013). Gut microbiota in children with type 1 diabetes differs from that in healthy children: a case-control study. *BMC Med.* 11:46. doi: 10.1186/1741-7015-11-46

Nguyen, A., and Bouscarel, B. (2008). Bile acids and signal transduction: role in glucose homeostasis. *Cell. Signal.* 20, 2180–2197. doi: 10.1016/j.cellsig.2008.06.014

Nie, Y. F., Hu, J., and Yan, X. H. (2015). Cross-talk between bile acids and intestinal microbiota in host metabolism and health. *J. Zhejiang Univ. Sci. B* 16, 436–446. doi: 10.1631/jzus.B1400327

Oh, P. L., Benson, A. K., Peterson, D. A., Patil, P. B., Moriyama, E. N., Roos, S., et al. (2010). Diversification of the gut symbiont *Lactobacillus reuteri* as a result of host-driven evolution. *ISME J.* 4, 377–387. doi: 10.1038/ismej.2009.123

Pallav, K., Dowd, S. E., Villafuerte, J., Yang, X., Kabbani, T., Hansen, J., et al. (2014). Effects of polysaccharopeptide from *Trametes versicolor* and amoxicillin on the gut microbiome of healthy volunteers: a randomized clinical trial. *Gut Microbes* 5, 458–467. doi: 10.4161/gmic.29558

Panda, S., El khader, I., Casellas, F., Lopez Vivancos, J., Garcia Cors, M., Santiago, A., et al. (2014). Short-term effect of antibiotics on human gut microbiota. *PLOS ONE* 9:e95476. doi: 10.1371/journal.pone.0095476

Pang, J., Rhodes, D. H., Pini, M., Akasheh, R. T., Castellanos, K. J., Cabay, R. J., et al. (2013). Increased adiposity, dysregulated glucose metabolism and systemic inflammation in Galectin-3 KO mice. *PLOS ONE* 8:e57915. doi: 10.1371/journal.pone.0057915

Parekh, P. J., Balart, L. A., and Johnson, D. A. (2015). The influence of the gut microbiome on obesity, metabolic syndrome and gastrointestinal disease. *Clin. Transl. Gastroenterol.* 6:e91. doi: 10.1038/ctg.2015.16

Pathak, P., Liu, H., Boehme, S., Xie, C., Krausz, K. W., Gonzalez, F., et al. (2017). Farnesoid X receptor induces Takeda G-protein receptor 5 cross-talk to regulate bile acid synthesis and hepatic metabolism. *J. Biol. Chem.* 292, 11055–11069. doi: 10.1074/jbc.M117.784322

Paulson, J. N., Stine, O. C., Bravo, H. C., and Pop, M. (2013). Differential abundance analysis for microbial marker-gene surveys. *Nat. Methods* 10, 1200–1202. doi: 10.1038/nmeth.2658

Perez-Cobas, A. E., Gosalbes, M. J., Friedrichs, A., Knecht, H., Artacho, A., Eismann, K., et al. (2013). Gut microbiota disturbance during antibiotic therapy: a multi-omic approach. *Gut* 62, 1591–1601. doi: 10.1136/gutjnl-2012-303184

Potthoff, M. J., Boney-Montoya, J., Choi, M., He, T., Sunny, N. E., Satapati, S., et al. (2011). FGF15/19 regulates hepatic glucose metabolism by inhibiting the CREB-PGC-1alpha pathway. *Cell Metab.* 13, 729–738. doi: 10.1016/j.cmet.2011.03.019

Prawitt, J., Caron, S., and Staels, B. (2011). Bile acid metabolism and the pathogenesis of type 2 diabetes. *Curr. Diab. Rep.* 11, 160–166. doi: 10.1007/s11892-011-0187-x

Qin, J., Li, Y., Cai, Z., Li, S., Zhu, J., Zhang, F., et al. (2012). A metagenome-wide association study of gut microbiota in type 2 diabetes. *Nature* 490, 55–60. doi: 10.1038/nature11450

Rakoff-Nahoum, S., Paglino, J., Eslami-Varzaneh, F., Edberg, S., and Medzhitov, R. (2004). Recognition of commensal microflora by toll-like receptors is required for intestinal homeostasis. *Cell* 118, 229–241. doi: 10.1016/j.cell.2004.07.002

Raymond, F., Ouameur, A. A., Deraspe, M., Iqbal, N., Gingras, H., Dridi, B., et al. (2016). The initial state of the human gut microbiome determines its reshaping by antibiotics. *ISME J.* 10, 707–720. doi: 10.1038/ismej.2015.148

Ritchie, M. E., Phipson, B., Wu, D., Hu, Y. F., Law, C. W., Shi, W., et al. (2015). limma powers differential expression analyses for RNA-sequencing and microarray studies. *Nucleic Acids Res.* 43:e47. doi: 10.1093/nar/gkv007

Rodrigues, R. R., Shulzhenko, N., and Morgun, A. (2017). Transkingdom networks: a systems biology approach to identify causal members of host-microbiota interactions. (accepted). Available at: https://arxiv.org/pdf/1709.05701.pdf

Saeidi, N., Meoli, L., Nestoridi, E., Gupta, N. K., Kvas, S., Kucharczyk, J., et al. (2013). Reprogramming of intestinal glucose metabolism and glycemic control in rats after gastric bypass. *Science* 341, 406–410. doi: 10.1126/science.1235103

Sanz, Y., Olivares, M., Moya-Perez, A., and Agostoni, C. (2015). Understanding the role of gut microbiome in metabolic disease risk. *Pediatr. Res.* 77, 236–244. doi: 10.1038/pr.2014.170

Sayin, S. I., Wahlstrom, A., Felin, J., Jantti, S., Marschall, H. U., Bamberg, K., et al. (2013). Gut microbiota regulates bile acid metabolism by reducing the levels of tauro-beta-muricholic acid, a naturally occurring FXR antagonist. *Cell Metab.* 17, 225–235. doi: 10.1016/j.cmet.2013.01.003

Schneeberger, M., Everard, A., Gomez-Valades, A. G., Matamoros, S., Ramirez, S., Delzenne, N. M., et al. (2015). *Akkermansia muciniphila* inversely correlates with the onset of inflammation, altered adipose tissue metabolism and metabolic disorders during obesity in mice. *Sci. Rep.* 5:16643. doi: 10.1038/srep16643

Shannon, P., Markiel, A., Ozier, O., Baliga, N. S., Wang, J. T., Ramage, D., et al. (2003). Cytoscape: a software environment for integrated models of biomolecular interaction networks. *Genome Res.* 13, 2498–2504. doi: 10.1101/gr.1239303

Shin, N. R., Lee, J. C., Lee, H. Y., Kim, M. S., Whon, T. W., Lee, M. S., et al. (2014). An increase in the Akkermansia spp. population induced by metformin treatment improves glucose homeostasis in diet-induced obese mice. *Gut* 63, 727–735. doi: 10.1136/gutjnl-2012-303839

Stenman, L. K., Burcelin, R., and Lahtinen, S. (2016). Establishing a causal link between gut microbes, body weight gain and glucose metabolism in humans - towards treatment with probiotics. *Benef. Microbes* 7, 11–22. doi: 10.3920/BM2015.0069

Suez, J., Shapiro, H., and Elinav, E. (2016). Role of the microbiome in the normal and aberrant glycemic response. *Clin. Nutr. Exp.* 6, 59–73. doi: 10.1016/j.yclnex.2016.01.001

Thomas, C., Pellicciari, R., Pruzanski, M., Auwerx, J., and Schoonjans, K. (2008). Targeting bile-acid signalling for metabolic diseases. *Nat. Rev. Drug Discov.* 7, 678–693. doi: 10.1038/nrd2619

Trabelsi, M. S., Lestavel, S., Staels, B., and Collet, X. (2016). Intestinal bile acid receptors are key regulators of glucose homeostasis. *Proc. Nutr. Soc.* 76, 192–202. doi: 10.1017/S0029665116002834

Trauner, M., Claudel, T., Fickert, P., Moustafa, T., and Wagner, M. (2010). Bile acids as regulators of hepatic lipid and glucose metabolism. *Dig. Dis.* 28, 220–224. doi: 10.1159/000282091

Tremaroli, V., and Backhed, F. (2012). Functional interactions between the gut microbiota and host metabolism. *Nature* 489, 242–249. doi: 10.1038/nature11552

Ubeda, C., Taur, Y., Jenq, R. R., Equinda, M. J., Son, T., Samstein, M., et al. (2010). Vancomycin-resistant Enterococcus domination of intestinal microbiota is enabled by antibiotic treatment in mice and precedes bloodstream invasion in humans. *J. Clin. Invest.* 120, 4332–4341. doi: 10.1172/JCI43918

Utzschneider, K. M., Kratz, M., Damman, C. J., and Hullarg, M. (2016). Mechanisms linking the gut microbiome and glucose metabolism. *J. Clin. Endocrinol. Metab.* 101, 1445–1454. doi: 10.1210/jc.2015-4251

Vazquez-Baeza, Y., Pirrung, M., Gonzalez, A., and Knight, R. (2013). EMPeror: a tool for visualizing high-throughput microbial community data. *Gigascience* 2:16. doi: 10.1186/2047-217X-2-16

Warnes, G. R., Bolker, B., Bonebakker, L., Gentleman, R., Liaw, W. H. A., Lumley, T., et al. (2016). *gplots: Various R Programming Tools for Plotting Data.* Available at: https://CRAN.R-project.org/package=gplots

Wickham, H. (2009). *ggplot2: Elegant Graphics for Data Analysis.* New York, NY: Springer-Verlag. doi: 10.1007/978-0-387-98141-3

Wu, H., Tremaroli, V., and Backhed, F. (2015). Linking microbiota to human diseases: a systems biology perspective. *Trends Endocrinol. Metab.* 26, 758–770. doi: 10.1016/j.tem.2015.09.011

Yamagata, K., Daitoku, H., Shimamoto, Y., Matsuzaki, H., Hirota, K., Ishida, J., et al. (2004). Bile acids regulate gluconeogenic gene expression via small heterodimer partner-mediated repression of hepatocyte nuclear factor 4 and Foxo1. *J. Biol. Chem.* 279, 23158–23165. doi: 10.1074/jbc.M314322200

Yambartsev, A., Perlin, M. A., Kovchegov, Y., Shulzhenko, N., Mine, K. L., Dong, X., et al. (2016). Unexpected links reflect the noise in networks. *Biol. Direct* 11:52. doi: 10.1186/s13062-016-0155-0

Zhang, X., Shen, D., Fang, Z., Jie, Z., Qiu, X., Zhang, C., et al. (2013). Human gut microbiota changes reveal the progression of glucose intolerance. *PLOS ONE* 8:e71108. doi: 10.1371/journal.pone.0071108

Zhang, Y., Lee, F. Y., Barrera, G., Lee, H., Vales, C., Gonzalez, F. J., et al. (2006). Activation of the nuclear receptor FXR improves hyperglycemia and hyperlipidemia in diabetic mice. *Proc. Natl. Acad. Sci. U.S.A.* 103, 1006–1011. doi: 10.1073/pnas.0506982103

Zhou, M., Luo, J., Chen, M., Yang, H., Learned, R. M., DePaoli, A. M., et al. (2017). Mouse species-specific control of hepatocarcinogenesis and metabolism by FGF19/FGF15. *J. Hepatol.* 66, 1182–1192. doi: 10.1016/j.jhep.2017.01.027

Zmora, N., Zeevi, D., Korem, T., Segal, E., and Elinav, E. (2016). Taking it personally: personalized utilization of the human microbiome in health and disease. *Cell Host Microbe* 19, 12–20. doi: 10.1016/j.chom.2015.12.016

7

PCR-Based Analysis of ColE1 Plasmids in Clinical Isolates and Metagenomic Samples Reveals their Importance as Gene Capture Platforms

Manuel Ares-Arroyo[1], *Cristina Bernabe-Balas*[1], *Alfonso Santos-Lopez*[1†], *Maria R. Baquero*[2], *Kashi N. Prasad*[3], *Dolores Cid*[1], *Carmen Martin-Espada*[1], *Alvaro San Millan*[4] *and Bruno Gonzalez-Zorn*[1*]

[1] Departamento de Sanidad Animal and Centro de Vigilancia Sanitaria Veterinaria (VISAVET), Facultad de Veterinaria, Universidad Complutense de Madrid, Madrid, Spain, [2] Departamento de Microbiología, Facultad de Veterinaria, Universidad Alfonso X el Sabio, Madrid, Spain, [3] Department of Microbiology, Sanjay Gandhi Postgraduate Institute of Medical Sciences, Lucknow, India, [4] Servicio de Microbiología Hospital Universitario Ramón y Cajal, Instituto de Investigación Sanitaria (IRYCIS), Madrid, Spain

Edited by:
Manuel Espinosa, Centro de Investigaciones Biológicas (CSIC), Spain

Reviewed by:
Fabián Lorenzo, Universidad de La Laguna, Spain
Raul Fernandez-Lopez, University of Cantabria, Spain
Ellen Lorraine Zechner, University of Graz, Austria
Antonio Juárez, Universitat de Barcelona, Spain

***Correspondence:**
Bruno Gonzalez-Zorn
bgzorn@ucm.es

ColE1 plasmids are important vehicles for the spread of antibiotic resistance in the Enterobacteriaceae and Pasteurellaceae families of bacteria. Their monitoring is essential, as they harbor important resistant determinants in humans, animals and the environment. In this work, we have analyzed ColE1 replicons using bioinformatic and experimental approaches. First, we carried out a computational study examining the structure of different ColE1 plasmids deposited in databases. Bioinformatic analysis of these ColE1 replicons revealed a mosaic genetic structure consisting of a host-adapted conserved region responsible for the housekeeping functions of the plasmid, and a variable region encoding a wide variety of genes, including multiple antibiotic resistance determinants. From this exhaustive computational analysis we developed a new PCR-based technique, targeting a specific sequence in the conserved region, for the screening, capture and sequencing of these small plasmids, either specific for Enterobacteriaceae or specific for Pasteurellaceae. To validate this PCR-based system, we tested various collections of isolates from both bacterial families, finding that ColE1 replicons were not only highly prevalent in antibiotic-resistant isolates, but also present in susceptible bacteria. In Pasteurellaceae, ColE1 plasmids carried almost exclusively antibiotic resistance genes. In Enterobacteriaceae, these plasmids encoded a large range of traits, including not only antibiotic resistance determinants, but also a wide variety of genes, showing the huge genetic plasticity of these small replicons. Finally, we also used a metagenomic approach in order to validate this technique, performing this PCR system using total DNA extractions from fecal samples from poultry, turkeys, pigs and humans. Using Illumina sequencing of the PCR products we identified a great diversity of genes encoded by ColE1 replicons, including different antibiotic resistance determinants, supporting the previous results achieved with the collections of bacterial isolates. In addition, we detected cryptic ColE1 plasmids in both families with no known genes in their variable region, which we have named sentinel plasmids. In conclusion, in this work we present a useful genetic tool for the detection and analysis of ColE1 plasmids, and confirm their important role in the dissemination of antibiotic resistance, especially in the Pasteurellaceae family of bacteria.

Keywords: antibiotic resistance, ColE1 plasmids, detection PCR, capture PCR, sentinel plasmids

INTRODUCTION

Plasmids are autonomously replicating fragments of extra-chromosomal DNA that can be transferred horizontally between bacteria. They usually harbor genes that confer a selective advantage under adverse conditions, becoming a major source of genetic variability in bacteria and playing a key role in their adaptation and evolution (Baquero, 2011; Wiedenbeck and Cohan, 2011). Antimicrobial resistance has become one of the most serious problems in public health, and the concern about the ability of plasmids to spread antimicrobial resistance determinants has greatly increased (Carattoli, 2013).

ColE1-type plasmids (hereafter ColE1 plasmids) are small, mobilizable, multi-copy replicons, found mainly in Enterobacteriaceae and Pasteurellaceae (Tomizawa et al., 1977; San Millan et al., 2007), although they have been described in other families of bacteria (Pan et al., 2010; Vincent et al., 2016). These plasmids have a distinctive theta replication mechanism, regulated by two small RNAs encoded close to the origin of replication (*oriV*) (Tomizawa, 1984; Lilly and Camps, 2015). According to the sequences of their mobilization genes, ColE1 replicons belong to the MOB_P family of plasmids (Garcillán-Barcia et al., 2009, 2015).

Several works have shown that ColE1 plasmids are carriers of resistance mechanisms of high clinical relevance. In Enterobacteriaceae, these plasmids are present in commensal (Pallecchi et al., 2010, 2011; Anantham and Hall, 2012; Moran and Hall, 2017) and pathogenic isolates (de Toro et al., 2013; Garbari et al., 2015; Stoesser et al., 2016, 2017; Albornoz et al., 2017), where they have been found to carry genes conferring resistance to fluoroquinolones, aminoglycosides, sulfonamides and β-lactams, including antibiotics of last resort such as carbapenems (Papagiannitsis et al., 2015) and even colistin (Borowiak et al., 2017). In Pasteurellaceae, several ColE1 plasmids conferring resistance to tetracyclines, aminoglycosides, sulfonamides and β-lactams have been described in human and animal pathogens of the genera *Pasteurella* (San Millan et al., 2009), *Haemophilus* (San Millan et al., 2010, 2011; Tristram et al., 2010; Moleres et al., 2015) and *Actinobacillus* (Blanco et al., 2007). Furthermore, in Pasteurellaceae species of animal origin, ColE1 plasmids are probably the main vehicle for the acquisition of antibiotic resistance determinants (Lancashire et al., 2005; Blanco et al., 2006, 2007; San Millan et al., 2007), being postulated as key strategy for multidrug resistance in this family (San Millan et al., 2009).

In this study, we carried out a computational analysis of the genetic structure of ColE1 plasmids in Enterobacteriaceae and Pasteurellaceae. With this data, we developed a novel PCR-based strategy for the detection, capture and study of ColE1 plasmids, and validated this technique using antibiotic susceptible and resistant isolates from Enterobacteriaceae and Pasteurellaceae. The results presented here highlight the importance of ColE1 plasmids as a source of antibiotic resistance, providing a useful tool for the study of these small replicons.

MATERIALS AND METHODS

Bacterial Strains, Sample Collection and Culture Conditions

In this study we used a total number of 135 Pasteurellaceae and 50 Enterobacteriaceae isolates. A collection of 44 *Pasteurella multocida* and 39 *Mannheimia haemolytica* isolates were obtained from the lungs of 3-month-old lambs and are described in Supplementary Table 1. These strains were isolated, identified on the basis of phenotype, this identification being confirmed by species-specific PCR, and then further characterized by serotyping (Fraser et al., 1983; Townsend et al., 1998, 2001; Angen et al., 2002). In addition, 52 Pasteurellaceae isolates from dogs and cats were recovered during 2009–2010, from oral samples cultured on Columbia 5% sheep blood agar plus 16 mg/l bacitracin, and characterized by Gram staining, oxidase tests and a lack of growth on MacConkey agar (BioMérieux, France). They were further characterized with the API 20NE microorganism identification test kit (BioMérieux, France) and by species-specific PCR and sequencing of the 16S rRNA gene (Król et al., 2011). *Haemophilus influenzae* and *Haemophilus parasuis* were cultured on chocolate agar PolyViteX plates (BioMérieux, France) and in *Haemophilus* Test Medium (HTM) broth (Wider, Francisco Soria Melguizo, S.A., Spain) at 37°C in microaerophilic conditions (5% CO_2) for 48 h. Moreover, 50 multidrug-resistant Enterobacteriaceae isolates were obtained from the Sanjay Gandhi Postgraduate Institute of Medical Sciences, in Lucknow, India. All the strains were isolated in 2010 and their antibiotic resistance and plasmid profiles are described in Supplementary Table 2. The remaining Enterobacteriaceae species and Pasteurellaceae species were cultured on Columbia agar +5% sheep blood plates and in BHI broth (BioMérieux, France) at 37°C for 24 h.

Additionally, we collected fecal samples from different origins: poultry, turkey, pig and human. Animal fecal samples were collected from three different farms located in the center area of Spain during 2015-2016. Each farm produced a different animal species: poultry, turkey and pig, respectively. Twenty-five individual samples of feces were taken from each animal species in each farm, and then pooled into one unique sample, stored at −80° until DNA extraction. Additionally, a fecal sample of a human from the same region was also collected during the same period and stored at the same temperature.

Antibiotic Susceptibility Testing

Antimicrobial susceptibility was determined by disk diffusion and microdilution methods, in accordance with CLSI guidelines (CLSI, 2013a,b). Commercially prepared dehydrated Sensititre panels (Trek Diagnostics, Inc., Westlake, OH) were used for MIC determination. Quality control of the panels was performed according to the manufacturer's instructions. Antibiotic disks were obtained from BioMérieux (BioMérieux, France) and Oxoid (Oxoid Ltd., Basingstoke, United Kingdom). Antibiotics were supplied by Merck (Merck KGaA, Darmstadt, Germany) and Sigma-Aldrich (Sigma Chemical Co. St Louis, Mo, USA).

Computational DNA Analysis

Data on complete plasmid sequences were obtained from GenBank. The structure of ColE1 from Pasteurellaceae was studied in the complete sequence of 24 wild type plasmids (**Figure 1**). The structure of ColE1 from Enterobacteriaceae was studied in the complete sequence of 37 wild type plasmids (**Figure 2**), excluding the numerous sequences from cloning vectors or other ColE1-based genetic tools. For the recovery of wild type ColE1 plasmids, short input sequence BLAST of different 20 bp segments including the origin of replication of ColE1 plasmids was performed (http://blast.ncbi.nlm.nih.gov/). The BLASTs were performed excluding all ColE1-derived cloning vectors, and only ColE1 plasmids from wild type strains were analyzed. The conserved region of plasmids was determined independently in Pasteurellaceae and Enterobacteriaceae by sequence alignments using Megablast. Plasmid sequence identity score of plasmid conserved regions was established using ColE1 (Accession no. J01566) and pB1000 (Accession no. DQ840517) as prototypes for ColE1 plasmids from Enterobacteriaceae and Pasteurellaceae families, respectively. Nucleotide BLASTs of the conserved region were performed to establish sequence identity using Nblast and Megablast. Conserved region of plasmid ColE1 was established from nucleotide 641 to 3,940, from the start of RNAII to the end of *mbeE* gene. The conserved region of plasmid pB1000 was established from nucleotide 3,584 to 2,665, excluding bla_{ROB-1} gene. Highly related sequences were selected to determine the GC content (%) in the conserved region using Serial Cloner 2.1 (Serial Basics, France). The regions with no homology were used to estimate Guanine plus Cytosine percentage in the variable region of the plasmids. Differences between variances were tested with the F test to compare the variances of two samples from normal populations. Different software were used for sequence analysis: 4Peaks 1.6 (Mek&Tosj, Netherlands) for the analysis of chromatograms, NIH online analysis tools (http://www.ncbi.nlm.nih.gov) for DNA alignments, and DNA Strider 1.4f13 (CEA, France) and CLC DNA workbench (CLC bio, Cambridge, MA) for general features like full integration of the data or analysis of the annotated sequences.

In Vitro DNA Analysis

The plasmids used in this study are shown in **Figures 1**, **2**, **4**, **6**. Plasmid DNA from bacteria isolates was extracted with Plasmid Midi and QIAprep Spin Miniprep (Qiagen, Inc., Chatworth, California, USA). PCR products were purified with Qiagen PCR Purification or Gel Extraction kits (Qiagen, Inc., Chatworth, California, USA). PCR was performed with Taq polymerase from Biotools (B&M Labs, Spain), Phusion high-fidelity DNA polymerase (Finnzymes, Woburn, MA, USA), AmpliTaq Gold DNA polymerase (Applied Biosystems, AB, Foster City, CA, USA) and Taq-Core (Qbiogene, Carlsbad, CA, USA). Automated Sanger sequencing of six complete plasmids was carried out with an Abi-Prism Apparatus (Perkin-Elmer) at Secugen S. L. (Madrid, Spain). The *P. multocida* isolates were identified by PCR detection of the *kmt* gene with the species-specific primers KMT1T7 and KMT1SP6 (Townsend et al., 1998). Capsular type was determined by the PCR method described by Townsend et al. (2001).

Detection PCR for ColE1 Plasmids

DNA-free polymerases, such as AmpliTaq Gold DNA polymerase (Applied Biosystems, Foster City, CA, USA) and Taq-Core (Qbiogene, Carlsbad, CA, USA), are required for this test in Enterobacteriaceae species to avoid false positive results (Supplementary Figure 1). Crude bacterial lysate can act as DNA template for this reaction. Detection PCR was performed according to the kit manufacturer's instructions, with primers ColE1 detF (tgaacggggggttcgtgca)/ColE1 detR (cgtttttccataggctccgcc) for Enterobacteriaceae, producing a PCR product of about 300 bp; and ColE1-P detF (gtctccgtttcgtgctacggt)/ColE1-P detR (aaatcagcggagccgataggc) for Pasteurellaceae, producing a PCR product of about 450 bp. The amplification conditions were: initial denaturation for 5 min at 94°C, followed by 25 cycles of denaturation for 30 s at 94°C, annealing for 30 s at 55°C and extension for 30 s at 72°C, with a final extension phase for 10 min at 72°C.

Capture PCR for ColE1 Plasmids

Capture PCR was performed with the Phusion high-fidelity DNA polymerase (Finnzymes, Woburn, MA, USA). No false positive results were observed using this polymerase for the capture PCR. This PCR has been shown to amplify pUC19 vectors with insertions generating products of up to 15 kb (**Figure 3**). Plasmid preparations using kits as the QIAprep Spin Miniprep (Qiagen, Inc., Chatworth, California, USA) are the recommended DNA template. PCR was performed with the ColE1 cap-1 (tgcacgaacccccgttca)/ColE1 cap-2 (ggcggagcctatggaaaaacg) primers for Enterobacteriaceae and the ColE1-P cap-1 (accgtagcacgaaacggagac)/ColE1-P cap-2 (gcctatcggctccgctgattt) primers for Pasteurellaceae, according to the kit manufacturer's instructions. The following conditions were used: initial denaturation for 30 s at 98°C, followed by 30 cycles of denaturation at 98°C for 10 s, annealing at 56°C for 10 s and extension at 72°C for 4 min, with a final extension phase for 10 min at 72°C.

Sequencing and Analysis of Metagenomic Samples

We extracted the total DNA from the four fecal samples with QIAmp Fast DNA Stool Mini Kit (Qiagen, Inc., Chatworth, California, USA), and then performed the Capture PCR from this

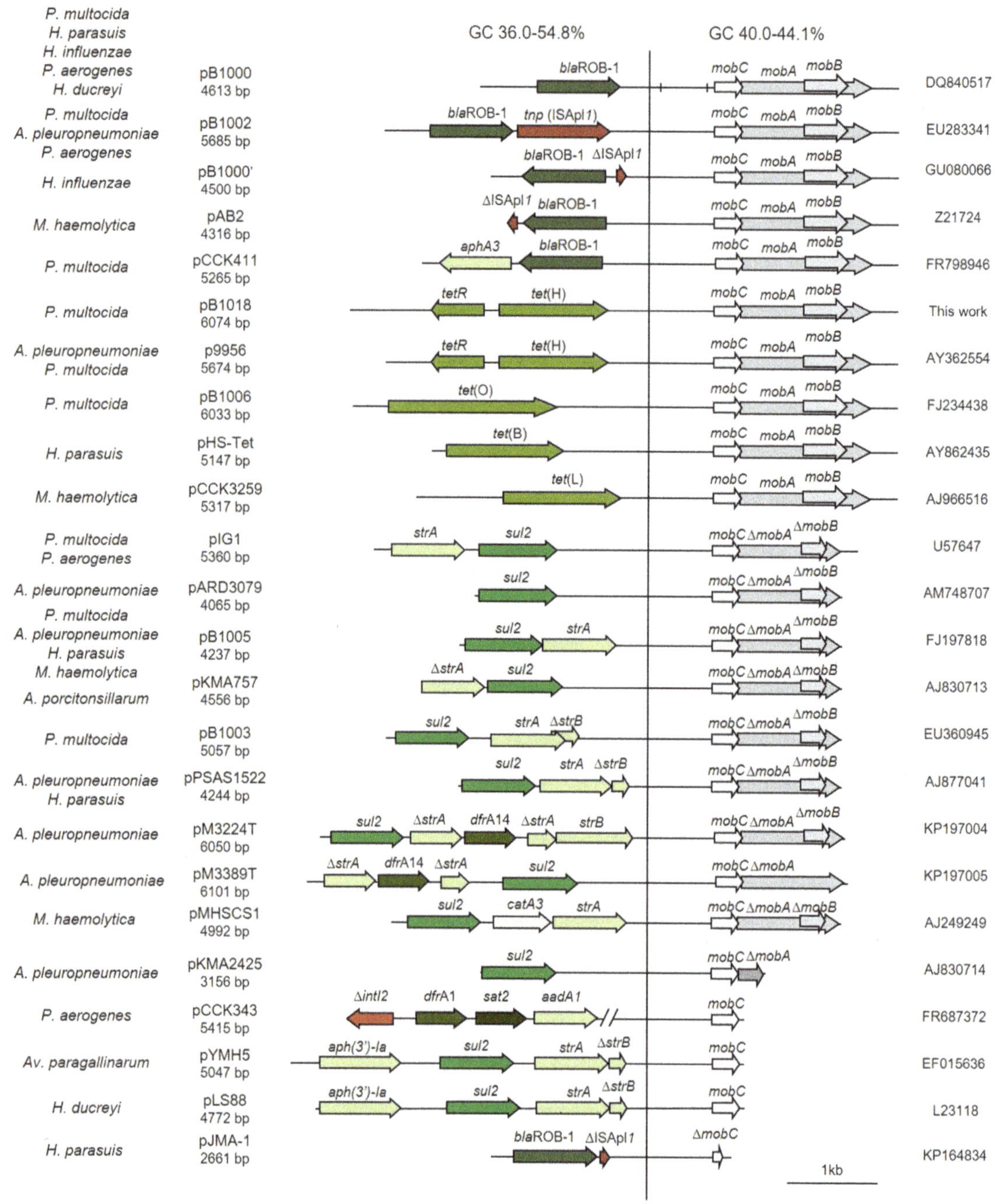

FIGURE 1 | Genetic structure of ColE1 plasmids from Pasteurellaceae. Schematic diagram of ColE1 plasmids from the Pasteurellaceae family. The reading frames for genes are shown as arrows, with the direction of transcription indicated by the arrowhead. Antimicrobial drug resistance determinants are shown in green whereas genes involved in genetic transposition or integration are shown in red. Genes encoding plasmid relaxases are shown in gray. In pB1000, two vertical bars bracket the region containing the putative origin of replication (*oriV*) and the putative origin of transfer (*oriT*). The large vertical bar separates the conserved region of the plasmid, to the right, from the variable region of the plasmid, to the left. Percentage ranges of GC content of variable and conserved regions of the plasmids are indicated in the top of the figure. The species in which the plasmid has been described and the name, size and accession number of plasmids are also indicated.

total DNA. After purifying the amplified DNA with the Qiagen PCR Purification kit (Qiagen, Inc., Chatworth, California, USA), it was sequenced and assembled at MicrobesNG (Birmingham, United Kingdom) by Next Generation Sequencing following their standard analysis pipeline.

DNA was quantified in triplicates with the Quantit dsDNA HS assay in an Ependorff AF2200 plate reader. Genomic DNA libraries were prepared using Nextera XT Library Prep Kit (Illumina, San Diego, USA) following the manufacturer's protocol with two modifications: two nanograms of DNA instead

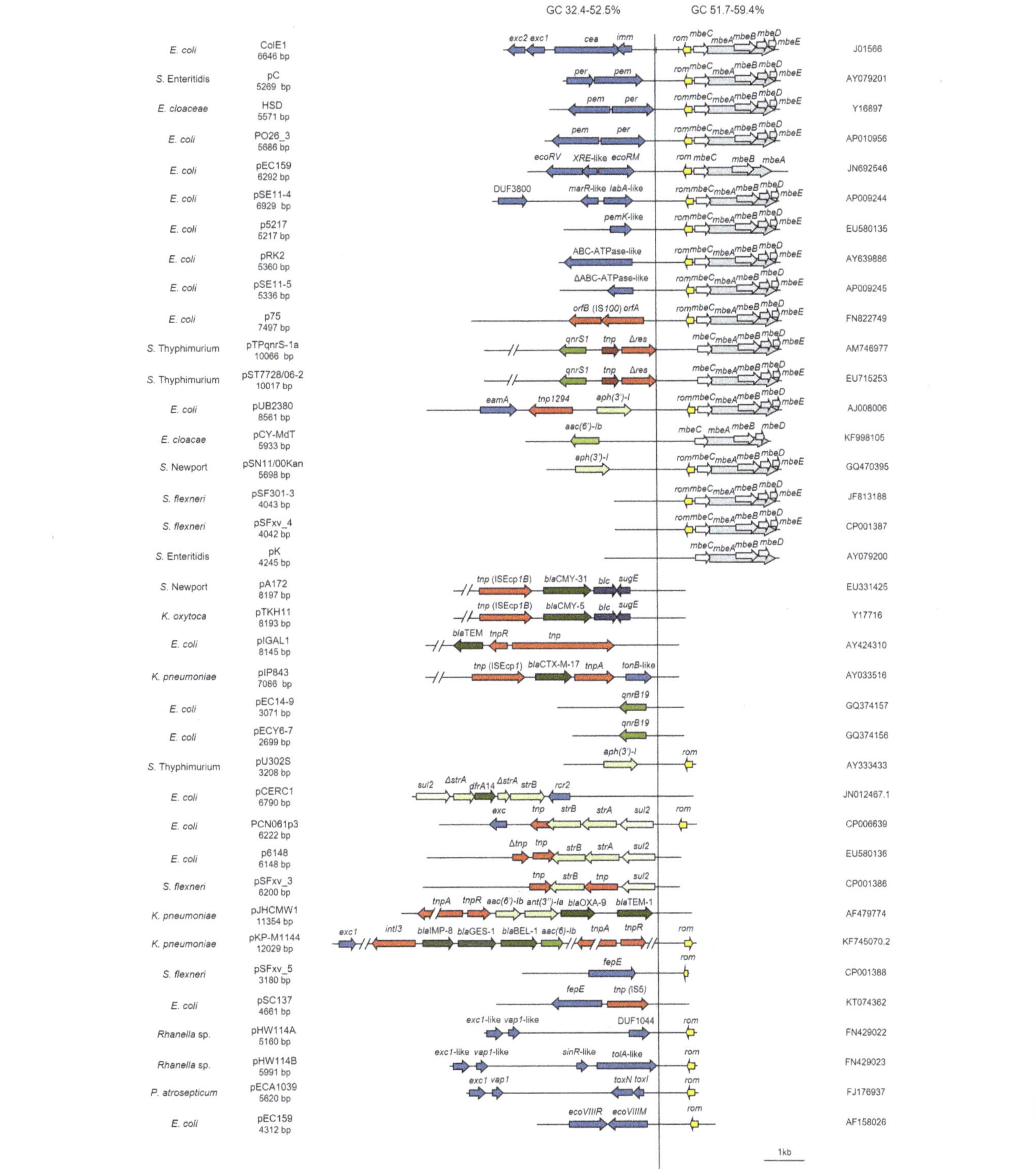

FIGURE 2 | Genetic structure of ColE1 plasmids from Enterobacteriaceae. Schematic diagram of the 37 ColE1 plasmids from the Enterobacteriaceae family studied in this work. The reading frames for genes are shown as arrows, with the direction of transcription indicated by the arrowhead. The names of the genes, or the names of the family of proteins they encode, are indicated. Antimicrobial drug resistance genes are shown in green and genes involved in genetic transposition or integration are shown in red. Genes encoding plasmid relaxases are shown in gray and the rom gene implicated in the regulation of plasmid replication is shown in yellow. The remaining ORFs are shown in blue. In ColE1, two vertical bars bracket the region containing the origin of replication (*oriV*) and the origin of transfer (*oriT*). The large vertical bar separates the conserved region of the plasmids, to the right, from the variable region of the plasmids, to the left. Percentage ranges of GC content of variable **(Left)** and conserved **(Right)** regions of the plasmids are indicated in the top of the figure. The species in which the plasmid has been described, and the name, size, and accession number of plasmids are also indicated.

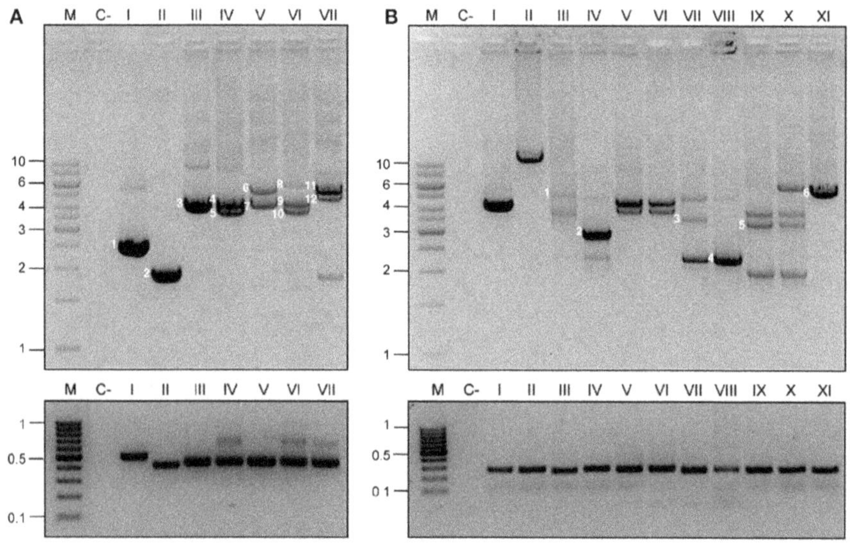

FIGURE 3 | Detection and capture PCR for ColE1 plasmids. ColE1 detection (bottom) and capture (top) PCRs in Pasteurellaceae **(A)** and Enterobacteriaceae **(B)**. M stands for molecular weight marker and numbers to the left of the panels indicate the size in kb. C- Stands for negative control. The lanes correspond to different strains and are indicated with roman numbers. **(A)** Pasteurellaceae family PCRs. Negative control (c-) corresponds to *H. influenzae* RdKW20. The products of the capture PCR, corresponding to the ColE1 plasmids carried by the strains, are indicated by numbers. Lane I, *P. stomatis* BB1086: pB000a (1). Lane II, *Frederiksenia canicola* BB1087: pB000b (2). Lane III, *H. influenzae* BB1059: pB1000 (3). Lane IV, *P. multocida* BB1035: pB1000 (4) and pB1005 (5). Lane V, *P. multocida* BB1041: p9956 (6) and pB1000 (7). Lane VI, *P. multocida* BB1044: pB1000 (9), pB1005 (10) and pB1006 (8). Lane VII, *P. multocida* BB1046: pB1002 (11) and pB1003 (12). **(B)** Enterobacteriaceae family PCRs. Negative control (c-) represents *E. coli* DH5α. Lanes I and II correspond to ColE1 based cloning vectors pTOPO and pUC19 (with the insertion of a ~13 kb DNA fragment), respectively. Lanes III to XI correspond to wild type strains from the Sanjay Gandhi Postgraduate Institute of Medical Sciences in India. Lanes III-VII, *K. pneumonia*. Lane VIII, *P. mirabilis*. Lanes IX and X, *E. cloacae*. Lane XI, *E. coli*. Six random plasmids were completely sequenced from these strains, and are indicated by numbers in the agarose gel: pB1019 (1), pB1020 (2), pB1022 (4), pB1023 (5), and pB1024 (6).

of one were used as input, and PCR elongation time was increased to 1 min from 30 s. DNA quantification and library preparation were carried out on a Hamilton Microlab STAR automated liquid handling system. Pooled libraries were quantified using the KapaBiosystems Library Quantification Kit for Illumina on a Roche light cycler 96 qPCR machine. Libraries were sequenced on the Illumina MiSeq using a 250 bp paired end protocol. Reads were adapter trimmed using Trimmomatic 0.30 with a sliding window quality cutoff of Q15 (Bolger et al., 2014). *De novo* assembly was performed on samples using SPAdes version 3.7 (Bankevich et al., 2012). Additional information about the data of this sequencing is presented in Supplementary Table 3.

An automated annotation of the contigs assembled was performed at MicrobesNG using Prokka (Seemann, 2014). Moreover, we developed an additional annotation of the sequences combining the analysis tools RAST (Aziz et al., 2008) and ResFinder (Zankari et al., 2012). In order to assure that the annotations were actually present in ColE1 replicons, we analyzed the genetic environment of the genes in the contigs assembled at MicrobesNG. We assumed that these genes were encoded actually on ColE1 replicons when their whole contig sequence upstream and downstream the gene corresponded to ColE1-like sequences according to the NIH online analysis tool Nucleotide BLAST (http://blast.ncbi.nlm.nih.gov/).

Nucleotide Sequence Accession Numbers

Nucleotide sequences of the plasmids obtained in this study have been deposited in GenBank under the following accession numbers: pB1018 from BB1253, JQ319774; pB000a from BB1086, JQ319773; pB000b from BB1087, JQ319771; pB1019 from BB1088, JQ319775; pB1020 from BB1089, JQ319772; pB1021 from BB1090, JQ319767; pB1022 from BB1091, JQ319766; pB1023 from BB1092, JQ319770; and pB1024 from BB1093, JQ319768.

RESULTS

Computational Analysis of ColE1 Plasmids

ColE1 Plasmids in Pasteurellaceae

The bioinformatic analysis of the ColE1 plasmids from Pasteurellaceae species led to the identification of two differentiated genetic regions (**Figure 1**): a conserved region carrying all the elements controlling replication and transfer, and a variable region encoding mainly antibiotic resistance genes. The conserved region of ColE1 plasmids from Pasteurellaceae had an average size of 2,513 bp (Standard Deviation, $SD = 718$ bp) and was highly similar among all of the plasmids, with a GC content between 40.0 and 44.1% and high nucleotide sequence identity (average= 97.62%, SD= 2.06%) (**Figure 1**). In contrast, the variable region of these ColE1 plasmids presented a high level of genetic divergence. As result of this variability, the GC content in this region varied between 36.0 and 54.8%. The variance of the GC content was significantly lower in the conserved region than in the variable region [$F_{(23)} = 40.22$, $P < 0.001$].

ColE1 Plasmids in Enterobacteriaceae

In this family, the ColE1 replicons also showed a conserved area involved in plasmid housekeeping functions and a variable region. However, in contrast to plasmids from Pasteurellaceae, ColE1 plasmids in Enterobacteriaceae encoded a wide variety of accessory genes, including also antibiotic resistance determinants (**Figure 2**). The conserved region of ColE1 plasmids from Enterobacteriaceae had an average size of 1,817 bp ($SD = 1{,}197$ bp) with an average nucleotide sequence identity of 82.59% ($SD = 10.79\%$). This region presented a GC content between 51.7 and 59.4%, whereas in the variable region GC content varied between 32.4 and 52.5%. Again, the variance of the GC content was significantly lower in the conserved region [$F_{(36)} = 14.11$, $P < 0.001$].

Development of a PCR-Based System for Detection and Capture of ColE1 Plasmids

We used the data obtained from the *in silico* analysis of the sequences of wild type ColE1 plasmids to develop a two PCR-based system for the detection and capture of ColE1 replicons in Enterobacteriaceae and Pasteurellaceae. Pairwise and multiple DNA alignments of the conserved sequences of the plasmids were performed in order to detect conserved regions suitable for the design of PCR primers (for detailed description of the PCRs conditions and primers see Material and Methods).

As the nucleotide sequence of ColE1 plasmids presents a highly conserved region among all the replicons belonging to the same bacterial family (82.59% in Enterobacteriaceae and 97.62% in Pasteurellaceae), but drastically different when compared against plasmids from the other family, we were forced to design two different set of primers, each one specific for Enterobacteriaceae ColE1 plasmids and Pasteurellaceae ColE1 plasmids, respectively.

Thus, we developed a two PCRs system for the specific analysis of ColE1 plasmids in each family of bacteria. First, a "Detection PCR" using a pair of universal primers was designed to amplify a small fragment from the conserved region of the replicons, close to the *oriV*. Second, we designed a "Capture PCR," using a pair of primers annealing to the exact same region as the primers form the detection PCR, but amplifying outwards, allowing the capture of the variable region of the plasmid. Using this technique, the whole plasmid sequence is available for further analysis.

Validation of the PCR-Based System for ColE1 Analysis

Validation in ColE1 Plasmids From Pasteurellaceae

In order to validate the PCR-based system in Pasteurellaceae, we used a well-characterized series of strains of *H. influenzae* (San Millan et al., 2010, 2011), *H. parasuis* (San Millan et al., 2007) and *P. multocida* (San Millan et al., 2009; Santos-Lopez et al., 2017). These strains, previously described by our group, carried one, two or three ColE1 plasmids. The Detection PCR was positive in every isolate and, most importantly, the Capture PCR was not only able to amplify a single ColE1 plasmid, but it gave rise to different PCR products, corresponding to each coexisting plasmid in those strains carrying multiple (up to three) plasmids. As negative controls we used the reference strains *H. influenzae* RdKW20, *P. multocida* ATCC 43137 and *H. parasuis* ATCC 19417, which do not carry ColE1 plasmids. As expected, no PCR product was observed in any of these strains. In **Figure 3A** we show the results of a representative group of Pasteurellaceae strains tested: *H. influenzae* RdKW20, *P. stomatis* BB1086, *Frederiksenia canicola* BB1087, *H. influenzae* BB1059, *P. multocida* BB1035, *P. multocida* BB1041, *P. multocida* BB1044, *P. multocida* BB1046.

Additionally, we decided to establish two new collections of Pasteurellaceae isolates to test the PCR-based system in a wider range of species. We constructed a first collection of 52 Pasteurellaceae isolates from oral samples collected from healthy dogs and cats, including strains from six different Pasteurellaceae species: *P. multocida*, *Pasteurella canis*, *Pasteurella pneumotropica*, *Pasteurella stomatis*, *Pasteurella dagmatis* and *F. canicola*. A second collection was collected from lungs of 3-month-old lambs counting with 44 *P. multocida* and 39 *M. haemolytica* strains (Supplementary Table 1). We performed antibiotic susceptibility testing for clinically relevant antibiotics in both collections, detecting eight tetracycline resistant *P. multocida* strains in the lamb collection (Supplementary Table 1). We carried out the ColE1 Detection PCR and detected 10 positive strains; the eight tetracycline resistant *P. multocida* isolates from the lamb collection and two susceptible isolates from the dog collection: a *P. stomatis* (BB1086) and a *F. canicola* (BB1087). The Capture PCR was also positive in these 10 strains. In the eight tetracycline resistant *P. multocida* the Capture PCR revealed the presence of a plasmid of about 6 kb in size in all the isolates. This plasmid, bearing the tetracycline resistance gene *tet(H)*, was completely sequenced and named pB1018 (**Figure 1**). The strains from the dog collection harbored two plasmids of 2,975 bp (BB1086) and a 2,319 bp (BB1087) in size, respectively. These cryptic plasmids carried no detectable known gene in their variable region, and we named them pB000a and pB000b (**Figure 4A**).

Validation in ColE1 Plasmids From Enterobacteriaceae

We first tested the PCR system using ColE1-based cloning vectors, such as pTOPO and pUC19. The Detection PCR was positive and the Capture PCR was able to amplify fragments of up to 15 kb in size from genetic constructions using the ColE1-based pUC19 plasmid (**Figure 3B**). Therefore, this reaction should be able to capture any ColE1 plasmid from wild type strains. As negative controls we used laboratorial strains carrying no ColE1 plasmids as *Escherichia coli* DH5α. In contrast to the case of Pasteurellaceae, we did not have access to a previously characterized collection of Enterobacteriaceae strains carrying ColE1 plasmids. Hence, we decided to analyse a new collection of 50 clinical isolates of Enterobacteriaceae, including six different species and displaying resistance to various antibiotics, recovered at the Sanjay Gandhi Postgraduate Institute of Medical Sciences in India (Supplementary Table 2). Thirty seven of fifty isolates gave positive results in the ColE1 Detection PCR (**Figure 3B**). The amplicons from a representative number of strains were

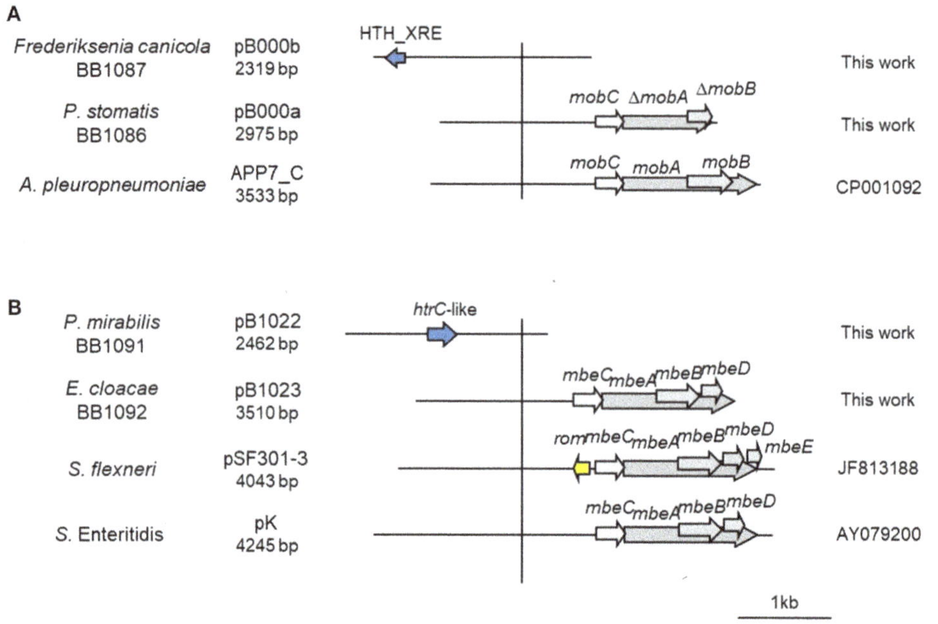

FIGURE 4 | Genetic structure of a selection of ColE1 cryptic plasmids. Schematic diagram of ColE1 cryptic plasmids from Pasteurellaceae **(A)** and Enterobacteriaceae **(B)**. The reading frames for genes are shown as arrows, with the direction of transcription indicated by the arrowhead. Genes encoding plasmid relaxases are shown in gray. The vertical bar separates the conserved region of the plasmids, to the right, from the variable region of the plasmids, to the left. The species in which the plasmid has been described, and the name, size, and accession number of plasmids are also indicated.

sequenced, and were confirmed to have originated from ColE1 plasmids. Thus, 74% of the Enterobacteriaceae studied isolates actually carried at least one ColE1 plasmid. The Capture PCR was then performed for the isolates bearing ColE1 plasmids, and it generated from one to three amplicons per isolate, with sizes ranging from 2 to 10 kb, corresponding to the various ColE1 plasmids present in the cell (**Figure 3B**). This result was confirmed by partial sequencing of the different PCR products after DNA purification from agarose gel (see Material and Methods).

We completely sequenced six random ColE1 plasmids (pB1019 to pB1024) from this collection of strains to confirm the results and the utility of the method developed here (**Figure 5**). These sequenced replicons had the typical characteristics of ColE1 plasmids, with both the conserved and the variable regions. Again, the GC content of the conserved region was very similar among these plasmids (54.8–57.7%), whereas the plasmid variable region presented diverse GC content (44.3–54.0%). We found different genes in the variable region of these replicons: toxin-antitoxin systems (pB1021), restriction-modification systems (pB1019) or transposases like Tn501 (pB1024). Interestingly, we also found two cryptic plasmids (pB1022 and pB1023). However, no antibiotic resistance gene was found in these plasmids despite the high level of resistance of the strains.

It is important to mention that contamination with DNA from ColE1 cloning vectors in some of the commercial DNA polymerases generated a false-positive reaction in the ColE1 Detection PCR in Enterobacteriaceae (Supplementary Figure 1). DNA-free polymerases, such as AmpliTaq Gold DNA polymerase (Applied Biosystems) and Taq-Core (Qbiogene), should therefore be used for this PCR reaction. Such contamination, leading to erroneous PCR results, has been described before and is of particular relevance in the case of the bla_{TEM-1} and $bla_{TEM-116}$ β-lactamases genes (Koncan et al., 2007; Jacoby and Bush, 2016).

Validation in ColE1 Plasmids From Intestinal Microbiota

In order to validate if the PCR system is useful for the analysis of ColE1 plasmids from metagenomic samples, we decided to test the Capture PCR directly on total DNA extracted from fecal samples using the primers specific for ColE1 plasmids from Enterobacteriaceae. Four pools of fecal samples were tested in this study, three of them collected from healthy animals of different species: poultry, turkey and pig; and a fourth one from human origin. We did not use the PCR-system for Pasteurellaceae given the limited presence of these bacteria in the gut microbiota of both animals and humans (Roto et al., 2015; Burrough et al., 2017; Gupta et al., 2017). After performing the Capture PCR and purifying the final product, we sequenced the amplified DNA resulting from all the ColE1 plasmids harbored in the Enterobacteriaceae cells present in the fecal samples using Illumina MiSeq. We analyzed the genes present on the amplicons, and their genetic environment, confirming that they were actually present in ColE1 plasmids. We found several mobilization genes, transposases, toxin-antitoxin systems, bacteriocins and restriction-modification

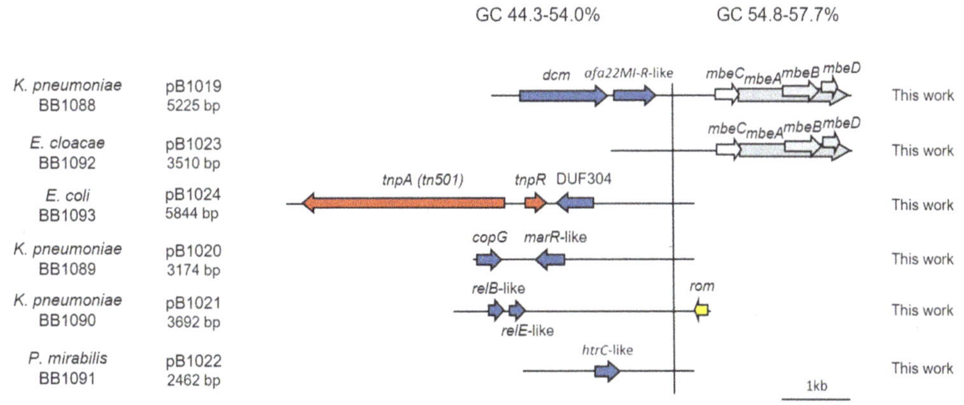

FIGURE 5 | Genetic structure of ColE1 plasmids from Enterobacteriaceae clinical isolates. Schematic diagram of the ColE1 plasmids from the Enterobacteriaceae family described here. The reading frames for genes are shown as arrows, with the direction of transcription indicated by the arrowhead. The names of the genes, or the names of the family of proteins they encode, are indicated. Genes involved in genetic transposition or integration are shown in red. Genes encoding plasmid relaxases are shown in gray and the *rom* gene involved in the regulation of plasmid replication is shown in yellow. The remaining ORFs are shown in blue. Percentage ranges of GC content of variable and conserved regions of the plasmids are indicated in the top of the figure. The vertical bar separates the conserved region of the plasmids, to the right, from the variable region of the plasmids, to the left. The strain in which the plasmid was described, and the name and size of the plasmids, are also indicated.

systems (**Table 1**), in addition to several hypothetical proteins ranging between 32 and 418 amino acids. Interestingly we also observed multiple antibiotic resistance genes conferring resistance to some of the most important antibiotic families such as ß-lactams, aminoglycosides and fluoroquinolones (**Table 1**).

DISCUSSION

In this work, we analyzed the structure and content of ColE1 plasmids described in Pasteurellaceae and Enterobacteriaceae up to date. Our results showed that although ColE1 plasmids are different in these two families, they presented the same genetic structure. We observed two differentiated regions in the plasmids, one highly conserved and another one highly variable. The conserved region harbored all the housekeeping functions of the plasmid, including the origins of replication (*oriV*) and transfer (*oriT*) and, in most of plasmids analyzed, relaxases genes. The GC content of this conserved region is similar to the GC content of the genomes in which ColE1 plasmids have been found (**Figure 6**): 37–44% in Pasteurellaceae (Supplementary Table 3) and 50–59% in Enterobacteriaceae (Supplementary Table 4), strongly suggesting a plasmid/host adaptation process. On the other hand, the variable region carries accessory genes from different origins. In this variable region we found a wide variety of antibiotic resistance determinants in both families. In Pasteurellaceae, we found genes conferring resistance to tetracycline, β-lactams, aminoglycosides, sulphonamides, trimethoprim and chloramphenicol (**Figure 1**). In contrast to the plasmids from Pasteurellaceae, ColE1 plasmids from Enterobacteriaceae carried a wide variety of genes apart from antibiotic resistance determinants: genes involved in resistance to phage infections such as abortive infection systems (Fineran et al., 2009) and restriction-modification systems (Gregorova et al., 2002), genes involved in ferric transport (Ye et al., 2010) or genes encoding bacteriocins, such as the colicin E1 (Tomizawa et al., 1977). However, despite the higher qualitative diversity in the genes encoded in their variable region, ColE1 plasmids from Enterobacteriaceae carried antibiotic resistance genes, mainly against β-lactams, aminoglycosides and sulphonamides (**Figure 2**).

Using the data obtained from our computational analysis, we developed a new PCR-based system able to detect and completely capture ColE1 plasmids in Enterobacteriaceae and Pasteurellaceae. This is, to the best of our knowledge, the first system of this nature developed for ColE1 plasmids in Pasteurellaceae. In Enterobacteriaceae, previous PCR-based tests (García-Fernández et al., 2009; Chen et al., 2010; Alvarado et al., 2012) have been described for the detection of ColE1 plasmids.

García-Fernández et al. (2009) designed a set of primers targeting a conserved region in the origin of replication of ColE1 plasmids while looking for plasmids harboring quinolone resistance genes in *Salmonella*. With these primers (**Table 2**) they successfully detected three different ColE1 replicons, demonstrating the efficacy of this PCR. However, by testing these primers *in silico* against the ColE1 represented in **Figure 2**, just 14 out of the 37 plasmids carried the complete sequence for primers hybridization. In addition, Chen et al. (2010) developed a PCR-based system for the detection of ColE1 plasmids in *Salmonella*, by using primers targeting a conserved region within the origin of replication and the *rom* gene (**Table 2**). However, we also tested these primers *in silico* and just 13 out of the 37 replicons did harbor the whole sequence complementary to these oligonucleotides. In parallel to the previous techniques targeting the origin of replication, Alvarado et al. (2012)

TABLE 1 | List of genes harbored by ColE1 plasmids in the fecal samples.

NAME	DESCRIPTION				
REPLICATION					
rom	RNAI modulator protein	*	*	*	*
CONJUGATION					
mbeA	Mobilization protein	*	*	*	*
mbeB-like	Mobilization protein	*			
mbeC	Mobilization protein	*	*	*	*
mbeD	Mobilization protein	*			
virB4	Type IV secretion system protein		*		*
ANTIBIOTIC RESISTANCE					
aph(3′)-IIa	Aminoglycoside phosphotransferase	*	*		*
$bla_{TEM-116}$	β-Lactamase	*	*	*	*
strA	Aminoglycoside phosphotransferase		*	*	
strB	Aminoglycoside phosphotransferase		*		
sul2	Sulfonamide-resistance dihydropteroate synthase		*		
qnrB19	Quinolone resistance protein			*	*
TRANSPOSASES					
IS91	Transposase		*		
IS5	Transposase			*	*
IS3	Transposase				*
Tn3	Transposon				*
TOXIN-ANTITOXIN SYSTEMS					
ccdB	CcdB/CcdA Toxin protein	*			*
higA	HigB/HigA Antitoxin protein	*		*	*
higB	HigB/HigA Toxin protein				*
lsoA	LsoA/LsoB Toxin protein				*
parE	ParE/ParD Toxin protein	*		*	
pemI	PemK/PemI Antitoxin protein				*
pemK	PemK/PemI Toxin protein				*
BACTERIOCINS					
caa	Colicin-A	*	*		
cai	Colicin-A immunity protein		*		
cea	Colicin-E1	*	*		
imm	Colicin-E1 immunity protein				*
lys	Colicin-E1 lysis protein				*
csa	Colicin-S4			*	
csi	Colicin-S4 immunity protein			*	
cta	Colicin-10		*	*	*
kba	Klebicin-B				*
kbi	Klebicin-B immunity protein				*
RESTRICTION-MODIFICATION SYSTEMS					
ecoRVR	Restriction endonuclease EcoRV	*	*	*	
ecoRVM	DNA-methyltransferase EcoRV	*			
ecoVIIIR	Restriction endonuclease EcoVIII			*	
ecoVIIIM	DNA-methyltransferase EcoVIII			*	
banIR	Restriction endonuclease BanI	*		*	*
banIM	DNA-methyltransferase BanI				*
styD4IR	Restriction endonuclease StyD4I	*			*
styD4IM	DNA-methyltransferase StyD4I	*			*

(Continued)

TABLE 1 | Continued

NAME	DESCRIPTION				
eco29kIR	Restriction endonuclease Eco29kI	*			
eco29kIM	DNA-methyltransferase Eco29kI	*			
eco034IR	Restriction Enzyme Eco034I	*			*
eco034IM	DNA-methyltransferase Eco034I	*			
ydiOR	Restriction endonuclease YdiO		*		
ydiOM	DNA-methyltransferase YdiO		*		
MISCELLANEOUS					
abi-like	Abortive infection bacteriophage protein		*	*	*
chrB	Membrane chromate resistance protein				*
copG family	Ribbon-helix-helix protein	*	*	*	
csp	Cold shock protein	*		*	*
era	GTPase Era			*	
exc1	Entry exclusion protein 1	*	*	*	*
exc2	Entry exclusion protein 2	*			*
escC family	Type III secretion system protein	*		*	*
fepE	Ferric enterobactin transport protein	*	*	*	*
Helix-turn-helix	Helix-turn-helix domain containing protein	*	*	*	*
htrC	Heat shock protein C	*	*	*	*
kdgF	Pectin degradation protein				*
nicB	Nicotine dehydrogenase subunit B	*			
nikA	Nickel ABC transporter substrate-binding protein	*	*	*	*
parA	Plasmid partitioning protein		*		
Phage integrase	Phage integrase family	*			
Phosphatase 2C	Type 2C protein phosphatase			*	*
pspF	Phage shock protein operon transcriptional activator			*	*
Reverse transcriptase motif	RNA-dependant DNA-polymerase	*		*	*
rihC	Ribonucleoside hydrolase	*			
tsr	Methyl-accepting chemotaxis protein		*		*
xre	Transcriptional regulator XRE family protein			*	
ydaM	Diguanylate cyclase, csgD regulator		*		
DIMER RESOLUTION					
xerD	Site-specific tyrosine recombinase	*			*
xerC	Site-specific tyrosine recombinase	*			*

The name and description of the genes is given on the left side of the table, while the asterisks (*) of the right side indicate the presence of each gene in the different samples, representing from left to the right: poultry, pig, turkey and human.

developed a Degenerate Primer MOB Typing (DPMT) technique, extremely useful to detect and classify plasmids present in gamma-proteobacteria by targeting their relaxases genes. This DPMT included different degenerate primers against the MOB_{P5} relaxases of ColE1 plasmids (**Table 2**). However, as these genes are actually absent in a substantial proportion of ColE1 replicons (**Figure 2**), a considerably part of these plasmids would not be detected by using only this set of primers. In summary, our bioinformatic analysis revealed that these prior methods, although scrupulously designed and useful in the particular studies in which they were employed, would fail to detect part of the wild type ColE1 plasmids described to date, either for lack of sensitivity of the primers or for targeting the mobilization genes. Nevertheless, in order to reach the most sensitivity as possible, we suggest the combination of all these primers to assure the detection of any ColE1 plasmid present in a sample.

In order to validate our technique, we used the PCR system in a range of bacterial collections from Enterobacteriaceae and Pasteurellaceae as well as in metagenomic samples from fecal origin. These experiments revealed interesting results. In Pasteurellaceae we confirmed the tight link between ColE1 plasmids and antibiotic resistance. Moreover, we discovered the presence of cryptic ColE1 plasmids in antibiotic susceptible *P. stomatis* BB1086 (pB000a) and *F. canicola* BB1087 (pB000b), which only encoded plasmid housekeeping genes. In Enterobacteriaceae, the PCR-based screening system showed that the prevalence of the ColE1 replicons in the antibiotic

TABLE 2 | Primers used to detect ColE1 plasmids in Enterobacteriaceae isolates in previous studies.

Primer name	Sequence (5′-3′)	Length	Reference
oricolE Fw	GTTCGTGCATACAGTCCA	18	García-Fernández et al., 2009
oricolE Rv	GGCGAAACCCGACAGGACT	19	
CC7059F	TTCGTGCACACAGCCCA	17	Chen et al., 2010
CC7062R	TGCGGTTATCCACAGAATCA	20	
CC7063F	GCGGACAGGTATCCGGTAA	19	
P51-f	TACCACGCCCTATGCGAARAARTAYAC	27	Alvarado et al., 2012
P52-f	GATAGCCTTGATTTTAATAACACCAAYACYTAYAC	35	
P5-r	CCCTTGTCCTGGTGYTSNACCCA	23	
P53-f	GGGCTCGCACGAYCAYACNGG	21	
P53-r	GCCCAGCCCTTTTCRTGRTTRTG	23	
ColE1 detF	TGAACGGGGGGTTCGTGCA	19	This work
ColE1 detR	CGTTTTTCCATAGGCTCCGCC	21	

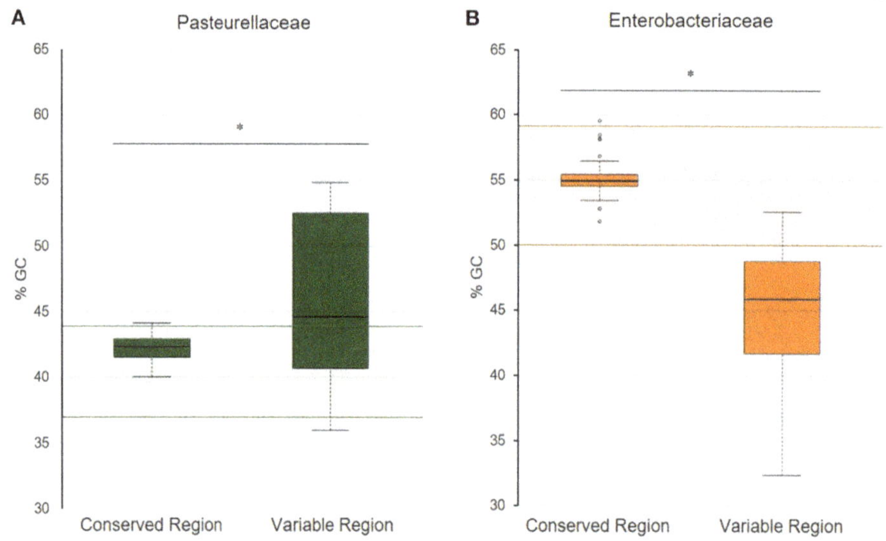

FIGURE 6 | GC content of the conserved and variable regions of ColE1 plasmids. Representation of the percentage of GC content in both the conserved and variable region of ColE1 plasmids in Pasteurellaceae **(A)** and Enterobacteriaceae **(B).** Colored boxplots represent the GC content of each of the ColE1 replicons studied. The shaded areas correspond to the %GC range of the genomes analyzed of these two bacteria families (Supplementary Tables S1, S2). The asterisks indicate the difference of the mean values of GC content between the conserved and variable region in both families of bacteria [$t_{(24)} = 2.91$, $P < 0.05$ and $t_{(41)} = 10.83$, $P < 0.05$, respectively].

resistance collection tested was especially high, with the 74% of the isolates carrying at least one plasmid. We also detected cryptic ColE1 plasmids (pB1022 and pB1023) with no evident genes in their variable region in this collection (**Figure 4B**). Other cryptic ColE1 plasmids, such as pB000a, pB000b, pB1022, and pB1023, have been previously described in human, animal and environmental isolates (Rozhon et al., 2006; Handford et al., 2009; Bleicher et al., 2013) (**Figure 4**) and, interestingly, some of the antibiotic resistance genes encoded in ColE1 replicons had been described in different bacterial genus and families (Miranda et al., 2003; Soge et al., 2006; Warburton et al., 2016). Hence, our hypothesis is that these unexpected prevalent cryptic replicons might act as "sentinel plasmids," capable of maintaining just the conserved region in bacteria due to their capacity of replication and conjugation (Burian et al., 1997), but able to acquire a wide variety of genes from heterogeneous origins, providing an increased genetic plasticity to their host.

In addition, our metagenomic approach confirmed the large diversity of genes that these small replicons can encode in the gut enterobacteria in healthy animals and humans (**Table 1**). We consider important to mention that we do not suggest a species distribution of the ColE1 genes based on our sample collection. However, we kept it separated in **Table 1**, firstly to show that there is no bias in our approach, and secondly, as it could be interesting in future works aiming to study the epidemiology of ColE1 plasmids. Many of the detected genes have well known functions in plasmids biology, as mobilization genes, toxin-antitoxin systems (Moran and Hall, 2017), the previously cited restriction-modification systems or the bacteriocins. However, the function of other genes found in this sample, and previously described in other ColE1 plasmids such as the *copG*-like genes (de Toro et al., 2013), are still unknown. Of especial relevance in these ColE1 amplicons from fecal samples was the detection of antibiotic resistance determinants against quinolones, aminoglycosides, sulphonamides and β-lactams. Previous works showed that the most prevalent antibiotic resistance genes in human and animal gut microbiomes are those conferring resistance to tetracycline, representing even the 90% of the gut resistome (Durso et al., 2012; Pal et al., 2016). However, although ColE1 plasmids frequently carry tetracycline resistance genes in Pasteurellaceae (**Figure 1**), they are not common in Enterobacteriaceae (**Figure 2**) and none has been detected in our approach. Other resistance genes are less represented in animal and human microbiomes, although new techniques with higher sensitivity (Lanza et al., 2018) might highlight the presence of genes that have been underrepresented to date in metagenomics samples. In our study, the genes detected with our ColE1 capture PCR were the *qnrB19* quinolone resistance gene in turkey and human, the *aph(3′)-IIa* aminoglycoside phosphotranspherases in poultry, pig and human, or the *strA* gene in turkey, which was found both alone and forming a *strA-strB-sul2* complex in the pig sample, being all these genes previously described in ColE1 replicons (**Figure 2**). The only antibiotic resistance determinant found in all the species was the $bla_{\text{TEM}-116}$ gene, which encodes an extended-spectrum β-lactamase (ESBL) (Lahlaoui et al., 2011). TEM-116 has been recently described as the central node of one of the two TEM clusters that group all the known TEM variants (Zeil et al., 2016), emphasizing its importance in antimicrobial resistance evolution and the role of ColE1 plasmids in its dissemination.

In conclusion, we have developed a simple system to screen and characterize ColE1 plasmids, which will allow monitoring the increasingly relevant role of these plasmids in the spread and evolution of antibiotic resistance in Pasteurellaceae and Enterobacteriaceae.

AUTHOR CONTRIBUTIONS

MA-A, CB-B, and AS-L have contributed with the design of the work, data collection, analysis and interpretation, drafting the article, and AS-L witch critical revision of the draft. MB, CM-E, DC, and KP have contributed with strains and critical revision of the article for final approval. AS has contributed with the design of the work, data collection and analysis, and critical revision of the draft. BG-Z has contributed with the conception and design of the work, interpretation of the data collection and analysis, critical revision of the draft and final approval of the article.

ACKNOWLEDGMENTS

We thank Natalia Montero for excellent technical support. We thank R. del Campo and J. L. San Millan for advice and technical assistance in the choice of DNA-free polymerases for ColE1 plasmid detection. We also thank to I. Cuesta for her bioinformatics support and advice with the sequencing data analysis. Finally, we would also like to thank the Veterinary Microbiology Students from the Universidad Alfonso X El Sabio of Madrid (term 2009–2010) for their assistance in the establishment of the collection of Pasteurellaceae isolates from dogs and cats. The authors would like to thank the reviewers for their careful and constructive comments.

REFERENCES

Albornoz, E., Lucero, C., Romero, G., Quiroga, M. P., Rapoport, M., Guerriero, L., et al. (2017). Prevalence of plasmid-mediated quinolone resistance genes in clinical enterobacteria from Argentina. *Microb. Drug Resist.* 23, 177–187. doi: 10.1089/mdr.2016.0033

Alvarado, A., Garcillan-Barcia, M. P., and de la Cruz, F. (2012). A degenerate primer MOB typing (DPMT) method to classify gamma-proteobacterial plasmids in clinical and environmental settings. *PLoS ONE* 7:e40438. doi: 10.1371/journal.pone.0040438

Anantham, S., and Hall, R. M. (2012). pCERC1, a small, globally disseminated plasmid carrying the dfrA14 cassette in the *strA* gene of the sul2-strA-strB gene cluster. *Microb. Drug Resist.* 18, 364–371. doi: 10.1089/mdr.2012.0008

Angen, O., Ahrens, P., and Bisgaard, M. (2002). Phenotypic and genotypic characterization of *Mannheimia* (*Pasteurella*) *haemolytica*-like strains isolated from diseased animals in Denmark. *Vet. Microbiol.* 84, 103–114. doi: 10.1016/S0378-1135(01)00439-4

Aziz, R. K., Bartels, D., Best, A. A., DeJongh, M., Diaz, T., Edwards, R. A., et al. (2008). The RAST Server: rapid annotations using subsystems technology. *BMC Genomics.* 9:75. doi: 10.1186/1471-2164-9-75

Bankevich, A., Nurk, S., Antipov, D., Gurevich, A., Dvorkin, M., Kulikov, A. S., et al. (2012). SPAdes: a new genome assembly algorithm and its applications to single-cell sequencing. *J. Comput. Biol.* 19, 455–477. doi: 10.1089/cmb.2012.0021

Baquero, F. (2011). The 2010 Garrod Lecture: the dimensions of evolution in antibiotic resistance: ex unibus plurum et ex pluribus unum. *J. Antimicrob. Chemother.* 66, 1659–1672. doi: 10.1093/jac/dkr214

Blanco, M., Gutiérrez-Martin, C. B., Rodriguez-Ferri, E. F., Roberts, M. C., and Navas, J. (2006). Distribution of tetracycline resistance genes in *Actinobacillus pleuropneumoniae* isolates from Spain. *Antimicrob. Agents Chemother.* 50, 702–708. doi: 10.1128/AAC.50.2.702-708.2006

Blanco, M., Kadlec, K., Gutierrez-Martin, C. B., de la Fuente, A. J., Schwarz, S., and Navas, J. (2007). Nucleotide sequence and transfer properties of two novel types of *Actinobacillus pleuropneumoniae* plasmids carrying the tetracycline resistance gene *tet*(H). *J. Antimicrob. Chemother.* 60, 864–867. doi: 10.1093/jac/dkm293

Bleicher, A., Schofl, G., Rodicio Mdel, R., and Saluz, H. P. (2013). The plasmidome of a *Salmonella enterica* serovar Derby isolated from pork meat. *Plasmid* 69, 202–210. doi: 10.1016/j.plasmid.2013.01.001

Bolger, A. M., Lohse, M., and Usadel, B. (2014). Trimmomatic: a flexible trimmer for illumina sequence data. *Bioinformatics* 30, 2114–2120. doi: 10.1093/bioinformatics/btu170

Borowiak, M., Fischer, J., Hammerl, J. A., Hendriksen, R. S., Szabo, I., and Malorny, B. (2017). Identification of a novel transposon-associated phosphoethanolamine transferase gene, *mcr-5* conferring resistance in d-tartrate fermenting *Salmonella enterica* subsp. enterica serovar Paratyphi B. *J. Antimicrob. Chemother.* 72, 3317–3324. doi: 10.1093/jac/dkx327

Burian, J., Guller, L., Macor, M., and Kay, W. W. (1997). Small crpytic plasmids of multiplasmid, clinical *Escherichia coli*. *Plasmid* 37, 2–14. doi: 10.1006/plas.1996.1273

Burrough, E. R., Arruda, B. L., and Plummer, P. J. (2017). Comparison of the luminal and mucosa-associated microbiota in the colon of pigs with and without swine dysentery. *Front. Vet. Sci.* 4:139. doi: 10.3389/fvets.2017.00139

Carattoli, A. (2013). Plasmids and the spread of resistance. *Int. J. Med. Microbiol.* 303, 298–304. doi: 10.1016/j.ijmm.2013.02.001

Chen, C. Y., Lindsey, R. L., Strobaugh, T. P. Jr., Frye, J. G., and Meinersmann, R. J. (2010). Prevalence of ColE1-like plasmids and kanamycin resistance genes in *Salmonella enterica serovars*. *Appl. Environ. Microbiol.* 76, 6707–6714. doi: 10.1128/AEM.00692-10

Clinical and Laboratory Standards Institute (CLSI). (2013a). *Performance Standards for Antimicrobial Disk and Dilution Susceptibility Tests for Bacteria Isolated from Animals, Approved Standards—4th Edn, VET01-A4.* Wayne, PA: CLSI.

Clinical and Laboratory Standards Institute (CLSI). (2013b). *Performance Standards for Antimicrobial Susceptibility Testing; 23rd Edn, Approved Standard M100-S23.* Wayne, PA: CLSI.

de Toro, M., Rodriguez, I., Rojo-Bezares, B., Helmuth, R., Torres, C., Guerra, B., et al. (2013). pMdT1, a small ColE1-like plasmid mobilizing a new variant of the *aac(6')-Ib-cr* gene in *Salmonella enterica* serovar Typhimurium. *J. Antimicrob. Chemother.* 68, 1277–1280. doi: 10.1093/jac/dkt001

Durso, L. M., Miller, D. N., and Wienhold, B. J. (2012). Distribution and quantification of antibiotic resistant genes and bacteria across agricultural and non-agricultural metagenomes. *PLoS ONE* 7:e48325. doi: 10.1371/journal.pone.0048325

Fineran, P. C., Blower, T. R., Foulds, I. J., Humphreys, D. P., Lilley, K. S., and Salmond, G. P. (2009). The phage abortive infection system, ToxIN, functions as a protein-RNA toxin-antitoxin pair. *Proc. Natl. Acad. Sci. U.S.A.* 106, 894–899. doi: 10.1073/pnas.0808832106

Fraser, J., Donachie, W., Quirie, M., and Gilmour, N. J. (1983). Rapid indirect hemaglutination test for serotyping *Pasteurella haemolytica*. *J. Clin. Microbiol.* 18, 206–207.

Garbari, L., Busetti, M., Dolzani, L., Petix, V., Knezevich, A., Bressan, R., et al. (2015). pKBuS13, a KPC-2-encoding plasmid from *Klebsiella pneumoniae* sequence type 833, carrying Tn4401b inserted into an Xer site-specific recombination locus. *Antimicrob. Agents Chemother.* 59, 5226–5231. doi: 10.1128/AAC.04543-14

García-Fernández, A., Fortini, D., Veldman, K., Mevius, D., and Carattoli, A. (2009). Characterization of plasmids harbouring *qnrS1*, *qnrB2* and *qnrB19* genes in *Salmonella*. *J. Antimicrob. Chemother.* 63, 274–281. doi: 10.1093/jac/dkn470

Garcillán-Barcia, M. P., Francia, M. V., and de la Cruz, F. (2009). The diversity of conjugative relaxases and its application in plasmid classification. *FEMS Microbiol. Rev.* 33, 657–687. doi: 10.1111/j.1574-6976.2009.00168.x

Garcillan-Barcia, M. P., Ruiz del Castillo, B., Alvarado, A., de la Cruz, F., and Martinez-Martinez, L. (2015). Degenerate primer MOB typing of multiresistant clinical isolates of *E. coli* uncovers new plasmid backbones. *Plasmid* 77, 17–27. doi: 10.1016/j.plasmid.2014.11.003

Gregorova, D., Pravcova, M., Karpiskova, R., and Rychlik, I. (2002). Plasmid pC present in *Salmonella enterica* serovar Enteritidis PT14b strains encodes a restriction modification system. *FEMS Microbiol. Lett.* 214, 195–198. doi: 10.1111/j.1574-6968.2002.tb11346.x

Gupta, V. K., Paul, S., and Dutta, C. (2017). Geography, ethnicity or subsistence-specific variations in human microbiome composition and diversity. *Front. Microbiol.* 8:1162. doi: 10.3389/fmicb.2017.01162

Handford, C. L., Stang, C. T., Raivio, T. L., and Dennis, J. J. (2009). The contribution of small cryptic plasmids to the antibiotic resistance of enteropathogenic *Escherichia coli* E2348/69. *Can. J. Microbiol.* 55, 1229–1239. doi: 10.1139/W09-079

Jacoby, G. A., and Bush, K. (2016). The Curious Case of TEM-116. *Antimicrob. Agents Chemother.* 60:7000. doi: 10.1128/AAC.01777-16

Koncan, R., Valverde, A., Morosini, M. I., Garcia-Castillo, M., Canton, R., Cornaglia, G., et al. (2007). Learning from mistakes: Taq polymerase contaminated with beta-lactamase sequences results in false emergence of Streptococcus pneumoniae containing TEM. *J. Antimicrob. Chemother.* 60, 702–703. doi: 10.1093/jac/dkm239

Król, J., Bania, J., Florek, M., Pliszczak-Krol, A., and Staroniewicz, Z. (2011). Polymerase chain reaction-based identification of clinically relevant Pasteurellaceae isolated from cats and dogs in Poland. *J. Vet. Diagn. Invest.* 23, 532–537. doi: 10.1177/1040638711403434

Lahlaoui, H., Dahmen, S., Moussa, M. B., and Omrane, B. (2011). First detection of TEM-116 extended-spectrum beta-lactamase in a *Providencia stuartii* isolate from a Tunisian hospital. *Indian J. Med. Microbiol.* 29, 258–261. doi: 10.4103/0255-0857.83909

Lancashire, J. F., Terry, T. D., Blackall, P. J., and Jennings, M. P. (2005). Plasmid-encoded Tet B tetracycline resistance in *Haemophilus parasuis*. *Antimicrob. Agents Chemother.* 49, 1927–1931. doi: 10.1128/AAC.49.5.1927-1931.2005

Lanza, V. F., Baquero, F., Martínez, J. L., Ramos-Ruíz, R., González-Zorn, B., Andremont, A., et al. (2018). In-depth resistome analysis by targeted metagenomics. *Microbiome* 6:11. doi: 10.1186/s40168-017-0387-y.

Lilly, J., and Camps, M. (2015). Mechanisms of theta plasmid replication. *Microbiol Spectr.* 3:Plas-0029-2014. doi: 10.1128/microbiolspec.PLAS-0029-2014

Miranda, C. D., Kehrenberg, C., Ulep, C., Schwarz, S., and Roberts, M. C. (2003). Diversity of tetracycline resistance genes in bacteria from Chilean salmon farms. *Antimicrob. Agents Chemother.* 47, 883–888. doi: 10.1128/AAC.47.3.883-888.2003

Moleres, J., Santos-Lopez, A., Lazaro, I., Labairu, J., Prat, C., Ardanuy, C., et al. (2015). Novel *blaROB-1*-bearing plasmid conferring resistance to beta-lactams in *Haemophilus parasuis* isolates from healthy weaning pigs. *Appl. Environ. Microbiol.* 81, 3255–3267. doi: 10.1128/AEM.03865-14

Moran, R. A., and Hall, R. M. (2017). Analysis of pCERC7, a small antibiotic resistance plasmid from a commensal ST131 *Escherichia coli*, defines a diverse group of plasmids that include various segments adjacent to a multimer resolution site and encode the same NikA relaxase accessory protein enabling mobilisation. *Plasmid* 89, 42–48. doi: 10.1016/j.plasmid.2016.11.001

Pal, C., Bengtsson-Palme, J., Kristiansson, E., and Joakim Larsson, D. G. (2016). The structure and diversity of human, animal and environmental resistomes. *Microbiome* 4, 54. doi: 10.1186/s40168-016-0199-5

Pallecchi, L., Riccobono, E., Mantella, A., Fernandez, C., Bartalesi, F., Rodriguez, H., et al. (2011). Small *qnrB*-harbouring ColE-like plasmids widespread in commensal enterobacteria from a remote Amazonas population not exposed to antibiotics. *J. Antimicrob. Chemother.* 66, 1176–1178. doi: 10.1093/jac/dkr026

Pallecchi, L., Riccobono, E., Sennati, S., Mantella, A., Bartalesi, F., Trifoso, C., et al. (2010). Characterization of small ColE-like plasmids mediating widespread dissemination of the *qnrB19* gene in commensal enterobacteria. *Antimicrob. Agents Chemother.* 54, 678–682. doi: 10.1128/AAC.01160-09

Pan, L., Leung, P. C., and Gu, J. D. (2010). A new ColE1-like plasmid group revealed by comparative analysis of the replication proficient fragments of Vibrionaceae plasmids. *J. Microbiol. Biotechnol.* 20, 1163–1178. doi: 10.4014/jmb.1003.03007

Papagiannitsis, C. C., Dolejska, M., Izdebski, R., Dobiasova, H., Studentova, V., Esteves, F. J., et al. (2015). Characterization of pKP-M1144, a novel ColE1-Like plasmid encoding IMP-8, GES-5, and BEL-1 beta-lactamases, from a *Klebsiella pneumoniae* sequence type 252 isolate. *Antimicrob. Agents Chemother.* 59, 5065–5068. doi: 10.1128/AAC.00937-15

Roto, S. M., Rubinelli, P. M., and Ricke, S. C. (2015). An introduction to the avian gut microbiota and the effects of yeast-based prebiotic-type compounds as potential feed additives. *Front. Vet. Sci.* 2:28. doi: 10.3389/fvets.2015.00028

Rozhon, W. M., Petutsching, E. K., and Jonak, C. (2006). Isolation and characterization of pHW15, a small cryptic plasmid from *Rahnella genomospecies* 2. *Plasmid* 56, 202–215. doi: 10.1016/j.plasmid.2006.05.007

San Millan, A., Escudero, J. A., Catalan, A., Nieto, S., Farelo, F., Gibert, M., et al. (2007). Beta-lactam resistance in *Haemophilus parasuis* is mediated by plasmid pB1000 bearing blaROB-1. *Antimicrob. Agents Chemother.* 51, 2260–2264. doi: 10.1128/AAC.00242-07

San Millan, A., Escudero, J. A., Guitierrez, B., Hidalgo, L., Garcia, N., Llagostera, M., et al. (2009). Multiresistance in *Pasteurella multocida* is mediated by coexistence of small plasmids. *Antimicrob. Agents Chemother.* 53, 3399–3404. doi: 10.1128/AAC.01522-08

San Millan, A., Garcia-Cobos, S., Escudero, J. A., Hidalgo, L., Gutierrez, B., Carrilero, L., et al. (2010). *Haemophilus influenzae* clinical isolates with plasmid pB1000 bearing blaROB-1: fitness cost and interspecies dissemination. *Antimicrob. Agents Chemother.* 54, 1506–1511. doi: 10.1128/AAC.01489-09

San Millan, A., Giufre, M., Escudero, J. A., Hidalgo, L., Gutierrez, B., Cerquetti, M., et al. (2011). Contribution of ROB-1 and PBP3 mutations to the resistance phenotype of a beta-lactamase-positive amoxicillin/clavulanic acid-resistant *Haemophilus influenzae* carrying plasmid pB1000 in Italy. *J. Antimicrob. Chemother.* 66, 96–99. doi: 10.1093/jac/dkq392

Santos-Lopez, A., Bernabe-Balas, C., Ares-Arroyo, M., Ortega-Huedo, R., Hoefer, A., San Millan, A., et al. (2017). A Naturally occurring single nucleotide polymorphism in a multicopy plasmid produces a reversible increase in antibiotic resistance. *Antimicrob. Agents Chemother.* 61:e01735-16. doi: 10.1128/AAC.01735-16.

Seemann, T. (2014). Prokka: rapid prokaryotic genome annotation. *Bioinformatics* 30, 2068–2069. doi: 10.1093/bioinformatics/btu153

Soge, O. O., Adeniyi, B. A., and Roberts, M. C. (2006). New antibiotic resistance genes associated with CTX-M plasmids from uropathogenic *Nigerian Klebsiella* pneumoniae. *J. Antimicrob. Chemother.* 58, 1048–1053. doi: 10.1093/jac/dkl370

Stoesser, N., Sheppard, A. E., Pankhurst, L., De Maio, N., Moore, C. E., Sebra, R., et al. (2016). Evolutionary history of the global emergence of the *Escherichia coli* epidemic clone ST131. *MBio* 7:e02162-15. doi: 10.1128/mBio.02162-15

Stoesser, N., Sheppard, A. E., Peirano, G., Anson, L. W., Pankhurst, L., Sebra, R., et al. (2017). Genomic epidemiology of global *Klebsiella pneumoniae* carbapenemase 334 (KPC)-producing *Escherichia coli*. *Sci. Rep.* 7:5917. doi: 10.1038/s41598-017-06256-2

Tomizawa, J. (1984). Control of ColE1 plasmid replication: the process of binding of RNA I to the primer transcript. *Cell* 38, 861–870. doi: 10.1016/0092-8674(84)90281-2

Tomizawa, J. I., Ohmori, H., and Bird, R. E. (1977). Origin of replication of colicin E1 plasmid DNA. *Proc. Natl. Acad. Sci. U.S.A.* 74, 1865–1869. doi: 10.1073/pnas.74.5.1865

Townsend, K. M., Boyce, J. D., Chung, J. Y., Frost, A. J., and Adler, B. (2001). Genetic organization of *Pasteurella multocida* cap Loci and development of a multiplex capsular PCR typing system. *J. Clin. Microbiol.* 39, 924–929. doi: 10.1128/JCM.39.3.924-929.2001

Townsend, K. M., Frost, A. J., Lee, C. W., Papadimitriou, J. M., and Hawkins, H. J. (1998). Development of PCR assays for species- and type-specific identification of *Pasteurella multocida* isolates. *J. Clin. Microbiol.* 36, 1096–1100.

Tristram, S. G., Littlejohn, R., and Bradbury, R. S. (2010). $bla_{\text{ROB-1}}$ presence on pB1000 in *Haemophilus influenzae* is widespread, and variable cefaclor resistance is associated with altered penicillin-binding proteins. *Antimicrob. Agents Chemother.* 54, 4945–4947. doi: 10.1128/aac.00263-10

Vincent, A. T., Emond-Rheault, J. G., Barbeau, X., Attere, S. A., Frenette, M., Lague, P., et al. (2016). Antibiotic resistance due to an unusual ColE1-type replicon plasmid in *Aeromonas salmonicida*. *Microbiology* 162, 942–953. doi: 10.1099/mic.0.000286

Warburton, P. J., Amodeo, N., and Roberts, A. P. (2016). Mosaic tetracycline resistance genes encoding ribosomal protection proteins. *J. Antimicrob. Chemother.* 71, 3333–3339. doi: 10.1093/jac/dkw304

Wiedenbeck, J., and Cohan, F. M. (2011). Origins of bacterial diversity through horizontal genetic transfer and adaptation to new ecological niches. *FEMS Microbiol. Rev.* 35, 957–976. doi: 10.1111/j.1574-6976.2011.00292.x

Ye, C., Lan, R., Xia, S., Zhang, J., Sun, Q., Zhang, S., et al. (2010). Emergence of a new multidrug-resistant serotype X variant in an epidemic clone of *Shigella flexneri*. *J. Clin. Microbiol.* 48, 419–426. doi: 10.1128/JCM.00614-09

Zankari, E., Hasman, H., Cosentino, S., Vestergaard, M., Rasmussen, S., Lund, O., et al. (2012). Identification of acquired antimicrobial resistance genes. *J. Antimicrob. Chemother.* 67, 2640–2644. doi: 10.1093/jac/dks261

Zeil, C., Widmann, M., Fademrecht, S., Vogel, C., and Pleiss, J. (2016). Network analysis of sequence-function relationships and exploration of sequence space of TEM beta-lactamases. *Antimicrob. Agents Chemother.* 60, 2709–2717. doi: 10.1128/AAC.02930-15

8

Molecular Mechanisms that Contribute to Horizontal Transfer of Plasmids by the Bacteriophage SPP1

*Ana Valero-Rello[1,2†], María López-Sanz[1†], Alvaro Quevedo-Olmos[1], Alexei Sorokin[2] and Silvia Ayora[1]**

[1] Department of Microbial Biotechnology, Centro Nacional de Biotecnología, Consejo Superior de Investigaciones Científicas, Madrid, Spain, [2] Micalis Institute, INRA, AgroParisTech, Universite Paris-Saclay, Jouy-en-Josas, France

Natural transformation and viral-mediated transduction are the main avenues of horizontal gene transfer in Firmicutes. *Bacillus subtilis* SPP1 is a generalized transducing bacteriophage. Using this lytic phage as a model, we have analyzed how viral replication and recombination systems contribute to the transfer of plasmid-borne antibiotic resistances. Phage SPP1 DNA replication relies on essential phage-encoded replisome organizer (G*38*P), helicase loader (G*39*P), hexameric replicative helicase (G*40*P), recombinase (G*35*P) and in less extent on the partially dispensable $5' \rightarrow 3'$ exonuclease (G*34.1*P), the single-stranded DNA binding protein (G*36*P) and the Holliday junction resolvase (G*44*P). Correspondingly, the accumulation of linear concatemeric plasmid DNA, and the formation of transducing particles were blocked in the absence of G*35*P, G*38*P, G*39*P, and G*40*P, greatly reduced in the G*34.1*P, G*36*P mutants, and slightly reduced in G*44*P mutants. In contrast, establishment of injected linear plasmid DNA in the recipient host was independent of viral-encoded functions. DNA homology between SPP1 and the plasmid, rather than a viral packaging signal, enhanced the accumulation of packagable plasmid DNA. The transfer efficiency was also dependent on plasmid copy number, and rolling-circle plasmids were encapsidated at higher frequencies than theta-type replicating plasmids.

Edited by:
Tatiana Venkova,
University of Texas Medical Branch, United States

Reviewed by:
Maite Muniesa,
University of Barcelona, Spain
Elisabeth Grohmann,
Beuth University of Applied Sciences, Germany

***Correspondence:**
Silvia Ayora
sayora@cnb.csic.es

† These authors have contributed equally to this work.

Keywords: horizontal gene transfer, plasmid transduction, SPP1, bacteriophages, antibiotic resistance

INTRODUCTION

Bacteriophage-mediated horizontal gene transfer enhances bacterial adaptive responses to environmental changes, and it is one of the mechanisms responsible for the rapid spread of antibiotic resistance, bacterial virulence and pathogenicity (Canchaya et al., 2003; Brussow et al., 2004; Brown-Jaque et al., 2015; Penades et al., 2015; Touchon et al., 2017). Bacteriophages, or simply phages, play active roles in the specialized mobilization of discrete chromosomal regions (specialized transduction), and also with significant efficiency can transfer any chromosomal segment or plasmid DNA (generalized transduction). The difference between these two transduction modes is that specialized transduction is the consequence of the faulty excision of the prophage from the bacterial chromosome, resulting into packaging of phage DNA as well as

Abbreviations: PFGE, pulsed field gel electrophoresis; RCR, rolling circle replication; SPP1, *B. subtilis* bacteriophage SPP1; *sus*, suppressor sensitive (mutation); TR, theta replication; ts, thermosensitive; wt, wild type.

adjacent DNA from the bacterial chromosome (Canchaya et al., 2003; Brussow et al., 2004; Penades et al., 2015; Touchon et al., 2017). In generalized transduction, phage DNA mispackaging occurs, and the viral packaging machinery uses chromosomal or plasmid DNA as a substrate for DNA packaging into the empty proheads instead of viral DNA (Ikeda and Tomizawa, 1965; Viret et al., 1991). Generalized transduction, which is recognized as a widespread mechanism for the transfer of any gene from one bacterium to another, was originally reported in γ-proteobacteria (Zinder and Lederberg, 1952; Lennox, 1955), and it has been also reported in many Gram-positive pathogens (Maslanova et al., 2013; Giovanetti et al., 2014; Winstel et al., 2015). The majority of generalized transducing phages package their DNA by the headful packaging mechanism (*pac* phages). One remarkable event related to this, is the encapsidation of pathogenicity islands, as it occurs with the *Staphylococcus aureus* pathogenicity islands (SaPIs). SAPIs have developed elegant strategies to hijack the phage machinery to use it for their own transfer (Penades et al., 2015). Most SaPI helper phages identified to date are *pac* phages, and many well-studied SaPIs are packaged by the headful mechanism (Ruzin et al., 2001). Despite its importance in spreading antibiotic resistances and virulence, the mechanisms that occur inside the cell and lead to the erroneous encapsidation of foreign DNA upon phage infection remain largely unexplored.

SPP1 is a 44-kb virulent *Bacillus subtilis* phage that can carry out generalized transduction (plasmid and chromosomal) with a significant frequency (Yasbin and Young, 1974; Ferrari et al., 1978; Canosi et al., 1982). The SPP1 replication and packaging machineries have been studied in deep (Alonso et al., 2006; Lo Piano et al., 2011; Oliveira et al., 2013). SPP1 DNA replication starts by the theta mode when the replisome organizer, G*38*P, binds to the replication origin, *ori*L (Pedre et al., 1994; Missich et al., 1997; Seco and Ayora, 2017). Then, the phage helicase loader (G*39*P) recruits the replicative haxameric helicase (G*40*P). The viral helicase recruits the host-encoded primase (DnaG) and DnaX, which is a subunit of the clamp loader (Pedre et al., 1994; Ayora et al., 1999; Martinez-Jimenez et al., 2002), so that a full replisome is loaded at the phage origin. SPP1 replication uses the host replicase holoenzyme and topoisomerases from the host (Seco et al., 2013; Seco and Ayora, 2017). After one or two rounds of theta-type replication (TR), it shifts to concatemeric (sigma-type) DNA replication in a process driven by recombination (Lo Piano et al., 2011). Two viral proteins may participate in this shift, the ATP-independent single-strand annealing recombinase (G*35*P) and its partner, the $5' \rightarrow 3'$ exonuclease (G*34.1*P) (Ayora et al., 2002; Martinez-Jimenez et al., 2005). In the shift to concatemeric DNA replication, G*38*P, bound to *ori*R, or working as a pre-primosome organizer (like the bacterial PriA enzyme), may restart DNA replication at stalled or paused replication forks (Seco et al., 2013). SPP1 codes for two other proteins involved in DNA replication and recombination: the G*36*P and G*44*P proteins. G*36*P is a single-stranded DNA binding protein (SSB), and G*44*P is a Holliday junction resolvase of the RusA family, which recognizes and cleaves a variety of recombination intermediates (Martinez-Jimenez et al., 2005; Zecchi et al., 2012). Biochemical assays showed that G*36*P is crucial for SPP1 DNA replication *in vitro*, but it can be substituted by host-encoded SSB (known as SsbA) (Seco et al., 2013). The role of G*44*P in SPP1 replication is thought to be the processing of the stalled replication fork, which may trigger the shift to the sigma-type or concatemeric DNA replication. This type of DNA replication is essential to generate the concatemeric DNA, which is the substrate for encapsidation. Viral replication and packaging are sequential and in some way coupled events. SPP1 encapsidates linear double-strand (ds) DNA into an empty prohead by a processive (~4 sequential packaging cycles) headful packaging mechanism, using the linear head-to-tail concatemer as a substrate (Oliveira et al., 2013). This is consistent with the observation that an *in vitro* DNA packaging system efficiently packaged mature SPP1 DNA as well as linear plasmid DNA, but no DNA packaging could be detected when circular DNA was the substrate for encapsidation (Oliveira et al., 2005). SPP1 packaging is initiated with the recognition of the specific *pac* region by the terminase small subunit, G*1*P, and the sequence specific cleavage at the *pac* sequence (CTATTGCGG↓C) by the terminase large subunit, G*2*P (Chai et al., 1992, 1995, 1997). This generates the first DNA end to be encapsidated (Chai et al., 1992; Gual et al., 2000; Camacho et al., 2003). A sequence independent cleavage, at 104% of the genome (headful cleavage), terminates one packaging round, generating a new starting point for another one (Chai et al., 1995; Camacho et al., 2003). Hence, the first cleavage in the concatemeric SPP1 DNA occurs specifically at *pac*, whereas the next ones do not (Gual et al., 2000).

In addition to package viral DNA, SPP1 is able to encapsidate chromosomal or plasmid DNA. However, some differences were observed with these two substrates. Rolling-circle replicating plasmids could be transduced at a frequency much higher than chromosomal DNA (Ferrari et al., 1978; Deichelbohrer et al., 1985), and an explanation for this could be that the copy number of plasmids in the cell is higher than that of the chromosome. Alternatively, another possibility could be that the replication mode influences the transduction frequency. It was also observed that the frequency of transduction of pUB110 and pC194 naturally occurring plasmids was enhanced 100- to 1000-fold by the presence of inserts homologous to the transducing phage DNA (Deichelbohrer et al., 1985). This homology-facilitated plasmid transduction was independent of the host RecA (Canosi et al., 1982; Deichelbohrer et al., 1985). In contrast, another report showed that SPP1 mediated chromosomal transduction was reduced 30-fold in cells having mutations in host functions involved in homologous recombination, such as RecA, RecU, and RecF (Ferrari et al., 1978). These differences, which were observed between plasmid and chromosomal transduction in the SPP1 system motivated us to analyze in deep and throughout the manuscript the influence of the replication mode and of the plasmid copy number in plasmid generalized transduction. In addition, we have analyzed the role of phage recombination and replication proteins. We show that in absence of G*35*P, G*38*P, G*39*P, or G*40*P linear plasmid transduction is blocked. In contrast, establishment of injected linear plasmid DNA in the recipient host was independent of viral-encoded functions. The transfer efficiency was found to be dependent on homology to phage DNA, plasmid copy number, and replication mechanism.

MATERIALS AND METHODS

Bacterial Strains and Plasmids

Bacillus subtilis BG214 (*trpCE metA5 amyE1 ytsJ1 rsbV37 xre1 xkdA1* $att^{SPß}$ att^{ICEBs1}) and its isogenic derivative BG295 (*sup3*) were used. They lack the ICE*Bs1* integrative conjugative element as well as prophage PBSX, and PBSX prohage cannot be induced (Kidane et al., 2009). The plasmids used are derivatives of pHP13, pUB110, pBT233 or pNDH33 (**Table 1**). To construct pBT233N, the pUB110 neomycin resistance gene was cloned into AvaI-linearized pBT233. Different regions of the SPP1 genome were cloned into the HpaI site of the pBT233N plasmid as indicated in **Table 1**. pHP13 derivatives were kindly provided by J. C. Alonso (CNB-CSIC). Plasmid pBT400 is a pHP13 derivative bearing an EcoRI-SalI fragment of SPP1 DNA. Different SPP1 DNA fragments were cloned into XbaI- or SmaI-cleaved pNDH33 DNA, rendering pNDH33-1300 and pNDH33-*pac* (**Table 1**).

SPP1 Phages

The SPP1 phages used in this work are listed in **Table 2**, including those (*sus19, sus53, sus109*, tsB3, and SPP1ΔA) previously described (Chai et al., 1992; Pedre et al., 1994; Zecchi et al., 2012).

The SPP1 tsI20F mutant was sequenced and it was found that the mutation that conferred thermosensitivity (ts), P159S, mapped in gene *35*, rather than in gene *34.1*, as it was previously suggested after genetic mapping (Burger and Trautner, 1978). This phage was used to construct the SPP1 *sus35* mutant. First, a lysine codon (the 10th codon in the gene *35*) was replaced by an ochre (UAA) stop codon by site-directed mutagenesis using plasmid pCB610 as template (a pHP13 derivative containing SPP1 genes *34.4* to *35*) and the Quickchange protocol. After sequencing confirmation the resulting plasmid (pHP13-G35P-ochre) was introduced into BG295 cells by transformation. BG295 cells bearing pHP13-G35P-ochre plasmid were infected with SPP1 tsI20F phage at 30°C for 2 h. The resulting phage lysate was used to infect BG295 cells at non-permissive temperature to

TABLE 1 | Plasmids used in this work.

Plasmids	Plasmid characteristics	Reference
pC194	Natural rolling circle replicating (RCR) plasmid, 2.9-kb	Horinouchi and Weisblum, 1982; Alonso and Trautner, 1985
pHP13	RCR plasmid derivative of pTA1060, 4.9-kb	Haima et al., 1987
pBT163 (pHP13-*pac*)	pHP13 derivative containing SPP1 DNA including *pac* (2675 bp cloned, coordinates 43778–44010 and 1–2439)	Chai et al., 1992
pBT271 (pHP13-*oriL*)	pHP13 derivative containing SPP1 DNA including *oriL* (2975 bp, coordinates 33875–36850)	Chai et al., 1993
pBT400 (pHP13-800)	pHP13 derivative containing SPP1 DNA (864 bp, coordinates 3225–4089)	This work
pUB110	Natural RCR plasmid, 4.5-kb	Leonhardt, 1990
pUB110-cop1	pUB110 derivative, lower copy number	Leonhardt, 1990
pBG55 (pUB110-3600)	pUB110 derivative containing SPP1 DNA (3639 bp, coordinates 23117–26756)	Deichelbohrer et al., 1985
pBT233	Theta replicating (TR) plasmid, 9-kb	Ceglowski et al., 1993a
pBT233N	pBT233 derivative containing the 1304 bp neomycin resistance gene (N) from pUB110	This work
pBT233N-400	pBT233N derivative containing SPP1 DNA (414 bp, coordinates 32562–32976)	This work
pBT233N-1300	pBT233N derivative containing SPP1 DNA (1340 bp, coordinates 25051–26391)	This work
pBT233N-*oriL*	pBT233N derivative containing SPP1 *oriL* DNA (350 bp, coordinates 35801–36151	This work
pBT233N-*pac*	pBT233N derivative containing SPP1 *pac* DNA (412 bp, coordinates 43689–44010 and 1–70)	This work
pNDH33	TR plasmid derivative of pBS72, 8.1-kb	Titok et al., 2003
pNDH33-1300	pNDH33 derivative containing SPP1 DNA (1340 bp, coordinates 25051–26391)	This work
pNDH33-*pac*	pNDH33 derivative containing SPP1 *pac* DNA (412 bp, coordinates 43689–44010 and 1–70)	This work

TABLE 2 | SPP1 phages used in this work.

Genotype	Name	Activity / type of mutant	Reference
wt	SPP1wt	Wild type	
34.1^{-}	*sus34.1*	Exonuclease, ochre mutant (OM)	This work
35^{-}	*sus35*	Recombinase, OM	This work
35^{-}	tsI20F	Recombinase/thermosensitive (ts) mutant	This work
36^{-}	*sus36*	ssDNA binding protein, OM	This work
38^{-}	tsB3	Replisome organizer/ts mutant	Pedre et al., 1994
39^{-}	*sus53*	Helicase loader, OM	Pedre et al., 1994
40^{-}	*sus109*	Helicase, OM	Pedre et al., 1994
44^{-}	SPP1ΔA	Deletion mutant lacking Holliday junction resolvase	Zecchi et al., 2012
2^{-}	*sus19*	Terminase large subunit, OM	Chai et al., 1992

obtain the recombinant phages. They were picked from Luria-Bertani (LB) plates supplemented with 10 mM $MgCl_2$ (LB-Mg^+) incubated at 50°C. The amplified phage was sequenced to confirm that phages had acquired the ochre mutation in gene *35*, and that it had reverted to wt the tsI20F mutation. The resulting mutant phage, containing the ochre codon, was named SPP1 *sus35*.

The 37th codon (Lys) in gene *36* was replaced by an ochre (UAA) stop codon in a pHP13 derivative containing SPP1 genes *34.4* to *37*. The SPP1 *sus34.1* mutant was generated by replacing, in a pHP13 derivative containing SPP1 genes *34.1* to *35*, the 31th codon (AAA) of gene *34.1* by an ochre (UAA) stop codon. The SPP1 *sus36* and *sus*34.1 mutants were then generated by homologous recombination between the SPP1 tsI20F phage and these plasmids carrying the stop ochre codon into the gene to be mutated, as described above. The accuracy of the resulting mutant phages was confirmed by sequencing.

SPP1wt, SPP1ΔA phages and the thermosensitive phages (tsI20F, and tsB3) were amplified in BG214 cells grown at 37°C or 30°C in LB-Mg^+, whereas the *sus* phages were routinely amplified in the suppressor strain BG295 (*sup3*) at 37°C.

Preparation of Transducing Lysates

Transducing lysates were obtained by infecting with the different SPP1 phages, at a multiplicity of infection (MOI) of 10, *B. subtilis* BG214 cells bearing the indicated plasmids, grown up to mid-exponential phase in LB-Mg^+ and appropriated antibiotics. Aliquots were taken at different post-infection times for DNA analysis and processed as described below. The cultures were centrifuged after 90 min of infection (14,000 rpm, 5 min), and the supernatants were filtered through 0.45 μm filters to remove donor cells. Under these growth conditions *B. subtilis* cells are not competent, so that DNAse I treatment was not required. Phage lysates were titrated on BG214 cells or BG295 cells before use and were stored at 4°C.

Plasmid Transduction

Exponentially growing recipient *B. subtilis* BG214 or BG295 cells ($OD_{560} = 0.4$) grown at 37°C in LB-Mg^+, were infected with the transducing phage lysate at MOI of 1. Phages were allowed to be absorbed for 5 min, and then the non-absorbed phages were removed by centrifugation. Cell pellets were washed and finally resuspended in 1 ml LB. Appropriate dilutions were plated in selective LB-agar plates containing the respective antibiotics, and incubated overnight at 37°C to quantify the number of transductants. As a control, 1 ml of the recipient host was plated to discard the appearance of spontaneous resistant colonies. In another LB-agar plate with antibiotic the same amount of the stock transducing lysate was plated without recipient cells, to discard a contamination with donor cells.

Analysis of Plasmid DNA Forms

B. subtilis BG214 cells bearing the different plasmids were grown at 37°C to an OD_{560} of 0.40 in LB-Mg^+ media supplemented with appropriate antibiotics, and infected with a MOI of 10. Phage addition marked the time zero of our experiments. At given times, aliquots of 1ml were collected, rapidly placed in a water-ice mixture and centrifuged for 5 min at 14,000 rpm and 4°C. The pellets were stored at −80°C. In experiments with thermosensitive phage mutants, the strains bearing plasmids were first grown at 30°C to an OD_{560} of 0.2, transferred to 50°C and then further grown to OD_{560} of 0.4. They were infected at 50°C, and the samples were processed as described above. Total DNA was isolated following a protocol described earlier (Viret and Alonso, 1987) with some minor modifications. Samples were resuspended in 200 μl of lysis buffer (25 mM Tris-HCl pH 8.0, 50mM glucose, 10 mM EDTA, 0.5 mg/ml lysozyme and 0.1 mg/ml RNase A). After 30 min of incubation at 30°C, Proteinase K (0.5 mg/ml) and SDS (0.8%) were added, and the mixture was further incubated for 30 min at 37°C. The lysate obtained was then treated twice with phenol and dialyzed against 20 mM Tris-HCl pH 8.0, 1 mM EDTA.

Pulsed field gel electrophoresis (PFGE) was performed on a Bio-Rad CHEF-DR II apparatus. 15 μl of samples were loaded on the 1% agarose gel. Running conditions were 5 V/cm, 0.5% TBE, 0.5–10 switch time for 20 h at 14°C. The molecular weight marker used was LW range PFG marker or λ DNA-HindIII digest, both from New England Biolabs. The probe used for Southern blot hybridization was a PCR product of 500 bp corresponding to neomycin or chloramphenicol resistance genes. Southern blots were performed with Hybond-N+ membranes as recommended by the manufacturer (GE Healthcare), and detection was done with the AlkPhos Direct Labeling kit (GE Healthcare).

RESULTS

Viral Replication and Recombination Proteins Are Responsible for the Generation of Plasmid Transducing Particles

To unravel the mechanisms that contribute to SPP1-mediated horizontal plasmid transfer we used *B. subtilis* BG214 strain, which is non-inducible for PBSX prophage and lacks prophage SPβ and the ICE*Bs1* integrative conjugative element. To analyze the role in antibiotic resistance transfer of SPP1 replication and recombination proteins, phages *sus34.1* and *sus36*, bearing mutations in genes *34.1* and *36* respectively, were constructed. SPP1 phage variants bearing mutations in the other genes were available in our phage collection (*sus19*, *sus53*, *sus109*, SPP1ΔA, tsB3). For comparison, a SPP1 *sus35* phage was also constructed, although a thermosensitive gene *35* mutant (the tsI20F phage) was available. The list of the bacteriophages used is shown in **Table 2**.

First we analyzed if G*34.1*P and G*36*P proteins, which were not yet studied *in vivo*, are essential for SPP1 replication (**Figure 1**). BG214 cells were grown until mid-exponential phase and then infected at MOI of 10 with the SPP1wt, SPP1ΔA, tsB3 (at restrictive temperature), or the different *sus* mutants (*sus34.1*, *sus35*, *sus36*, and *sus53*, a phage with a mutation in gene *39*). After 90 min of infection, the phage lysates were collected and titrated. As previously observed, deletion of gene *44* reduced the

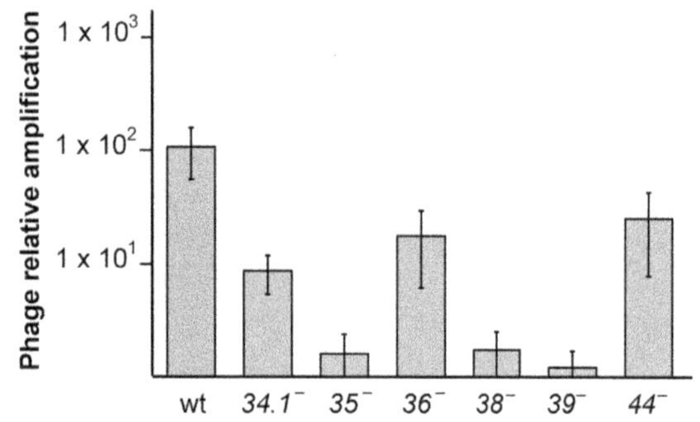

FIGURE 1 | Effect of the different SPP1 mutations on phage titer. *B. subtilis* BG214 cells were infected with the different phages at a MOI of 10, and after 2 h of infection PFU/ml was calculated. The number of phages, relative to the number of phages initially added, is indicated. The values are the mean of at least five independent assays and error bars indicate SD.

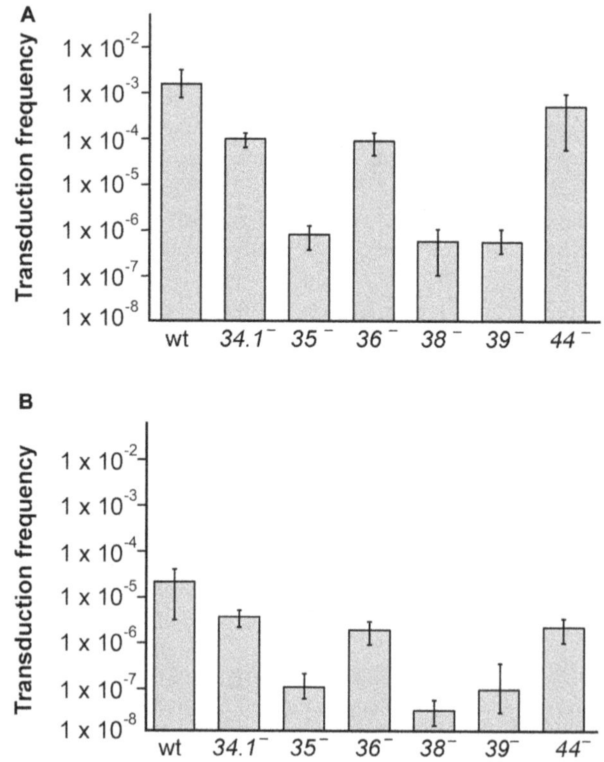

FIGURE 2 | The generation of transducing lysates bearing plasmids with (pBG55) or without (pUB110) sequence homology with SPP1 is affected by mutations in viral replication and recombination genes. **(A)** Generation of pBG55 transducing particles after phage infections, expressed as frequency of transductants/CFU. **(B)** Generation of pUB110 transducing particles. BG214 cells bearing plasmids were infected with the different phage mutants and lysates were used to infect the BG295 *sup3* strain to have the *sus* mutation only in the donor cells. The values are the mean of at least four independent assays and error bars indicate SD.

phage titer only 5-fold (Zecchi et al., 2012), whereas the mutation in gene *35, 38* or *39* completely abolished SPP1 amplification (Pedre et al., 1994; Ayora et al., 2002). The mutation in gene *36* reduced SPP1 titer only 6-fold, in agreement with the biochemical data showing that G*36*P can be replaced by the host SsbA during SPP1 DNA replication (Seco et al., 2013; Seco and Ayora, 2017). Deletion of the *34.1* gene reduced the phage titer 10-fold, and the size of the phage plaques was considerably smaller compared to the wt phage (Supplementary Figure S1). These results show that both, G*36*P, and G*34.1*P are not essential for phage amplification, although their defects reduce phage development.

To analyze if SPP1 replication and recombination proteins are involved in the generation of the transducing particle, the different *sus* mutant phages were used to infect BG214 cells bearing plasmid pBG55, a rolling circle replicating (RCR) plasmid with high-frequency of transduction (see **Table 1** for more description). The lysates were collected after 90 min of infection, filtered and used to infect the BG295 *sup3* strain, to have the effect of phage *sus* mutation only in the donor and not in the recipient strain. The frequency of pBG55 transfer (Neomycin resistants [Nm^R]/CFU) for the wt phage was similar to previously published results obtained using the BG214 strain, both as donor and as recipient (Deichelbohrer et al., 1985). These results show that the *sup3* genotype does not affect the transduction frequency. In parallel, infections with the thermosensitive phage mutants were performed at 50°C for 90 min. The lysates were then collected, filtered and used to infect BG214 cells at 30°C to have the effect of the thermosensitive mutation only in the donor, and not in the recipient strain. Mutations in genes *35, 38,* or *39* blocked the transfer of the plasmid with homology (pBG55), with more than 1000-fold reduction in the transduction frequency (**Figure 2A**). A similar result was obtained with *sus109,* bearing a mutation in gene *40* (data not shown). Mutations in the exonuclease (G*34.1*P) or in the viral SSB (G*36*P) reduced the transduction frequency by ~12-fold, whereas the mutation in G*44*P only reduced it by ~4-fold.

To analyze if these proteins are also involved in the transfer of plasmids having no homology with the SPP1 phage, or just very short homologous regions (sequences of 11–16 bp complementary to SPP1 DNA, see Supplementary Table S1) we performed transduction assays with the natural occurring pUB110 plasmid and the different phage mutants (**Figure 2B**). As already observed the transduction frequency of this plasmid was reduced by a factor of ~100-fold compared to the frequency of pBG55 transduction. The transduction frequencies were reduced in all of the SPP1 mutants, and similarly to the results obtained with the plasmid having homology, mutations in the recombinase or in replication proteins drastically reduced the phage-mediated transfer of pUB110, whereas mutations in the exonuclease, the SSB, or the HJ resolvase reduced the number of transductants/ml to a lesser extent.

SPP1 Replication and Recombination Proteins Are Essential for the Generation of Plasmid Concatemeric DNA

Concatemeric plasmid DNA synthesis was observed with RCR plasmids after phage infection (Alonso et al., 1986; Bravo and Alonso, 1990). The results obtained in the previous section

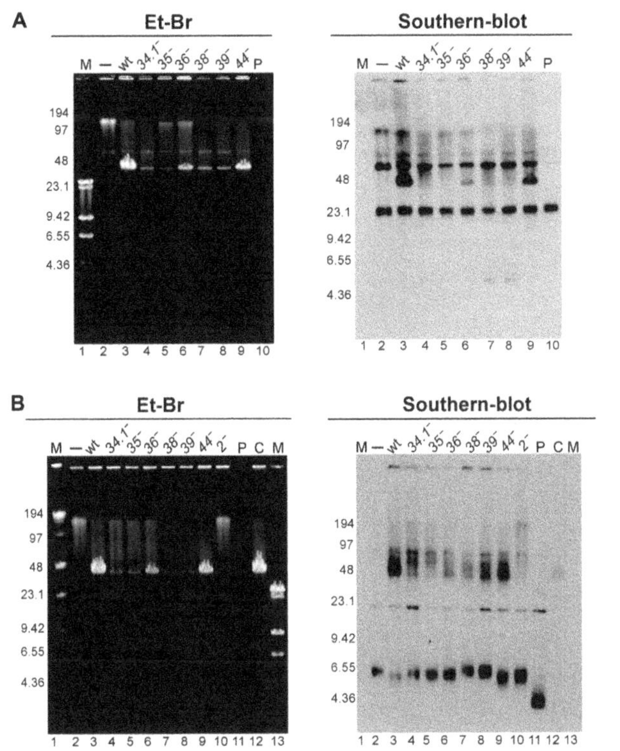

FIGURE 3 | Analysis of the appearance of the transducing particles after infection with different SPP1 mutants of cells bearing RCR replicating plasmids with (pBG55 infections, **A**) or without regions homologous to SPP1 DNA (pUB110 infections, **B**). To unravel the appearance of *hmw* plasmid DNA 30 min after phage infections samples were analyzed by PFGE and Et-Br staining (left panel) followed by Southern-blot (right panel). M, LW and λ-HindIII markers. P, purified plasmid DNA 15 ng (pBG55 in **A**, and pUB110 in **B**); C, control, a SPP1 infection of BG214 cells without plasmid.

suggest that the essential viral recombination (G35P) and replication (G38P, G39P, and G40P) proteins could be responsible for the generation of this linear concatemeric plasmid DNA. To test this, we infected BG214 cells bearing pBG55 with the different phage mutants. After 30 min of infection, the infected cells were collected, total DNA was extracted, and separated by PFGE and Southern blotted to detect the production of concatemeric plasmid DNA forms. After infection with the wt phage the appearance of plasmid DNA that migrates with the bulk of SPP1 DNA (i.e., a multimeric plasmid DNA band of 44-kb) was observed (**Figure 3A**). In the absence of *G35P*, *G38P* or *G39P*, the production of this concatemeric band was not observed, consistent with the above result that mutations in these proteins block plasmid transduction. In agreement with its minor role in plasmid transfer, the 44-kb plasmid DNA band was observed after infection with phages bearing mutations in *G34.1P*, *G36P*, or in *G44P*. Moreover, the amount of 44-kb pBG55 DNA observed by PFGE and Southern blot correlated in these mutants with their transduction frequencies.

We also observed the appearance of a similar 44-kb plasmid band after infection with the wt SPP1 phage of cells bearing the natural pUB110 plasmid (**Figure 3B**). In concordance with observations using the plasmid with extensive homology, the appearance of this 44-kb plasmid DNA band was clearly observed after infections with SPP1wt and SPP1 ΔA phages, which showed the highest transduction frequencies.

Viral Replication and Recombination Proteins Are Not Involved in the Establishment of the Transduced Plasmid

The results presented above and in earlier reports (Deichelbohrer et al., 1985; Bravo and Alonso, 1990) indicate that a concatemeric ~44-kb plasmid DNA is encapsidated into the viral capsids. Once this concatemeric plasmid DNA (5.4 plasmid copies in the case of pBG55 plasmid) is injected into a recipient cell, it needs to circularize and monomerize to prepare the plasmid for correct replication and segregation cycles. The duplicated regions present in the concatemer could be used for monomerization, through a homologous recombination event, as it occurs during natural plasmid transformation (Kidane et al., 2009). In order to analyze if the viral replication and recombination machinery is involved in this monomerization and plasmid establishment process, we performed transduction assays with *sup3* as donor and wt as recipient cells (**Table 3**). It appeared that none of the viral proteins were required for the establishment of the transduced plasmid in the recipient cells.

The Influence of Plasmid Copy Number and Replication Mode in Transduction

Plasmid-borne genes are transduced at much higher frequency than chromosomal-borne genes (Ferrari et al., 1978; Deichelbohrer et al., 1985), suggesting that copy number of plasmids could account for such differences. However, there is no tight correlation. As an example, it was published that the transduction frequency of plasmid pUB110, which has ~50 copies per cell (Viret and Alonso, 1988) is lower than that of pC1943 with ~15 copies per cell (Deichelbohrer et al., 1985). We confirmed these results (**Table 4**). This suggests that plasmid copy number is not the major determining factor, or not the only one. Other factors such as the presence of pseudo-*pac* sites, or of single-stranded (ssDNA) plasmid forms (recombinogenic particles, see below) could be the cause of this increased transduction frequency. Both plasmids, pUB110 and pC194, are RCR plasmids, but it was found that pC194 is more prone to formation of ssDNA than pUB110 (te Riele et al., 1986; Viret and Alonso, 1987).

To elucidate the influence of copy number, we compared the transduction efficiency of plasmid pUB110 (48 ± 4 copies/cell) and its derivative pUB110-copl (9 ± 1 copies/cell). pUB110-copl results from a single mutation in pUB110 plasmid, and therefore it has the same amount of ssDNA as the parental plasmid, but its copy number is reduced by 5-fold (Leonhardt, 1990). Both plasmids should have similar rates of circularization and establishment when they are injected into the recipient cell. As shown in **Table 4**, the transduction efficiency of pUB110-copl was proportionally reduced 4.6 times. In parallel we compared also the transduction frequencies of two other plasmids that accumulate ssDNA, pC194 (15 ± 2 copies per cell, Alonso and Trautner, 1985) and pHP13 (a pTA1060 derivative, 7 ± 2 copies per cell, Wang et al., 2004). Here also the transduction efficiency decreased by lowering the copy number of the plasmids. Nevertheless in all cases the transduction frequencies were higher for the plasmids accumulating ssDNA intermediates (**Table 4**).

Previous studies of the plasmid transduction by the SPP1 phage were done only with RCR plasmids. To determine the transduction frequency of theta replicating (TR) plasmids we used two such plasmids: pBT233 and pNDH33, which have a copy number similar to that of pHP13 plasmid (**Table 1**). Plasmid pBT233 is a pSM19035 derivative (erythromycin resistant), which has a copy number of ~8 ± 2, and replicates unidirectionally by a DNA polymerase I (PolI)-dependent theta mechanism (Ceglowski et al., 1993a,b,c). Plasmid pNDH33 is a derivative of pBS72 (chloramphenicol resistant) with a copy number of ~6 ± 1 plasmids/cell (Nguyen et al., 2005; Phan et al., 2006). pNDH33 is thought to replicate by a DnaA-dependent and DNA PolI-independent theta type mechanism (Titok et al., 2003; Schumann, 2007). To compare TR and RCR plasmids, and to eliminate any resistance marker effects, the neomycin gene of the pUB110 was cloned into plasmid pBT233, to render plasmid pBT233N. The transduction frequency of the TR plasmid pBT233N was about 70-fold lower than that of pHP13. We measured also the transduction frequency of the second TR plasmid, pNDH33. This appeared to be also low, but only ~10-fold lower than that of pHP13 plasmid (**Table 4**). This higher transduction could be due to the occasional presence in the pNDH33 plasmid of a *pseudo-pac* site or because of a 16 bp stretch of homology (**Table 4** and Supplementary Table S1). When analyzing the fate of TR plasmids in infected cells, it was observed that, as with RCR plasmids, the infection with wt SPP1 phage produced the accumulation of a 44-kb plasmid DNA band, which was not observed after infection with a *sus35* mutant (**Figure 4**).

The Presence of Homology to Phage Enhances the Transduction of TR Plasmids

When the phage packaging signal (*pac*) was integrated into the host chromosome, SPP1 mediated the transduction of chromosomal genes located close to the region of integration of the *pac* signal (Bravo et al., 1990). It was not tested if the presence of other SPP1 regions also increases the transduction frequencies of chromosomal DNA. To test this, we used the pBT233N derivative conferring Nm^R, which replicates via the theta-type mechanism as the chromosome. Different regions of SPP1 were cloned into pBT233N in order to evaluate whether the presence of *pac* sequence or the replication origin (*ori*L) results in higher transduction than simply homology to the phage (**Table 5**). Overall, the presence of a homologous region increased the transduction frequency of pBT233N plasmid by more than 1000-fold, and this increase was observed independently of the homologous region cloned (*pac*, *ori*L, or a 400 bp or 1000 bp region unrelated to replication and packaging processes).

TABLE 3 | Viral replication and recombination proteins are not involved in the establishment of transduced plasmids.

Donor strain[a] and plasmid	Recipient strain	Phage	Transduction Frequency[b]	SD[c]	TF^M/TF^{wt}
BG295 pBG55	BG214	SPP1 wt	4.3×10^{-3}	$\pm 2.0 \times 10^{-3}$	1.00[d]
BG295 pBG55	BG214	*34.1*⁻	3.4×10^{-3}	$\pm 1.5 \times 10^{-3}$	7.9×10^{-1}
BG295 pBG55	BG214	*35*⁻	2.1×10^{-3}	$\pm 1.1 \times 10^{-3}$	4.9×10^{-1}
BG295 pBG55	BG214	*36*⁻	3.4×10^{-3}	$\pm 1.8 \times 10^{-3}$	7.9×10^{-1}
BG214 pBG55[e]	BG214	*38*⁻	4.3×10^{-3}	$\pm 2.1 \times 10^{-3}$	1.0×10^{0}
BG295 pBG55	BG214	*39*⁻	2.9×10^{-3}	$\pm 2.0 \times 10^{-3}$	6.7×10^{-1}

[a]*BG214 is the wild type strain and BG295 is the isogenic sup3 strain.* [b]*The transduction frequency (Neo^R/CFU) is the average of at least three independent experiments.* [c]*SD: standard deviation.* [d]*The frequency of pBG55 plasmid transduction with the phage mutants (TF^M) with respect to the wt phage (TF^{wt}) is presented.* [e]*The 38⁻ mutant is a thermosensitive phage (tsB3), therefore the infection was done at permissive temperature (30°C) and the transduction at non-permissive temperature (50°C) to have the mutation only in the recipient strain.*

TABLE 4 | Transduction frequency of theta and rolling circle replicating plasmids without sequence homology with SPP1.

Plasmid	Ab^R marker	Replication mechanism[a]	Copy number[b]	ssDNA production[c]	*pseudo-pac* site[d]	Transduction Frequency[e]	$CI_{0.95}$[f]
pUB110	Nm	RCR	50	+	–	2.1×10^{-5}	$\pm 1.3 \times 10^{-5}$
pUB110-cop1	Nm	RCR	15	+	–	4.3×10^{-6}	$\pm 2.2 \times 10^{-6}$
pC194	Cm	RCR	15	+++	–	5.2×10^{-5}	$\pm 4.7 \times 10^{-5}$
pHP13	Cm	RCR	5	+++	–	5.9×10^{-6}	$\pm 2.9 \times 10^{-6}$
pBT233N	Nm	TR	8	–	–	8.2×10^{-8}	$\pm 7.1 \times 10^{-8}$
pNDH33	Cm	TR	6	–	1	6.4×10^{-7}	$\pm 3.2 \times 10^{-7}$

[a]*RCR, rolling circle replication; TR, theta replication.* [b]*Plasmid copy numbers were reported in the literature and are presented here for comparison.* [c]*ssDNA production was reported in the literature and is presented here for comparison.* [d]*The pac motif (5′-CTATTGCGG⇓C-3′) is absent in all of the plasmids. Here the presence of a shorter motif that we call pseudo-pac site 5′-TTGCGG⇓CW-3′ is indicated.* [e]*The transduction frequency (transductans/CFU) is the mean of at least five independent experiments.* [f]*CI, confidence interval.*

Similarly, cloning into a RCR plasmid (pHP13) one of the phage origins of replication of SPP1 did not further increase the transduction frequency (Supplementary Table S2). Using of other TR-type replicon, pNDH33, provided similar results (**Table 5**). Furthermore, the accumulation of the 44-kb plasmid band was higher in the TR plasmids derivatives having homology with the phage (**Figure 4**, and data not shown).

DISCUSSION

Until recently, it was thought that generalized transduction occurred at low frequency. However, recent single-cell analyses observed transduction rates close to 1% per plaque forming units when natural communities were used as recipients (Kenzaka et al., 2010). Therefore the study of the transduction mechanisms is essential to prevent this highly frequent horizontal gene transfer process, to avoid the spread of antibiotic resistance among bacteria. In this aspect, the SPP1 bacteriophage is a valuable model, because its replication, recombination, and packaging machineries haven been studied in deep for many years. Furthermore, it was recently reported that SPP1 can occasionally infect resistant cells when combined with sensitive cells, providing new routes for horizontal gene transfer (Tzipilevich et al., 2017). Previous biochemical studies assigned a role to SPP1 proteins *G34.1*P, *G35*P, *G36*P, *G38*P, *G39*P, *G40*P, and *G44*P in replication and recombination, but their contribution to generalized plasmid transduction remained unknown. Here we show that all SPP1 replication proteins contribute to horizontal plasmid transfer, although to a different extent. The origin binding protein (*G38*P), helicase loader (*G39*P), and helicase (*G40*P) are essential to produce concatemeric plasmid DNA, which is synthesized after phage infection. Infections with the *sus36* mutants show only a 10-fold reduction in the transduction frequency, probably due to potential complementation of the *G36*P function by cellular SsbA protein (Seco et al., 2013; Seco and Ayora, 2017). The SPP1 recombination proteins contribute to plasmid transfer to a different extent. The exonuclease *G34.1*P and the Holliday junction resolvase *G44*P only contribute partially to plasmid transduction, with a reduction of the

TABLE 5 | Transduction frequency of theta replicating plasmids bearing different SPP1 DNA regions.

Plasmid	Length of homologous region	Special features	Transduction Frequency[a]	$CI_{0.95}$[b]
pBT233N	–	–	8.2×10^{-8}	$\pm 7.1 \times 10^{-8}$
pBT233N-400	414 bp	homology	3.1×10^{-5}	$\pm 1.9 \times 10^{-5}$
pBT233N-1300	1340 bp	homology	4.5×10^{-4}	$\pm 3.0 \times 10^{-4}$
pBT233N-*pac*	412 bp	*pac*	1.7×10^{-4}	$\pm 2.8 \times 10^{-4}$
pBT233N-*oriL*	360 bp	*oriL*	1.5×10^{-4}	$\pm 1.6 \times 10^{-4}$
pNDH33	–	–	5.6×10^{-7}	$\pm 2.9 \times 10^{-7}$
pNDH33-1300	1340 bp	homology	2.0×10^{-4}	$\pm 8.5 \times 10^{-5}$
pNDH33-*pac*	412 bp	*pac*	3.9×10^{-4}	$\pm 3.2 \times 10^{-4}$

[a]*The transduction frequency (transductans/CFU) is the mean of at least five independent experiments.* [b]*CI, confidence interval.*

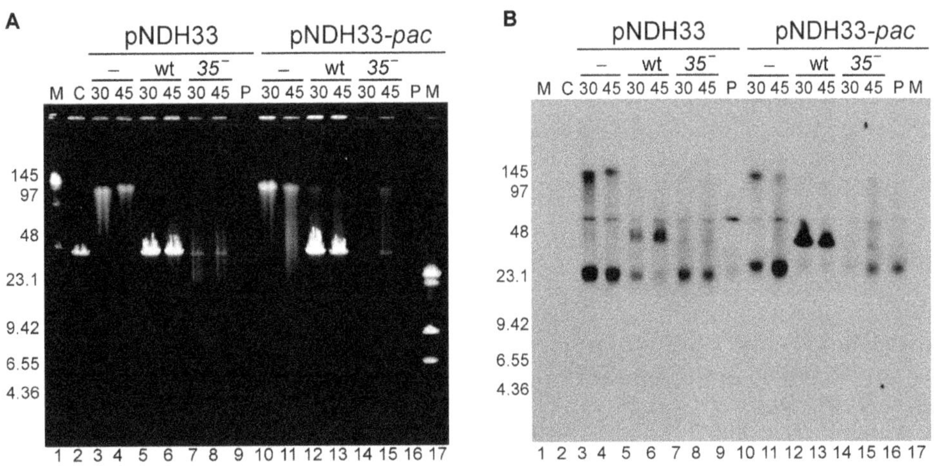

FIGURE 4 | Southern-blot analysis of the appearance of the transducing particles after infection with SPP1 or with *sus35* phage of cells bearing TR plasmids having (pNHD33-*pac*) or lacking (pNDH33) homologous regions to the phage. **(A)** Ethidium bromide stain and **(B)** Southern blot of the same gel developed with a chloramphenicol probe to visualize plasmid DNA. Lanes: 1 and 17: LW and λ-HindIII markers. Lane 2: C, control SPP1 infection of BG214 cells without plasmid. Lanes 3–4 and 10–11: control, non-infected BG214 cells bearing pNDH33 or pNDH33-*pac* plasmid. Lanes 5–6 and 12–13: SPP1 infection of BG214 cells bearing pNDH33 or pNDH33-*pac* plasmid, after 30 and 45 min infection. Lanes 7–8 and 14–15: BG214 cells bearing pNDH33 or pNDH33-*pac*, after 30 and 45 min infection with *sus35* phage. Lane 9 and lane 16: P, 15 ng of purified pNDH33 or pNDH33-*pac* respectively.

transduction frequency of 12- and 5-fold in their mutants, respectively. The G35P recombinase is essential, with its inactivation leading to a >100-fold decrease.

Previous studies with SPP1 and RCR plasmids showed that: (i) the transduction of pUB110 and pC194 plasmids was enhanced 100- to 1000-fold when there was any homology between the plasmid and the SPP1 genome rather than with the specific *pac* signal; (ii) pUB110 and pC194 plasmid transduction was independent on RecA (Deichelbohrer et al., 1985), and (iii) linear plasmid concatemeric DNA (or high-molecular-weight [*hmw*] DNA) accumulated during phage infection, and in certain genetic backgrounds (Viret and Alonso, 1987; Viret et al., 1991). The synthesis of *hmw* DNA and its independence of the host-encoded recombinase (RecA) strongly suggests that the formation of transducing particles may rely on viral replication and/or recombination functions. In this work we show that the synthesis of this *hmw* DNA, and consequently transduction of RCR plasmids requires an active G35P protein. Biochemical analysis shows that G35P is an ATP-independent single-strand annealing enzyme, similar to the RecT enzyme encoded by the Rac prophage (Ayora et al., 2002). Both, G35P and RecT, belong to the Redβ family of viral single strand annealing proteins. To date, five different single strand annealing recombinase families have been identified in phages: Sak, Redβ, Erf, Sak4 and Gp2.5 (Lopes et al., 2010). These recombinases have gained increased attention in recent years because of their abundance in phage genomes (Lopes et al., 2010; Delattre et al., 2016), and also due to their wide use in recombineering systems (Datta et al., 2008; Sun et al., 2015). Many of these recombinases, including G35P, are essential for the phage life cycle (Zecchi et al., 2012; Neamah et al., 2017).

In this work we found that variations in copy-number affect the transduction frequency. Since the transduction is a stochastic process, it is expected that the more plasmid DNA in the cell the more generalized transducing phage particles should carry a plasmid copy and therefore the chances of transduction increase. The plasmids replicating in *B. subtilis* cells are either of the TR (circle-to-circle) type or RCR (sigma) type, and the products of both replication modes are usually covalently closed circular monomers (Khan, 2005). Comparing plasmids with similar copy number we observed that the frequency of transduction for RCR plasmids is ~60-fold higher than that for TR plasmids. This result suggests that the type of DNA replication also determines the transduction frequency. In the small RCR plasmids leading and lagging strand replication are uncoupled, and they contain two modules: the Rep protein with its cognate double-strand origin (DSO), and a single strand origin (SSO), which functions as the major initiation site for lagging-strand synthesis (Alonso et al., 1988; Espinosa et al., 1995; Khan, 2005). All RCR plasmids accumulate ssDNA although to a different extent: pUB110 accumulates traces and pC194 accumulates circular ssDNA (te Riele et al., 1986; Viret and Alonso, 1988). In contrast, the large low-copy-number TR plasmids, such as pBT233, which replicates via an unidirectional mechanism, do not accumulate circular ssDNA intermediates (Ceglowski et al., 1993b,c). We propose that the high transfer frequencies of some RCR plasmids may be correlated with the high accumulation of recombinogenic ssDNA intermediates in these plasmids. Such ssDNA intermediates may constitute the substrates for formation of the transducing particles, through a recombination catalyzed by the G35P protein. This is in agreement with recent results observed with viral recombinases: when analyzing their recombineering activity *in vivo*, it was found that they catalyze single-strand annealing preferentially on the lagging strand (van Kessel and Hatfull, 2008; Mosberg et al., 2010; Lajoie et al., 2012; Fricker and Peters, 2014; Ander et al., 2015). We propose that all the phages encoding recombinases will transduce RCR plasmids with high efficiency by the mechanism of viral recombinase-mediated generalized transduction. Furthermore, we also observed that the transduction of the pUB110 and pNDH33 plasmids, which do not have an extensive region of homology, was strongly reduced in infections with the *sus35* mutant (**Figures 2, 4**). All phage recombinases studied so far are single-strand annealing proteins that promote genetic recombination under more permissive conditions than RecA (Scaltriti et al., 2011; De Paepe et al., 2014; Menouni et al., 2015). Our results suggest that G35P contributes to the transfer of natural plasmids by catalyzing a recombination reaction using small stretches of homology found in many plasmids (Supplementary Table S1).

The different contributions of the SPP1 recombination proteins to plasmid transduction, together with the high recombinogenic nature of the RCR plasmids, suggest that the initial DNA substrate, used for the production of transducing particles by recombination, is indeed ssDNA. This is consistent with the result that the G34.1P exonuclease, which resects the dsDNA ends to generate the appropriate substrate for the recombinase (Martinez-Jimenez et al., 2005), has a minor role in plasmid transfer. Similarly, we found that the SPP1 SSB protein, G36P, only slightly contributes to the mechanisms of plasmid transduction. However, in some phages the recombinases require the activity of their cognate SSB proteins to perform their function (Neamah et al., 2017).

It was previously observed with RCR plasmids that any SPP1 DNA segment larger than 50 bp, cloned into such plasmids, greatly increased the transduction frequency (Deichelbohrer et al., 1985; Alonso et al., 1986). We extend this observation to TR plasmids, where the transduction frequency was highly increased, independently of what is the region of homology cloned, whether it was the packaging sequence, a phage origin of replication, or any other region of homology. Similarly, the cloning of the origin of replication of SPP1 (*oriL*) into a RCR-type plasmid did not further increase its transduction frequency (Supplementary Table S2). We conclude that any DNA region homologous to the phage genome increases the frequency of horizontal transfer of plasmids, independently of their replication mechanism. Enhanced transduction of plasmids bearing homology with phage DNA has been also observed with phage T4, which codes for a different recombinase, the UvsX protein (Kreuzer et al., 1988), and with *Salmonella typhimurium* phage P22, which codes for the Erf recombinase (Orbach and Jackson, 1982).

How is the plasmid substrate for generalized transduction generated? Three different mechanisms could account for the generation of a concatemeric plasmid DNA with high frequency of transduction. In the first model, the multiple tandem repeats of plasmid DNA might be produced by intermolecular recombination, as proposed for P22 plasmid transduction (Mann and Slauch, 1997). This mechanism resembles phage T4 generation of concatemeric DNA during its replication (Kreuzer, 2000; Mosig et al., 2001). Here, multiple strand invasions catalyzed by the ATP-dependent RecA-like recombinase encoded by this phage, UvsX, and the resolution of the Holliday junction intermediates by its Holliday junction resolvase Gp49 (also called EndoVII), produce the concatemeric DNA, as well as the transducing particle (Kreuzer et al., 1988; Kreuzer, 2000; Mosig et al., 2001). We do not favor this hypothesis in the SPP1 system, because we found that the Holliday junction resolvase G*44*P has only a minor role in plasmid pBG55 and pUB110 transduction. In the second model, plasmid over-replication leads to the accumulation of linear concatemeric *hmw* DNA (Cohen and Clark, 1986; Viret and Alonso, 1987; Viret et al., 1991). The accumulation of linear head-to-tail multigenome-length plasmid DNA (*hmw* DNA) in the absence of RecBCD/AddAB was documented in both *Escherichia coli* and *B. subtilis* cells (Silberstein and Cohen, 1987; Viret and Alonso, 1987). Indeed, upon infection, many bacteriophages directly or indirectly inactivate end-resection catalyzed by this host encoded multi-subunit helicase-nuclease enzyme (Szczepanska, 2009). It was observed that the synthesis of pC194 or pUB110 *hmw* plasmid DNA occurred in the absence of plasmid-encoded Rep protein, and required DNA PolI, RecA and pre-primosomal proteins (e.g., DnaB) (Viret and Alonso, 1987; Leonhardt et al., 1991; Viret et al., 1991). Analysis of this *hmw* plasmid DNA by electron microscopy displayed linear DNA molecules up to 100 kb in size, which were either single-stranded, double-stranded or duplex DNA with single-stranded tailed ends (Leonhardt et al., 1991). This *hmw* DNA can be encapsidated into a viral prohead by a headful packaging mechanism (Schmidt and Schmieger, 1984; Schmieger, 1984). If this model is correct, the presence of a *pac* signal will significantly increase the encapsidation of the plasmid *hmw* DNA, and we found that there was not an increase in the transduction frequency when plasmids contained the *pac* signal. In the third model, phage infection arrests host and plasmid replication. Then SPP1-dependent replication restarts, and the linear plasmid concatemer is synthesized. This is consistent with the result that the phage G*38*P protein may act as a PriA-like enzyme, restarting DNA replication outside form a replication origin (Seco et al., 2013; Seco and Ayora, 2017). In this *de novo* synthesis of plasmid DNA, a viral *pac* site might be gained by recombination and recognized by the viral packaging machinery (Alonso et al., 1986; Bravo et al., 1990; Viret et al., 1991). In this model, the phage might form a phage-plasmid chimera and the plasmid hijacks the viral replication machinery to promote *de novo* synthesis of linear plasmid concatemeric DNA. The concatemeric plasmid DNA is then packaged into an empty prohead by the headful mechanism, indistinguishable of viral DNA, provided that the packaged substrate is larger than mature phage DNA. Our data support the third model, because we found that in infections with a phage bearing a mutation in the terminase (*sus19* infections), plasmid concatemers up to 200-kb long are produced after phage infection (**Figure 3B** and Supplementary Figure S2). This model explains also the requirement of viral replication proteins for the formation of the transducing particles. However, we were unable to detect the phage-plasmid chimeras, which might be rapidly processed to produce the plasmid head-to-tail concatemers.

Our results show that the establishment of the transduced concatemeric plasmid in the host is independent of phage encoded recombination functions, which only participate in the generation of the transducing particle. We propose that the injected linear concatemer can be converted into a circular form by the homologous recombination machinery of the recipient cells. In this respect, transduction of plasmids might have similar host requirements as the resolution of phage-plasmid chimeras analyzed in the P22 and SPP1 systems (Orbach and Jackson, 1982; Alonso et al., 1992). In the former case, the plasmid integrated into the phage genome has to be excised from the genome of the defective phage prior to establishment, whereas in the latter case the head-to-tail plasmid concatemer has to recombine intramolecularly to facilitate plasmid establishment. This process was found to be RecA-independent but dependent on host RecO and RecR functions that also catalyze single-strand annealing (Alonso et al., 1992; Manfredi et al., 2008).

AUTHOR CONTRIBUTIONS

AV-R, ML-S, AQ-O, and SA: performed the experiments; AV-R, AS, and SA: analyzed data; SA: conceived the project, integrated the results and wrote the paper.

ACKNOWLEDGMENTS

We thank J. C. Alonso (CNB-CSIC, Spain) for providing us with pUB110 and pHP13 plasmid derivatives, and for critically reading this manuscript. Plasmid pNDH33 was kindly provided by Wolfgang Schumann (University of Bayreuth, Germany).

REFERENCES

Alonso, J. C., Luder, G., and Trautner, T. A. (1986). Requirements for the formation of plasmid-transducing particles of *Bacillus subtilis* bacteriophage SPP1. *EMBO J.* 5, 3723–3728.

Alonso, J. C., Luder, G., and Trautner, T. A. (1992). Intramolecular homologous recombination in *Bacillus subtilis* 168. *Mol. Gen. Genet.* 236, 60–64.

Alonso, J. C., Stiege, C. A., Tailor, R. H., and Viret, J. F. (1988). Functional analysis of the dna (Ts) mutants of *Bacillus subtilis*: plasmid pUB110 replication as a model system. *Mol. Gen. Genet.* 214, 482–489. doi: 10.1007/BF00330484

Alonso, J. C., Tavares, P., Lurz, R., and Trautner, T. A. (2006). "Bacteriophage SPP1," in *The Bacteriophages*, ed. R. Calendar (New York, NY: Oxford University Press), 18.

Alonso, J. C., and Trautner, T. A. (1985). A gene controlling segregation of the *Bacillus subtilis* plasmid pC194. *Mol. Gen. Genet.* 198, 427–431. doi: 10.1007/BF00332934

Ander, M., Subramaniam, S., Fahmy, K., Stewart, A. F., and Schaffer, E. (2015). A single-strand snnealing protein clamps DNA to detect and secure homology. *PLOS Biol.* 13:e1002213. doi: 10.1371/journal.pbio.1002213

Ayora, S., Missich, R., Mesa, P., Lurz, R., Yang, S., Egelman, E. H., et al. (2002). Homologous-pairing activity of the *Bacillus subtilis* bacteriophage SPP1 replication protein G35P. *J. Biol. Chem.* 277, 35969–35979. doi: 10.1074/jbc.M20446720

Ayora, S., Stasiak, A., and Alonso, J. C. (1999). The *Bacillus subtilis* bacteriophage SPP1 G39P delivers and activates the G40P DNA helicase upon interacting with the G38P-bound replication origin. *J. Mol. Biol.* 288, 71–85. doi: 10.1006/jmbi.1999.2662

Bravo, A., and Alonso, J. C. (1990). The generation of concatemeric plasmid DNA in *Bacillus subtilis* as a consequence of bacteriophage SPP1 infection. *Nucleic Acids Res.* 18, 4651–4657. doi: 10.1093/nar/18.16.4651

Bravo, A., Alonso, J. C., and Trautner, T. A. (1990). Functional analysis of the *Bacillus subtilis* bacteriophage SPP1 pac site. *Nucleic Acids Res.* 18, 2881–2886. doi: 10.1093/nar/18.10.2881

Brown-Jaque, M., Calero-Caceres, W., and Muniesa, M. (2015). Transfer of antibiotic-resistance genes via phage-related mobile elements. *Plasmid* 79, 1–7. doi: 10.1016/j.plasmid.2015.01.001

Brussow, H., Canchaya, C., and Hardt, W. D. (2004). Phages and the evolution of bacterial pathogens: from genomic rearrangements to lysogenic conversion. *Microbiol. Mol. Biol. Rev.* 68, 560–602. doi: 10.1128/MMBR.68.3.560-602.2004

Burger, K. J., and Trautner, T. A. (1978). Specific labelling of replicating SPP1 DNA: analysis of viral DNA synthesis and identification of phage DNA-genes. *Mol. Gen. Genet.* 166, 277–285.

Camacho, A. G., Gual, A., Lurz, R., Tavares, P., and Alonso, J. C. (2003). *Bacillus subtilis* bacteriophage SPP1 DNA packaging motor requires terminase and portal proteins. *J. Biol. Chem.* 278, 23251–23259. doi: 10.1074/jbc.M301805200

Canchaya, C., Fournous, G., Chibani-Chennoufi, S., Dillmann, M. L., and Brussow, H. (2003). Phage as agents of lateral gene transfer. *Curr. Opin. Microbiol.* 6, 417–424. doi: 10.1016/S1369-5274(03)00086-9

Canosi, U., Luder, G., and Trautner, T. A. (1982). SPP1-mediated plasmid transduction. *J. Virol.* 44, 431–436.

Ceglowski, P., Boitsov, A., Chai, S., and Alonso, J. C. (1993a). Analysis of the stabilization system of pSM19035-derived plasmid pBT233 in *Bacillus subtilis*. *Gene* 136, 1–12.

Ceglowski, P., Boitsov, A., Karamyan, N., Chai, S., and Alonso, J. C. (1993b). Characterization of the effectors required for stable inheritance of *Streptococcus* pyogenes pSM19035-derived plasmids in *Bacillus subtilis*. *Mol. Gen. Genet.* 241, 579–585.

Ceglowski, P., Lurz, R., and Alonso, J. C. (1993c). Functional analysis of pSM19035-derived replicons in *Bacillus subtilis*. *FEMS Microbiol. Lett.* 109, 145–150.

Chai, S., Bravo, A., Luder, G., Nedlin, A., Trautner, T. A., and Alonso, J. C. (1992). Molecular analysis of the *Bacillus subtilis* bacteriophage SPP1 region encompassing genes 1 to 6. The products of gene 1 and gene 2 are required for pac cleavage. *J. Mol. Biol.* 224, 87–102. doi: 10.1016/0022-2836(92)90578-8

Chai, S., Lurz, R., and Alonso, J. C. (1995). The small subunit of the terminase enzyme of *Bacillus subtilis* bacteriophage SPP1 forms a specialized nucleoprotein complex with the packaging initiation region. *J. Mol. Biol.* 252, 386–398. doi: 10.1006/jmbi.1995.0505

Chai, S., Szepan, U., and Alonso, J. C. (1997). *Bacillus subtilis* bacteriophage SPP1 terminase has a dual activity: it is required for the packaging initiation and represses its own synthesis. *Gene* 184, 251–256. doi: 10.1016/S0378-1119(96)00609-9

Chai, S., Szepan, U., Luder, G., Trautner, T. A., and Alonso, J. C. (1993). Sequence analysis of the left end of the *Bacillus subtilis* bacteriophage SPP1 genome. *Gene* 129, 41–49. doi: 10.1016/0378-1119(93)90694-X

Cohen, A., and Clark, A. J. (1986). Synthesis of linear plasmid multimers in *Escherichia coli* K-12. *J. Bacteriol.* 167, 327–335. doi: 10.1128/jb.167.1.327-335.1986

Datta, S., Costantino, N., Zhou, X., and Court, D. L. (2008). Identification and analysis of recombineering functions from Gram-negative and Gram-positive bacteria and their phages. *Proc. Natl. Acad. Sci. U.S.A.* 105, 1626–1631. doi: 10.1073/pnas.0709089105

De Paepe, M., Hutinet, G., Son, O., Amarir-Bouhram, J., Schbath, S., and Petit, M. A. (2014). Temperate phages acquire DNA from defective prophages by relaxed homologous recombination: the role of Rad52-like recombinases. *PLOS Genet.* 10:e1004181. doi: 10.1371/journal.pgen.1004181

Deichelbohrer, I., Alonso, J. C., Luder, G., and Trautner, T. A. (1985). Plasmid transduction by *Bacillus subtilis* bacteriophage SPP1: effects of DNA homology between plasmid and bacteriophage. *J. Bacteriol.* 162, 1238–1243.

Delattre, H., Souiai, O., Fagoonee, K., Guerois, R., and Petit, M. A. (2016). Phagonaute: a web-based interface for phage synteny browsing and protein function prediction. *Virology* 496, 42–50. doi: 10.1016/j.virol.2016.05.007

Espinosa, M., del Solar, G., Rojo, F., and Alonso, J. C. (1995). Plasmid rolling circle replication and its control. *FEMS Microbiol. Lett.* 130, 111–120. doi: 10.1111/j.1574-6968.1995.tb07707.x

Ferrari, E., Canosi, U., Galizzi, A., and Mazza, G. (1978). Studies on transduction process by SPP1 phage. *J. Gen. Virol.* 41, 563–572. doi: 10.1099/0022-1317-41-3-563

Fricker, A. D., and Peters, J. E. (2014). Vulnerabilities on the lagging-strand template: opportunities for mobile elements. *Annu. Rev. Genet.* 48, 167–186. doi: 10.1146/annurev-genet-120213-092046

Giovanetti, E., Brenciani, A., Morroni, G., Tiberi, E., Pasquaroli, S., Mingoia, M., et al. (2014). Transduction of the *Streptococcus pyogenes* bacteriophage Phim46.1, carrying resistance genes mef(A) and tet(O), to other *Streptococcus* species. *Front. Microbiol.* 5:746. doi: 10.3389/fmicb.2014.00746

Gual, A., Camacho, A. G., and Alonso, J. C. (2000). Functional analysis of the terminase large subunit, G2P, of *Bacillus subtilis* bacteriophage SPP1. *J. Biol. Chem.* 275, 35311–35319. doi: 10.1074/jbc.M004309200

Haima, P., Bron, S., and Venema, G. (1987). The effect of restriction on shotgun cloning and plasmid stability in *Bacillus subtilis* Marburg. *Mol. Gen. Genet.* 209, 335–342. doi: 10.1007/BF00329663

Horinouchi, S., and Weisblum, B. (1982). Nucleotide sequence and functional map of pC194, a plasmid that specifies inducible chloramphenicol resistance. *J. Bacteriol.* 150, 815–825.

Ikeda, H., and Tomizawa, J. I. (1965). Transducing fragments in generalized transduction by phage P1. I. Molecular origin of the fragments. *J. Mol. Biol.* 14, 85–109. doi: 10.1016/S0022-2836(65)80232-7

Kenzaka, T., Tani, K., and Nasu, M. (2010). High-frequency phage-mediated gene transfer in freshwater environments determined at single-cell level. *ISME J.* 4, 648–659. doi: 10.1038/ismej.2009.145

Khan, S. A. (2005). Plasmid rolling-circle replication: highlights of two decades of research. *Plasmid* 53, 126–136. doi: 10.1016/j.plasmid.2004.12.008

Kidane, D., Carrasco, B., Manfredi, C., Rothmaier, K., Ayora, S., Tadesse, S., et al. (2009). Evidence for different pathways during horizontal gene transfer in competent *Bacillus subtilis* cells. *PLOS Genet.* 5:e1000630. doi: 10.1371/journal.pgen.1000630

Kreuzer, K. N. (2000). Recombination-dependent DNA replication in phage T4. *Trends Biochem. Sci.* 25, 165–173. doi: 10.1016/S0968-0004(00)01559-0

Kreuzer, K. N., Yap, W. Y., Menkens, A. E., and Engman, H. W. (1988). Recombination-dependent replication of plasmids during bacteriophage T4 infection. *J. Biol. Chem.* 263, 11366–11373.

Lajoie, M. J., Gregg, C. J., Mosberg, J. A., Washington, G. C., and Church, G. M. (2012). Manipulating replisome dynamics to enhance lambda Red-mediated multiplex genome engineering. *Nucleic Acids Res.* 40:e170. doi: 10.1093/nar/gks751

Lennox, E. S. (1955). Transduction of linked genetic characters of the host by bacteriophage P1. *Virology* 1, 190–206. doi: 10.1016/0042-6822(55)90016-7

Leonhardt, H. (1990). Identification of a low-copy-number mutation within the pUB110 replicon and its effect on plasmid stability in *Bacillus subtilis*. *Gene* 94, 121–124. doi: 10.1016/0378-1119(90)90477-9

Leonhardt, H., Lurz, R., and Alonso, J. C. (1991). Physical and biochemical characterization of recombination-dependent synthesis of linear plasmid multimers in *Bacillus subtilis*. *Nucleic Acids Res.* 19, 497–503. doi: 10.1093/nar/19.3.497

Lo Piano, A., Martinez-Jimenez, M. I., Zecchi, L., and Ayora, S. (2011). Recombination-dependent concatemeric viral DNA replication. *Virus Res.* 160, 1–14. doi: 10.1016/j.virusres.2011.06.009

Lopes, A., Amarir-Bouhram, J., Faure, G., Petit, M. A., and Guerois, R. (2010). Detection of novel recombinases in bacteriophage genomes unveils Rad52, Rad51 and Gp2.5 remote homologs. *Nucleic Acids Res.* 38, 3952–3962. doi: 10.1093/nar/gkq096

Manfredi, C., Carrasco, B., Ayora, S., and Alonso, J. C. (2008). *Bacillus subtilis* RecO nucleates RecA onto SsbA-coated single-stranded DNA. *J. Biol. Chem.* 283, 24837–24847. doi: 10.1074/jbc.M802002200

Mann, B. A., and Slauch, J. M. (1997). Transduction of low-copy number plasmids by bacteriophage P22. *Genetics* 146, 447–456.

Martinez-Jimenez, M. I., Alonso, J. C., and Ayora, S. (2005). *Bacillus subtilis* bacteriophage SPP1-encoded gene 34.1 product is a recombination-dependent DNA replication protein. *J. Mol. Biol.* 351, 1007–1019. doi: 10.1016/j.jmb.2005.06.064

Martinez-Jimenez, M. I., Mesa, P., and Alonso, J. C. (2002). *Bacillus subtilis* tau subunit of DNA polymerase III interacts with bacteriophage SPP1 replicative DNA helicase G40P. *Nucleic Acids Res.* 30, 5056–5064. doi: 10.1093/nar/gkf650

Maslanova, I., Doskar, J., Varga, M., Kuntova, L., Muzik, J., Maluskova, D., et al. (2013). Bacteriophages of *Staphylococcus aureus* efficiently package various bacterial genes and mobile genetic elements including SCCmec with different frequencies. *Environ. Microbiol. Rep.* 5, 66–73. doi: 10.1111/j.1758-2229.2012.00378.x

Menouni, R., Hutinet, G., Petit, M. A., and Ansaldi, M. (2015). Bacterial genome remodeling through bacteriophage recombination. *FEMS Microbiol. Lett.* 362, 1–10. doi: 10.1093/femsle/fnu022

Missich, R., Weise, F., Chai, S., Lurz, R., Pedre, X., and Alonso, J. C. (1997). The replisome organizer (G38P) of *Bacillus subtilis* bacteriophage SPP1 forms specialized nucleoprotein complexes with two discrete distant regions of the SPP1 genome. *J. Mol. Biol.* 270, 50–64. doi: 10.1006/jmbi.1997.1060

Mosberg, J. A., Lajoie, M. J., and Church, G. M. (2010). Lambda red recombineering in *Escherichia coli* occurs through a fully single-stranded intermediate. *Genetics* 186, 791–799. doi: 10.1534/genetics.110.120782

Mosig, G., Gewin, J., Luder, A., Colowick, N., and Vo, D. (2001). Two recombination-dependent DNA replication pathways of bacteriophage T4, and their roles in mutagenesis and horizontal gene transfer. *Proc. Natl. Acad. Sci. U.S.A.* 98, 8306–8311. doi: 10.1073/pnas.131007398

Neamah, M. M., Mir-Sanchis, I., Lopez-Sanz, M., Acosta, S., Baquedano, I., Haag, A. F., et al. (2017). Sak and Sak4 recombinases are required for bacteriophage replication in *Staphylococcus aureus*. *Nucleic Acids Res.* 45, 6507–6519. doi: 10.1093/nar/gkx308

Nguyen, H. D., Nguyen, Q. A., Ferreira, R. C., Ferreira, L. C., Tran, L. T., and Schumann, W. (2005). Construction of plasmid-based expression vectors for *Bacillus subtilis* exhibiting full structural stability. *Plasmid* 54, 241–248. doi: 10.1016/j.plasmid.2005.05.001

Oliveira, L., Alonso, J. C., and Tavares, P. (2005). A defined in vitro system for DNA packaging by the bacteriophage SPP1: insights into the headful packaging mechanism. *J. Mol. Biol.* 353, 529–539. doi: 10.1016/j.jmb.2005.08.063

Oliveira, L., Tavares, P., and Alonso, J. C. (2013). Headful DNA packaging: bacteriophage SPP1 as a model system. *Virus Res.* 173, 247–259. doi: 10.1016/j.virusres.2013.01.021

Orbach, M. J., and Jackson, E. N. (1982). Transfer of chimeric plasmids among *Salmonella* typhimurium strains by P22 transduction. *J. Bacteriol.* 149, 985–994.

Pedre, X., Weise, F., Chai, S., Luder, G., and Alonso, J. C. (1994). Analysis of cis and trans acting elements required for the initiation of DNA replication in the *Bacillus subtilis* bacteriophage SPP1. *J. Mol. Biol.* 236, 1324–1340. doi: 10.1016/0022-2836(94)90061-2

Penades, J. R., Chen, J., Quiles-Puchalt, N., Carpena, N., and Novick, R. P. (2015). Bacteriophage-mediated spread of bacterial virulence genes. *Curr. Opin. Microbiol.* 23, 171–178. doi: 10.1016/j.mib.2014.11.019

Phan, T. T., Nguyen, H. D., and Schumann, W. (2006). Novel plasmid-based expression vectors for intra- and extracellular production of recombinant proteins in *Bacillus subtilis*. *Protein Expr. Purif.* 46, 189–195. doi: 10.1016/j.pep.2005.07.005

Ruzin, A., Lindsay, J., and Novick, R. P. (2001). Molecular genetics of SaPI1–a mobile pathogenicity island in *Staphylococcus aureus*. *Mol. Microbiol.* 41, 365–377. doi: 10.1046/j.1365-2958.2001.02488.x

Scaltriti, E., Launay, H., Genois, M. M., Bron, P., Rivetti, C., Grolli, S., et al. (2011). Lactococcal phage p2 ORF35-Sak3 is an ATPase involved in DNA recombination and AbiK mechanism. *Mol. Microbiol.* 80, 102–116. doi: 10.1111/j.1365-2958.2011.07561.x

Schmidt, C., and Schmieger, H. (1984). Selective transduction of recombinant plasmids with cloned pac sites by *Salmonella* phage P22. *Mol. Gen. Genet.* 196, 123–128. doi: 10.1007/BF00334103

Schmieger, H. (1984). pac sites are indispensable for in vivo packaging of DNA by phage P22. *Mol. Gen. Genet.* 195, 252–255. doi: 10.1007/BF00332755

Schumann, W. (2007). Production of recombinant proteins in *Bacillus subtilis*. *Adv. Appl. Microbiol.* 62, 137–189. doi: 10.1016/S0065-2164(07)62006-1

Seco, E. M., and Ayora, S. (2017). *Bacillus subtilis* DNA polymerases, PolC and DnaE, are required for both leading and lagging strand synthesis in SPP1 origin-dependent DNA replication. *Nucleic Acids Res.* 45, 8302–8313. doi: 10.1093/nar/gkx493

Seco, E. M., Zinder, J. C., Manhart, C. M., Lo Piano, A., McHenry, C. S., and Ayora, S. (2013). Bacteriophage SPP1 DNA replication strategies promote viral and disable host replication in vitro. *Nucleic Acids Res.* 41, 1711–1721. doi: 10.1093/nar/gks1290

Silberstein, Z., and Cohen, A. (1987). Synthesis of linear multimers of OriC and pBR322 derivatives in *Escherichia coli* K-12: role of recombination and replication functions. *J. Bacteriol.* 169, 3131–3137. doi: 10.1128/jb.169.7.3131-3137.1987

Sun, Z., Deng, A., Hu, T., Wu, J., Sun, Q., Bai, H., et al. (2015). A high-efficiency recombineering system with PCR-based ssDNA in *Bacillus subtilis* mediated by the native phage recombinase GP35. *Appl. Microbiol. Biotechnol.* 99, 5151–5162. doi: 10.1007/s00253-015-6485-5

Szczepanska, A. K. (2009). Bacteriophage-encoded functions engaged in initiation of homologous recombination events. *Crit. Rev. Microbiol.* 35, 197–220. doi: 10.1080/10408410902983129

te Riele, H., Michel, B., and Ehrlich, S. D. (1986). Single-stranded plasmid DNA in *Bacillus subtilis* and *Staphylococcus aureus*. *Proc. Natl. Acad. Sci. U.S.A.* 83, 2541–2545. doi: 10.1073/pnas.83.8.2541

Titok, M. A., Chapuis, J., Selezneva, Y. V., Lagodich, A. V., Prokulevich, V. A., Ehrlich, S. D., et al. (2003). *Bacillus subtilis* soil isolates: plasmid replicon analysis and construction of a new theta-replicating vector. *Plasmid* 49, 53–62. doi: 10.1016/S0147-619X(02)00109-9

Touchon, M., Moura de Sousa, J. A., and Rocha, E. P. (2017). Embracing the enemy: the diversification of microbial gene repertoires by phage-mediated horizontal gene transfer. *Curr. Opin. Microbiol.* 38, 66–73. doi: 10.1016/j.mib.2017.04.010

Tzipilevich, E., Habusha, M., and Ben-Yehuda, S. (2017). Acquisition of phage sensitivity by bacteria through exchange of phage receptors. *Cell* 168, 186e–199e. doi: 10.1016/j.cell.2016.12.003

van Kessel, J. C., and Hatfull, G. F. (2008). Efficient point mutagenesis in mycobacteria using single-stranded DNA recombineering: characterization of antimycobacterial drug targets. *Mol. Microbiol.* 67, 1094–1107. doi: 10.1111/j.1365-2958.2008.06109.x

Viret, J. F., and Alonso, J. C. (1987). Generation of linear multigenome-length plasmid molecules in *Bacillus subtilis*. *Nucleic Acids Res.* 15, 6349–6367. doi: 10.1093/nar/15.16.6349

Viret, J. F., and Alonso, J. C. (1988). A DNA sequence outside the pUB110 minimal replicon is required for normal replication in *Bacillus subtilis*. *Nucleic Acids Res.* 16, 4389–4406. doi: 10.1093/nar/16.10.4389

Viret, J. F., Bravo, A., and Alonso, J. C. (1991). Recombination-dependent concatemeric plasmid replication. *Microbiol. Rev.* 55, 675–683.

Wang, J. D., Rokop, M. E., Barker, M. M., Hanson, N. R., and Grossman, A. D. (2004). Multicopy plasmids affect replisome positioning in *Bacillus subtilis*. *J. Bacteriol.* 186, 7084–7090. doi: 10.1128/JB.186.21.7084-7090.2004

Winstel, V., Kuhner, P., Krismer, B., Peschel, A., and Rohde, H. (2015). Transfer of plasmid DNA to clinical coagulase-negative staphylococcal pathogens by using a unique bacteriophage. *Appl. Environ. Microbiol.* 81, 2481–2488. doi: 10.1128/AEM.04190-14

Yasbin, R. E., and Young, F. E. (1974). Transduction in *Bacillus subtilis* by bacteriophage SPP1. *J. Virol.* 14, 1343–1348.

Zecchi, L., Lo Piano, A., Suzuki, Y., Canas, C., Takeyasu, K., and Ayora, S. (2012). Characterization of the Holliday junction resolving enzyme encoded by the *Bacillus subtilis* bacteriophage SPP1. *PLOS ONE* 7:e48440. doi: 10.1371/journal.pone.0048440

Zinder, N. D., and Lederberg, J. (1952). Genetic exchange in *Salmonella*. *J. Bacteriol.* 64, 679–699.

9

The Intriguing Evolutionary Journey of Enteroinvasive *E. coli* (EIEC) toward Pathogenicity

Martina Pasqua[1], Valeria Michelacci[2], Maria Letizia Di Martino[1†], Rosangela Tozzoli[2], Milena Grossi[1], Bianca Colonna[1], Stefano Morabito[2] and Gianni Prosseda[1]*

[1] *Istituto Pasteur Italia, Department of Biology and Biotechnology "C. Darwin", Sapienza Università di Roma, Rome, Italy,* [2] *European Union Reference Laboratory for Escherichia coli, Department of Veterinary Public Health and Food Safety, Istituto Superiore di Sanità, Rome, Italy*

Edited by:
Tatiana Venkova,
Fox Chase Cancer Center,
United States

Reviewed by:
David A. Rasko,
University of Maryland, Baltimore,
United States
Alessandra Polissi,
Università degli Studi di Milano, Italy
Antonio Juárez,
University of Barcelona, Spain

***Correspondence:**
Gianni Prosseda
gianni.prosseda@uniroma1.it

Among the intestinal pathogenic *Escherichia coli*, enteroinvasive *E. coli* (EIEC) are a group of intracellular pathogens able to enter epithelial cells of colon, multiplicate within them, and move between adjacent cells with a mechanism similar to *Shigella*, the ethiological agent of bacillary dysentery. Despite EIEC belong to the same pathotype of *Shigella*, they neither have the full set of traits that define *Shigella* nor have undergone the extensive gene decay observed in *Shigella*. Molecular analysis confirms that EIEC are widely distributed among *E. coli* phylogenetic groups and correspond to bioserotypes found in many *E. coli* serogroups. Like *Shigella*, also in EIEC the critical event toward a pathogenic life-style consisted in the acquisition by horizontal gene transfer of a large F-type plasmid (pINV) containing the genes required for invasion, intracellular survival, and spreading through the intestinal mucosa. In *Shigella*, the ample gain in virulence determinants has been counteracted by a substantial loss of functions that, although important for the survival in the environment, are redundant or deleterious for the life inside the host. The pathoadaptation process that has led *Shigella* to modify its metabolic profile and increase its pathogenic potential is still in infancy in EIEC, although maintenance of some features typical of *E. coli* might favor their emerging relevance as intestinal pathogens worldwide, as documented by recent outbreaks in industrialized countries. In this review, we will discuss the evolution of EIEC toward *Shigella*-like invasive forms going through the epidemiology, including the emergence of new virulent strains, their genome organization, and the complex interactions they establish with the host.

Keywords: pathogenic *E. coli*, enteroinvasive *E. coli* (EIEC), *Shigella*, bacterial evolution, emerging EIEC

Abbreviations: DAMP, damage-associated molecular pattern; DEC, diarrheagenic *E. coli*; HGT, horizontal gene transfer; H-NS, heat-stable nucleoid-structuring protein; IL, interleukin; IS, insertion sequence; MAPK, mitogen-activated protein kinase; NLR, Nod-like receptor; PAI, pathogenicity island; PAMP, pathogen-associated molecular pattern; pINV, virulence plasmid; PMNL, polymorphonuclear leukocytes; PRR, pattern recognition receptor; SHI, *Shigella* pathogenicity island; SRL, *Shigella* resistance locus; sRNA, small RNA; T2SS, type II secretion system; T3SS, type III secretion system; TLR, Toll-like receptor; TNF, tumor necrosis factor.

INTRODUCTION

Escherichia coli is not only a harmless commensal of the human and animal intestine but also a major cause of morbidity and mortality (Kaper et al., 2004; Wirth et al., 2006). Indeed, many pathogenic *E. coli* have been described as cause of diseases both in healthy and immunocompromised individuals. Based on the specific virulence factors and pathogenicity processes, pathogenic *E. coli* have been subdivided into different pathogroups, that can be broadly grouped as DEC (or intestinal) or extraintestinal *E. coli* (ExPEC) (Kaper et al., 2004; Croxen and Finlay, 2009; Gomes et al., 2016). DEC include at least six major pathotypes differing in virulence mechanisms, infectious processes, and damages provoked to the target cells: enteropathogenic *E. coli* (EPEC), Shiga toxin-producing *E. coli* (STEC), enterotoxigenic *E. coli* (ETEC), enteroinvasive *E. coli* (EIEC), enteroaggregative *E. coli* (EAEC), diffusely adherent *E. coli* (DAEC), as well as adherent invasive *E. coli* (AIEC), a recently identified pathotype. As for ExPEC, the most common strains belong to two different pathotypes targeting different body compartments: uropathogenic *E. coli* (UTI) and neonatal meningitis *E. coli* (NMEC).

The presence of so many different pathotypes exemplifies the remarkable plasticity of *E. coli* genome, which is characterized by an extremely large pangenome of approximately 20,000 genes in contrast to a common core of about 1700 genes (Rasko et al., 2008; Touchon et al., 2009). Those that vary among different pathogenic *E. coli* strains have been acquired by intense HGT and are often conveyed by mobile genetic elements (Touchon et al., 2009; Dobrindt et al., 2010; van Elsas et al., 2011).

Among the DEC pathotypes, EIEC are etiological agents of bacillary dysentery in humans, particularly in low-income countries (Croxen et al., 2013; Gomes et al., 2016). The pathogenesis of EIEC infection is characterized by the ability of bacteria to invade the human colonic mucosa, conferred by the expression of chromosomal and plasmid-borne genes (Harris et al., 1982; Sansonetti et al., 1982; Hale et al., 1983; Kaper et al., 2004). Following penetration into colonic epithelial cells, EIEC replicate intracellularly and spread to adjacent cells causing the inflammatory destruction of the intestinal epithelial barrier. This provokes the characteristic dysentery syndrome, usually self-limiting, characterized by the presence of blood, mucus, and leukocytes in stools (DuPont et al., 1971; O'Brien et al., 1979; Taylor et al., 1988). The clinical illness caused by EIEC is similar to that induced by *Shigella* spp. (Formal and Hornick, 1978; Small and Falkow, 1988), with whom they are closely related in their virulence and other phenotypic properties (Kopecko et al., 1985; Lan et al., 2004). Notwithstanding the similarities in the invasion mechanisms, the infectious dose of EIEC has been observed to be much higher than that of *Shigella* and the diseases caused by EIEC appear in some cases to be milder (DuPont et al., 1971).

Despite several studies, whether EIEC are precursors of the "full blown" pathogen *Shigella*, or not is still under debate. In this review, we will attempt at tracing the evolutionary pathway of EIEC considering their epidemiology, the complex mechanisms of their interaction with host cells, the key steps that could have characterized their evolution from a commensal life style toward pathogenicity, and the organization of their genome, including the description of the major traits of emerging EIEC clones.

EPIDEMIOLOGY OF ENTEROINVASIVE *E. coli* (EIEC)

The first report of an EIEC strain dates back at 1947 (Ewing and Gravatti, 1947). At that time, it was defined as "paracolon bacillus" but the strain was later identified as an O124 *E. coli*. In the 1950s and 1960s, other *E. coli* strains, isolated from dysentery and initially classified as *Shigella manolovi*, *S. sofia*, *Shigella* strain 13, and *S. metadysenteriae*, due to their ability to cause experimental keratoconjunctivitis in guinea pigs, were later renamed as EIEC (Manolov, 1959; Rowe et al., 1977; Edwards and Ewing, 1986). Their biochemical characters were first described in 1967 (Sakazaki et al., 1967; Trabulsi et al., 1967).

Enteroinvasive *E. coli* and *Shigella* spp. share several phenotypic and genotypic characteristics, often making the discrimination between the two genera challenging (Silva et al., 1980; Toledo and Trabulsi, 1983; Bando et al., 1998; Lan and Reeves, 2002; Pavlovic et al., 2011; van den Beld and Reubsaet, 2012), especially in case of shared serogroups. This difficulty biases the interpretation of the epidemiological information available, hindering the evaluation of the real burden of EIEC infections. As a matter of fact, both EIEC and *Shigella* spend much of their life cycle within the eukaryotic cells, possessing the ability to use nutrients coming from the host environment. Similarly to *Shigella*, most EIEC strains are unable to decarboxylate lysine, lack the ability to ferment lactose, and are generally non-motile, with the exception of strains belonging to a few serogroups (Silva et al., 1980; Farmer et al., 1985; Bando et al., 1998; Casalino et al., 2003; Tozzoli and Scheutz, 2014).

A limited set of serotypes have been assigned to EIEC, namely O28ac:H-, O29:H-, O112ac:H-, O115:H-, O121:H-, O124:H-, O124:H7, O124:H30, O124:H32, O135:H-, O136:H-, O143:H-, O144:H-, O144:H25,O152:H-, O159:H-, O159:H2, O164:H-, O167:H-, O167:H4, O167:H5, O173:H-, and recently O96:H19 (Voeroes et al., 1964; Silva et al., 1980; Gomes et al., 1987, 2016; Orskov et al., 1991; Matsushita et al., 1993; Escher et al., 2014; Tozzoli and Scheutz, 2014; Michelacci et al., 2016; Newitt et al., 2016). Some of these EIEC-associated O antigens, such as O28, O112ac, O121, O124, O143, O144, O152, and O167, are identical to O antigens present in *Shigella* spp. (Cheasty and Rowe, 1983; Tozzoli and Scheutz, 2014).

Enteroinvasive *E. coli*-infected humans seem to be the major source of infection, as no animal reservoirs have been identified, and transmission uses mainly the oral–fecal route. Although EIEC infections occur worldwide, these are particularly common in low-income countries where poor general hygiene favors their spreading (Chatterjee and Sanyal, 1984; Beutin et al., 1997; Kaper et al., 2004; Vieira et al., 2007).

Enteroinvasive *E. coli* incidence has been estimated in several countries, and it differs depending on the region (Gomes et al., 2016). Discrepancies among some of the reports can be observed, probably due to the difficulty in discriminating between *Shigella* and EIEC. In certain countries of Latin America and Asia, namely Chile, Thailand, India, and Brazil, EIEC were found to be common diarrheagenic pathogens (Chatterjee and Sanyal, 1984; Faundez et al., 1988; Echeverria et al., 1992; Blake et al., 1993; Levine et al., 1993), with frequent reports of asymptomatic infected subjects excreting the pathogen (Beutin et al., 1997). In industrialized countries, EIEC infections have been mainly described as travel-related, being reported in returning travelers from high-incidence countries (Wanger et al., 1988; Beutin et al., 1997; Svenungsson et al., 2000). Occasionally, food and water sources have been identified as vehicles of infection, but usually as a secondary contamination by a human source (Tozzoli and Scheutz, 2014).

Enteroinvasive *E. coli* cause sporadic cases of infection but have been implicated in outbreaks as well, sometimes involving large numbers of cases. In the 1970s a huge outbreak, affecting 387 patients and linked to cheese contaminated with an O124 *E. coli* strain, occurred in United States (Marier et al., 1973). Recently, an increase of cases of infections linked to an emerging EIEC clone has been observed in Europe, where in 2012 a large and severe outbreak of bloody diarrhea in Italy involving more than 100 individuals was reported (Escher et al., 2014; Pettengill et al., 2015). An EIEC O96:H19 strain, a serotype never described before for EIEC, was isolated and the suspected source of infection was traced to cooked vegetables (Escher et al., 2014). During the outbreak investigation an EIEC O96:H19 strain was also isolated from two asymptomatic food handlers working in the canteen linked with the outbreak, supporting the hypothesis of a secondary contamination of the vegetables during post-cooking handling procedures (Escher et al., 2014). In 2014, two linked outbreaks of gastrointestinal disease occurred in the United Kingdom, involving more than 100 cases of infection. One of the episodes was associated to the consumption of contaminated salad vegetables and, again, an O96:H19 EIEC was isolated from some of the patients and from vegetable samples (Newitt et al., 2016). Finally, an EIEC belonging to the same serotype was isolated in a case of traveler's diarrhea in Spain in 2013 (Michelacci et al., 2016). Pheno-genotypic characterization of the strains involved in the three episodes suggests that the EIEC O96:H19 could be emerged as a result of the recent acquisition of the invasion plasmid by an *E. coli* strain (Michelacci et al., 2016).

THE INVASIVE PROCESS

Similarly to *Shigella*, EIEC are responsible of bacillary dysentery (Taylor et al., 1988). However, the disease caused by EIEC is usually less severe than that induced by *Shigella* (DuPont et al., 1971). Following the discovery that EIEC strains carry a pINV plasmid identical to that of *Shigella* (Harris et al., 1982; Sansonetti et al., 1982; Hale et al., 1983) and that they can display a *Shigella*-like invasive behavior (Hale et al., 1985; Small and Falkow, 1988; Taylor et al., 1988), *in vitro* and *in vivo* studies have been extensively focused on *Shigella*, providing in-depth knowledge about its pathogenicity/virulence mechanisms. In recent years, the pathogenicity of EIEC has gained new interest and comparative analyses between EIEC and *Shigella* have been performed, aimed at understanding the different clinical outcome severity of the two infections (Moreno et al., 2009; Bando et al., 2010; Sanchez-Villamil et al., 2016). Here we first present the invasive process as it has been inferred from studies on *S. flexneri*. Then, we address what it is known about the difference between these two enteroinvasive bacteria.

In order to gain access to intestinal epithelia, bacteria first transit from the lumen to the submucosa by preferentially entering M cells in Peyer's patches. After endocytosis by M cells bacteria are transcytosed toward the M cell pocket, where they meet, and are phagocytosed by resident macrophages. *Shigella* infection of macrophages is accompanied by the release of T3SS effectors and components that are recognized as PAMPs by NLRs, ultimately leading to pyroptosis with the release of proinflammatory cytokines, IL-1β and IL-18 (Ashida et al., 2011). The induction of macrophage cell death is pivotal for bacteria to invade enterocytes, though pyroptosis is a form of cell death that induces a massive inflammatory response. Once released from dying macrophages, invasive bacteria infect the neighboring enterocytes by entering through the basolateral surface. Here they are enclosed into a vacuole that is rapidly disrupted freeing them into the cytosol. Subsequently, the bacteria multiply and, using actin-based motility, spread to adjacent cells (Schroeder and Hilbi, 2008).

Inside epithelial cells, bacterial PAMPs and DAMPs are detected by various PRRs, including TLRs and NLRs, which stimulate host defense signal pathways such as those involving MAPKs and NF-κB leading to the secretion of proinflammatory cytokines (e.g., IL-8 and TNF-α) (Takeuchi and Akira, 2010). These molecules induce the recruitment of phagocytic cells to the infection site, initially facilitating the invasion process and eventually clearing the bacterial pathogens. In order to maximize invasion and permanence and save the replicative niche in epithelial cells, invading *Shigella* modulate host cell responses throughout the infection process by secreted effectors (Killackey et al., 2016). Induction of a very early inflammatory response upon invasion of epithelial cells is functional to bacterial spreading as it results in recruitment of polymorphonuclear leucocytes (PMNL), which migrate across the epithelium destabilizing the intercellular junctions and increasing the surface available for bacterial entry into target cells (Ashida et al., 2011). Several T3SS effectors, such as OspB, OspC1, and OspZ (Zurawski et al., 2009; Ambrosi et al., 2015; Mattock and Blocker, 2017), contribute to promote inflammation at early stages of the infection process. They mainly act by enhancing activation of MAPK and NF-κB pathways, which are involved in the control of the production of PMNL chemoattractants, including IL-8, whose secretion triggers PMNL migration in a basolateral to apical direction causing epithelial barrier disruption. However, though this early inflammatory response is essential to initiate infection, it would also contribute toward rapidly clearing the infecting agents. Thus, to establish infection, at later stages *Shigella* must overcome the host innate response. This is achieved by delivering T3SS effectors, whose function is aimed mainly at inhibiting MAPK and NF-κB signaling pathways with the consequent decrease of inflammatory chemokine and cytokine production (Killackey et al., 2016; Mattock and Blocker, 2017).

An important obstacle *Shigella* must tackle during the invasion of the epithelial tissue is host cell targeting and degradation by autophagy. Several studies have demonstrated that *Shigella* are particularly exposed to autophagy targeting only when they are associated to cell membranes. Two bacterial factors, IcsB and VirA, have been implicated in bacterial evasion of autophagy targeting by interfering with LC3 recruitment and by allowing bacteria to escape from LC3-positive vacuoles (Ogawa et al., 2005; Baxt and Goldberg, 2014; Campbell-Valois et al., 2015).

Typically, intracellular pathogens need to save their host to establish a successful infection. As part of their pathogenic mechanism *Shigella* employ several countermeasures to avoid premature cell death to maintain their epithelial replicative niche. The early stage of infection is characterized by induction of DNA damage and genotoxic stress, which lead to activation of p53 and stimulation of apoptosis. Apoptotic cell death is prevented by the activity of the T3SS effectors VirA and IpgD, which promote p53 degradation and activate the PI3K/Akt pro-survival pathway, respectively, and by the pilus component protein FimA, which inhibits cytochrome c release by mitochondria (Mattock and Blocker, 2017).

As discussed above, EIEC share many aspects of the *Shigella* infection process that involves crossing of intestinal epithelial barrier, killing of resident macrophage cells, invasion of enterocytes, intra-cellular replication, and dissemination from cell to cell without extracellular steps (Croxen and Finlay, 2009). Moreover, EIEC express the same virulence factors found in *Shigella* (Parsot, 2005). However, the infectious dose required for EIEC to cause disease is higher than that of *Shigella* and the disease caused by EIEC appears to be milder (DuPont et al., 1971), suggesting differences between EIEC and *Shigella* in sensing and shaping the host environment, which, in turn, would influence the pathways toward virulence. To date only few studies have investigated the differences in the infectiveness between EIEC and *Shigella*. Moreno et al. (2009) detailed for the first time the relationship between the expression of some genes crucial for the infection process and the reduced ability of EIEC to cause disease. This is well supported by their Serény tests in guinea pigs, showing how the signs of keratoconjunctivitis induced by *Shigella* appear earlier and are more severe as compared to those caused by EIEC. Using an epithelial cell model, the authors also demonstrate that, although *Shigella* and EIEC display similar invasion ability, EIEC disseminates less efficiently, producing smaller plaques in plaque assays. As compared to *Shigella* the overall behavior of EIEC apparently reflects a reduced expression of key virulence genes, during both invasion and cell-to-cell spreading, except for *virF* that is expressed at higher levels by intracellular EIEC than *Shigella* during the dissemination step. This apparent discrepancy may be explained in the light of recent results showing that *Shigella virF* is transcribed into two mRNAs, with the shortest one encoding a smaller protein that negatively regulates transcription of full-length mRNA and, consequently, the expression of the VirF regulator (Di Martino et al., 2016b). Since in the real-time PCR experiments carried out by Moreno et al. (2009) *virF* expression was assayed by using primers that did not discriminate between the two mRNAs, comparative *virF* expression studies between *Shigella* and EIEC deserve further investigations to deeper analyze potential differences.

A more recent work has compared the host cell response to infection by different *E. coli* pathotypes, including EIEC, and by *Shigella*. The kinetic of NF-κB and ERK1/2 activation in HT-29 epithelial cells shows only a slightly higher p65 phosphorylation after 4 h of infection with *Shigella* as compared with EIEC. Conversely, although following a similar kinetics, the accumulation of phosphorylated ERK1/2 is much higher in cells infected with EIEC at 4 h post-infection. Despite these differences, HT-29 cells infected with EIEC or *Shigella* release comparable amounts of cytokines, as IL-8 and TNF-α with similar kinetics (Sanchez-Villamil et al., 2016). The phosphorylation of ERK1/2 and p38 is controlled by the phosphothreonine lyase activity of OspF, to which both anti-inflammatory (Arbibe et al., 2007) and pro-inflammatory roles (Reiterer et al., 2011) have been attributed. Since both *Shigella* and EIEC express OspF, it is reasonable that additional factors are involved in determining the different ERK1/2 phosphorylation profile and the outcome of MAPK activation.

The key step in invasion of the epithelial cells resides in the ability of EIEC and *Shigella* to escape from macrophages after phagocytosis by induction of caspase 1-dependent cell death. It has been reported that, as compared to *Shigella*, EIEC have a decreased capacity to escape from murine J774 macrophages and are less efficient in cell killing during the first 4 h of infection (Bando et al., 2010). This likely depends on differences in the expression of some virulence genes. In particular, as compared to *Shigella* the expression of the *ipaC* gene is reduced in intracellular EIEC at all the time points after infection. As for the release of pro- and anti-inflammatory cytokines (as TNF-α, IL-1, and IL-10) by infected cells, contrasting results exist. While no significant differences between EIEC and *Shigella*-infected J774 cells (Bando et al., 2010) have been reported, other studies carried out using human THP-1 cells differentiated into macrophages (Sanchez-Villamil et al., 2016) have shown that *Shigella* infection results in higher secretion of both pro-inflammatory and anti-inflammatory cytokines.

To date, banking on the modest amount of data available from *in vitro* infection of macrophage-like cells and epithelial cells, the milder disease caused by EIEC appears to be mainly associated to a lower expression of key virulence genes involved in phagosomal escape inside host cells and in dissemination among epithelial cells (Moreno et al., 2009; Bando et al., 2010). There are no obvious differences in the inflammatory response by epithelial cells, at least as far as the secretion of IL-8 and TNF-α is concerned, neither at early nor at late times of infection. Despite this cytokine profile, the activation state of ERK1/2 MAPK seems to be more elevated in epithelial cells infected with EIEC than in those infected with *Shigella* (Sanchez-Villamil et al., 2016). Deeper investigations will clarify to what extent this may depend on differences in manipulating certain cell signaling pathways and on differences in the activity of bacterial factors involved therein.

THE MAJOR VIRULENCE TRAIT OF EIEC: THE LARGE VIRULENCE PLASMID pINV

The evolution of *E. coli* toward pathogenic phenotypes has been determined, as in many other bacterial pathogens, mainly by two mechanisms: the acquisition of virulence genes by HGT as parts of plasmids, phages, transposons, or PAI and the loss or modification of genes of the core genome. While the first mechanism plays a crucial role in the colonization of a new host environment, the latter, known as pathoadaptation, strongly contributes to drive the evolution of bacteria toward a more pathogenic phenotype (Kaper et al., 2004; Dobrindt et al., 2010).

It is widely acknowledged that, as in *Shigella*, in EIEC the critical event in the transition toward a pathogenic lifestyle has been the acquisition of a large F-type plasmid (pINV) which encodes the molecular machinery required for invasion, survival, and diffusion of the bacterium within the host (Harris et al., 1982; Sansonetti et al., 1982; Hale et al., 1983; Small and Falkow, 1988; **Figure 1**). The pINV plasmid has been found only in the *Shigella*/EIEC pathotype and its loss is a very rare event, which determines an avirulent phenotype.

The genetic organization of the pINV is very complex (Johnson and Nolan, 2009). As a matter of fact these plasmids are made up of a mosaic of genes of various origins and harbor traces of four different plasmids (Buchrieser et al., 2000; Venkatesan et al., 2001; Escobar-Páramo et al., 2003). pINV isolated from EIEC share wide regions of high structural and functional homology and are interchangeable with those isolated from *Shigella* strains (Hale et al., 1983; Lan et al., 2001; Johnson and Nolan, 2009). pINV share with IncFIIA plasmids high homology in the regions involved in replication (*rep*) and conjugation (*tra*) (Makino et al., 1988) and stable inheritance of pINV is ensured by the presence of several plasmid segregation and maintenance systems (Lan et al., 2001). Due to large deletions in the *tra* region, pINV are not capable of self-transfer by conjugation, but they can be mobilized by other conjugative plasmids. All over the plasmid genome, an astonishing number of ISs is present as a mixture of complete and incomplete IS elements repeated several times, confirming the relevant role played by ISs in pINV assembly and evolution (Buchrieser et al., 2000; Venkatesan et al., 2001). Most ISs are related to known elements while others represent novel ISs. Among the latter, ISEc11, an IS belonging to the IS1111 family, is widespread and functional in pINV from EIEC while only defective copies are present in the *Shigella* pINV plasmids (Prosseda et al., 2006).

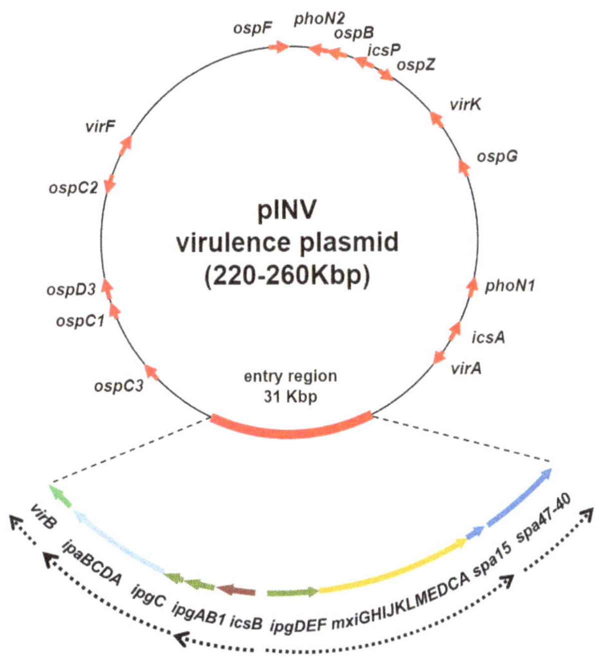

FIGURE 1 | Genetic map of the pINV of *Shigella* and EIEC strains. The red arrows indicate major virulence determinants. Due to the variability in position and number, the *ipaH* genes are not shown. The genetic organization of the entry region is shown in detail, with dashed arrow lines indicating known transcriptional units. The entry region organization is based on the sequence of plasmid pWR100 (Venkatesan et al., 2001) while the entire plasmid is freely drawn to provide the layout of a typical pINV plasmid (the figure is not to scale).

In the pINV there is only one large (31 kb) region, which does not host any IS elements. This is the so-called entry region, which displays a PAI-like structure (Buchrieser et al., 2000; Venkatesan et al., 2001). It is composed by two large, divergently transcribed gene clusters coding for a T3SS apparatus (Mxi and Spa), for most of its effectors (IpaB, IpaC, and IpaD) with their chaperons (IpgA, IpgC, IpgE, and Spa15), and for two transcriptional regulators (VirB and MxiE), both required for the activation of most virulence genes (Schroeder and Hilbi, 2008; **Figure 1**). The entry region is extremely conserved among *Shigella* and EIEC pINV plasmids (Lan et al., 2001). Albeit it had been initially proposed as a PAI, likely acquired in a single recombination event, it lacks the presence of flanking tRNA sequences and at least remnants of a recombinase-encoding gene. It is therefore unclear if the acquisition of the entry region has occurred independently from its insertion into tRNA sequences or if the absence of tRNA genes may have resulted from rearrangement following gene transfer. The latter hypothesis is supported by the fact that the entry region is flanked by truncated IS elements, suggesting that rearrangements may have occurred after its acquisition *en bloc* by the plasmid (Buchrieser et al., 2000). The T3SS encoded by the entry region plays a critical role in the bacterial invasive process, since it delivers a large number of effectors involved in the reorganization of the host cell actin cytoskeleton and in the modulation of cell signaling pathways to evade the host immune response (Mattock and Blocker, 2017). With the exception of few proteins of the IpaH family, which are chromosomally encoded, all T3SS effectors are encoded by pINV genes located within or outside the entry region. Since the entry region is highly conserved, the phylogenetic analysis of three of its genes (*ipgD*, *mxiC*, and *mxiA*) has allowed differentiating pINV from *Shigella* spp. and EIEC into two forms, A and B, with the first one predominantly associated with EIEC strains (Lan et al., 2001, 2004).

Besides the large PAI-like region, a small islet carries the genes coding for IcsA (a protein responsible for the bacterial motility inside the cytoplasm), VirA (a GTPase-activating protein), and RnaG (a regulatory sRNA negatively controlling *icsA* expression) (Giangrossi et al., 2010; Tran et al., 2011; Dong et al., 2012). Other genes encoding proteins crucial for the invasive process cover the pINV plasmid including the OspG and OspF proteins which interfere with the host innate immune response (Kim et al., 2005; Arbibe et al., 2007), the PhoN2 protein required for IcsA localization (Scribano et al., 2014), and the IpaH proteins which interfere with the host protein degradation (Ashida and Sasakawa, 2015; **Figure 1**). Moreover, in contrast to the other two virulence regulatory genes (*virB* and *mxiE*), the *virF* gene, coding for the primary virulence regulator, is located on a "desert island" surrounded by several IS sequences and far away from all other virulence genes, including those under its direct control, *virB* and *icsA* (Di Martino et al., 2016a). While the CG content of *virF* is only slightly lower as compared to that of the entry region (Buchrieser et al., 2000), its position suggests that it has been acquired independently to promote the expression of the virulence genes. VirF is also involved in the activation of some chromosomal genes, indicating that it acts as global regulator and that its acquisition by HGT has contributed to a reshaping of the core genome, easing the adaptation of bacteria to the host environment (Barbagallo et al., 2011; Leuzzi et al., 2015).

The mechanisms involved in the activation of the pINV virulence genes have been extensively studied both in EIEC and in *Shigella* (Dagberg and Uhlin, 1992; Prosseda et al., 2002). They rely on a sophisticated regulatory cascade involving global and specific regulators, encoded by both, pINV and the chromosome. Outside the human host, the nucleoid-associated protein H-NS represses each of the key promoters of the pINV virulence genes (Dorman, 2004). In response to environmental conditions found in the human intestine, the transcriptional activation of the invasive operons is triggered by an increased level of VirF counteracting H-NS repression at the *icsA* and *virB* promoters (Prosseda et al., 2004). Then VirB activates most operons within the entry region, including the gene for the last regulator (*mxiE*), as well as all other virulence genes scattered along the pINV genome, except *icsA*. Finally, MxiE, assisted by IpgC, activates the transcription of genes encoding the late effectors (Schroeder and Hilbi, 2008).

As in other pathogenic *E. coli*, also in EIEC the virulence genes are stably maintained on an extrachromosomal element (Johnson and Nolan, 2009). Nevertheless, it has been reported that the pINV of EIEC strain HN280 is able to integrate into the host chromosome and that integration results in silencing of all pINV-encoded virulence genes also under host temperature conditions (Zagaglia et al., 1991). Silencing was shown to depend on a severe reduction of *virB* transcription, likely dependent on the inability of VirF to counteract the negative control of H-NS at the *virB* promoter when it is chromosomally located (Colonna et al., 1995). This has led to the hypothesis that the presence of virulence genes on the pINV is the result of an evolutionary pathway toward the optimization of gene expression.

EVOLUTION OF EIEC

The studies on the evolutionary origin of the *Shigella*/EIEC pathovar have led to two major hypotheses. The pINV could have been acquired only once by an ancestral *E. coli* that subsequently gave rise to the different *Shigella*/EIEC lineages (Escobar-Páramo et al., 2003; Zuo et al., 2013), as suggested by the inability of the plasmid to autonomously undergo horizontal transmission. Alternatively, the different *Shigella*/EIEC strains could have arisen from different *E. coli* that had acquired the pINV independently, e.g., from an unknown donor or from other *Shigella*/EIEC that already harbored it. This view is supported by the diversity of the genotypes within the *Shigella*/EIEC pathovar, revealed by phylogenetic analyses of chromosomal genes and by genome comparison (Pupo et al., 2000; Hazen et al., 2016; Pettengill et al., 2016). Besides the large pINV, several virulence genes have been acquired on the chromosome of *Shigella* and EIEC as part of PAIs (**Figure 2**). The PAIs described so far for *Shigella* (SHI islands) carry genes encoding different traits, including an enterotoxin and a cytotoxic protease (SHI-1) and systems involved in iron uptake and evasion of immune response (SHI-2 and SHI-3 in *S. boydii*), in the modification of O antigens (SHI-O) or in multi-drug resistance (SRL) (Schroeder and Hilbi, 2008). Recently, 20 genomes from EIEC belonging to different serotypes have been compared with those of reference strains belonging to diverse *E. coli* pathovars and *Shigella* species. This comparison highlights the existence of at least three distinct lineages containing only EIEC strains and suggests a convergent evolution of non-pathogenic *E. coli* toward invasive phenotype (Hazen et al., 2016). An *in silico* search for protein-encoding genes of SHI-1, SHI-2, SHI-3, SHI-O, and SRL indicates that, with

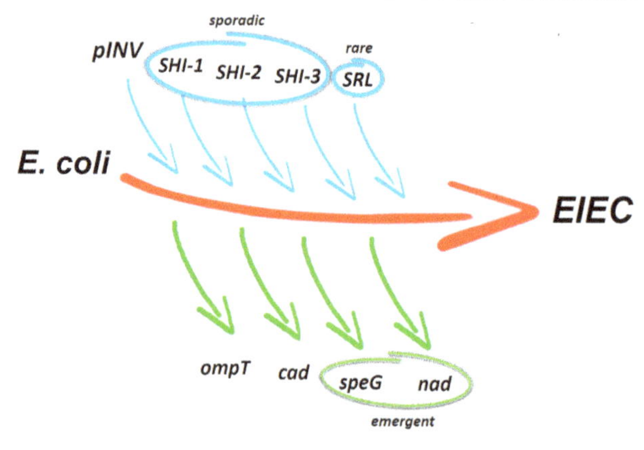

FIGURE 2 | Genetic events contributing to the evolution of EIEC from ancestral commensal *E. coli*. The acquisition of the pINV by HGT is a major evolutionary event toward pathogenicity. This can be accompanied by the sporadic acquisition of entire or incomplete SHI-1 PAI and incomplete SHI-2 and SHI-3 PAIs. Rarely, also incomplete SRL PAI are acquired by EIEC genomes. The absence of *ompT* and the loss of cadaverine synthesis (usually resulting from *cadC* silencing) counterbalance the gain of virulence-associated determinants. The inactivation of *speG* (involved in spermidine acetylation) and *nad* (involved in NAD biosynthesis) is regarded as emergent pathoadaptive mutations in EIEC.

the exception of SHI-O, portions of the other PAIs are present in EIEC genomes in a lineage-specific manner (Hazen et al., 2016). Interestingly, while a whole SHI-1 Island has never been detected in EIEC, SHI-1 fragments of different length have been found in all EIEC genomes. However, the ShET1 toxin genes, typically harbored by SHI-1 in *S. flexneri* genomes, were found only in EIEC strains of lineage 2. In the case of virulence genes associated with SHI-2, the *shi*A gene, involved in the reduction of the host inflammatory response, is absent in all EIEC lineages, while *shiD*, which provides immunity to colicins, is present in all EIEC of lineages 1 and 2. An entire SHI-3 PAI, typically associated with *S. boydii* strains, has been detected only in few EIEC strains of lineages 1 and 2, while portions of it, including the genes encoding for aerobactin-mediated iron uptake, are found in all three lineages. As for the large SRL PAI, widely distributed among *Shigella* spp. and containing a cluster of multiple antibiotic resistance determinants (Turner et al., 2003), only a few of its genes are present in EIEC genomes.

The variable presence of the PAIs in EIEC confirms the phylogenetic diversity among EIEC and *Shigella* and further supports the hypothesis that the EIEC pathovar has not a single origin but rather stems from multiple lineages (Hazen et al., 2016; Michelacci et al., 2016; Pettengill et al., 2016).

A significant complementary step toward the pathogenic lifestyle has been pathoadaptation, the inactivation, or loss of several chromosomal genes, which negatively interfere with the expression of virulence factors required for survival within the host. The antivirulence loci identified encode a broad spectrum of functions, confirming that adaptation to the new host environments is the result of long and ordered process targeting core genome determinants (Casalino et al., 2003; Di Martino et al., 2013b; Campilongo et al., 2014).

Despite the close similarity of the *Shigella* and EIEC pathogenicity process, it is well known that EIEC have a metabolic activity more similar to *E. coli* and have not undergone the intense gene decay observed in *Shigella* (Silva et al., 1980; Pettengill et al., 2016). It is therefore not surprising that the pathoadaptation in EIEC has not reached a level comparable to *Shigella* (Prosseda et al., 2012) and that most of the antivirulence loci characterized in *Shigella* are still encoding functional products in EIEC. One of the pathoadaptive mutations conserved both in EIEC and in *Shigella* is the deletion of the *ompT* gene, located within the defective lambdoid prophage DLP12 (Nakata et al., 1993; **Figure 2**). The OmpT protease triggers the degradation of IcsA protein and therefore negatively interferes with host cell invasion by drastically reducing the ability of *Shigella* to spread into adjacent epithelial cells. Considering that the loss of OmpT is widespread, it is as yet unclear if *E. coli* lineages that gave rise to the *Shigella*/EIEC pathovar have not hosted DLP12 *ab initio* or if the prophage has been excised during the pathoadaption process (Bliven and Maurelli, 2012; Leuzzi et al., 2017).

Another typical pathoadaptive mutation of *Shigella* spp. is the inability to catabolise lysine, due to the silencing of lysine decarboxylase (LDC) activity (Prosseda et al., 2007). The LDC$^-$ phenotype, which is found also in most EIEC, is determined by mutations in the *cad* locus, which hamper the synthesis of cadaverine. Cadaverine is a polyamine that interferes with pathogenicity by blocking the release of *Shigella* into the cytoplasm of the infected cells and inhibiting the migration of PMNL across the intestinal epithelium (Bliven and Maurelli, 2012). A detailed analysis of the molecular rearrangements occurred in the *cad* operon of several EIEC strains belonging to different serotypes (Casalino et al., 2003) has shown that, similarly to *Shigella*, the silencing of the *cad* locus has been accomplished through convergent evolution. In contrast to *Shigella*, in EIEC the *cad* region is colinear with the *E. coli* K12 chromosome and the lack of cadaverine synthesis is mainly due to the inactivation of the gene encoding the CadC transcriptional regulator (Casalino et al., 2010). By comparing the *cad* loci of EIEC and *Shigella*, it appears that the rearrangements occurred in EIEC are less severe compared to the complete erosion of the locus observed in several *Shigella* strains (Casalino et al., 2005; Prosseda et al., 2007; **Figure 2**). Indeed, despite the antivirulence role played by cadaverine (Fernandez et al., 2001), emerging O96:H19 EIEC strains still maintains an integer *cad* operon and exhibits a LDC$^+$ phenotype (Michelacci et al., 2016).

As compared to the commensal *E. coli* the polyamine profile of *Shigella* is affected not only by the lack of cadaverine but also by the marked accumulation of spermidine and by the loss of *N*-acetyl spermidine, the inert form of spermidine (Di Martino et al., 2013a). The increased spermidine content depends on the loss of the spermidine acetyltransferase (SAT), the enzyme encoded by the *speG* gene and responsible for the conversion of spermidine into *N*-acetylspermidine. In *Shigella* it has been demonstrated that a higher level of spermidine increases survival within macrophages and confers bacteria a higher resistance to oxidative stress (Barbagallo et al., 2011). Similarly to how observed for the *cad* locus, also *speG* silencing is the result of convergent evolution. A comparison of the polyamine profiles of several EIEC strains with those of *Shigella* and *E. coli* K12 has revealed that in EIEC major polyamines attain levels in-between those observed in *E. coli* and *Shigella*. Indeed, as compared to commensal *E. coli*, in EIEC intracellular putrescine is significantly increased and spermidine tends to be higher. Nevertheless, in contrast to *Shigella*, *N*-acetylspermidine is still present in most EIEC strains (Campilongo et al., 2014), indicating that the loss of *speG* is an emerging trait. However, when spermidine accumulation is induced in EIEC through deletion of the *speG* gene, survival within macrophages, as well as resistance to oxidative stress are increased (Campilongo et al., 2014). This confirms that the absence of SAT activity confers to intracellular bacteria like EIEC and *Shigella* an increased capability to defy antagonistic host environment. Moreover, the analysis of the polyamine profiles has revealed that the higher level of putrescine in EIEC is determined by increased transcription of *speC*, promoted by the lack of cadaverine. The *speC* gene encodes the enzyme converting L-ornithine into putrescine. On the basis of these observations it has been suggested (Campilongo et al., 2014) that during the transition toward the pathogenic phenotype, the modification of the polyamine profile might have been triggered by the loss of cadaverine, with the double effect of favoring the invasive process and increasing the putrescine level. Since putrescine is an important intermediate in the synthesis of spermidine and, consequently, of *N*-acetylspermidine, its increase may in turn have caused higher levels of both polyamines. In this scenario the silencing of *speG*, which appears completed in *Shigella* but can be regarded as an ongoing process in EIEC, would represent the last step favoring further accumulation of spermidine and the disappearance of *N*-acetylspermidine.

Another noteworthy pathoadaptive mutation in *Shigella* is the requirement for exogenous nicotinic acid (NAD) due to inactivation of the *nad* genes (Prunier et al., 2007), required for *de novo* synthesis of NAD. Also in this case the inability to synthesize NAD is not a generalized feature among EIEC strains (Di Martino et al., 2013b). In those EIEC strains requiring NAD it has been shown that the preferential target in the pathoadaptation process is the *nadB* gene, inactivated through diverse strategies, involving point mutations or IS insertions.

Altogether, the picture emerging from the observations on pathoadaptive mutations suggests that EIEC might represent intermediates in the evolution toward a full-blown phenotype, with some mutational events still confined to *Shigella* (**Figure 2**). However, a recent whole-genome comparative analysis (Pettengill et al., 2016), performed on a large number of *Shigella* and EIEC genomes, indicates that *Shigella* and EIEC evolved independently. Nevertheless, the same authors proposed that, while EIEC as a group cannot be considered the ancestor to *Shigella*, some EIEC lineages might have been the *Shigella* ancestor.

EMERGING ENTEROINVASIVE *Escherichia coli*

The recent outbreaks occurred in Europe caused by the EIEC O96:H19 led the scientific community to reconsider the role of EIEC infection in industrialized countries (Escher et al., 2014; Michelacci et al., 2016; Newitt et al., 2016). Such EIEC serotype had never been reported before 2012 and represents a new virulent emergent clone. The EIEC O96:H19 isolated from two outbreaks occurred in Italy and United Kingdom and from a sporadic case of disease reported in Spain were studied by whole genome sequencing (Pettengill et al., 2015; Michelacci et al., 2016). The genomic analysis confirmed that all the isolates belonged not only to the same unprecedented EIEC serotype, but also to the same sequence type (ST-99), never observed before in EIEC strains (Michelacci et al., 2016). The analysis of the distribution of virulence genes typical of EIEC and *Shigella* highlighted the presence in the three strains of the plasmid genes encoding the T3SS system and its effectors, as well as the master transcriptional regulators genes *virF* and *virB*. As for the chromosomally located virulence genes, the three isolates showed the presence of the genetic determinants of a T2SS and were all negative for those encoding the aerobactin system involved in iron uptake. Interestingly, none of the O96:H19 isolates was found to have undergone the process of pathoadaptation through accumulation of the mutations described in the literature for EIEC and *Shigella* (Bliven and Maurelli, 2012; Prosseda et al., 2012). Nevertheless, the three isolates were shown to display minor differences. The plasmid profiles obtained through the genomic analysis highlighted the presence of five plasmids in the strains isolated in Spain and United Kingdom and three plasmids in that responsible of the Italian outbreak, with three plasmids in common in the three strains. Altogether, these observations strengthen the hypothesis of the emergence of a new virulent EIEC clone circulating in Europe.

Phenotypic analysis also highlighted peculiar properties of this EIEC clone, when compared to reference EIEC and *Shigella* strains. Biochemical characterization showed that the isolates displayed the LDC activity, confirming the lack of the related pathoadaptive mutations observed through genome analysis, and interestingly showed that the isolates retained the ability to ferment lactose (Michelacci et al., 2016), usually lacking in *Shigella* and in the majority of EIEC strains (Tozzoli and Scheutz, 2014). Generally, a better fitness was observed for the O96:H19 strains when comparing the growth curves with those of *Shigella* and reference EIEC strains (Michelacci et al., 2016). Moreover, swimming motility was observed for the strains from Italian and Spanish cases, which was instead completely absent in the strain from United Kingdom and in all the other EIEC and *Shigella* strains tested. Such phenotypic traits are not typical of intracellular pathogens such as EIEC and *Shigella*, while they are more common in *E. coli* strains, contributing to their great ability in surviving and adapting in different ecological niches.

These findings support the hypothesis of the evolution of EIEC and *Shigella* after the acquisition of the pINV by multiple lineages of commensal *E. coli*, followed by a multi-step adaptation process. Such an evolutionary pathway could be exemplified by EIEC ST-280 Clonal Complex, which could have been generated with the acquisition of the pINV plasmid by a commensal *E. coli* eventually evolving toward *Shigella* belonging to related clonal complexes (ST-149, 152, 243, 245, 250) (Wirth et al., 2006; Michelacci et al., 2016). The mechanism could have involved multiple events of pathoadaptive mutations, giving origin to the existing *Shigella* clones, specialized for intracellular survival with detriment of the ability to persist outside the host. A similar paradigm could also explain the emergence of other EIEC clones following the acquisition of pINV by other commensal *E. coli*. This event in some cases could be followed by the accumulation of pathoadaptive mutation, as it is the case of the EIEC strains belonging to ST-6 clonal complex, while some other clones could have maintained all the functions granting an efficient extracellular persistence, such as the EIEC O96:H19 belonging to ST-99 (Michelacci et al., 2016). The observed better fitness of EIEC O96:H19 in comparison with that of the other reference EIEC and *Shigella* strains could have favored its survival in the extracellular environment and allowed its overgrowth in the food vehicles, granting it a high potential as a foodborne pathogen, as demonstrated in the two large episodes occurred in Italy and United Kingdom (Michelacci et al., 2016; Newitt et al., 2016).

CONCLUSION AND PERSPECTIVES

Genomics approaches in combination with phenotypic analyses have a strong potential toward the formulation of new intriguing hypotheses on the ongoing evolution of EIEC. Currently available comparisons between EIEC and *Shigella* genomes support the need for a taxonomical revision moving the *Shigella* genus back within the *E. coli* species (Michelacci et al., 2016; Pettengill et al., 2016). As a matter of fact, *Shigella* clades are interspersed in clusters of *E. coli* genomes regardless of the bioinformatics approach used for the phylogenetic analysis (Sahl et al., 2015; Pettengill et al., 2016). In the light of recent studies, the organization of the EIEC genome appears to have been originated from multiple independent events (Hazen et al., 2016; Pettengill et al., 2016). This hypothesis finds even stronger evidence in the emergence of a novel EIEC clone belonging to O96:H19 serotype, which exhibits phenotypic traits more typical of *E. coli* than of reference EIEC or *Shigella* (Michelacci et al., 2016).

The acquisition of the plasmid may represent the first step in the emergence of new EIEC clones, but it is well known to be not sufficient for establishing the full pathogenicity (Sansonetti et al., 1983). In this context, it is of great interest to deeper investigate on the role and relevance of functions that *Shigella* has lost in its route toward an intracellular life-style but that are still retained by most EIEC strains.

AUTHOR CONTRIBUTIONS

BC, VM, SM, and GP proposed the idea of the review; MP, VM, MG, and RT wrote the review draft; MP, MM, and GP design the figures; BC, MM, GP, MG, and SM wrote the final version of the review. The final text has been read and approved by all the authors of the review.

REFERENCES

Ambrosi, C., Pompili, M., Scribano, D., Limongi, D., Petrucca, A., Cannavacciuolo, S., et al. (2015). The *Shigella flexneri* OspB effector: an early immunomodulator. *Int. J. Med. Microbiol.* 305, 75–84. doi: 10.1016/j.ijmm.2014.11.004

Arbibe, L., Kim, D. W., Batsche, E., Pedron, T., Mateescu, B., Muchardt, C., et al. (2007). An injected bacterial effector targets chromatin access for transcription factor NF-κB to alter transcription of host genes involved in immune responses. *Nat. Immunol.* 8, 47–56. doi: 10.1038/ni1423

Ashida, H., Ogawa, M., Kim, M., Suzuki, S., Sanada, T., Punginelli, C., et al. (2011). *Shigella* deploy multiple countermeasures against host innate immune responses. *Curr. Opin. Microbiol.* 14, 16–23. doi: 10.1016/j.mib.2010.08.014

Ashida, H., and Sasakawa, C. (2015). *Shigella* IpaH family effectors as a versatile model for studying pathogenic bacteria. *Front. Cell. Infect. Microbiol.* 5:100. doi: 10.3389/fcimb.2015.00100

Bando, S. Y., do Valle, G. R., Martinez, M. B., Trabulsi, L. R., and Moreira-Filho, C. A. (1998). Characterization of enteroinvasive *Escherichia coli* and *Shigella* strains by RAPD analysis. *FEMS Microbiol. Lett.* 165, 159–165. doi: 10.1111/j.1574-6968.1998.tb13141.x

Bando, S. Y., Moreno, A. C. R., Albuquerque, J. A. T., Amhaz, J. M. K., Moreira-Filho, C. A., and Martinez, M. B. (2010). Expression of bacterial virulence factors and cytokines during in vitro macrophage infection by enteroinvasive *Escherichia coli* and *Shigella flexneri*: a comparative study. *Mem. Inst. Oswaldo Cruz* 105, 786–791. doi: 10.1590/S0074-02762010000600009

Barbagallo, M., Di Martino, M. L., Marcocci, L., Pietrangeli, P., De Carolis, E., Casalino, M., et al. (2011). A new piece of the *Shigella* pathogenicity puzzle: spermidine accumulation by silencing of the speG gene [corrected]. *PLOS ONE* 6:e27226. doi: 10.1371/journal.pone.0027226

Baxt, L. A., and Goldberg, M. B. (2014). Host and bacterial proteins that repress recruitment of LC3 to *Shigella* early during infection. *PLOS ONE* 9:e94653. doi: 10.1371/journal.pone.0094653

Beutin, L., Gleier, K., Kontny, I., Echeverria, P., and Scheutz, F. (1997). Origin and characteristics of enteroinvasive strains of *Escherichia coli* (EIEC) isolated in Germany. *Epidemiol. Infect.* 118, 199–205. doi: 10.1017/S0950268897007413

Blake, P. A., Ramos, S., MacDonald, K. L., Rassi, V., Gomes, T. A., Ivey, C., et al. (1993). Pathogen-specific risk factors and protective factors for acute diarrheal disease in urban Brazilian infants. *J. Infect. Dis.* 167, 627–632. doi: 10.1093/infdis/167.3.627

Bliven, K. A., and Maurelli, A. T. (2012). Antivirulence genes: insights into pathogen evolution through gene loss. *Infect. Immun.* 80, 4061–4070. doi: 10.1128/IAI.00740-12

Buchrieser, C., Glaser, P., Rusniok, C., Nedjari, H., D'Hauteville, H., Kunst, F., et al. (2000). The virulence plasmid pWR100 and the repertoire of proteins secreted by the type III secretion apparatus of *Shigella flexneri*. *Mol. Microbiol.* 38, 760–771. doi: 10.1046/j.1365-2958.2000.02179.x

Campbell-Valois, F.-X., Sachse, M., Sansonetti, P. J., and Parsot, C. (2015). Escape of actively secreting *Shigella flexneri* from ATG8/LC3-positive vacuoles formed during cell-to-cell spread is facilitated by IcsB and VirA. *mBio* 6:e02567-14. doi: 10.1128/mBio.02567-14

Campilongo, R., Di Martino, M. L., Marcocci, L., Pietrangeli, P., Leuzzi, A., Grossi, M., et al. (2014). Molecular and functional profiling of the polyamine content in enteroinvasive *E. coli*: looking into the gap between commensal *E. coli* and harmful *Shigella*. *PLOS ONE* 9:e106589. doi: 10.1371/journal.pone.0106589

Casalino, M., Latella, M. C., Prosseda, G., Ceccarini, P., Grimont, F., and Colonna, B. (2005). Molecular evolution of the lysine decarboxylase-defective phenotype in *Shigella sonnei*. *Int. J. Med. Microbiol.* 294, 503–512. doi: 10.1016/j.ijmm.2004.11.001

Casalino, M., Latella, M. C., Prosseda, G., and Colonna, B. (2003). CadC is the preferential target of a convergent evolution driving enteroinvasive *Escherichia coli* toward a lysine decarboxylase-defective phenotype. *Infect. Immun.* 71, 5472–5479. doi: 10.1128/IAI.71.10.5472-5479.2003

Casalino, M., Prosseda, G., Barbagallo, M., Iacobino, A., Ceccarini, P., Carmela Latella, M., et al. (2010). Interference of the CadC regulator in the arginine-dependent acid resistance system of *Shigella* and enteroinvasive *E. coli*. *Int. J. Med. Microbiol.* 300, 289–295. doi: 10.1016/j.ijmm.2009.10.008

Chatterjee, B. D., and Sanyal, S. N. (1984). Is it all shigellosis? *Lancet* 2, 574. doi: 10.1016/S0140-6736(84)90783-9

Cheasty, T., and Rowe, B. (1983). Antigenic relationships between the enteroinvasive *Escherichia coli* O antigens O28ac, O112ac, O124, O136, O143, O144, O152, and O164 and *Shigella* O antigens. *J. Clin. Microbiol.* 17, 681–684.

Colonna, B., Casalino, M., Fradiani, P. A., Zagaglia, C., Naitza, S., Leoni, L., et al. (1995). H-NS regulation of virulence gene expression in enteroinvasive *Escherichia coli* harboring the virulence plasmid integrated into the host chromosome. *J. Bacteriol.* 177, 4703–4712. doi: 10.1128/jb.177.16.4703-4712.1995

Croxen, M. A., and Finlay, B. B. (2009). Molecular mechanisms of *Escherichia coli* pathogenicity. *Nat. Rev. Microbiol.* 8, 26–38. doi: 10.1038/nrmicro2265

Croxen, M. A., Law, R. J., Scholz, R., Keeney, K. M., Wlodarska, M., and Finlay, B. B. (2013). Recent advances in understanding enteric pathogenic *Escherichia coli*. *Clin. Microbiol. Rev.* 26, 822–880. doi: 10.1128/CMR.00022-13

Dagberg, B., and Uhlin, B. E. (1992). Regulation of virulence-associated plasmid genes in enteroinvasive *Escherichia coli*. *J. Bacteriol.* 174, 7606–7612. doi: 10.1128/jb.174.23.7606-7612.1992

Di Martino, M. L., Campilongo, R., Casalino, M., Micheli, G., Colonna, B., and Prosseda, G. (2013a). Polyamines: emerging players in bacteria-host interactions. *Int. J. Med. Microbiol.* 303, 484–491. doi: 10.1016/j.ijmm.2013.06.008

Di Martino, M. L., Falconi, M., Micheli, G., Colonna, B., and Prosseda, G. (2016a). The multifaceted activity of the VirF regulatory protein in the *Shigella* lifestyle. *Front. Mol. Biosci.* 3:61. doi: 10.3389/fmolb.2016.00061

Di Martino, M. L., Fioravanti, R., Barbabella, G., Prosseda, G., Colonna, B., and Casalino, M. (2013b). Molecular evolution of the nicotinic acid requirement within the *Shigella*/EIEC pathotype. *Int. J. Med. Microbiol.* 303, 651–661. doi: 10.1016/j.ijmm.2013.09.007

Di Martino, M. L., Romilly, C., Wagner, E. G. H., Colonna, B., and Prosseda, G. (2016b). One gene and two proteins: a leaderless mRNA supports the translation of a shorter form of the *Shigella* VirF regulator. *mBio* 7:e01860-16. doi: 10.1128/mBio.01860-16

Dobrindt, U., Chowdary, M. G., Krumbholz, G., and Hacker, J. (2010). Genome dynamics and its impact on evolution of *Escherichia coli*. *Med. Microbiol. Immunol.* 199, 145–154. doi: 10.1007/s00430-010-0161-2

Dong, N., Zhu, Y., Lu, Q., Hu, L., Zheng, Y., and Shao, F. (2012). Structurally distinct bacterial TBC-like GAPs link Arf GTPase to Rab1 inactivation to counteract host defenses. *Cell* 150, 1029–1041. doi: 10.1016/j.cell.2012.06.050

Dorman, C. J. (2004). H-NS: a universal regulator for a dynamic genome. *Nat. Rev. Microbiol.* 2, 391–400. doi: 10.1038/nrmicro883

DuPont, H. L., Formal, S. B., Hornick, R. B., Snyder, M. J., Libonati, J. P., Sheahan, D. G., et al. (1971). Pathogenesis of *Escherichia coli* diarrhea. *N. Engl. J. Med.* 285, 1–9. doi: 10.1056/NEJM197107012850101

Echeverria, P., Sethabutr, O., Serichantalergs, O., Lexomboon, U., and Tamura, K. (1992). *Shigella* and enteroinvasive *Escherichia coli* infections in households of children with dysentery in Bangkok. *J. Infect. Dis.* 165, 144–147. doi: 10.1093/infdis/165.1.144

Edwards, P. R., and Ewing, W. H. (1986). *Edwards and Ewing's Identification of Enterobacteriaceae*, 4th Edn. New York, NY: Elsvier Science.

Escher, M., Scavia, G., Morabito, S., Tozzoli, R., Maugliani, A., Cantoni, S., et al. (2014). A severe foodborne outbreak of diarrhoea linked to a canteen in Italy caused by enteroinvasive *Escherichia coli*, an uncommon agent. *Epidemiol. Infect.* 142, 2559–2566. doi: 10.1017/S0950268814000181

Escobar-Páramo, P., Giudicelli, C., Parsot, C., and Denamur, E. (2003). The evolutionary history of *Shigella* and enteroinvasive *Escherichia coli* revised. *J. Mol. Evol.* 57, 140–148. doi: 10.1007/s00239-003-2460-3

Ewing, W. H., and Gravatti, J. L. (1947). *Shigella* types encountered in the mediterranean area. *J. Bacteriol.* 53, 191–195.

Farmer, J. J., Davis, B. R., Hickman-Brenner, F. W., McWhorter, A., Huntley-Carter, G. P., Asbury, M. A., et al. (1985). Biochemical identification of new species and biogroups of Enterobacteriaceae isolated from clinical specimens. *J. Clin. Microbiol.* 21, 46–76.

Faundez, G., Figueroa, G., Troncoso, M., and Cabello, F. C. (1988). Characterization of enteroinvasive *Escherichia coli* strains isolated from children with diarrhea in Chile. *J. Clin. Microbiol.* 26, 928–932.

Fernandez, I. M., Silva, M., Schuch, R., Walker, W. A., Siber, A. M., Maurelli, A. T., et al. (2001). Cadaverine prevents the escape of *Shigella flexneri* from the phagolysosome: a connection between bacterial dissemination and neutrophil transepithelial signaling. *J. Infect. Dis.* 184, 743–753. doi: 10.1086/323035

Formal, S. B., and Hornick, R. B. (1978). Invasive *Escherichia coli*. *J. Infect. Dis.* 137, 641–644. doi: 10.1093/infdis/137.5.641

Giangrossi, M., Prosseda, G., Tran, C. N., Brandi, A., Colonna, B., and Falconi, M. (2010). A novel antisense RNA regulates at transcriptional level the virulence gene icsA of *Shigella flexneri*. *Nucleic Acids Res.* 38, 3362–3375. doi: 10.1093/nar/gkq025

Gomes, T. A., Toledo, M. R., Trabulsi, L. R., Wood, P. K., and Morris, J. G. (1987). DNA probes for identification of enteroinvasive *Escherichia coli*. *J. Clin. Microbiol.* 25, 2025–2027.

Gomes, T. A. T., Elias, W. P., Scaletsky, I. C. A., Guth, B. E. C., Rodrigues, J. F., Piazza, R. M. F., et al. (2016). Diarrheagenic *Escherichia coli*. *Braz. J. Microbiol.* 47, 3–30. doi: 10.1016/j.bjm.2016.10.015

Hale, T. L., Oaks, E. V., and Formal, S. B. (1985). Identification and antigenic characterization of virulence-associated, plasmid-coded proteins of *Shigella* spp. and enteroinvasive *Escherichia coli*. *Infect. Immun.* 50, 620–629.

Hale, T. L., Sansonetti, P. J., Schad, P. A., Austin, S., and Formal, S. B. (1983). Characterization of virulence plasmids and plasmid-associated outer membrane proteins in *Shigella flexneri*, *Shigella sonnei*, and *Escherichia coli*. *Infect. Immun.* 40, 340–350.

Harris, J. R., Wachsmuth, I. K., Davis, B. R., and Cohen, M. L. (1982). High-molecular-weight plasmid correlates with *Escherichia coli* enteroinvasiveness. *Infect. Immun.* 37, 1295–1298.

Hazen, T. H., Leonard, S. R., Lampel, K. A., Lacher, D. W., Maurelli, A. T., and Rasko, D. A. (2016). Investigating the relatedness of enteroinvasive *Escherichia coli* to other *E. coli and Shigella* isolates by using comparative genomics. *Infect. Immun.* 84, 2362–2371. doi: 10.1128/IAI.00350-16

Johnson, T. J., and Nolan, L. K. (2009). Pathogenomics of the virulence plasmids of *Escherichia coli*. *Microbiol. Mol. Biol. Rev.* 73, 750–774. doi: 10.1128/MMBR.00015-09

Kaper, J. B., Nataro, J. P., and Mobley, H. L. T. (2004). Pathogenic *Escherichia coli*. *Nat. Rev. Microbiol.* 2, 123–140. doi: 10.1038/nrmicro818

Killackey, S. A., Sorbara, M. T., and Girardin, S. E. (2016). Cellular aspects of *Shigella* pathogenesis: focus on the manipulation of host cell processes. *Front. Cell. Infect. Microbiol.* 6:38. doi: 10.3389/fcimb.2016.00038

Kim, D. W., Lenzen, G., Page, A.-L., Legrain, P., Sansonetti, P. J., and Parsot, C. (2005). The *Shigella flexneri* effector OspG interferes with innate immune responses by targeting ubiquitin-conjugating enzymes. *Proc. Natl. Acad. Sci. U.S.A.* 102, 14046–14051. doi: 10.1073/pnas.0504466102

Kopecko, D. J., Baron, L. S., and Buysse, J. (1985). Genetic determinants of virulence in *Shigella* and dysenteric strains of *Escherichia coli*: their involvement in the pathogenesis of dysentery. *Curr. Top. Microbiol. Immunol.* 118, 71–95. doi: 10.1007/978-3-642-70586-1_5

Lan, R., Alles, M. C., Donohoe, K., Martinez, M. B., and Reeves, P. R. (2004). Molecular evolutionary relationships of enteroinvasive *Escherichia coli* and *Shigella* spp. *Infect. Immun.* 72, 5080–5088. doi: 10.1128/IAI.72.9.5080-5088.2004

Lan, R., Lumb, B., Ryan, D., and Reeves, P. R. (2001). Molecular evolution of large virulence plasmid in *Shigella* clones and enteroinvasive *Escherichia coli*. *Infect. Immun.* 69, 6303–6309. doi: 10.1128/IAI.69.10.6303-6309.2001

Lan, R., and Reeves, P. R. (2002). Escherichia coli in disguise: molecular origins of *Shigella*. *Microbes Infect.* 4, 1125–1132. doi: 10.1016/S1286-4579(02)01637-4

Leuzzi, A., Di Martino, M. L., Campilongo, R., Falconi, M., Barbagallo, M., Marcocci, L., et al. (2015). Multifactor regulation of the MdtJI polyamine transporter in *Shigella*. *PLOS ONE* 10:e0136744. doi: 10.1371/journal.pone.0136744

Leuzzi, A., Grossi, M., Di Martino, M. L., Pasqua, M., Micheli, G., Colonna, B., et al. (2017). Role of the SRRz/Rz1 lambdoid lysis cassette in the pathoadaptive evolution of *Shigella*. *Int. J. Med. Microbiol.* 307, 268–275. doi: 10.1016/j.ijmm.2017.03.002

Levine, M. M., Ferreccio, C., Prado, V., Cayazzo, M., Abrego, P., Martinez, J., et al. (1993). Epidemiologic studies of *Escherichia coli* diarrheal infections in a low socioeconomic level peri-urban community in Santiago. *Chile. Am. J. Epidemiol.* 138, 849–869. doi: 10.1093/oxfordjournals.aje.a116788

Makino, S., Sasakawa, C., and Yoshikawa, M. (1988). Genetic relatedness of the basic replicon of the virulence plasmid in shigellae and enteroinvasive *Escherichia coli*. *Microb. Pathog.* 5, 267–274. doi: 10.1016/0882-4010(88)90099-X

Manolov, D. G. (1959). A new type of the genus *Shigella* –"*Shigella* 13". *J. Hyg. Epidemiol. Microbiol. Immunol.* 3, 184–190.

Marier, R., Wells, J., Swanson, R., Callahan, W., and Mehlman, I. (1973). An outbreak of enteropathogenic *Escherichia coli* foodborne disease traced to imported French cheese. *Lancet* 302, 1376–1378. doi: 10.1016/S0140-6736(73)93335-7

Matsushita, S., Yamada, S., Kai, A., and Kudoh, Y. (1993). Invasive strains of *Escherichia coli* belonging to serotype O121:NM. *J. Clin. Microbiol.* 31, 3034–3035.

Mattock, E., and Blocker, A. J. (2017). How do the virulence factors of *Shigella* work together to cause disease? *Front. Cell. Infect. Microbiol.* 7:64. doi: 10.3389/fcimb.2017.00064

Michelacci, V., Prosseda, G., Maugliani, A., Tozzoli, R., Sanchez, S., Herrera-León, S., et al. (2016). Characterization of an emergent clone of enteroinvasive *Escherichia coli* circulating in Europe. *Clin. Microbiol. Infect.* 22, 287.e11–287.e19. doi: 10.1016/j.cmi.2015.10.025

Moreno, A. C. R., Ferreira, L. G., and Martinez, M. B. (2009). Enteroinvasive *Escherichia coli* vs. *Shigella flexneri*: how different patterns of gene expression affect virulence. *FEMS Microbiol. Lett.* 301, 156–163. doi: 10.1111/j.1574-6968.2009.01815.x

Nakata, N., Tobe, T., Fukuda, I., Suzuki, T., Komatsu, K., Yoshikawa, M., et al. (1993). The absence of a surface protease, OmpT, determines the intercellular spreading ability of *Shigella*: the relationship between the ompT and kcpA loci. *Mol. Microbiol.* 9, 459–468. doi: 10.1111/j.1365-2958.1993.tb01707.x

Newitt, S., MacGregor, V., Robbins, V., Bayliss, L., Chattaway, M. A., Dallman, T., et al. (2016). Two linked enteroinvasive *Escherichia coli* outbreaks, Nottingham, UK, June 2014. *Emerg. Infect. Dis.* 22, 1178–1184. doi: 10.3201/eid2207.152080

O'Brien, A. D., Gentry, M. K., Thompson, M. R., Doctor, B. P., Gemski, P., and Formal, S. B. (1979). Shigellosis and *Escherichia coli* diarrhea: relative importance of invasive and toxigenic mechanisms. *Am. J. Clin. Nutr.* 32, 229–233.

Ogawa, M., Yoshimori, T., Suzuki, T., Sagara, H., Mizushima, N., and Sasakawa, C. (2005). Escape of intracellular *Shigella* from autophagy. *Science* 307, 727–731. doi: 10.1126/science.1106036

Orskov, I., Wachsmuth, I. K., Taylor, D. N., Echeverria, P., Rowe, B., Sakazaki, R., et al. (1991). Two new *Escherichia coli* O groups: O172 from "Shiga-like" toxin II-producing strains (EHEC) and O173 from enteroinvasive *E. coli* (EIEC). *APMIS* 99, 30–32. doi: 10.1111/j.1699-0463.1991.tb05114.x

Parsot, C. (2005). *Shigella* spp. and enteroinvasive *Escherichia coli* pathogenicity factors. *FEMS Microbiol. Lett.* 252, 11–18. doi: 10.1016/j.femsle.2005.08.046

Pavlovic, M., Luze, A., Konrad, R., Berger, A., Sing, A., Busch, U., et al. (2011). Development of a duplex real-time PCR for differentiation between *E. coli* and *Shigella* spp. *J. Appl. Microbiol.* 110, 1245–1251. doi: 10.1111/j.1365-2672.2011.04973.x

Pettengill, E. A., Pettengill, J. B., and Binet, R. (2015). Phylogenetic analyses of *Shigella* and enteroinvasive *Escherichia coli* for the identification of molecular epidemiological markers: whole-genome comparative analysis does not support distinct genera designation. *Front. Microbiol.* 6:1573. doi: 10.3389/fmicb.2015.01573

Pettengill, E. A., Pettengill, J. B., and Binet, R. (2016). Phylogenetic analyses of *Shigella* and enteroinvasive *Escherichia coli* for the identification of molecular epidemiological markers: whole-genome comparative analysis does not support distinct genera designation. *Front. Microbiol.* 6:1573. doi: 10.3389/fmicb.2015.01573

Prosseda, G., Carmela Latella, M., Barbagallo, M., Nicoletti, M., Al Kassas, R., Casalino, M., et al. (2007). The two-faced role of cad genes in the virulence of pathogenic *Escherichia coli*. *Res. Microbiol.* 158, 487–493. doi: 10.1016/j.resmic.2007.05.001

Prosseda, G., Di Martino, M. L., Campilongo, R., Fioravanti, R., Micheli, G., Casalino, M., et al. (2012). Shedding of genes that interfere with the pathogenic lifestyle: the *Shigella* model. *Res. Microbiol.* 163, 399–406. doi: 10.1016/j.resmic.2012.07.004

Prosseda, G., Falconi, M., Giangrossi, M., Gualerzi, C. O., Micheli, G., and Colonna, B. (2004). The virF promoter in *Shigella*: more than just a curved DNA stretch. *Mol. Microbiol.* 51, 523–537. doi: 10.1046/j.1365-2958.2003.03848.x

Prosseda, G., Falconi, M., Nicoletti, M., Casalino, M., Micheli, G., and Colonna, B. (2002). Histone-like proteins and the *Shigella* invasivity regulon. *Res. Microbiol.* 153, 461–468. doi: 10.1016/S0923-2508(02)01346-3

Prosseda, G., Latella, M. C., Casalino, M., Nicoletti, M., Michienzi, S., and Colonna, B. (2006). Plasticity of the P junc promoter of ISEc11, a new insertion sequence of the IS1111 family. *J. Bacteriol.* 188, 4681–4689. doi: 10.1128/JB.00332-06

Prunier, A.-L., Schuch, R., Fernandez, R. E., Mumy, K. L., Kohler, H., McCormick, B. A., et al. (2007). nadA and nadB of *Shigella flexneri* 5a are antivirulence loci responsible for the synthesis of quinolinate, a small molecule inhibitor of *Shigella* pathogenicity. *Microbiology* 153, 2363–2372. doi: 10.1099/mic.0.2007/006916-0

Pupo, G. M., Lan, R., and Reeves, P. R. (2000). Multiple independent origins of *Shigella* clones of *Escherichia coli* and convergent evolution of many of their characteristics. *Proc. Natl. Acad. Sci. U.S.A.* 97, 10567–10572. doi: 10.1073/pnas.180094797

Rasko, D. A., Rosovitz, M. J., Myers, G. S. A., Mongodin, E. F., Fricke, W. F., Gajer, P., et al. (2008). The pangenome structure of *Escherichia coli*: comparative genomic analysis of *E. coli* commensal and pathogenic isolates. *J. Bacteriol.* 190, 6881–6893. doi: 10.1128/JB.00619-08

Reiterer, V., Grossniklaus, L., Tschon, T., Kasper, C. A., Sorg, I., and Arrieumerlou, C. (2011). *Shigella flexneri* type III secreted effector OspF reveals new crosstalks of proinflammatory signaling pathways during bacterial infection. *Cell. Signal.* 23, 1188–1196. doi: 10.1016/j.cellsig.2011.03.006

Rowe, B., Gross, R. J., and Woodroof, D. P. (1977). Proposal to recognise seroovar 145/46 (synonyms: 147, *Shigella* 13, *Shigella sofia*, and *Shigella manolovii*) as a New *Escherichia coli* O group, O164. *Int. J. Syst. Bacteriol.* 27, 15–18. doi: 10.1099/00207713-27-1-15

Sahl, J. W., Morris, C. R., Emberger, J., Fraser, C. M., Ochieng, J. B., Juma, J., et al. (2015). Defining the phylogenomics of *Shigella* species: a pathway to diagnostics. *J. Clin. Microbiol.* 53, 951–960. doi: 10.1128/JCM.03527-14

Sakazaki, R., Tamura, K., and Saito, M. (1967). Enteropathogenic *Escherichia coli* associated with diarrhea in children and adults. *Jpn. J. Med. Sci. Biol.* 20, 387–399. doi: 10.7883/yoken1952.20.387

Sanchez-Villamil, J., Tapia-Pastrana, G., and Navarro-Garcia, F. (2016). Pathogenic lifestyles of *E. coli* pathotypes in a standardized epithelial cell model influence inflammatory signaling pathways and cytokines secretion. *Front. Cell. Infect. Microbiol.* 6:120. doi: 10.3389/fcimb.2016.00120

Sansonetti, P. J., Hale, T. L., Dammin, G. J., Kapfer, C., Collins, H. H., and Formal, S. B. (1983). Alterations in the pathogenicity of *Escherichia coli* K-12 after transfer of plasmid and chromosomal genes from *Shigella flexneri*. *Infect. Immun.* 39, 1392–1402.

Sansonetti, P. J., Kopecko, D. J., and Formal, S. B. (1982). Involvement of a plasmid in the invasive ability of *Shigella flexneri*. *Infect. Immun.* 35, 852–860.

Schroeder, G. N., and Hilbi, H. (2008). Molecular pathogenesis of *Shigella* spp.: controlling host cell signaling, invasion, and death by type III secretion. *Clin. Microbiol. Rev.* 21, 134–156. doi: 10.1128/CMR.00032-07

Scribano, D., Petrucca, A., Pompili, M., Ambrosi, C., Bruni, E., Zagaglia, C., et al. (2014). Polar localization of PhoN2, a periplasmic virulence-associated factor of *Shigella flexneri*, is required for proper IcsA exposition at the old bacterial pole. *PLOS ONE* 9:e90230. doi: 10.1371/journal.pone.0090230

Silva, R. M., Toledo, M. R., and Trabulsi, L. R. (1980). Biochemical and cultural characteristics of invasive *Escherichia coli*. *J. Clin. Microbiol.* 11, 441–444.

Small, P. L., and Falkow, S. (1988). Identification of regions on a 230-kilobase plasmid from enteroinvasive *Escherichia coli* that are required for entry into HEp-2 cells. *Infect. Immun.* 56, 225–229.

Svenungsson, B., Lagergren, A., Ekwall, E., Evengård, B., Hedlund, K. O., Kärnell, A., et al. (2000). Enteropathogens in adult patients with diarrhea and healthy control subjects: a 1-year prospective study in a Swedish clinic for infectious diseases. *Clin. Infect. Dis.* 30, 770–778. doi: 10.1086/313770

Takeuchi, O., and Akira, S. (2010). Pattern recognition receptors and inflammation. *Cell* 140, 805–820. doi: 10.1016/j.cell.2010.01.022

Taylor, D. N., Echeverria, P., Sethabutr, O., Pitarangsi, C., Leksomboon, U., Blacklow, N. R., et al. (1988). Clinical and microbiologic features of *Shigella* and enteroinvasive *Escherichia coli* infections detected by DNA hybridization. *J. Clin. Microbiol.* 26, 1362–1366.

Toledo, M. R., and Trabulsi, L. R. (1983). Correlation between biochemical and serological characteristics of *Escherichia coli* and results of the Serény test. *J. Clin. Microbiol.* 17, 419–421.

Touchon, M., Hoede, C., Tenaillon, O., Barbe, V., Baeriswyl, S., Bidet, P., et al. (2009). Organised genome dynamics in the *Escherichia coli* species results in highly diverse adaptive paths. *PLOS Genet.* 5:e1000344. doi: 10.1371/journal.pgen.1000344

Tozzoli, R., and Scheutz, F. (2014). "Diarrheagenic *Escherichia coli* infections in humans," in *Pathogenic Escherichia coli: Molecular and Cellular Microbiology*, ed. S. Morabito (Norfolk: Caister Academic Press).

Trabulsi, L. R., Fernandes, M. R., and Zuliani, M. E. (1967). New intestinal bacteria pathogenic to man. *Rev. Inst. Med. Trop. Sao Paulo* 9, 31–39.

Tran, C. N., Giangrossi, M., Prosseda, G., Brandi, A., Di Martino, M. L., Colonna, B., et al. (2011). A multifactor regulatory circuit involving H-NS, VirF and an antisense RNA modulates transcription of the virulence gene icsA of *Shigella flexneri*. *Nucleic Acids Res.* 39, 8122–8134. doi: 10.1093/nar/gkr521

Turner, S. A., Luck, S. N., Sakellaris, H., Rajakumar, K., and Adler, B. (2003). Molecular epidemiology of the SRL pathogenicity island. *Antimicrob. Agents Chemother.* 47, 727–734. doi: 10.1128/AAC.47.2.727-734.2003

van den Beld, M. J. C., and Reubsaet, F. A. G. (2012). Differentiation between *Shigella*, enteroinvasive *Escherichia coli* (EIEC) and noninvasive *Escherichia coli*. *Eur. J. Clin. Microbiol. Infect. Dis.* 31, 899–904. doi: 10.1007/s10096-011-1395-7

van Elsas, J. D., Semenov, A. V., Costa, R., and Trevors, J. T. (2011). Survival of *Escherichia coli* in the environment: fundamental and public health aspects. *ISME J.* 5, 173–183. doi: 10.1038/ismej.2010.80

Venkatesan, M. M., Goldberg, M. B., Rose, D. J., Grotbeck, E. J., Burland, V., and Blattner, F. R. (2001). Complete DNA sequence and analysis of the large virulence plasmid of *Shigella flexneri*. *Infect. Immun.* 69, 3271–3285. doi: 10.1128/IAI.69.5.3271-3285.2001

Vieira, N., Bates, S. J., Solberg, O. D., Ponce, K., Howsmon, R., Cevallos, W., et al. (2007). High prevalence of enteroinvasive *Escherichia coli* isolated in a remote region of northern coastal Ecuador. *Am. J. Trop. Med. Hyg.* 76, 528–533.

Voeroes, S., Redey, B., and Csizmazia, F. (1964). Antigenic structure of a new enteropathogenic *E. coli* strain. *Acta Microbiol. Acad. Sci. Hung.* 11, 125–129.

Wanger, A. R., Murray, B. E., Echeverria, P., Mathewson, J. J., and DuPont, H. L. (1988). Enteroinvasive *Escherichia coli* in travelers with diarrhea. *J. Infect. Dis.* 158, 640–642. doi: 10.1093/infdis/158.3.640

Wirth, T., Falush, D., Lan, R., Colles, F., Mensa, P., Wieler, L. H., et al. (2006). Sex and virulence in *Escherichia coli*: an evolutionary perspective. *Mol. Microbiol.* 60, 1136–1151. doi: 10.1111/j.1365-2958.2006.05172.x

Zagaglia, C., Casalino, M., Colonna, B., Conti, C., Calconi, A., and Nicoletti, M. (1991). Virulence plasmids of enteroinvasive *Escherichia coli* and *Shigella flexneri* integrate into a specific site on the host chromosome: integration greatly reduces expression of plasmid-carried virulence genes. *Infect. Immun.* 59, 792–799.

Zuo, G., Xu, Z., and Hao, B. (2013). *Shigella* strains are not clones of *Escherichia coli* but sister species in the genus *Escherichia*. *Genomics Proteomics Bioinformatics* 11, 61–65. doi: 10.1016/j.gpb.2012.11.002

Zurawski, D. V., Mumy, K. L., Faherty, C. S., McCormick, B. A., and Maurelli, A. T. (2009). *Shigella flexneri* type III secretion system effectors OspB and OspF target the nucleus to downregulate the host inflammatory response via interactions with retinoblastoma protein. *Mol. Microbiol.* 71, 350–368. doi: 10.1111/j.1365-2958.2008.06524.x

10

Environmental Origin of the Genus *Bordetella*

Illiassou Hamidou Soumana[1,2†], *Bodo Linz*[2,3*†] *and Eric T. Harvill*[1,2,3*]

[1] Department of Infectious Diseases, University of Georgia, Athens, GA, USA, [2] Center for Vaccines and Immunology, University of Georgia, Athens, GA, USA, [3] Department of Veterinary and Biomedical Sciences, Pennsylvania State University, University Park, PA, USA

Edited by:
Tatiana Venkova,
The University of Texas Medical Branch at Galveston, USA

Reviewed by:
Nikolai Ravin,
Research Center for Biotechnology (RAS), Russia
Louise Temple,
James Madison University, USA

***Correspondence:**
Bodo Linz
bodo.linz@uga.edu
Eric T. Harvill
harvill@uga.edu

† These authors have contributed equally to this work.

Members of the genus *Bordetella* include human and animal pathogens that cause a variety of respiratory infections, including whooping cough in humans. Despite the long known ability to switch between a within-animal and an extra-host lifestyle under laboratory growth conditions, no extra-host niches of pathogenic *Bordetella* species have been defined. To better understand the distribution of *Bordetella* species in the environment, we probed the NCBI nucleotide database with the 16S ribosomal RNA (16S rRNA) gene sequences from pathogenic *Bordetella* species. Bacteria of the genus *Bordetella* were frequently found in soil, water, sediment, and plants. Phylogenetic analyses of their 16S rRNA gene sequences showed that *Bordetella* recovered from environmental samples are evolutionarily ancestral to animal-associated species. Sequences from environmental samples had a significantly higher genetic diversity, were located closer to the root of the phylogenetic tree and were present in all 10 identified sequence clades, while only four sequence clades possessed animal-associated species. The pathogenic bordetellae appear to have evolved from ancestors in soil and/or water. We show that, despite being animal-adapted pathogens, *Bordetella bronchiseptica*, and *Bordetella hinzii* have preserved the ability to grow and proliferate in soil. Our data implicate soil as a probable environmental origin of *Bordetella* species, including the animal-pathogenic lineages. Soil may further constitute an environmental niche, allowing for persistence and dissemination of the bacterial pathogens. Spread of pathogenic bordetellae from an environmental reservoir such as soil may potentially explain their wide distribution as well as frequent disease outbreaks that start without an obvious infectious source.

Keywords: *Bordetella*, environmental strains, ecological niches, extra-host adaptation, environmental origin

INTRODUCTION

Bacteria of the genus *Bordetella* are of primary importance in human and veterinary medicine because of their ability to colonize the respiratory tract, causing a wide range of pulmonary and bronchial infections. The common human- and animal-adapted pathogens *B. pertussis*, *B. parapertussis*, and *B. bronchiseptica* are known as the "classical" *Bordetella* species. *B. pertussis* is a strictly human pathogen, but *B. parapertussis* consists of two lineages, one infecting humans and the other infecting sheep (Mattoo and Cherry, 2005). In contrast to these examples of adaptation to a single host, *B. bronchiseptica* colonizes a variety of animals and even humans (Register et al., 2015), resulting in a broad array of respiratory diseases, from asymptomatic colonization to lethal pneumonia (Goodnow, 1980). Phylogenetic analyses (Musser et al., 1986; Diavatopoulos et al., 2005) and genome comparisons (Parkhill et al., 2003) have revealed that *B. pertussis* and *B. parapertussis* represent human-adapted forms of *B. bronchiseptica* that have evolved independently from a *B. bronchiseptica*-like ancestor. The genus also contains a number of additional, more recently described species. For example, *B. avium* (Kersters et al., 1984) causes respiratory disease in birds. *B. hinzii* (Vandamme et al., 1995) is generally regarded as a non-pathogenic colonizer of the respiratory tract of poultry but some strains were shown to cause disease in turkeys when experimentally inoculated (Register and Kunkle, 2009). Meanwhile, the closely related species *B. pseudohinzii* colonizes laboratory mice (Ivanov et al., 2015, 2016). *B. holmesii* (Weyant et al., 1995) causes pertussis-like disease and septicemia in humans (Shepard et al., 2004), and *B. bronchialis*, *B. flabilis*, and *B. sputigena* (Vandamme et al., 2015) were also isolated from human respiratory specimens. In contrast to other bordetellae, *B. trematum* (Vandamme et al., 1996) and *B. ansorpii* (Ko et al., 2005) are not associated with respiratory problems but were isolated from human wound infection.

B. petrii, a species originally isolated from a dechlorinating bioreactor enriched by river sediment, represents the first described environmental species within the *Bordetella* genus (von Wintzingerode et al., 2001). *B. petrii* strains were also found in marine sponges (Wang et al., 2007), grass root consortia (Wang et al., 2007), and in other environmental samples as members of microbial communities involved in the degradation of aromatic hydrocarbons, such as benzenes (Bianchi et al., 2005; Wang et al., 2007). In apparent conflict with its environmental source, the *B. petrii* genome contains genes that allow for the synthesis and secretion of factors specifically associated with the virulence of the pathogenic *Bordetella* sp., including the BvgAS master regulon and filamentous hemagglutinin (Gross et al., 2008). In addition to these environmental sources, *B. petrii* was also isolated from immunocompromised patients with ear infection, cystic fibrosis and chronic pulmonary disease (Fry et al., 2005; Biederman et al., 2015; Nagata et al., 2015), suggesting broad adaptability of this bacterial species to both environmental conditions and as an opportunistic pathogen of humans and possibly other animals.

Other *Bordetella* species have been obtained from environments not associated with animal hosts. Ten different bacterial strains were cultured from cotton swabs taken from the plaster wall surface of 1300-year-old mural paintings inside the stone chamber of the Takamatsuzuka Tomb, an ancient circular burial mound in Japan. Taxonomic classification of these isolates revealed three novel species that were then named *B. muralis*, *B. tumulicola*, and *B. tumbae* (Tazato et al., 2015). Isolation of the bacteria from the paintings, but not from the surrounding stone walls, suggested that these species might be involved in the observed biodeterioration of the colorful paintings (Kigawa et al., 2013).

According to their 16S rRNA gene sequences, other environmental bacteria from soil also belong to the genus *Bordetella*. Interestingly, the majority of those samples originated from contaminated sites such as soil polluted with chlorinated benzenes (Wang et al., 2007), from arctic soils contaminated with polycyclic aromatic hydrocarbons such as oil, diesel fuel or tar (Eriksson et al., 2003), from the sediment of a municipal wastewater plant (Nisola et al., 2010) and from arsenic polluted soils (Cavalca et al., 2010; Bachate et al., 2012). All these observations suggest that members of the *Bordetella* genus may have the potential for biodegradation of a great variety of organic compounds.

Although these anecdotal findings suggest that members of the *Bordetella* genus may be found in nature, there is currently no systematic analysis of the occurrence of *Bordetella* outside human or animal hosts, and the potential impact of environmental isolates on human and animal health is uncertain. Environmental niches of pathogenic *Bordetella* species have been proposed but not identified. Yet, the ability of *Bordetella* to survive and persist outside mammalian hosts would allow for its greater dissemination and persistence, and could contribute to a wide distribution of infections and disease, often without an obvious infectious source.

Here, we search the NCBI nucleotide database for 16S ribosomal RNA gene sequences of *Bordetella*-like microorganisms from various environments and compare them to those of the described species, including the classical bordetellae, to determine their phylogenetic relatedness. We identified 10 clades of related strains, all of which contained samples isolated from environmental sources, though only four clades also contained sequences from animal-associated species. Sequences from environmental samples had a significantly higher genetic diversity and were located closer to the root of the phylogenetic tree than those from animal-associated isolates, suggesting an environmental origin of the genus *Bordetella*. In addition, we show that the animal-adapted pathogens *B. bronchiseptica* and *B. hinzii* grow efficiently in soil extract, indicating that diverse pathogenic bordetellae may have retained the ability to proliferate in the environment.

MATERIALS AND METHODS

Search for *Bordetella* 16S rRNA Gene Sequences in the NCBI nt Database

The16S ribosomal gene sequences of *Bordetella hinzii* strain LMG 13501 (GenBank accession number NR_027537.1); *Bordetella holmesii* strain ATCC 51541 (NR_121717.1); and *Bordetella pertussis* strain Tohama I (AF142326.1) were each used as queries for BLAST search (blastn) against the NCBI nr/nt database using the default search parameters with a hitlist size of 5000. From the numerous hits, we excluded sequences of isolates from the known species that are mentioned in the introduction and selected only those that showed higher percentage of similarity to known *Bordetella* species than to bacteria from any other genus, including *Achromobacter*. As a control, we ran blastn searches with each of the identified sequences against the NCBI nr/nt database to remove potential false positives. The remaining sequences, all of which were from bacteria obtained from environmental sources, were considered for further analysis. All three searches using 16S rRNA sequences of

B. pertussis, *B. hinzii*, and *B. holmesii* as queries, respectively, gave consistent results. The accession numbers were then explored for details on sample source, year of isolation, and associated publications. Most sequences were described as *Bordetella* sp. in the gene description, but some were designated as "uncultured bacterium."

Phylogenetic Analysis and Tree Construction

All 16S rRNA sequences were aligned in Clustal Omega (http://www.ebi.ac.uk/Tools/msa/clustalo/), and the alignment was checked manually for consistency. Only sequences containing a 1376 bp gene fragment (near full-length) were used for further analyses. In order to identify the closely related species of environmental *Bordetella* isolates, the 16S ribosomal RNA gene sequences of members of 16 named *Bordetella* species were used as references; namely *B. pertussis* Tohama I, *B. parapertussis* BPP5, *B. bronchiseptica* RB50, *B. avium* 197N, *B. hinzii* LMG 13501, *B. pseudohinzii* 8-296-03, *B. holmesii* ATCC 51541, *B. trematum* DSM 11334, *B. ansorpii* SMC-8986, *B. bronchialis* LMG 28640, *B. sputigena* LMG 28641, *B. flabilis* LMG 28642, *B. petrii* Se-1111R, *B. muralis* T6220-3-2b, *B. tumulicola* T6517-1-4b, and *B. tumbae* T6713-1-3b (**Table 1**). The 16S rRNA gene sequences of *Burkholderia pseudomallei* NCTC13179 and *Ralstonia solanacearum* YP-01 were used as outgroups. The aligned and trimmed sequences (one per unique sequence) were used to generate a Neighbor-joining tree using the Maximum Composite Likelihood algorithm in MEGA (Tamura et al., 2007), and bootstrap support was estimated running 100,000 replications. Nucleotide diversity (Π) within environmental samples and within animal-associated samples were estimated in DnaSP (Librado and Rozas, 2009), and 95% confidence limits (Π_{95}) were estimated using an online confidence limit calculator (https://www.allto.co.uk/tools/statistic-calculators/confidence-interval-for-mean-calculator/).

Soil Sample Collection and *Bordetella* Growth in Soil Extract

Soils were sampled in April 2016 at two random sites in State College, Pennsylvania, near a suburban park (40°48′40.7″ N 77°53′06.1″ W and 40°48′38.2″ N 77°53′04.2″ W). Each sample was collected to a depth of 20 cm and thoroughly mixed. Fifty grams of each soil sample (100 g total) was placed in a bottle which was filled to 500 ml with sterile PBS. The sample was homogenized by shaking for 10 min, then left to settle for 1 h at room temperature and carefully decanted. The soil-PBS suspension was filter sterilized. Single colonies of *B. bronchiseptica* strain RB50, *B. hinzii* strain L60, and *B. petrii* strain DSMZ12804 were picked from Bordet-Gengou (BG) agar (Difco) plates supplemented with 10% defibrinated Sheep's blood (HemoStat Laboratories, Dixon, CA, USA) and were cultured in liquid Stainer-Scholte medium (Stainer and Scholte, 1970) overnight at 37°C. The *Bordetella* inocula were prepared as follows. The cultures were centrifuged, resuspended in 1 ml PBS, and the optical density (OD_{600}) was determined. Following five consecutive 10-fold dilutions in 1 ml PBS, 100 μl (=10^6-fold dilution) containing ~150 (*B. petrii*) or 240 bacterial cells (*B. hinzii* or *B. bronchiseptica*) were added to 5 ml of the soil extract resulting in starting concentrations of ~30 bacterial cells/ml (*B. petrii*) and 48 bacterial cells/ml (*B. hinzii*, *B. bronchiseptica*). Bacterial numbers were determined by plating an aliquot of each inoculum. The culture tubes were incubated at room temperature (25°C) with shaking. After 24, 48, and 72 h, 100 μl of each culture was plated on BG agar supplemented with

TABLE 1 | Reference strains of named *Bordetella* species.

Strain name	References	GenBank accession number	16S rRNA sequence length (bp)
Burkholderia pseudomallei NCTC 13179	Johnson et al., 2015	CP003976.1	1487
Ralstonia solanacearum YP-01	NCBI	FJ494776.1	1500
Bordetella avium 197N	Sebaihia et al., 2006	NR_074639.1	1487
Bordetella bronchiseptica RB50	Parkhill et al., 2003	BX640447.1	1487
Bordetella hinzii LMG 13501	Kattar et al., 2000	NR_027537.1	1487
Bordetella parapertussis BPP5	Park et al., 2012	HE965803.1	1489
Bordetella holmesii ATCC 51541	NCBI	NR_121717.1	1487
Bordetella pertussis Tohama I	Parkhill et al., 2003	AF142326.1	1487
Bordetella trematum DSM 11334	von Wintzingerode et al., 2001	NR_025404.1	1521
Bordetella flabilis LMG 28642	Vandamme et al., 2015	EU082162.1	1376
Bordetella bronchialis LMG 28640	Vandamme et al., 2015	EU082135.1	1416
Bordetella sputigena LMG 28641	Vandamme et al., 2015	KF601914.1	1376
Bordetella ansorpii SMC-8986	Ko et al., 2005	AY594190.1	1424
Bordetella pseudohinzii 8-296-03	Ivanov et al., 2016	JHEP02000033.1	1542
Bordetella petrii DSMZ12804	Gross et al., 2008	NC_010170	1487
Bordetella muralis T6220-3-2b	Tazato et al., 2015	LC053647.1	1456
Bordetella tumbae T6713-1-3b	Tazato et al., 2015	LC053656.1	1456
Bordetella tumulicola T6517-1-4b	Tazato et al., 2015	LC053650.1	1456

10% defibrinated sheep's blood to determine bacterial numbers. Each experiment was carried out in triplicate. The mean and ± standard error as well as analysis of variance (ANOVA) were conducted using Graphpad Prism version 6.04. The bacterial doubling time was calculated by the formula: doubling time = $\ln(2)/\ln(N(t)/N(0))/t$, where $N(t)$ is the number of bacterial cells at time t, $N(0)$ is the number of bacteria at time 0 and t is the time in hours.

RESULTS

Bordetella in the Environment

We mined the NCBI nucleotide databases for *Bordetella* spp. 16S rRNA gene sequences. The search resulted in a total of 71 *Bordetella* spp. 16S rRNA gene sequences (**Table 2**) in addition to those from the named species (**Table 1**) *B. bronchiseptica*, *B. parapertussis*, *B. pertussis*, *B. hinzii*, *B. pseudohinzii*, *B. holmesii*, *B. avium*, *B. trematum*, *B. ansorpii*, *B. flabilis*, *B. bronchealis*, *B. sputigena* (isolated from samples of human/animal origin), *B. petrii*, *B. tumbae*, *B. muralis*, and *B. tumulicola* (isolated from environmental samples). The corresponding strains were recovered from different environmental niches (**Table 2**), including soil (52 strains) and water (11 strains), and from 8 strains associated with plants. The soil samples were of diverse origin, including compost, cave rocks, and metal mines, but the majority were sampled at sites contaminated with oil and several halogenated cyclic hydrocarbons such as chlorinated benzenes or hexachlorocyclohexane. The samples from aquatic environments were also of diverse origin, namely industrial wastewater, a sulfur spring, lake water, surface sea water, and river biofilms. Several samples from plants were isolated from roots and thus at the plant-soil interface (**Table 2**). Thus, members of the genus *Bordetella* appear to be widespread across different environmental niches.

16S rRNA Gene Sequence Clades are Associated with Particular Environmental Niches

To relate the environmental isolates to known *Bordetella* species, we aligned the 16S rRNA gene sequences and constructed a Neighbor-joining tree using the Maximum-likelihood algorithm implemented in MEGA (Tamura et al., 2007). Forty-eight sequences from environmental samples were of sufficient length and used for further analyses (**Table 2**). Of those, 36 originated from soil (27 haplotypes), eight from aquatic environments (7 haplotypes), and four from plants (4 haplotypes). The tree was rooted with sequences of *Burkholderia pseudomallei* and *Ralstonia solanacearum* as outgroups. The *Bordetella* sequences formed 10 distinct clusters (**Figure 1**). While most clusters contained at least one described species, such as *B. petrii* in cluster VI or *B. tumbae*/*B. muralis* in cluster V, several *Bordetella* sequences did not cluster with any described species but rather occupied distinct branches of the tree. These include the two isolates in cluster IV, the isolates from soil samples in clusters VII and X and strains *B.* sp. CC-PW-55 and *B.* sp. TS-T34 (cluster IX) isolated from surface sea water and lake water, respectively (**Figure 1**).

Superimposing the origin of the *Bordetella* spp. isolates revealed that most of the identified clusters were dominated by sequences of similar environmental/host origin. Thus, cluster I was composed of sequences of *B. holmesii* and the classical bordetellae (*B. bronchiseptica*, *B. parapertussis*, and *B. pertussis*), all of which were isolated from human and animal infection, but also contained *B.* sp. HT38 isolated from a river biofilm in China (**Figure 1**, **Table 2**). Cluster III contained sequences of species isolated from human respiratory specimen (*B. sputigena*, *B. bronchialis*, and *B. flabilis*) plus an isolate from soil in India. Other clusters either contained, or were dominated by, sequences of environmental origin such as cluster IV (water and soil), cluster V (soil), including the three species recovered from mural paintings *B. tumbae*, *B. tumulicola*, and *B. muralis*; but also *B.* sp. CCBAU from a maize rhizosphere and *B. ansorpii* from infection of an immunocompromised patient, and cluster VI (soil, including the environmental species *B. petrii*). The prominent exception to this pattern, cluster II, contained sequences from animal/human infection (*B. avium*, *B. hinzii*, *B. pseudohinzii*, and *B. trematum*) as well as from water (*B.* sp. MT-E1, *B.* sp. HF27), plant root (*B.* sp. R8-804, *B.* sp. R8-551), and soil samples (*B.* sp. BAB-4396). However, the other clusters were either dominated by animal-associated samples (clusters I and III) or samples of environmental origin (all other clusters).

If the genus *Bordetella* were of environmental origin, samples isolated from soil and water would be expected to be more diverse and would appear widespread across the tree. Indeed, environmental samples were present in all sequence clusters. In contrast, sequences from animal-associated samples were confined to four clusters, all of which also contained environmental isolates. Three of those four clusters formed a single super clade which originated from one of several clades among sequences from environmental isolates. In contrast, all clusters near the tree root exclusively contained environmental samples, but no animal associated samples (**Figure 1**). The phylogenetic analyses showed that the genetic diversity was significantly higher in sequences from environmental samples (Π_{95} = 2.02–2.13%) than in sequences from animal-associated samples (Π_{95} = 1.30–1.53%). The sequence of branching events within the phylogenetic tree is consistent with an environmental origin of *Bordetella* and subsequent adaptation of some lineages to animal hosts.

Bordetella bronchiseptica and *Bordetella hinzii* are Capable of Growing in Soil Extract

Since most environmental *Bordetella* samples were recovered from soil (and water), we hypothesized that pathogenic, animal-associated species may have retained the ability to thrive in soil as an environmental niche. Therefore, we assessed the ability of *B. bronchiseptica* strain RB50, *B. hinzii* strain L60, and *B. petrii* strain DSMZ12804, to grow in a sterile, homogenized suspension made from soil. Instead of growing pathogenic bordetellae directly on solid soil, we prepared a soil suspension

TABLE 2 | *Bordetella* strains for which the 16S ribosomal RNA sequences were recovered from environmental samples.

Bordetella strains	Isolation source	Country	References	GenBank accession No.	Sequence length (bp)	Duplicated sequences
SOIL ORIGIN						
***Bordetella* sp. F2**	Chlorinated benzenes polluted soil	Germany	Wang et al., 2007	DQ453689.1	1527	4
***Bordetella* sp. E3**	Chlorinated benzenes polluted soil	Germany	Wang et al., 2007	DQ453688.1	1527	4
***Bordetella* sp. QJ2–5**	Chlorinated benzenes polluted soil	China	NCBI	DQ152013.1	1393	4
***Bordetella* sp. 2b05**	HCH-contaminated soil	India	NCBI	JF979304.1	1523	
***Bordetella* sp. 2f06**	HCH-contaminated soil	India	NCBI	JF979347.1	1523	
***Bordetella* sp. 2e11**	HCH-contaminated soil	India	NCBI	JF979341.1	1522	
***Bordetella* sp. 1h08**	HCH-contaminated soil	India	NCBI	JF979288.1	1519	
***Bordetella* sp. 1c11**	HCH-contaminated soil	India	NCBI	JF979241.1	1521	
***Bordetella* sp. 2c11**	HCH-contaminated soil	India	NCBI	JF979320.1	1523	
***Bordetella* sp. 2a09**	HCH-contaminated soil	India	NCBI	JF979298.1	1519	
***Bordetella* sp. ud1**	1,2,4-TCB contaminated soil	Germany	NCBI	FJ529833.1	1523	4
***Bordetella* sp. ud29**	1,2,4-TCB contaminated soil	Germany	NCBI	FJ529848.1	1523	
***Bordetella* sp. ud3b**	1,2,4-TCB contaminated soil	Germany	NCBI	FJ529835.1	1523	4
***Bordetella* sp. ud13a**	1,2,4-TCB contaminated soil	Germany	NCBI	FJ529840.1	1525	
***Bordetella* sp. IITR02**	1,2,4-TCB contaminated soil	India	NCBI	EU752498.1	1422	
***Bordetella* sp. CTN-10**	Chemical factory soil	China	NCBI	FJ598334.1	1398	2
***Bordetella* sp. 2–12**	Chemical factory soil	China	NCBI	FJ598328.1	1410	2
***Bordetella* sp. CTN-16**	Chemical factory soil	China	NCBI	FJ598326.1	1412	4
***Bordetella* sp. C16-Siri112**	Oil-contaminated soil	Iran	NCBI	JX500276.1	1397	5
***Bordetella* sp. p23(2011)**	Magnetite drainage, Iron mine	China	NCBI	HQ652588.1	1518	
***Bordetella* sp. e3(2011)**	Magnetite drainage, Iron mine	China	NCBI	HQ652587.1	1501	3
***Bordetella* sp. d16(2011)**	Magnetite drainage, Iron mine	China	NCBI	HQ652589.1	1507	3
***Bordetella* sp. f17(2011)**	Magnetite drainage, Iron mine	China	NCBI	HQ652590.1	1520	
***Bordetella* sp. FB-8**	Creek sediment from former uranium-mining area	Germany	NCBI	JN885794.1	1385	
***Bordetella* sp. A2–436**	Uranium mine	Portugal	NCBI	KF441609.1	1528	
***Bordetella* sp. J4**	Acid mine drainage	France	Delavat et al., 2013	HF568988.1	1410	
***Bordetella* sp. BAB-4396**	Soil	India	NCBI	KM289182.1	1499	
***Bordetella* sp. B4**	Paddy field by yellow river	China	NCBI	EU140499.1	1523	
***Bordetella* sp. MCYF11**	Lake Taihu sediment	China	Yang et al., 2014	KC734882.1	1385	
***Bordetella* sp. PTG4–17**	Sediment of the Indian ocean	India	NCBI	EU603444.1	1496	
***Bordetella* sp. RCC3**	Caves rock	India	NCBI	KC119149.1	1476	
***Bordetella* sp. RCC4**	Caves rock	India	NCBI	KC119150.1	1464	
***Bordetella* sp. M1–6**	Compost	China	Kato et al., 2004	AB039335.1	1531	
***Bordetella* sp. FS1413**	Compost	Finland	Partanen et al., 2010	FN667145.1	1464	
***Bordetella* sp. SMG22**	Compost	China	Guo et al., 2015	AM930282.1	1491	6
***Bordetella* sp. OT-2-E7**	Compost	China	Tian et al., 2013	JQ337611.1	1397	6
Bordetella sp. strain 2ABA4	Solid waste dumpsites	Nigeria	Sanuth et al., 2013	HE858274.1	1168	
Bordetella sp. OS17	Benten-Cho station soil	Japan	Matsumura et al., 2009	AB453298.1	980	
Bordetella sp. VVAR	Soil	Japan	NCBI	FJ588707.1	1451	
Bordetella sp. Ds-4	Cultivated soil	India	NCBI	HQ857791.1	727	
Bordetella sp. R-8	Garden soil	India	NCBI	JX130378.1	1319	
Bordetella sp. SPB-24	Garden soil	India	Bachate et al., 2012	JN208922.1	1403	
Bordetella sp. As3–3	Arsenic contaminated soil	Italy	Cavalca et al., 2010	FN392624.2	544	
Bordetella sp. AGO-03	Arsenic contaminated rice field	India	NCBI	AB696982.1	979	
Bordetella sp. ADP-18	Arsenic contaminated rice fields	India	NCBI	AB697485.1	674	
Bordetella sp. C16-Siri108	Oil-contaminated soil	Iran	NCBI	JX500272.1	1069	
Bordetella sp. C16-Siri113	Oil-contaminated soil	Iran	NCBI	JX500277.1	1295	
Bordetella sp. BF07B02	Agricultural soil	Burkina Faso	Colinon et al., 2013	KC195878.1	1381	

(Continued)

TABLE 2 | Continued

Bordetella strains	Isolation source	Country	References	GenBank accession No.	Sequence length (bp)	Duplicated sequences
Bordetella sp. HPC772	Activated sludge of an effluent treatment plant	India	NCBI	AY838357.1	580	
Bordetella sp. PH21	Phenolic compounds-contaminated sediment	China	NCBI	JN171686.1	721	
Bordetella sp. PH22	Phenolic compounds-contaminated sediment	China	NCBI	JN171687.1	721	
Bordetella sp. VKRKCd3	Seashore surface sediment	India	NCBI	GQ262759.1	363	
PLANT ORIGIN						
***Bordetella* sp. CCBAU 10842**	Maize rhizosphere	China	NCBI	JF772555.1	1369	
***Bordetella* sp. R8–804**	*Jatropha curcas* L, plant root	Singapore	NCBI	JQ659985.1	1487	
***Bordetella* sp. R8–551**	*Jatropha curcas* L, plant root	Singapore	NCBI	JQ659951.1	1486	
***Bordetella* sp. S2–5-CL23**	velvetleaf seed	USA	NCBI	EU769148.1	1492	
Bordetella sp. S318(2010)	*M. sinensis* × giganteus internal stem tissue	Ireland	NCBI	HM102497.1	600	
Bordetella sp. Juv992	Lupine cluster roots	Switzerland	Weisskopf et al., 2011	JN590346.1	1302	
Bordetella sp. PnB 4	Pepper	India	NCBI	JQ886795.1	370	
Bordetella sp. RS-CIW-47	Maize rhizosphere	Pakistan	NCBI	KC430988.1	950	
WATER ORIGIN						
***Bordetella* sp. MT-I2**	Industrial wastewater	Germany	Toups et al., 2010	EU727195.1	1526	1
***Bordetella* sp. MT-E1**	Industrial wastewater	Germany	Toups et al., 2010	EU727194.1	1525	1
***Bordetella* sp. TS-T34**	Lake water	China	NCBI	KC762319.1	1398	
***Bordetella* sp. CC-PW-55**	Surface seawater	Taiwan	NCBI	KF851340.1	1500	
***Bordetella* sp. 13.1 KSS**	Mineral oil-based metalworking fluid	Germany	Lodders and Kämpfer, 2012	HE575910.1	1398	
***Bordetella* sp. HT19**	Sulfur spring	India	NCBI	FJ969843.1	1404	
***Bordetella* sp. HF38**	River biofilms	China	NCBI	KR188914.1	1523	
***Bordetella* sp. HF72**	River biofilms	China	NCBI	KR188948.1	1523	
Bordetella sp. MMJ09	Distillery wastewater	China	NCBI	GU244378.1	813	
Bordetella sp. Sulf-8	Municipal wastewater	South Korea	Nisola et al., 2010	GU812430.1	1314	
Bordetella sp. IPJ1	Rusted iron pipe in freshwater lake	India	NCBI	HM593901.1	1100	

In bold are the strains for which the length of the 16S ribosomal RNA sequence were at least 1376 bp, and were included in the phylogenetic tree construction.

to extract possible nutrients but to avoid solid matter which allowed visual monitoring of bacterial growth and selection of appropriate sampling time points. All three isolates were cultured at room temperature (25°C) with shaking in either liquid soil extract or in Stainer-Scholte (SS) medium as a control. All three species grew fast in SS medium with doubling times of 1.8 ± 0.02 h (*B. bronchiseptica*), 1.9 ± 0.01 h (*B. hinzii*), and 1.9 ± 0.02 h (*B. petrii*), and reached the stationary phase prior to 48 h post-inoculation (**Figure 2**). As expected from an environmental bacterium, *B. petrii* strain DSMZ12804 thrived when inoculated into a soil extract, with a doubling time of 7.25 ± 0.24 h (**Figure 2**). Surprisingly, both *B. hinzii* strain L60 with a doubling time of 6.4 ± 0.09 h and *B. bronchiseptica* strain RB50 with a doubling time of 4.0 ± 0.04 h grew in the soil extract faster than *B. petrii*. Thus, all three species can grow efficiently at 25°C on filter-sterilized soil extract, even though the growth rate was slower than in Stainer-Scholte medium.

DISCUSSION

Bacteria of the genus *Bordetella* occupy remarkably diverse ecological niches, ranging from soil, water, and plants, to the respiratory tracts of a wide variety of animals including humans. Several environmental *Bordetella* strains were isolated from soils polluted with oil and oil derivatives (**Table 2**), including halogenated polycyclic hydrocarbons (Eriksson et al., 2003; Bianchi et al., 2005; Wang et al., 2007). Other strains were found in garden soil, compost, and various sediments suggesting these organisms are quite adaptable to diverse sites. The only sequenced and analyzed genome of an environmental isolate, *B. petrii* strain DSMZ 12804, revealed a possible genomic basis for substantial metabolic versatility (Gross et al., 2008). The genome encodes multiple auxiliary pathways for the utilization of a variety of nutrients, including pectate, numerous sugar derivatives from degraded plant products and various aromatic

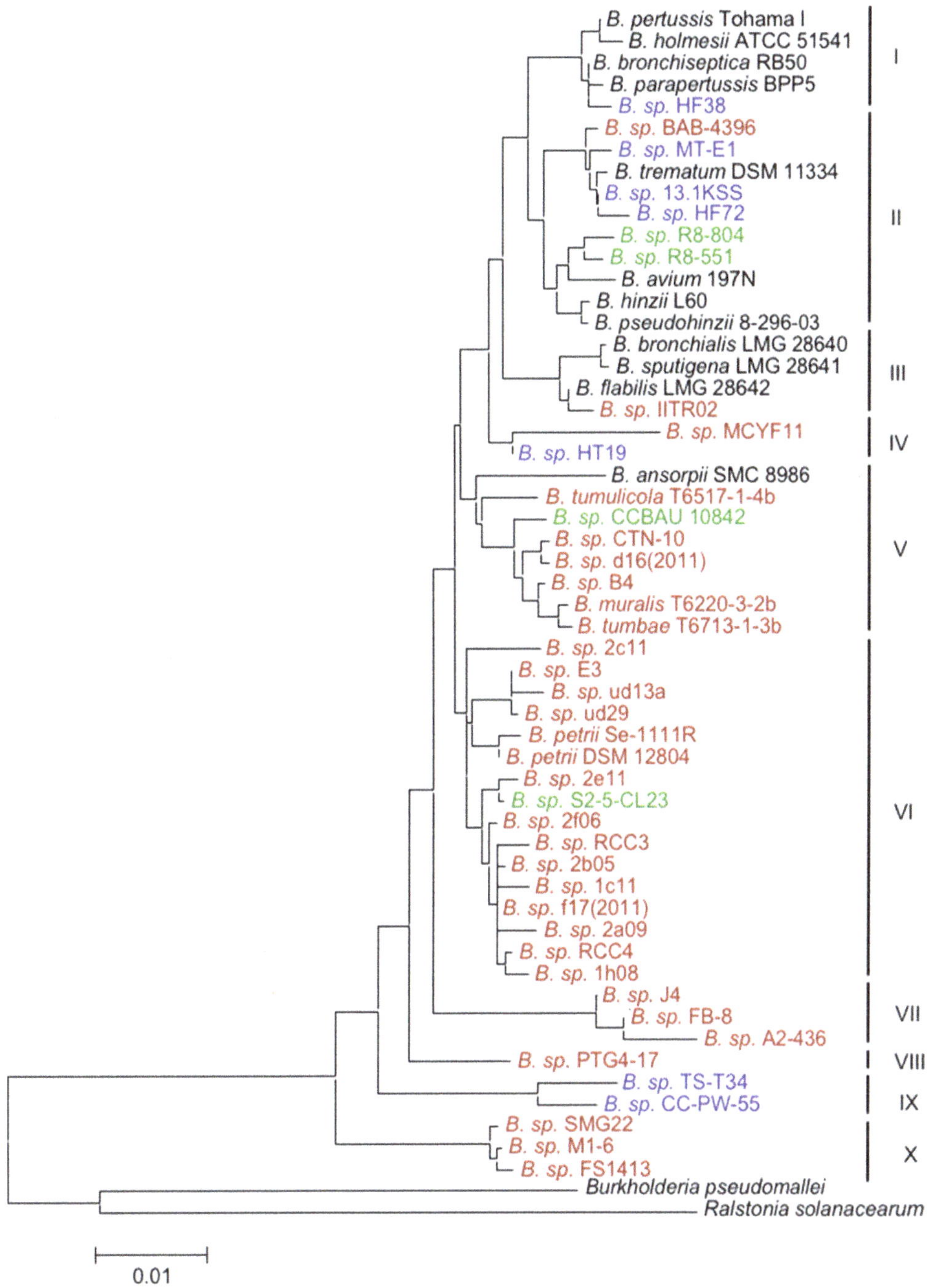

FIGURE 1 | Neighbor-Joining tree based on 16S rRNA gene sequences of animal-associated and environmental strains of *Bordetella*. The 52 near full-length sequences (1376 bp) formed 10 clades (I–X) of phylogenetically closely related *Bordetella* isolates/species recovered from soil (brown), water (blue), plants (green) and animals (black). The 16S rRNA gene sequences of the beta-proteobacteria *Burkholderia pseudomallei* and *Ralstonia solanacearum* were used as outgroups.

compounds. Five of the eight genomic islands that have been identified in this genome contain genes coding for enzymes for the metabolism of aromatic compounds, particularly clusters of genes encoding enzymes of the chlorocatechol pathway, including gene clusters that show high similarity to genes in a 1,2,4-trichlorobenzene-degrading *Pseudomonas* strain (Gross et al., 2008). The presence of multiple chlorocatechol gene clusters in addition to several different central pathways for aromatic metabolism may provide a competitive advantage for growth in contaminated environments.

Another striking feature of environmental *Bordetella* isolates is their resistance to heavy metals (Cavalca et al., 2010). Ten out

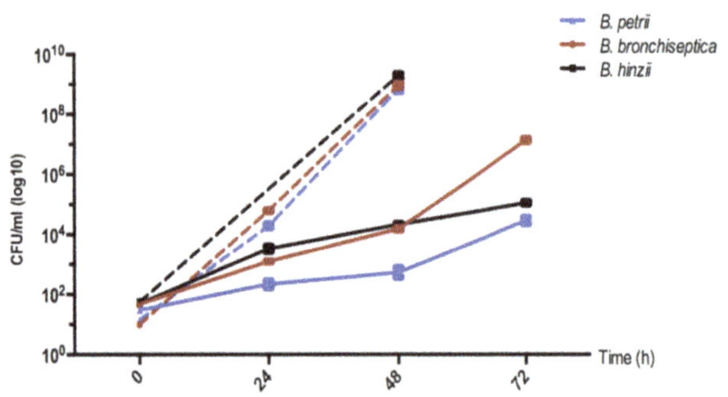

FIGURE 2 | Growth of *B. bronchiseptica* strain RB50, *B. hinzii* strain L60, and *B. petrii* strain DSMZ12804 in soil extract (solid lines) and in Stainer-Scholte medium (dashed lines). All three bacterial species efficiently grow in a sterile-filtered soil suspension suggesting that soil may represent an environmental niche for pathogenic *Bordetella* species.

of 52 soil samples (**Table 2**) were isolated from iron mines (e.g., *B.* sp. d16, *B.* sp. f17), from uranium mines (*B.* sp. FB-8, *B.* sp. A2–436), or from soil polluted with arsenic (e.g., *B.* sp. As3–3). Such remarkable metal tolerance is most likely conferred by heavy metal resistance systems. Indeed, the genome of *B. petrii* strain DSMZ 12804 contains several heavy metal resistance operons on a genomic island absent from the genomes of other sequenced bordetellae, whereas other strains contain different islands of genes. Ultimately, the presence of multiple heavy metal resistance systems may allow environmental *Bordetella* isolates to thrive in metal rich environments.

Most plant-associated *Bordetella* strains were recovered from roots (*B.* sp. R8–804, *B.* sp. R8–551, *B.* sp. Juv992) and the rhizosphere at the plant-soil interface (*B.* sp. CCBAU 10842). Thus, these isolates may in fact represent soil samples or, alternatively, may be involved in interactions with plants at the plant-soil interface. The resemblance between plant responses to bacterial virulence factors and the responses of mammalian immune cells (Berg et al., 2005) serve as evidence that bacteria-plant interactions may have paved the way for bacterial adaptation to animals. In this regard, plant-root isolates *B.* sp. R8–804 and *B.* sp. R8–551 from plant roots are closely related to bird pathogens, *B. hinzii* and *B. avium*, supporting the view that plants cells could serve as a "training ground" for environmental strains that eventually gain the ability to colonize animal hosts (Berg et al., 2005).

In addition to these plant root isolates, several other environmental isolates were also found to be very closely related to animal-associated pathogens (**Figure 1**). Interestingly, those strains were isolated from very diverse sources, namely (polluted) soil in India (*B.* sp. BAB-4396, *B.* sp. IITR02), industrial waste water (*B.* sp. MT-E1), and oil-based metal-working emulsion in Germany (*B.* sp. 13.1 KSS), as well as from river biofilms in China (*B.* sp. HF38 and *B.* sp. HF72). The two isolates from a river biofilm in China are of particular interest. The 16S rRNA sequence of one of those (strain HF72) showed 99.56% sequence similarity to that of the human pathogen *B. trematum* (6 SNPs). According to 16S rRNA gene sequence, the other isolate (*B.* sp. HF38) is even more closely related (99.78%, three SNPs) to the animal pathogen *B. bronchiseptica* strain RB50 and the human pathogen *B. parapertussis* strain 12822, which share an identical sequence in this gene. By this measure, isolate HF38 is as closely related to *B. bronchiseptica* strain RB50 and *B. parapertussis* strain 12822 as it is to *B. pertussis*. This exceptionally close phylogenetic relatedness makes several evolutionary scenarios conceivable. First, isolate *B.* sp. HF38 may be an environmental, non-pathogenic strain closely related to the animal/human pathogens among the classical bordetellae. Second, this isolate might be a descendant or relative of an ancestor of the classical bordetellae which later became pathogenic after acquisition of several virulence-associated factors, such as pertussis toxin, adenylate cyclase toxin, and dermonecrotic toxin. Third, this isolate may in fact represent a *B. bronchiseptica* or *B. parapertussis* strain that naturally survives and/or grows within an environmental reservoir. Although the classical bordetellae have not yet been isolated from outside a mammalian host, our results suggest that animal-pathogenic *Bordetella* species retain the ability to grow in soil as an environmental niche. This implies that *B. bronchiseptica* and other species might be found (at least transiently) in soil, for example at farms with suitable animal hosts such as cattle, pig, sheep and horse, or near dog kennels. Interestingly, even fastidious *B. pertussis* bacteria remained able to be cultured for up to 5 days when spread onto various hospital-setting surfaces such as fabrics, plastics, glass, and paper, and also in several infant foods (Ocklitz and Milleck, 1967). Fourth, *B.* sp. HF38 as well as other isolates from water and soil may be protected internally by a non-vertebrate host. For example, amoebae are known to host bacteria such as *Legionella pneumophila* (Molmeret et al., 2005), and amoeba-grown *L. pneumophila* exhibited radically increased resistance to harsh environmental conditions such as fluctuations in temperature, osmolarity, acidity, as well as to biocides that may facilitate bacterial survival and persistence in the environment (Barker et al., 1995; Abu Kwaik et al., 1997, 1998; Winiecka-Krusnell and Linder, 1999). Amoebae are ubiquitously found in most environments, and shared habitats between amoeba and *Bordetella* could be an important factor for the persistence of the bacteria. Indeed, our group has shown that the animal-adapted *B. bronchiseptica* is able to survive and multiply intracellularly in the trophozoites and sori of the amoeba *Dictyostelium discoideum* before being disseminated with the amoeba spores to novel geographical locations (Bendor et al., in revision). Thus, in addition to our recent data demonstrating that *B. bronchiseptica* can circulate and efficiently transmit amongst mammals, these data demonstrate that this species can also grow and disseminate efficiently in association with amoebae. These independent but interconnected *Bordetella* lifecycles allow for disease propagation, transmission, and re-emergence in the absence of an infected animal host.

Strains included in this study were identified as *Bordetella* spp. based on their 16S rRNA gene sequence. Currently, there are no data available regarding potential pathogenicity of these species. Whole genome sequencing will provide valuable insights into the evolution and ecology of environmental vs. animal-pathogenic bordetellae. Of special interest are environmental isolates closely related to animal pathogens, particularly isolate *B.* sp. HF38, and analysis of their genomes will reveal whether they are non-pathogenic relatives of known animal pathogens or if they in fact represent environmental reservoirs of *B. bronchiseptica* or *B. parapertussis*.

Finally, the majority of environmental *B.* sp. were recovered from soil samples indicating that soil could be the most frequent natural habitat of bordetellae. Indeed, sequences identified from soil samples were found in 8 of 10 sequence clusters, including samples from compost in cluster X at the root of the tree (**Figure 1**). The sequence of branching events within the phylogenetic tree, the significantly higher sequence diversity in samples from soil and water than in those from animals, as well as the preserved ability of animal pathogens to grow in soil, suggest an environmental, likely soil-based, origin of the genus *Bordetella*. Thus, similar to bacteria of the closely related genus *Achromobacter*, which are of environmental origin but also contain opportunistic pathogens (Li et al., 2013), *Bordetella* appears to be a bacterium of environmental origin that adapted and became pathogenic via the acquisition of factors mediating specific interactions with animal hosts.

AUTHOR CONTRIBUTIONS

IHS, BL, and ETH conceived and designed the experiments. IHS and BL performed the experiments and analyzed the data. IHS, BL and ETH wrote the paper.

ACKNOWLEDGMENTS

We thank Holly Vuong, Monica Cartelle Gestal, and Israel Rivera from the Harvill lab for helpful discussions. This work was supported by grants GM113681 and AI116186 by the National Institutes of Health (to ETH).

REFERENCES

Abu Kwaik, Y., Gao, L. Y., Harb, O. S., and Stone, B. J. (1997). Transcriptional regulation of the macrophage-induced gene (*gspA*) of *Legionella pneumophila* and phenotypic characterization of a null mutant. *Mol. Microbiol.* 24, 629–642. doi: 10.1046/j.1365-2958.1997.3661739.x

Abu Kwaik, Y., Gao, L. Y., Stone, B. J., Venkataraman, C., and Harb, O. S. (1998). Invasion of protozoa by *Legionella pneumophila* and its role in bacterial ecology and pathogenesis. *Appl. Environ. Microbiol.* 64, 3127–3133.

Bachate, S. P., Khapare, R. M., and Kodam, K. M. (2012). Oxidation of arsenite by two ß-proteobacteria isolated from soil. *Appl. Microbiol. Biotechnol.* 93, 2135–2145. doi: 10.1007/s00253-011-3606-7

Barker, J., Scaife, H., and Brown, M. R. (1995). Intraphagocytic growth induces an antibiotic-resistant phenotype of *Legionella pneumophila*. *Antimicrob. Agents Chemother.* 39, 2684–2688. doi: 10.1128/AAC.39.12.2684

Berg, G., Eberl, L., and Hartmann, A. (2005). The rhizosphere as a reservoir for opportunistic human pathogenic bacteria. *Environ. Microbiol.* 7, 1673–1685. doi: 10.1111/j.1462-2920.2005.00891.x

Bianchi, F., Careri, M., Mustat, L., Malcevschi, A., and Musci, M. (2005). Bioremediation of toluene and naphthalene: development and validation of a GC-FID method for their monitoring. *Ann. Chim.* 95, 515–524. doi: 10.1002/adic.200590061

Biederman, L., Rosen, M. R., Bobik, B. S., and Roberts, A. L. (2015). Bordetella petrii recovered from chronic pansinusitis in an adult with cystic fibrosis. *IDCases* 2, 97–98. doi: 10.1016/j.idcr.2015.09.004

Cavalca, L., Zanchi, R., Corsini, A., Colombo, M., Romagnoli, C., Canzi, E., et al. (2010). Arsenic-resistant bacteria associated with roots of the wild *Cirsium arvense* (L.) plant from an arsenic polluted soil, and screening of potential plant growth-promoting characteristics. *Syst. Appl. Microbiol.* 33, 154–164. doi: 10.1016/j.syapm.2010.02.004

Colinon, C., Deredjian, A., Hien, E., Brothier, E., Bouziri, L., Cournoyer, B., et al. (2013). Detection and enumeration of *Pseudomonas aeruginosa* in soil and manure assessed by an ecfX qPCR assay. *J. Appl. Microbiol.* 114, 1734–1749. doi: 10.1111/jam.12189

Delavat, F., Lett, M. C., and Lièvremont, D. (2013). Yeast and bacterial diversity along a transect in an acidic, As-Fe rich environment revealed by cultural approaches. *Sci. Total Environ.* 463–464, 823–828. doi: 10.1016/j.scitotenv.2013.06.023

Diavatopoulos, D. A., Cummings, C. A., Schouls, L. M., Brinig, M. M., Relman, D. A., and Mooi, F. R. (2005). *Bordetella pertussis*, the causative agent of whooping cough, evolved from a distinct, human-associated lineage of *B. bronchiseptica*. *PLoS Pathog.* 1:e45. doi: 10.1371/journal.ppat.0010045

Eriksson, M., Sodersten, E., Yu, Z., Dalhammar, G., and Mohn, W. W. (2003). Degradation of polycyclic aromatic hydrocarbons at low temperature under aerobic and nitrate-reducing conditions in enrichment cultures from northern soils. *Appl. Environ. Microbiol.* 69, 275–284. doi: 10.1128/AEM.69.1.275-284.2003

Fry, N. K., Duncan, J., Malnick, H., Warner, M., Smith, A. J., Jackson, M. S., et al. (2005). *Bordetella petrii* clinical isolate. *Emerging Infect. Dis.* 11, 1131–1133. doi: 10.3201/eid1107.050046

Goodnow, R. A. (1980). Biology of *Bordetella bronchiseptica*. *Microbiol. Rev.* 44, 722–738.

Gross, R., Guzman, C. A., Sebaihia, M., dos Santos, V. A., Pieper, D. H., Koebnik, R., et al. (2008). The missing link: *Bordetella petrii* is endowed with both the metabolic versatility of environmental bacteria and virulence traits of pathogenic Bordetellae. *BMC Genomics* 9:449. doi: 10.1186/1471-2164-9-449

Guo, Y., Zhang, J., Yan, Y., Wu, J., Zhu, N., and Deng, C. (2015). Molecular phylogenetic diversity and spatial distribution of bacterial communities in cooling stage during swine manure composting. *Asian-Australas. J. Anim Sci.* 28, 888–895. doi: 10.5713/ajas.14.0882

Ivanov, Y. V., Linz, B., Register, K. B., Newman, J. D., Taylor, D. L., Boschert, K. R., et al. (2016). Identification and taxonomic characterization of *Bordetella pseudohinzii* sp. nov. isolated from laboratory-raised mice. *Int. J. Syst. Evol. Microbiol.* 66, 5452–5459. doi: 10.1099/ijsem.0.001540

Ivanov, Y. V., Shariat, N., Register, K. B., Linz, B., Rivera, I., Hu, K., et al. (2015). A newly discovered Bordetella species carries a transcriptionally active CRISPR-Cas with a small Cas9 endonuclease. *BMC Genomics* 16:863. doi: 10.1186/s12864-015-2028-9

Johnson, S. L., Bishop-Lilly, K. A., Ladner, J. T., Daligault, H. E., Davenport, K. W., Jaissle, J., et al. (2015). Complete genome sequences for 59 Burkholderia isolates, both pathogenic and near neighbor. *Genome Announc.* 3:e00159-15. doi: 10.1128/genomeA.00159-15

Kato, S., Haruta, S., Cui, Z. J., Ishii, M., and Igarashi, Y. (2004). Effective cellulose degradation by a mixed-culture system composed of a cellulolytic Clostridium and aerobic non-cellulolytic bacteria. *FEMS Microbiol. Ecol.* 51, 133–142. doi: 10.1016/j.femsec.2004.07.015

Kattar, M. M., Chavez, J. F., Limaye, A. P., Rassoulian-Barrett, S. L., Yarfitz, S. L., Carlson, L. C., et al. (2000). Application of 16S rRNA gene sequencing to identify *Bordetella hinzii* as the causative agent of fatal septicemia. *J. Clin. Microbiol.* 38, 789–794.

Kersters, K., Hinz, K. H., Hertle, A., Segers, P., Lievens, A., Siegmann, O., et al. (1984). *Bordetella avium* sp. nov., isolated from the respiratory tracts of turkeys and other birds. *Int. J. Syst. Bacteriol.* 34, 56–70. doi: 10.1099/00207713-34-1-56

Kigawa, R., Sano, C., Nishijima, M., Tazato, N., Kiyuna, T., Hayakawa, N., et al. (2013). Investigation of acetic acid bacteria isolated from the Kitora tumulus in Japan and their involvement in the deterioration of the plaster of the mural paintings. *Stud. Conserv.* 58, 30–40. doi: 10.1179/2047058412Y.0000000040

Ko, K. S., Peck, K. R., Oh, W. S., Lee, N. Y., Lee, J. H., and Song, J. H. (2005). New species of Bordetella, *Bordetella ansorpii* sp. nov., isolated from the purulent exudate of an epidermal cyst. *J. Clin. Microbiol.* 43, 2516–2519. doi: 10.1128/JCM.43.5.2516-2519.2005

Li, X., Hu, Y., Gong, J., Zhang, L., and Wang, G. (2013). Comparative genome characterization of Achromobacter members reveals potential genetic determinants facilitating the adaptation to a pathogenic lifestyle. *Appl. Microbiol. Biotechnol.* 97, 6413–6425. doi: 10.1007/s00253-013-5018-3

Librado, P., and Rozas, J. (2009). DnaSP v5: a software for comprehensive analysis of DNA polymorphism data. *Bioinformatics* 25, 1451–1452. doi: 10.1093/bioinformatics/btp187

Lodders, N., and Kämpfer, P. (2012). A combined cultivation and cultivation-independent approach shows high bacterial diversity in water-miscible metalworking fluids. *Syst. Appl. Microbiol.* 35, 246–252. doi: 10.1016/j.syapm.2012.03.006

Matsumura, Y., Hosokawa, C., Sasaki-Mori, M., Akahira, A., Fukunaga, K., Ikeuchi, T., et al. (2009). Isolation and characterization of novel bisphenol-A-degrading bacteria from soils. *Biocontrol Sci.* 14, 161–169. doi: 10.4265/bio.14.161

Mattoo, S., and Cherry, J. D. (2005). Molecular pathogenesis, epidemiology, and clinical manifestations of respiratory infections due to *Bordetella pertussis* and other *Bordetella* subspecies. *Clin. Microbiol. Rev.* 18, 326–382. doi: 10.1128/CMR.18.2.326-382.2005

Molmeret, M., Horn, M., Wagner, M., Santic, M., and Abu Kwaik, Y. (2005). Amoebae as training grounds for intracellular bacterial pathogens. *Appl. Environ. Microbiol.* 71, 20–28. doi: 10.1128/AEM.71.1.20-28.2005

Musser, J. M., Hewlett, E. L., Peppler, M. S., and Selander, R. K. (1986). Genetic diversity and relationships in populations of *Bordetella* spp. *J. Bacteriol.* 166, 230–237. doi: 10.1128/jb.166.1.230-237.1986

Nagata, J. M., Charville, G. W., Klotz, J. M., Wickremasinghe, W. R., Kann, D. C., Schwenk, H. T., et al. (2015). *Bordetella petrii* sinusitis in an immunocompromised adolescent. *Pediatr. Infect. Dis. J.* 34, 458. doi: 10.1097/INF.0000000000000564

Nisola, G. M., Tuuguu, E., Farnazo, D. M., Han, M., Kim, Y., Cho, E., et al. (2010). Hydrogen sulfide degradation characteristics of *Bordetella* sp. Sulf-8 in a biotrickling filter. *Bioprocess. Biosyst. Eng.* 33, 1131–1138. doi: 10.1007/s00449-010-0440-8

Ocklitz, H. W., and Milleck, J. (1967). Die überlebenszeit von pertussisbakterien außerhalb des kranken. experimentelle untersuchungen zur keuchhustenepidemiologie. *Zentralblatt Bakteriologie Parasitenkunde Infektionskrankheiten* 203, 79–91.

Park, J., Zhang, Y., Buboltz, A. M., Zhang, X., Schuster, S. C., Ahuja, U. et al. (2012). Comparative genomics of the classical Bordetella subspecies: the evolution and exchange of virulence-associated diversity amongst closely related pathogens. *BMC Genomics* 13:545. doi: 10.1186/1471-2164-13-545

Parkhill, J., Sebaihia, M., Preston, A., Murphy, L. D., Thomson, N., Harris, D. E., et al. (2003). Comparative analysis of the genome sequences of *Bordetella pertussis*, *Bordetella parapertussis* and *Bordetella bronchiseptica*. *Nat. Genet.* 35, 32–40. doi: 10.1038/ng1227

Partanen, P., Hultman, J., Paulin, L., Auvinen, P., and Romantschuk, M. (2010). Bacterial diversity at different stages of the composting process. *BMC Microbiol.* 10:94. doi: 10.1186/1471-2180-10-94

Register, K. B., Ivanov, Y. V., Jacobs, N., Meyer, J. A., Goodfield, L. L., Muse, S. J. et al. (2015). Draft genome sequences of 53 genetically distinct isolates of *Bordetella bronchiseptica* representing 11 terrestrial and aquatic hosts. *Genome Announc.* 3:e00152-15. doi: 10.1128/genomeA.00152-15

Register, K. B., and Kunkle, R. A. (2009). Strain-specific virulence of *Bordetella hinzii* in poultry. *Avian Dis.* 53, 50–54. doi: 10.1637/8388-070108-Reg.1

Sanuth, H. A., Yadav, A., Fagade, O. E., and Shouche, Y. (2013). epsilon-caprolactam utilization by *Proteus* sp. and *Bordetella* sp. Isolated from solid waste dumpsites in Lagos State, Nigeria, first *report. Indian J. Microbiol.* 53, 221–226. doi: 10.1007/s12088-013-0356-5

Sebaihia, M., Preston, A., Maskell, D. J., Kuzmiak, H., Connell, T. D., King, N. D. et al. (2006). Comparison of the genome sequence of the poultry pathogen Bordetella avium with those of B. bronchiseptica, B. pertussis, and B. parapertussis reveals extensive diversity in surface structures associated with host interaction. *J. Bacteriol.* 188, 6002–6015. doi: 10.1128/JB.01927-05

Shepard, C. W., Daneshvar, M. I., Kaiser, R. M., Ashford, D. A., Lonsway, D., Patel, J. B., et al. (2004). *Bordetella holmesii* bacteremia: a newly recognized clinical entity among asplenic patients. *Clin. Infect. Dis.* 38, 799–804. doi: 10.1086/381888

Stainer, D. W., and Scholte, M. J. (1970). A simple chemically defined medium for the production of phase I *Bordetella pertussis*. *J. Gen. Microbiol.* 63, 211–220. doi: 10.1099/00221287-63-2-211

Tamura, K., Dudley, J., Nei, M., and Kumar, S. (2007). MEGA4: Molecular Evolutionary Genetics Analysis (MEGA) software version 4.0. *Mol. Biol. Evol.* 24, 1596–1599. doi: 10.1093/molbev/msm092

Tazato, N., Handa, Y., Nishijima, M., Kigawa, R., Sano, C., and Sugiyama, J. (2015). Novel environmental *Bordetella* species isolated from the plaster wall surface of mural paintings in the Takamatsuzuka tumulus: *Bordetella muralis* sp. nov., *Bordetella tumulicola* sp. nov. and *Bordetella tumbae* sp. nov. *Int. J. Syst. Evol. Microbiol.* 65, 4830–4838. doi: 10.1099/ijsem.0.000655

Tian, W., Sun, Q., Xu, D. B., Zhang, Z. H., Chen, D., Li, C. Y., et al. (2013). Succession of bacterial communities during composting process as detected by 16S rRNA clone libraries analysis. *Int. Biodeterior. Biodegradation* 78, 58–66. doi: 10.1016/j.ibiod.2012.12.008

Toups, M., Wübbeler, J. H., and Steinbüchel, A. (2010). Microbial utilization of the industrial wastewater pollutants 2-ethylhexylthioglycolic acid and iso-octylthioglycolic acid by aerobic gram-negative bacteria. *Biodegradation* 21, 309–319. doi: 10.1007/s10532-009-9302-y

Vandamme, P. A., Peeters, C., Cnockaert, M., Inganäs, E., Falsen, E., Moore, E. R., et al. (2015). *Bordetella bronchialis* sp. nov., *Bordetella flabilis* sp. nov. and *Bordetella sputigena* sp. nov., isolated from human respiratory specimens, and reclassification of *Achromobacter sediminum* Zhang et al. 2014 as *Verticia sediminum* gen. nov., comb. nov. *Int. J. Syst. Evol. Microbiol.* 65, 3674–3682. doi: 10.1099/ijsem.0.000473

Vandamme, P., Heyndrickx, M., Vancanneyt, M., Hoste, B., De Vos, P., Falsen, E., et al. (1996). *Bordetella trematum* sp. nov., isolated from wounds and ear infections in humans, and reassessment of *Alcaligenes denitrificans* Ruger and Tan 1983. *Int. J. Syst. Bacteriol.* 46, 849–858. doi: 10.1099/00207713-46-4-849

Vandamme, P., Hommez, J., Vancanneyt, M., Monsieurs, M., Hoste, B., Cookson, B., et al. (1995). *Bordetella hinzii* sp. nov., isolated from poultry and humans. *Int. J. Syst. Bacteriol.* 45, 37–45. doi: 10.1099/00207713-45-1-37

von Wintzingerode, F., Schattke, A., Siddiqui, R. A., Rösick, U., Göbel, U. B., and Gross, R. (2001). *Bordetella petrii* sp. nov., isolated from an anaerobic bioreactor, and emended description of the genus Bordetella. *Int. J. Syst. Evol. Microbiol.* 51, 1257–1265. doi: 10.1099/00207713-51-4-1257

Wang, F., Grundmann, S., Schmid, M., Dörfler, U., Roherer, S., Charles Munch, J., et al. (2007). Isolation and characterization of 1,2,4-trichlorobenzene mineralizing *Bordetella* sp. and its bioremediation potential in soil. *Chemosphere* 67, 896–902. doi: 10.1016/j.chemosphere.2006.11.019

Weisskopf, L., Heller, S., and Eberl, L. (2011). *Burkholderia* species are major inhabitants of white lupin cluster roots. *Appl. Environ. Microbiol.* 77, 7715–7720. doi: 10.1128/AEM.05845-11

Weyant, R. S., Hollis, D. G., Weaver, R. E., Amin, M. F., Steigerwalt, A. G., O'Connor, S. P., et al. (1995). *Bordetella holmesii* sp. nov., a new gram-negative species associated with septicemia. *J. Clin. Microbiol.* 33, 1–7.

Winiecka-Krusnell, J., and Linder, E. (1999). Free-living amoebae protecting Legionella in water: the tip of an iceberg? *Scand. J. Infect. Dis.* 31, 383–385. doi: 10.1080/00365549950163833

Yang, F., Zhou, Y., Yin, L., Zhu, G., Liang, G., and Pu, Y. (2014). Microcystin-degrading activity of an indigenous bacterial strain *Stenotrophomonas acidaminiphila* MC-LTH2 isolated from Lake Taihu. *PLoS ONE* 9:e86216. doi: 10.1371/journal.pone.0086216

11

In-Depth Characterization and Functional Analysis of Clonal Variants in a *Mycobacterium tuberculosis* Strain Prone to Microevolution

Yurena Navarro[1,2,3,4,5], *Laura Pérez-Lago*[1,2,3], *Marta Herranz*[1,2,3], *Olalla Sierra*[1,2], *Iñaki Comas*[6,7], *Javier Sicilia*[2,8], *Emilio Bouza*[1,2,3,9] *and Darío García de Viedma*[1,2,3,4]*

[1] Servicio Microbiología Clínica y Enfermedades Infecciosas, Hospital General Universitario Gregorio Marañón, Madrid, Spain, [2] Instituto de Investigación Sanitaria Gregorio Marañón, Madrid, Spain, [3] CIBER Enfermedades Respiratorias, CIBERES, Madrid, Spain, [4] CEI Campus Moncloa, UCM-UPM, Madrid, Spain, [5] Centro de Vigilancia Sanitaria Veterinaria, Universidad Complutense Madrid, Madrid, Spain, [6] Unidad Mixta Genómica y Salud, Centro Superior de Investigación en Salud Pública (FISABIO)-Universitat de València, Valencia, Spain, [7] CIBER en Epidemiología y Salud Pública, Madrid, Spain, [8] Unidad de Medicina y Cirugía Experimental, Hospital General Universitario Gregorio Marañón, Madrid, Spain, [9] Departamento de Medicina, Facultad de Medicina, Universidad Complutense de Madrid, Madrid, Spain

Edited by:
Manuel Espinosa,
Centro de Investigaciones Biológicas (CSIC), Spain

Reviewed by:
Anders Norman,
Statens Serum Institut, Denmark
Igor Mokrousov,
Saint Petersburg Pasteur Institute, Russia

***Correspondence:**
Darío García de Viedma
dgviedma2@gmail.com

The role of clonal complexity has gradually been accepted in infection by *Mycobacterium tuberculosis* (MTB), although analyses of this issue are limited. We performed an in-depth study of a case of recurrent MTB infection by integrating genotyping, whole genome sequencing, analysis of gene expression and infectivity in *in vitro* and *in vivo* models. Four different clonal variants were identified from independent intrapatient evolutionary branches. One of the single-nucleotide polymorphisms in the variants mapped in *mce*3R, which encodes a repressor of an operon involved in virulence, and affected expression of the operon. Competitive *in vivo* and *in vitro* co-infection assays revealed higher infective efficiency for one of the clonal variants. A new clonal variant, which had not been observed in the clinical isolates, emerged in the infection assays and showed higher fitness than its parental strain. The analysis of other patients involved in the same transmission cluster revealed new clonal variants acquired through novel evolutionary routes, indicating a high tendency toward microevolution in some strains that is not host-dependent. Our study highlights the need for integration of various approaches to advance our knowledge of the role and significance of microevolution in tuberculosis.

Keywords: microevolution, *Mycobacterium tuberculosis*, functional analysis, *in vitro* infections, *in vivo* infections, whole genome sequencing

INTRODUCTION

The idea that *Mycobacterium tuberculosis* (MTB) infection of a single case is caused by a single strain is increasingly debated. Genotyping has enabled us to describe cases of co-infection by ≥2 different strains (mixed infection) and coexistence of clonal variants of the same strain (polyclonal infection) (García de Viedma et al., 2005; Shamputa et al., 2006; Al-Hajoj et al., 2010; Navarro et al., 2011; Huyen et al., 2012; Muwonge et al., 2013). In the latter, clonal variants emerge through microevolution phenomena, which make it possible to detect subtle changes when applying standard fingerprinting strategies (IS*6110*-RFLP or mycobacterial interspersed repetitive units–variable number of tandem repeats [MIRU-VNTR]). These changes can have a functional impact on the expression of neighboring genes and can increase the variability of antigenic proteins (Akhtar et al., 2009; Olsen et al., 2009; Yindeeyoungyeon et al., 2009; Tantivitayakul et al., 2010; Perez-Lago et al., 2011, 2013).

Most studies on microevolution in infection by MTB or on the functional characterization of MTB clonal variants merely describe the genotypic changes involved. If they include a functional analysis, then this is only partially based on gene expression assays or standard infection models. In our study, we attempted to overcome this fragmented view of clonal complexity of MTB infection by performing an in-depth analysis of a representative case of clonally complex MTB infection by integrating the following approaches: (i) complete characterization of the variants by both standard genotyping and whole genome sequencing (WGS); (ii) gene expression analysis; and (iii) evaluation of infectivity by analyzing the behavior of clonal variants not only in standard infections, but also in co-infections in broader experimental conditions than usual, i.e., applying both cellular and animal models. We show how integrating various approaches can advance our knowledge of the role and significance of microevolution in MTB infection.

MATERIALS AND METHODS

Mycobacterium tuberculosis Clonal Variants

The clonal variants (A and B) were isolated from a patient with recurrent tuberculosis and identified using 15-locus MIRU-VNTR, as described in Martín et al. (2011).

Expanded Characterization of the Clonal Variants

24-Locus MIRU-VNTR

Clonal variants were further characterized using 24-locus MIRU-VNTR (Oelemann et al., 2007), as described in Navarro et al. (2011), and IS*6110*-restriction fragment length polymorphism (RFLP), following international standardization guidelines (van Embden et al., 1993).

Ligation-Mediated PCR (LM-PCR)

IS*6110* sequences were mapped using LM-PCR, as described in Perez-Lago et al. (2011).

Alelle-Specific-PCR (ASO-PCR)

Alelle-specific PCRs were designed to analyze single colonies to assign the alelles present for several SNPs identified in the following Rvs: Rv1201, 1527, 1553, 1963, 2209, 2579.

The reactions were carried out in 50 μl including $MgCl_2$ (1.2, 0.9, 1.5, 1.3, 1.3, and 1.3 mM; for the different SNPs in the order indicated above), 0.4 μl dNTPs (10 μM); 0.5 μl DMSO, 1.5 μl s of each primer (Supplementary Table; 10 μM), 0.4 μl AmpliTaq Gold enzyme. The PCR run was constituted by 30 rounds of the cycle: 10′ 95°C, 1′ 95°C, 1′ Tm (64, 60, 63, 64, 64, 64°C, for the different SNPs in the order indicated above), 10′ 72°C and a final tail of 4°C.

For each SNP, two independent PCRs were run. The first one using the selective primer complementary for the allele in the SNP and, as a control, a second PCR targeting the complementary allele. The expected sizes for the products were 239, 402, 239, 207, 234, and 147 bp (for the different SNPs in the order indicated above).

Whole Genome Sequencing

WGS was performed as detailed elsewhere (Perez-Lago et al., 2014) using a HiSeq 2000 device and a Miseq device (Illumina), which generated 101-51–bp paired-end reads. We mapped the reads for each strain using the Burrows-Wheeler Aligner and the ancestral MTB genome as detailed elsewhere (Comas et al., 2013). SNP calls were made with SAMtools and VarScan (coverage of at least 10×, mean SNP mapping quality of 20). The genome was compared between strains using an in-house script written in R.

Gene Expression Assay

Relative quantification assays based on reverse transcriptase polymerase chain reaction (RT-PCR) were performed to examine expression of the gene involved in the microevolution events. The clonal variants were grown in 7H9 liquid medium (Difco) supplemented with 10% ADC (Becton Dickinson) and 1% Tween 80 (Merck) for 3 weeks. Cell lysis and RNA extraction were performed as previously described (Perez-Lago et al., 2013). RNA was reversed-transcribed using the High Capacity RNA-to-cDNA Kit (Invitrogen, Life Technologies, CA, USA). This step was followed by qRT-PCR amplification (preincubation at 95°C for 10 min and 45 cycles of 95°C for 10 s, 60°C for 10 s, and 72°C for 20 s) using the LightCycler® FastStart DNA Master SYBR Green I kit (Roche) and the primers FyrbE3A (5′-GGTGTTTCTCATGCACGTCT-3′) and RyrbE3A (5′-CCG ACCGACATGCCCTTATA-3′). The results of the assay were expressed as the ratio of the values obtained for the variant harboring the SNP to the values from the wild-type variant. Three independent quantitative RT-PCR assays were performed using 2 independent RNA extractions. A 1-sample *t*-test was used to determine whether the average expression ratio was statistically different from 1 ($p < 0.05$).

Infection Assays

In vivo Infection Model

Specific pathogen–free 8-week-old female Balb/c mice were obtained from Charles River Laboratories (L'Arbresle, France) and from our animal experimental laboratory (Instituto de Investigación Sanitaria Gregorio Marañón, Madrid, Spain). The mice were shipped under appropriate conditions, with the corresponding certificate of health and origin. All the animals were kept under controlled conditions in a biosafety level 3 facility with food and water *ad libitum*.

Mice were anesthetized by intraperitoneal injection with xylazine (0.75 mg/g) and ketamine (0.1 mg/g) and subsequently infected with 200 μl of inoculum (1–5 × 10^5 bacilli) by intravenous inoculation in the lateral tail vein.

Monitoring of infections by each clonal variant was based on colony forming unit (CFU) counts. The lungs and spleen of the infected mice were individually homogenized in PBS supplemented with 0.05% Tween 80. Serial 10-fold dilutions of the homogenate were plated on Middlebrook 7H11 agar (Difco). The growth rates for each clonal variant were calculated by linear regression of the CFU logarithm. Eighteen mice were infected, and 3 mice were analyzed at each point for each strain (day 1, week 1, and week 3).

The results of the assays in which the mice were simultaneously co-infected with the 2 clonal variants were analyzed by comparing the proportion of co-infecting strains before infection (adjusting to a 1:1 proportion) and after infection. The lungs and spleen of the infected mice were individually homogenized and plated at each time point (day 1 and week 5) to analyze the allelic value of locus MIRU 42 in 40 single colonies. Ten mice were co-infected, and standard deviations were calculated on the basis of the results obtained from 5 mice at each time point.

In vitro Infection Model

THP-1 cells were differentiated to macrophages and simultaneously co-infected with both clonal variants following the protocol described by Alonso et al. (2010) at a multiplicity of infection of 3 bacteria per cell. The proportion of co-infecting variants before and after infection was calculated at 3 h and day 7 by plating 10-fold serial dilutions of lysates on Middlebrook 7H11 agar (Difco). Seventy colonies were analyzed at each time point using simplex PCR of locus MIRU 42.

Fitness Assay

Clonal variants were subcultured on Mycobacteria Growth Indicator Tubes (BACTEC MGIT 960 System; Becton Dickinson) supplemented with BBL™ PANTA™ and BACTEC MGIT 960 Growth Supplement, as indicated by the manufacturer. Inocula were obtained from positive tubes (3 days) following the protocol used for antimicrobial susceptibility testing. Three MGITs were inoculated after ensuring equivalent bacterial concentrations in the 3 aliquots of inoculum used by plating them on Middlebrook 7H11 agar (Difco). Growth curves were obtained by monitoring the growth units (GU) every hour using BD EpiCenter™. The fitness of the clonal variants was compared based on 2 parameters taken from the growth curves: (i) lag phase (time to a positive threshold [75 GU]) and (ii) rate growth (time required for the 4,000–6,000–GU increase).

The means and standard deviations were determined, and one-way analysis of variance with repeated measures was used to determine *P*-values, which were adjusted using a Bonferroni correction.

RESULTS

Complete Characterization of Clonal Variants

Four isolates from the patient with recurrent infection were available for analysis, 1 for the first episode (July 2000) and 3 for the second one (August-November 2001). Available genotyping data for the clonal variants using 15-locus MIRU-VNTR were further analyzed using 24-locus MIRU-VNTR. Differences in the VNTR analysis were restricted to a single-locus variant in locus MIRU 42 (424/Mtub04) (3 repetitions in the first isolate [VNTR variant A] and 1 repetition in the last isolate [VNTR variant B]) (**Figure 1A**). A mixture of VNTR variants A and B (detected by the observations of the corresponding double alleles in MIRU 42) was detected in the 2 intermediate specimens (August), thus leading us to select the first and last isolates for further genotyping.

According to RFLP, the differential IS*6110* band detected in one of the 2 variants (**Figure 1A**) was localized using LM-PCR and mapped at the location described for one of the 16 insertions of H37Rv (3551227-3552587).

WGS revealed 11 different SNPs between the 2 variants. After comparing the WGS data with the most recent common ancestor of MTB as a reference, we obtained 6 specific SNPs from variant A and 5 from variant B (**Table 1**). Of the 6 specific SNPs from variant A, 3 were non-synonymous and 3 were synonymous. Of the 5 SNPs from variant B, 4 were non-synonymous, and the remaining one was intergenic.

Analysis of Functional Relevance

The next step in the characterization of these clonal variants was to determine the functional relevance of the subtle modifications identified. The IS*6110* variation mapped in a hotspot for

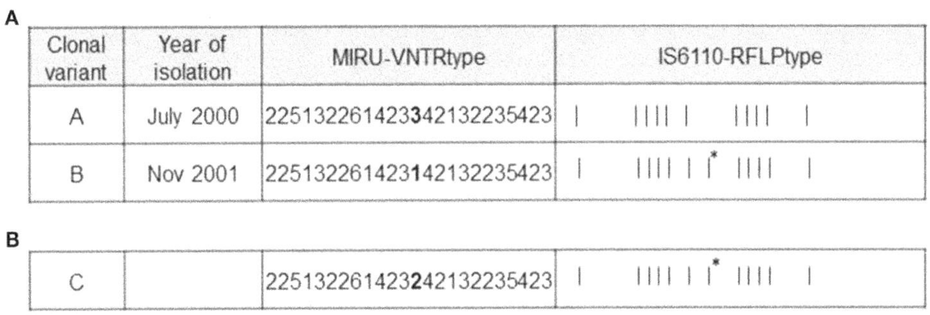

A

Clonal variant	Year of isolation	MIRU-VNTRtype	IS6110-RFLPtype
A	July 2000	22513226142**3**3**4**2132235423	
B	Nov 2001	22513226142**3**1**4**2132235423	

B

C		225132261423**2**42132235423	

FIGURE 1 | Genotypic features of clonal variants A, B, and C. (A) Genotypes of variants A and B. MIRU-VNTR loci differing between the variants are highlighted in bold. The asterisks indicate the additional IS*6110* band in variant B. **(B)** Genotypes of variant C. MIRU-VNTR loci differing between variants A and B are highlighted in bold. The asterisk indicates the same additional IS*6110* band obtained from variant B.

TABLE 1 | Specific SNPs from variants obtained by WGS.

Variant A	Ancestor	S/NS/1	Gene
T	C	NS	Rv1963c (*mce3R*)
C	A	NS	Rv2579 (*dhaA*)
T	C	NS	Rv2921c (*ftsY*)
T	C	s	Rv0836c
G	T	s	Rv1487
G	C	s	Rv1497 (*lipL*)
Variant B	**Ancestor**	**S/NS/1**	**Gene**
T	C	NS	Rv1201c
T	G	NS	Rv1527c (*pks5*)
A	G	NS	Rv1553 (*frdB*)
G	C	NS	Rv2209
C	G	I	Rv2644c–Rv2645

S, synonymous SNP; NS, Non-synonymous SNP; I, Intergenic SNP.

the IS*6110* insertion sequence, thus minimizing the potential functional significance of this variation. The MIRU-VNTR modifications mapped in a region between 2 stop codons for 2 adjacent genes (Rv0353 and Rv0354), again reducing the likelihood of a functional impact for these changes. In contrast, the fact that some of the SNP-based variability revealed by WGS corresponded to non-synonymous changes in relevant genes made it worthwhile to evaluate their functional impact.

We evaluated the potential impact of SNP-based variability by focusing on the SNP in Rv1963c (*mce3R*), because the SNP maps in a repressor of several genes included in the yrbE3A-Rv1971 operon (Santangelo et al., 2008; **Figure 2**). We compared the expression of the first gene in the operon, *yrbE3A,* in the 2 variants. A 0.7910-fold ratio (0.6936–0.8884; $p < 0.05$) was observed in the expression of *yrbE3A* in variant A compared with variant B, indicating that the SNP detected in variant A increased the efficiency of the repressor Mce3R.

Infectivity of Clonal Variants

The infectivity of the clonal variants was measured in both murine and cellular models. In the murine model, no significant differences were found in the growth rates of each variant in the lung (0.8039 ± 0.1707 and 0.8658 ± 0.1683 for variants A and B) or in the spleen (0.2328 ± 0.2996 and 0.5926 ± 0.1911 for variants A and B).

To evaluate the presence of subtle differences between the infectivity of the variants that were not revealed in standard infection assays, we performed competitive assays by simultaneously infecting mice with the 2 clonal variants. We did not detect differences in the proportion of the variants at day 1 compared with the proportion in the inoculum (both in lung and in spleen **Figure 3A**). However, the representativeness of variant B between day 1 and week 5 increased both in the lung (from 0.3095 ± 0.0336 at day 1 to 0.5058 ± 0.0950 at week 5) and in the spleen (from 0.2019 ± 0.0624 at day 1 to 0.3638 ± 0.0779 at week 5) ($p < 0.05$) (**Figure 3A**).

In order to obtain new evidence of the likely higher infectivity of clonal variant B, we performed a new co-infection experiment. On this occasion, we used macrophages and modified the inoculum to force underrepresentation of clonal variant B (96:4). Even at this unbalanced proportion, more efficient uptake was observed for variant B than for variant A, meaning that the 96:4 proportion in the inoculum became 61:39 at 3 h (**Figure 3B**). Measurements could not be taken appropriately at day 7 owing to an unexpected finding, namely, the emergence of a previously undetected clonal variant (variant C: 2 repetitions at locus MIRU 42, **Figure 1B**). The 3 variants were now found at a proportion 62:10:28 (A:B:C). Consequently, the representativeness of variant A remained constant compared with that observed at 3 h after infection, whereas the presence of variant B decreased.

Characterization of the New Clonal Variant

MIRU-VNTR analysis indicated that variant C only harbored a difference in locus MIRU 42 (allelic value 2). The IS*6110*-RFLP type was identical to that of clonal variant B, suggesting that variant B was its parental strain (**Figure 1B**). WGS of the new variant revealed no SNPs with respect to variant B, confirming that variant C derived from variant B.

The fitness of variant C was compared with that of the other 2 variants and proved to be higher, as indicated by its shorter lag phase (219 ± 5.292 h vs. 260 ± 12.12 h [variant B] and 255.3 ± 12.5 [variant A]; $p < 0.01$) (**Figure 4A**) and rate of growth (25.62 ± 1.22 vs. 71.23 ± 16.32 h [variant B] and 56.84 ± 5.569 h [variant A]; $p < 0.05$) (**Figure 4B**).

Analysis of Intrapatient Microevolution Dynamics

In order to determine the dynamics by which variant A (July 2000) was replaced by variant B (November 2001), we recovered the 2 intermediate specimens (August 2001, specimens 1 and 2) in which the 2 variants coexisted to evaluate whether it was possible to track the progressive elimination of variant A. Taking advantage of the differential SNPs found between the 2 variants, we selected the seven non-synonymous SNPs to assess the representativeness of each variant by analyzing multiple single colonies. For this aim we tailored allele specific PCRs to directly track the two alternative alleles in the seven SNPs. The ASO-PCR targeted to the analysis of the SNP in Rv2921 could not be optimized. The analysis of the remaining six SNPs on 30 colonies (12 and 18 from specimens 1 and 2, respectively) revealed that in the specimen 1 coexisted the variant A with two novel variants (new variant 1 and 2), sharing 4 and five alleles out of six, respectively, with variant B (**Table 2**). In the specimen 2 we observed the coexistence of variant A with the new variant 2.

The detailed analysis of the intermediate isolates allowed us to rule out our firstly assumed hypothesis of a chronological substitution of variant A by variant B. Instead, there was a coexistence of new evolutionary intermediates acquiring sequentially the variant B alleles (**Figure 5A**). In summary we could identify 4 independent clonal variants involved at different stages of the microevolutionary process along the infection.

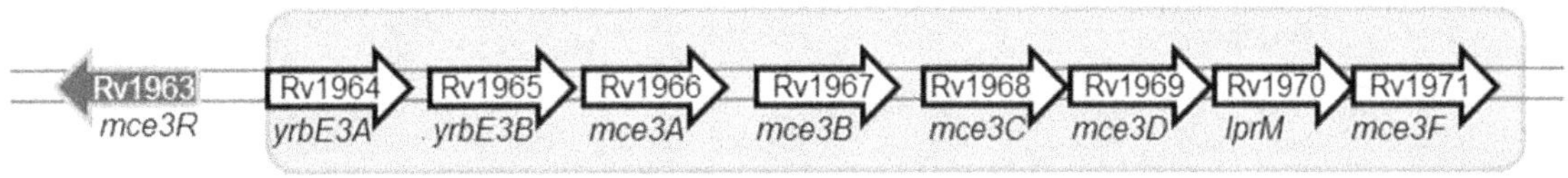

FIGURE 2 | Diagram showing the location of the gene *mce3R* within the *mce3R* operon.

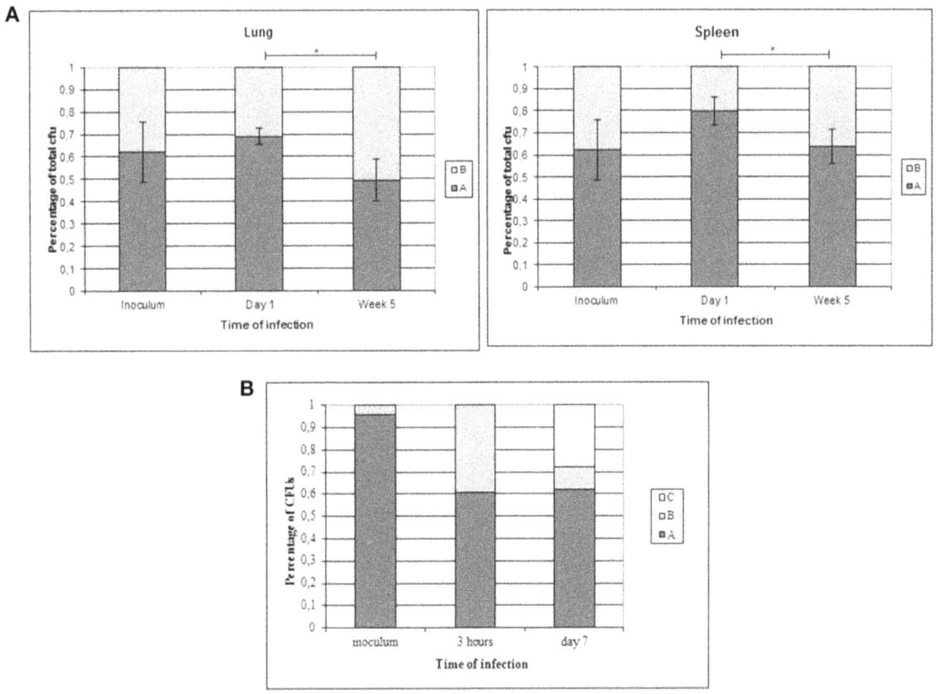

FIGURE 3 | Co-infection assays with both clonal variants. Asterisks indicate statistical significance ($p < 0.05$). **(A)** Proportion of co-infecting variants obtained in the lung (left panel) and spleen (right panel) from 5 mice at each time point. **(B)** Co-infection of THP-1 cells with the clonal variants.

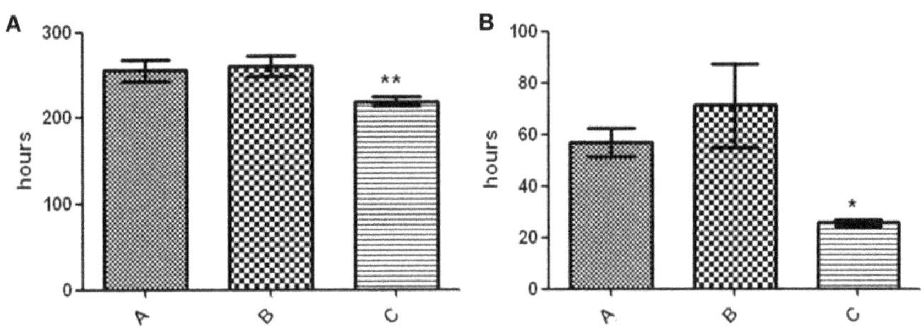

FIGURE 4 | Fitness assay of the 3 variants. (A) Lag phase (time to positive threshold [75 GU]). The asterisks indicate statistical significance ($p < 0.01$). **(B)** Growth rate (time required for the 4,000–6,000-GU increase). The asterisk indicates statistical significance ($p < 0.05$).

Analysis of Interpatient Microevolution Dynamics in Clustered Patients

Our final step was to search for other cases in the population involved in a transmission cluster with the case we report. We searched for cases infected by each of the initial and final variants A and B. For variant A, we identified 2 cases (C2 and C4, years 2003 and 2004 respectively) and for variant B none.

We performed WGS analysis with these 2 new isolates. Neither shared the alleles found in the SNPs of variants A and B with respect to the ancestor, indicating that they had

TABLE 2 | Distribution of SNPs (in bold, after comparing to the ancestor reference) identified in the clonal variants by analyzing single colonies.

		Episode 1	Episode 2					
		July 2000	August 2001:Specimen 1 (12 colonies)			August 2001:Specimen 2 (20 colonies)		November 2001
SNP:Gene	Ancestor	Variant A	Variant A (3)	New variant 1 (2)	New variant 2 (7)	Variant A (6)	New variant 2 (14)	Variant B
SNP1:Rv1201	G	G	G	**T**	**T**	G	**T**	**T**
SNP2:Rv1527	G	G	G	**A**	**A**	G	**A**	**A**
SNP3:Rv1553	C	C	C	C	**G**	C	**G**	**G**
SNP4:Rv1963	C	**T**	**T**	C	C	**T**	C	C
SNPS:Rv2209	G	G	G	G	G	G	G	**C**
SNP6:Rv2579	A	**C**	**C**	A	A	**C**	A	A
SNP7:Rv2921	C	**T**						C

The number between brackets indicates the number of colonies in which that composition of SNPs was identified.

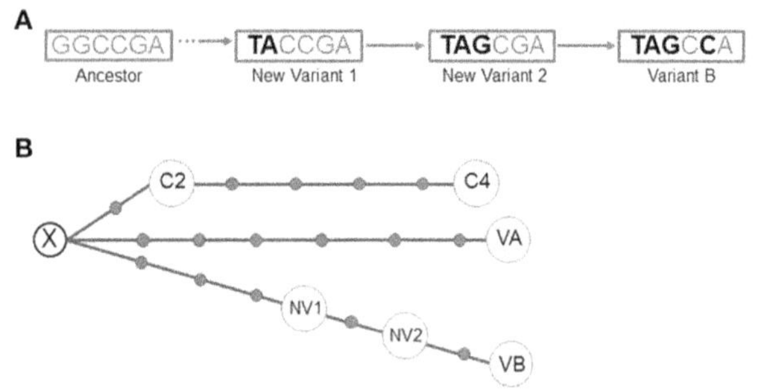

FIGURE 5 | (A) Schematic representation of the pathway of SNP acquisition for variant B. The dashed arrow indicates the lack of information to determine the potential intermediates between the ancestor and the new variant 1. **(B)** Median-joining network for all the variants from the 3 clustered cases (NETWORK 5.0). The circles correspond to the different variants (VA, VB, NV1, NV2, C2, and C4). Each dot along the lines linking the circles corresponds to a single-nucleotide polymorphism detected between the linked variants. A hypothetical unsampled variant (X) is postulated by the algorithm. VA (variant A), VB (variant B), NV1 (New variant 1), NV2 (New variant 2), C2 (clustered case 2), C4 (clustered case 4).

followed an independent microevolutionary branch (**Figure 5B**) from a common unsampled node (X). C2 and C4 were sequential steps of that branch. The SNP acquired in C2 was synonymous and mapped in Rv0037 (conserved membrane protein). Three of the 4 SNPs from C4 mapped in Rv0015 (non-synonymous, transmembrane serine/threonine-protein kinase A), Rv1821 (synonymous, preprotein translocase ATPase secA2), and Rv2934 (non-synonymous, phenolpthiocerol synthesis type-I polyketide synthase). The fourth SNP was intergenic.

DISCUSSION

We performed an in-depth study of the clonal variants that emerged in a patient with recurrent tuberculosis. Initial genotypic characterization confirmed that the strains isolated from each episode were clonal variants because they only differed in a single-locus variant in 24 MIRU loci and in a single IS*6110* copy. Similarly subtle differences have been reported for clonal variants in various studies (Shamputa et al., 2006; Al-Hajoj et al., 2010; Navarro et al., 2011).

WGS was applied to obtain a more exhaustive description of the degree of variability between the clonal variants, and 11 SNPs were identified. These figures are higher than expected according to the variability observed in transmission chains (similarity threshold of 12 SNPs before isolates can be considered clustered; Walker et al., 2013). Our findings indicate that intrapatient variability itself could be close to this threshold. Also, given previously reported mutation rates in MTB, usually within the range of 0.25–0.5 SNPs per genome per year, the accumulation of SNPs in the intrapatient variants is much higher than expected. It suggests a higher tendency to acquire variability by this strain, but we must also leave open the possibility that part of that variability could have been acquired during the latency period, as it has been suggested elsewhere (Lillebaek et al., 2016).

We compared the alleles found in the 11 SNPs with those found in the most recent common ancestor of MTB as a reference (Comas et al., 2010, 2013) and found that variant B did not evolve directly from variant A, because 6 and 5 alleles found in the 11 SNPs were specific to variants A and B, respectively. Therefore, each variant represented an independent evolutionary path from a common parental strain that had not been sampled in the specimens analyzed. The results obtained from this new analysis fitted better with the consensus variability thresholds (Walker et al., 2013), thus illustrating that it is essential to integrate allelic data from the ancestor in order to interpret SNP data appropriately and that this step is needed to establish the true phylogenetic relationships between clonal variants.

Our naïve initial interpretation of 2 intrapatient variants emerging one from the other became a more complex picture of 2 independent evolutionary branches from an unsampled common ancestor. This complexity increased after we found that the variants from the intermediate specimens (August), which were initially interpreted as co-infection of variants A and B, corresponded to mixtures of variant A with 2 novel intermediate variants. The detection of these new variants allowed us to infer partially the pathway followed in the emergence of variant B. Our findings are consistent with those reported for macaque models (Lin et al., 2014), where different clonal variants in different lesions evolved independently, that is, some finished in a *cul de sac*, whereas others progressed. The finding of different clonal variants in different lesions could lead to a heterogeneous drain of variants into respiratory samples and differential identification of one variant or another.

The variability of the evolutionary routes explored by our strain was also illustrated by the analysis of a further 2 cases of tuberculosis in the population involved in the same transmission cluster as the first study patient. The isolates from these 2 cases shared the same evolutionary branch, although this was independent of that of the intrapatient variants A and B.

Taken together, the SNP data and the subtle differences identified by RFLP and MIRU-VNTR suggest that the strain in question had a higher than expected tendency to microevolve. This is supported not only by its higher than expected accumulation of SNPs, but also by the variations according to IS6110 distribution and by the variations in the number of repetitions for certain VNTR loci. This tendency is not restricted to a specific host, because acquisition of variability was observed in all 3 clustered TB cases analyzed. The most obvious alert is that clusters involving strains of this kind are likely to exceed the similarity thresholds established to define clusters; therefore, related cases could be misinterpreted as unclustered.

The WGS analysis also provided clues about the potential functional significance of some of the above mentioned evolutionary routes followed by the clonal variants. In the case of specific SNPs for variant A, the most remarkable polymorphism is a non-synonymous substitution mapping in the essential gene *mce3R*, a transcriptional repressor of the *mce*3R regulon, which is involved in lipid metabolism and redox reactions (de la Paz Santangelo et al., 2009). The expression of the first gene in the *mce*3 operon, *yrbE3A,* in variant A was lower than in variant B, indicating that the SNP found in variant A leads to higher efficiency of the repressor, which would likely have a functional effect. At some stage in the infection process, more marked repression of this operon might have been advantageous, since both variant A and one of the intermediate variants following a different evolutionary branch shared this SNP.

The *in silico* analysis of the SNPs detected in the clonal variants revealed the involvement of enzymatic, membrane, and cell division proteins. Together with differential expression for variants showing a SNP in *mce*3R and their differences in TNFα production when infecting macrophages (Martín et al., 2011), this involvement indicated the potential functional impact of several of the specific SNPs acquired for each clonal variant.

These findings led us to explore in detail the infective behavior of the clonal variants. Based on a competitive strategy when assaying infectivity in infection models developed in our group (Martín et al., 2011; Navarro et al., 2013) we revealed that clonal variant B was more successful than variant A. This advantageous behavior was observed even when underrepresented (Barczak et al., 2005), and, surprisingly, likely owing to the marked imbalance forced in the co-infection assay, a new variant with a higher fitness emerged, namely variant C, which had not been identified in the clinical specimens. This finding suggested a marked tendency of this strain to microevolve.

In summary, we present an in-depth study of multiple clonal variants which emerged in sequential stages in a patient with recurrent tuberculosis through several independent exploratory branches in the same microevolution event. In addition, other microevolutionary branch leading to different variants was detected in other hosts sharing the same transmission cluster, and finally a novel variant not sampled from the clinical specimens emerged in an infection assay in the laboratory. We also observed differential gene expression and differential infectivity between some of the emerged variants.

These observations emphasize how complex and functional the microevolutionary phenomena in the infection by *M tuberculosis* can be and indicate that some strains are especially prone to microevolution, which could impact on the inference of clusters based on WGS if strict thresholds are applied.

ETHICS STATEMENT

The Institutional Animal Care and Use Committee (IACUC) of the Gregorio Marañón General Hospital (ES 280790000087) (Madrid, Spain) reviewed and approved the experimental protocol. The procedures followed were in agreement with the current Spanish Legislation (RD 53/2013), the European Directive 2010/63/UE [which follows the guidelines and recommendations approved by the Federation of Laboratory Animal Science Associations (FELASA)] and the ethical rules which are applied in this center.

AUTHOR CONTRIBUTIONS

DG, LP-L, JS, IC, MH, YN, and OS: Made substantial contributions to the concept and design of the work through acquisition, analysis and interpretation of data. YN, LP-L, and DG: Drafted the work and provided critical revision for important intellectual content. All: Approved the final version of the article.

ACKNOWLEDGMENTS

We are grateful to Thomas O'Boyle for proofreading the text. This study was funded by the Ministry of Economy and Competitiveness ISCIII-FIS (PI15/01554) and co-funded by ERDF (FEDER) funds from the European Commission, "A way of making Europe." Research by YN was supported by a PICATA pre-doctoral fellowship (BE55/11) from the Moncloa Campus of International Excellence (UCM-UPM, Instituto de Investigación Sanitaria Gregorio Marañón). LP-L was supported by a Miguel Servet contract (MS15/00075-CP15/00075). IC lab is financed by Ministerio de Economía y Competitividad (Spanish Government) research grant SAF2013-43521-R, SAF2016-77346-R, and the European Research Council (ERC) (638553-TB-ACCELERATE).

REFERENCES

Akhtar, P., Singh, S., Bifani, P., Kaur, S., Srivastava, B. S., and Srivastava, R. (2009). Variable-number tandem repeat 3690 polymorphism in Indian clinical isolates of *Mycobacterium tuberculosis* and its influence on transcription. *J. Med. Microbiol.* 58, 798–805. doi: 10.1099/jmm.0.002550-0

Al-Hajoj, S. A., Akkerman, O., Parwati, I., al-Gamdi, S., Rahim, Z., van Soolingen, D., et al. (2010). Microevolution of *Mycobacterium tuberculosis* in a tuberculosis patient. *J. Clin. Microbiol.* 48, 3813–3816. doi: 10.1128/JCM.00556-10

Alonso, M., Alonso Rodriguez, N., Garzelli, C., Garzelli, C., Martínez Lirola, M., Herranz, M., Samper, S., et al. (2010). Characterization of *Mycobacterium tuberculosis* Beijing isolates from the Mediterranean area. *BMC Microbiol.* 10:151. doi: 10.1186/1471-2180-10-151

Barczak, A. K., Domenech, P., Boshoff, H. I., Reed, M. B., Manca, C., Kaplan, G., et al. (2005). *In vivo* phenotypic dominance in mouse mixed infections with *Mycobacterium tuberculosis* clinical isolates. *J. Infect. Dis.* 192, 600–606. doi: 10.1086/432006

Comas, I., Chakravartti, J., Small, P. M., et al. (2010). Human T cell epitopes of *Mycobacterium tuberculosis* are evolutionarily hyperconserved. *Nat. Genet.* 42, 498–503. doi: 10.1038/ng.590

Comas, I., Coscolla, M., Luo, T., Borrell, S., Holt, K. E., Kato-Maeda, M., et al. (2013). Out-of-Africa migration and Neolithic coexpansion of *Mycobacterium tuberculosis* with modern humans. *Nat. Genet.* 45, 1176–1182. doi: 10.1038/ng.2744

de la Paz Santangelo, M., Klepp, L., Nuñez-García, J., Blanco, F. C., Soria, M., García-Pelayo, M. C., et al. (2009). Mce3R, a TetR-type transcriptional repressor, controls the expression of a regulon involved in lipid metabolism in *Mycobacterium tuberculosis. Microbiology* 155, 2245–2255. doi: 10.1099/mic.0.027086-0

García de Viedma, D., Alonso Rodriguez, N., Andrés, S., Ruiz Serrano, M. J., and Bouza, E. (2005). Characterization of clonal complexity in tuberculosis by mycobacterial interspersed repetitive unit-variable-number tandem repeat typing. *J. Clin. Microbiol.* 43, 5660-5664. doi: 10.1128/JCM.43.11.5660-5664.2005

Huyen, M. N., Kremer, K., Lan, N. T., Cobelens, F. G, Buu, T. N., Dung, N. H., et al. (2012). Mixed tuberculosis infections in rural South Vietnam. *J. Clin. Microbiol.* 50, 1586–1592. doi: 10.1128/JCM.00434-12

Lillebaek, T., Norman, A., Rasmussen, E. M., Marvig, R. L., Folkvardsen, D. B., Andersen, Å. B., et al. (2016). Substantial molecular evolution and mutation rates in prolonged latent *Mycobacterium tuberculosis* infection in humans. *Int. J. Med. Microbiol.* 306, 580–585. doi: 10.1016/j.ijmm.2016.05.017

Lin, P. L., Ford, C. B., Coleman, M. T., Myers, A. J., Gawande, R., Ioerger, T., et al. (2014). Sterilization of granulomas is common in active and latent tuberculosis despite within-host variability in bacterial killing. *Nat. Med.* 20, 75–79. doi: 10.1038/nm.3412

Martín, A., Herranz, M., Navarro, Y., Lasarte, S., Ruiz Serrano, M. J., Bouza, E., et al. (2011). Evaluation of the inaccurate assignment of mixed infections by *Mycobacterium tuberculosis* as exogenous reinfection and analysis of the potential role of bacterial factors in reinfection. *J. Clin. Microbiol.* 49, 1331–1338. doi: 10.1128/JCM.02519-10

Muwonge, A., Kankya, C., Olea-Popelka, F., Biffa, D., Ssengooba, W., Berit, D., et al. (2013). Molecular investigation of multiple strain infections in patients with tuberculosis in Mubende district, Uganda. *Infect. Genet. Evol.* 17, 16–22. doi: 10.1016/j.meegid.2013.03.039

Navarro, Y., Herranz, M., Perez-Lago, L., Martínez Lirola, M., INDAL-TB, Ruiz-Serrano, M. J., et al. (2011). Systematic survey of clonal complexity in tuberculosis at a populational level and detailed characterization of the isolates involved. *J. Clin. Microbiol.* 49, 4131–4137. doi: 10.1128/JCM.05203-11

Navarro, Y., Pérez-Lago, L., Sislema, F., Herranz, M., de Juan, L., Bouza, E., et al. (2013). Unmasking subtle differences in the infectivity of microevolved *Mycobacterium tuberculosis* variants coinfecting the same patient. *Int. J. Med. Microbiol.* 303, 693–696. doi: 10.1016/j.ijmm.2013.10.002

Oelemann, M. C., Diel, R., Vatin, V., Haas, W., Rüsch-Gerdes, S., Locht, C., et al. (2007). Assessment of an optimized mycobacterial interspersed repetitive- unit-variable-number tandem-repeat typing system combined with spoligotyping for population-based molecular epidemiology tudies of tuberculosis. *J. Clin. Microbiol.* 45, 691–697. doi: 10.1128/JCM.01393-06

Olsen, I., Balasingham, S. V., Davidsen, T., Debebe, E., Rødland, E. A., van Soolingen, D., et al. (2009). Characterization of the major formamidopyrimidine-DNA glycosylase homolog in *Mycobacterium tuberculosis* and its linkage to variable tandem repeats. *FEMS Immunol. Med. Microbiol.* 56, 151–161. doi: 10.1111/j.1574-695X.2009.00562.x

Perez-Lago, L., Comas, I., Navarro, Y., González-Candelas, F., Herranz, M., Bouza, E., et al. (2014). Whole genome sequencing analysis of intrapatient microevolution in *Mycobacterium tuberculosis*: potential impact on the inference of tuberculosis transmission. *J. Infect. Dis.* 209, 98–108. doi: 10.1093/infdis/jit439

Perez-Lago, L., Herranz, M., Lirola, M. M., Group, I.-T., Bouza, E., and Garcia de Viedma, D. (2011). Characterization of microevolution events in *Mycobacterium tuberculosis* strains involved in recent transmission clusters. *J. Clin. Microbiol.* 49, 3771–3776. doi: 10.1128/JCM.01285-11

Perez-Lago, L., Navarro, Y., Herranz, M., Bouza, E., and Garcia-de-Viedma, D. (2013). Differences in gene expression between clonal variants of *Mycobacterium tuberculosis* emerging as a result of microevolution. *Int. J. Med. Microbiol.* 303, 674–677. doi: 10.1016/j.ijmm.2013.09.010

Santangelo, M. P., Blanco, F. C., Bianco, M. V., Klepp, L. I., Zabal, O., Cataldi, A. A., et al. (2008). Study of the role of Mce3R on the transcription of mce genes of *Mycobacterium tuberculosis. BMC Microbiol.* 8:38. doi: 10.1186/1471-2180-8-38

Shamputa, I. C., Jugheli, L., Sadradze, N., Willery, E., Portaels, F., Supply, P., et al. (2006). Mixed infection and clonal representativeness of a single sputum sample in tuberculosis patients from a penitentiary hospital in Georgia. *Respir. Res.* 7:99. doi: 10.1186/1465-9921-7-99

Tantivitayakul, P., Panapruksachat, S., Billamas, P., and Palittapongarnpim, P. (2010). Variable number of tandem repeat sequences act as regulatory elements in *Mycobacterium tuberculosis. Tuberculosis (Edinb)* 90, 311–318. doi: 10.1016/j.tube.2010.08.003

van Embden, J. D., Cave, M. D., Crawford, J. T., Dale, J. W., Eisenach, K. D., Gicquel, B., et al. (1993). Strain identification of *Mycobacterium tuberculosis* by DNA fingerprinting: recommendations for a standardized methodology. *J. Clin. Microbiol.* 31, 406–409.

Walker, T. M., Ip, C. L., Harrell, R. H., Evans, J. T., Kapatai, G., Dedicoat, M. J., et al. (2013). Whole-genome sequencing to delineate *Mycobacterium tuberculosis* outbreaks: a retrospective observational study. *Lancet Infect. Dis.* 13, 137–146. doi: 10.1016/S1473-3099(12)70277-3

Yindeeyoungyeon, W., Likitvivatanavong, S., and Palittapongarnpim, P. (2009). Characterization of alpha-isopropylmalate synthases containing different copy numbers of tandem repeats in *Mycobacterium tuberculosis. BMC Microbiol.* 9:122. doi: 10.1186/1471-2180-9-122

12

A Novel Antimicrobial Coating Represses Biofilm and Virulence-Related Genes in Methicillin-Resistant *Staphylococcus aureus*

*Ankita Vaishampayan[1], Anne de Jong[2], Darren J. Wight[3], Jan Kok[2] and Elisabeth Grohmann[1,4]**

[1] Life Sciences and Technology, Beuth University of Applied Sciences Berlin, Berlin, Germany, [2] Department of Molecular Genetics, University of Groningen, Groningen, Netherlands, [3] Institute of Virology, Free University of Berlin, Berlin, Germany, [4] Division of Infectious Diseases, University Medical Center Freiburg, Freiburg, Germany

Edited by:
Manuel Espinosa,
Centro de Investigaciones Biológicas (CSIC), Spain

Reviewed by:
Mirian Moscoso,
Instituto de Investigación Biomédica de A Coruña (INIBIC), Spain
Maria Victoria Francia,
Marqués de Valdecilla University Hospital, Spain

***Correspondence:**
Elisabeth Grohmann
elisabeth.grohmann@beuth-hochschule.de

Methicillin-resistant *Staphylococcus aureus* (MRSA) has become an important cause of hospital-acquired infections worldwide. It is one of the most threatening pathogens due to its multi-drug resistance and strong biofilm-forming capacity. Thus, there is an urgent need for novel alternative strategies to combat bacterial infections. Recently, we demonstrated that a novel antimicrobial surface coating, AGXX®, consisting of micro-galvanic elements of the two noble metals, silver and ruthenium, surface-conditioned with ascorbic acid, efficiently inhibits MRSA growth. In this study, we demonstrated that the antimicrobial coating caused a significant reduction in biofilm formation (46%) of the clinical MRSA isolate, *S. aureus* 04-02981. To understand the molecular mechanism of the antimicrobial coating, we exposed *S. aureus* 04-02981 for different time-periods to the coating and investigated its molecular response via next-generation RNA-sequencing. A conventional antimicrobial silver coating served as a control. RNA-sequencing demonstrated down-regulation of many biofilm-associated genes and of genes related to virulence of *S. aureus*. The antimicrobial substance also down-regulated the two-component quorum-sensing system *agr* suggesting that it might interfere with quorum-sensing while diminishing biofilm formation in *S. aureus* 04-02981.

Keywords: antimicrobial surface, MRSA, virulence, biofilm, quorum-sensing, RNA sequencing

INTRODUCTION

Staphylococcus aureus is an opportunistic pathogen commonly found in the human respiratory tract, nasal areas and skin. It colonizes the anterior nares of approximately 20–25% of the healthy adult population, while 60% are intermittently colonized (Kluytmans et al., 1997; Ellis et al., 2014). Methicillin-resistant *Staphylococcus aureus* (MRSA) is a crucial human pathogen causing infections ranging from skin and soft tissue infections to fatal sepsis (Marathe et al., 2015). It is one of the leading pathogens that cause nosocomial infections (Paniagua-Contreras et al., 2012; Lister and Horswill, 2014); it is resistant to methicillin and many other antibiotics (Marathe et al., 2015), and it is also known to produce thick biofilm (Paniagua-Contreras et al., 2012; Qin et al., 2014). MRSA was shown to cause catheter-associated and other medical devices-related infections (Arciola et al., 2001; Paniagua-Contreras et al., 2012). Eighty percent of prosthetic infections are caused by Staphylococci (Kirmusaoglu, 2016). Its firm attachment to medical devices and host tissues, and its ability to form robust biofilms makes it a cause of chronic infections (Yarwood et al., 2004). *S. aureus* biofilms cause numerous infections in which the accessory gene regulator (*agr*) quorum-sensing system (QS) plays an important role (Yarwood et al., 2004). Around 90% of the infections caused by the bacterium are skin and soft tissue infections, and the *agr*QS system is associated with these infections (Sully et al., 2014).

Multiple drug resistance combined with a thick biofilm makes the treatment and eradication of *S. aureus* infections even more difficult. This entails the urge of development of novel antimicrobials, which could also be potential biofilm inhibitors. Virulence factors of *S. aureus* serve as targets for the newly developed class of biological anti-staphylococcal agents. These targets include, surface bound adhesins, immunoglobulin-binding proteins, surface-associated and secreted proteases, a family of immune-stimulatory exotoxins called 'superantigens' (SAgs), and potent leukocidal toxins (Sause et al., 2015).

Metals like copper and silver have been used as antimicrobials since a long time. The use of copper in human civilization is known since the 5th and 6th millennia B.C. Silver was officially approved for use as an antimicrobial agent in the 20th century (Chopra, 2007; Grass et al., 2011; Schäberle and Hack, 2014; Guridi et al., 2015). Copper and copper alloys have also been used as antimicrobials (Warnes and Keevil, 2013). These metals are known to kill bacteria and fungi by a phenomenon called contact killing (Grass et al., 2011) and can be used to coat medical devices as they inhibit biofilm formation of pathogens (Baker et al., 2010). In the 17th century, silver was described as an essential multipurpose medicinal product and the first scientific documentation of its medical use dates from 1901 (Maillard and Hartemann, 2013). However, in 1975, several patients died from a silver resistant *Salmonella* Typhimurium isolate in the Massachusetts General Hospital; this was the first report of silver resistant bacteria (Gupta et al., 1999). Excessive use of silver is questioned due to its toxicity to the environment as well as to the human body (Landsdown, 2010). Silver resistance, like antibiotic resistance in bacteria, prompts us to develop new strategies to control bacterial infections. One such novel, broad-spectrum antimicrobial agent is AGXX®.

AGXX® (Largentec GmbH, Berlin, Germany) is a combination of two transition metals, silver and ruthenium which can be galvanically electroplated on various carriers like V2A steel, silver sheets, Polydimethylsiloxane (PDMS), fleece, etc. The coating is conditioned by ascorbic acid and is active against many Gram-positive and Gram-negative bacteria (Guridi et al., 2015). It is not only an efficient antibacterial but also kills yeasts, viruses, and fungi (Landau et al., 2017a,b). The coating was used successfully for the decontamination of industrial cooling and process water (Landau, 2013). As it is only slightly cytotoxic (Bouchard, 2011), it can be incorporated into various medical applications. Although, the exact mode of action of the antimicrobial activity of the coating is not fully understood, it is known that the generation of reactive oxygen species (ROS) plays an important role in making it a potent antimicrobial. The formation of hydrogen peroxide and hydroxyl radicals has been detected by spectroscopic methods (Clauss-Lendzian et al., 2017). Putative formation of other ROS is under investigation. ROS can damage cellular components, including, DNA, lipids and proteins. Superoxide dismutase and catalase are involved in detoxification of ROS (Paraje, 2011).

In this study, we performed total RNA-sequencing of *S. aureus* 04-02981 (MRSA) to investigate differential gene expression after different times of exposure of the pathogen to the antimicrobials AGXX® or Ag. Our data demonstrate that AGXX® likely reduces biofilm formation and virulence in *S. aureus* 04-02981 by interfering with the QS, by down-regulating the expression of toxins like leukocidins (*lukE*) and gamma-hemolysins (*hlgA*), and of genes associated with surface adhesins and capsular polysaccharide.

MATERIALS AND METHODS

Preparation of Antimicrobial Metal Sheets

Silver sheets of 0.125 mm thickness were used as a base material to prepare the antimicrobial metal sheets. Both sides of the silver sheets were etched by immersing them in half-concentrated nitric acid, for 60 s. The silver sheets were cleaned with de-ionized water and galvanically plated with a 0.16 μm ruthenium coating on both sides for 40 s. Then, the sheets were cleaned with de-ionized water, conditioned with ascorbic acid, rinsed with de-ionized water and dried with a paper towel. Prior to use, AGXX®, and Ag sheets, used as reference material, were autoclaved at 121°C for 20 min.

Bacterial Strain and Culture Conditions

Staphylococcus aureus 04-02981 (Nuebel et al., 2010) was grown at 37°C in Tryptic Soy Broth [TSB] (Carl Roth GmbH & Co. KG, Karlsruhe, Germany) with constant agitation at 150 rpm or on Tryptic Soy Agar [TSA] (Carl Roth GmbH & Co. KG, Karlsruhe, Germany). Growth inhibition tests on agar surface were performed according to CLSI guidelines for disk diffusion test (Naas et al., 2006). For this assay, 0.25 cm^2 sheets of Ag and AGXX® were used.

For generation of growth curves, bacteria were pre-cultured overnight, diluted in TSB to an optical density at 600 nm (OD_{600}) of 0.05 and incubated for further 8 h either in presence of AGXX® or in the presence of silver (Ag), 24 cm^2 each in 30 mL medium to obtain a sheet surface to medium volume ratio (A: V) of 0.8. Cultures grown in the absence of a metal sheet served as controls. OD_{600} of the cultures was measured using the Genesys 10S UV-Vis spectrophotometer (Thermo Scientific, China). Colony forming units (CFU) per mL were determined hourly from 0 to 8 h post inoculation. Growth experiments were performed in triplicate with independent biological replicates.

Biofilm Screening Assay

To study the effect of Ag, and AGXX® on biofilm formation of *S. aureus* 04-02981, the Crystal Violet Assay was performed without any metal sheet, in presence of Ag (24 cm^2 uncoated silver sheet) and in presence of AGXX® (24 cm^2 silver sheet coated with ruthenium for 40 s). The sheet surface: medium volume ratio (A: V) was 0.8 (24 cm^2 metal sheet: 30 mL medium). The overnight culture of *S. aureus* 04-02981was diluted to an initial OD_{600} of 0.05. The culture was incubated at 37°C and 150 rpm for 4 h (mid- exponential phase, OD_{600}~1.5). Then, it was transferred to the transparent 96-well plate (Carl Roth GmbH & Co. KG, Karlsruhe, Germany) containing Ag, or AGXX®. The plate was incubated at 37°C for 24 h, then the antimicrobial metal sheets were carefully removed and OD_{600} of the cultures was measured. In addition, at this stage, the CFU per mL of the planktonic cultures and the biofilms in presence as well as in absence of the metal sheets were determined. Means of five values each and two biological replicates are given. The biofilm assay was performed according to Schiwon et al. (2013). *Enterococcus faecalis* 12030, a strong biofilm former was used as a positive control (Huebner et al., 1999), and Tryptic Soy Broth (TSB) as a negative control (Schiwon et al., 2013). Biofilm formation was measured in EnSpire Multimode Plate Reader 2300-0000 (PerkinElmer, Turku, Finland) at 570 nm. Normalized biofilm formation was calculated by dividing the biofilm measure at OD_{570} by the bacterial growth at OD_{600}. Following criteria were used for the interpretation of the results, ODc = negative control; OD $\leq$ ODc = non-adherent, ODc $\leq$ OD $\leq$ (2 × ODc) = weakly adherent, (2 × ODc) < OD $\leq$ (4 × ODc) = moderately adherent, (4 × ODc) < OD = strongly adherent, as described in Nyenje et al. (2013). Biofilm inhibitory rates of AGXX® and Ag were calculated using the following equation, as described by Qin et al. (2014).

$$\text{Inhibitory rate (\%)} = \frac{OD_{570}\,(\text{Control}) - OD_{570}\,(\text{Sample})^{*}100}{OD_{570}\,(\text{Control})}$$

Student's *t*- test was used to check if biofilm inhibition was statistically significant, using SigmaPlot version 11.0 (Systat software, Inc., San Jose, CA, United States[1]) (Wass, 2009).

Spinning Disk Confocal Microscopy

Staphylococcus aureus 04-02981 was grown in TSB overnight at 37°C, 150 rpm, then it was diluted to an OD_{600} of 0.05 and further incubated at 37°C for 4 h (mid-exponential phase, OD_{600} ~1.5). Then, the culture was transferred to a μ-Dish (μ-Dish 35 mm, low, from ibidi GmbH, Martinsried, Germany) containing Ag, or AGXX® (sheet surface: medium volume ratio = 0.8) and incubated at 37°C for 24 h. The culture was removed from the μ-Dish, and the biofilm on the μ-Dish was washed three times with phosphate buffered saline (PBS). The biofilm was stained for 10 min in the dark with Hoechst 33342 (5 μg/mL) and propidium iodide (1 μg/mL) (Thermo Fisher, Eugene, OR, United States). The staining solution was then replaced with 50% glycerol to prevent movement of bacteria during imaging. Imaging was performed with a Nikon TiE-based Visitron spinning disk confocal microscope using a 100× NA1.45 objective. Fluorescent dyes were excited using 405 nm (Hoechst 33342) and 561 nm (propidium iodide) laser lines and fluorescent emission captured through appropriate filters onto an iXon888 EMCCD detector (Andor, Belfast, United Kingdom). Images were subsequently analyzed using Fiji (ImageJ) version 3.2.0.2.

Metal Stress and RNA Extraction

Overnight cultures of *S. aureus* 04-02981 were diluted as described above and grown until mid-exponential growth phase (4 h post dilution, OD_{600}~1.5). The cultures were then subjected to metal stress by exposure to AGXX® or Ag sheets (sheet-surface to medium-volume ratio of 0.8) followed by further incubation for 6, 12, 24, 80, and 120 min at 37°C with constant agitation at 150 rpm. As a control, no metal sheet was added to the culture. Cells from 30 mL culture were harvested by centrifugation for 1 min at 10,000 rpm and 4°C in a Heraeus Multifuge X3R Centrifuge (Thermo Electron LED GmbH, Osterode am Harz, Germany). Cell pellets were immediately frozen in liquid nitrogen and stored at −80°C or directly used for RNA extraction using the ZR Fungal/Bacterial RNA MiniPrep™ Kit (ZymoResearch, Freiburg, Germany) following the manufacturer's instructions. To recover total RNA including small RNAs, 1.5 volumes of absolute ethanol were added in step 5. Finally, total RNA was eluted with 50 μl DNase- and RNase-free water and stored at −80°C. RNA quantity and quality were assessed with a NanoDrop 2000c UV-Vis Spectrophotometer (Thermo Scientific, Osterode am Harz, Germany) as well as on bleach agarose gels. Residual contaminating DNA was eliminated with TURBO DNA-free™ Kit Ambion (Life Technologies, Darmstadt, Germany).

RNA Sequencing

Total RNA sequencing was done by PrimBio Research Institute, Exton, PA, United States. The protocol was performed in five steps; rRNA removal was done using the Ribo-Zero rRNA Removal Kit (Bacteria) (Illumina, Cat# MRZMB126), followed by library preparation, and templating, enrichment and sequencing.

RNA-Sequencing Data Analysis

Raw sequencing reads were aligned to the reference genome of *S. aureus* 04-02981, using Bowtie2 (Langmead and Salzberg, 2012) version 2.2.3 with optimal settings for the IonProton™ Sequence. Post-processing of the SAM files into sorted BAM files was carried out with SAMtools (Li et al., 2009, version 1.2-207). The samples AGXX®, and Ag were normalized (AGXX®-Control, Ag-Control) against the control of the respective time-points. Length normalized confidence interval RPKM (=Reads per Kilobase of transcript per Million mapped reads) values were obtained with Cufflinks (Trapnell et al., 2010). Finally, statistical analysis was carried out using the T-REx RNA-Seq analysis pipeline (de Jong et al., 2015). A gene was considered significantly differentially expressed when the fold change was $\geq|2.0|$ and the false discovery rate (FDR) adjusted *p*-value $\leq$ 0.05. The data presented in this paper have been deposited at NCBI, and are accessible through GSE103064[2].

[1] http://www.systatsoftware.com

[2] https://www.ncbi.nlm.nih.gov/geo/query/acc.cgi?acc=GSE103064

Reverse Transcription Quantitative PCR (RT-qPCR)

To verify the results obtained from RNA-sequencing, RT-qPCR was performed on five genes detected as highly differentially expressed via RNA-seq. To this end, RNA extracted from *S. aureus* 04-02981 cultures exposed to Ag or AGXX® for 24, and 80 min, was used. First strand cDNA was synthesized with RevertAid™ First Strand cDNA Synthesis kit (Thermo Fisher Scientific Inc., Walham, Germany) as per the manufacturer's instructions using 120 ng total RNA as template and random hexamer primers. cDNA was diluted with DNase- and RNase-free water and amplified in a LightCycler®480 II (Roche Diagnostics GmbH, Mannheim, Germany).

The *agrC, lukE, sdrC, srrA*, and *cap5A* genes were selected to verify the data obtained through RNA-seq. The gene *gyrB* was used as a control. These genes were amplified using TaqMan chemistry according to the instructions provided in LightCycler® 480 Probes Master Kit (Roche Diagnostics). All RT-qPCR reactions were carried out in a total volume of 20 μL. The amplification step was performed with 'Quantification' analysis mode at 95°C for 10 s, with a ramp rate of 4.4°C/s, followed by annealing at the respective annealing temperature for 50 s, with a ramp rate of 2.2°C/s and finally an extension at 72°C for 1 s, with a ramp rate of 4.4°C/s. The amplification step was performed 45 times. All primers and probes used in the study are listed in **Supplementary Table S1**. All RT-qPCR experiments were done in triplicate and each experiment was repeated at least twice. Data were analyzed by LightCycler® 480 Software release 1.5.0 by using the 'Relative Standard Curve' method; the standard curves were constructed using genomic DNA from *S. aureus* 04-02981. Data represent expression ratios, calculated by normalizing to the *gyrB* gene and relative to the untreated culture of *S. aureus* 04-02981 which served as the calibrator, as described in 'Guide to performing Relative Quantitation of Gene expression using real time-quantitative PCR' by Applied Biosystems. Means of five Cp values each were used to calculate the relative expression ratio.

Statistical Analysis

Statistical tests were performed to analyze the significance of the obtained data. Student's *t*-test was applied to the normalized target, and normalized control values (normalized concentration). The tests were performed and analyzed using SigmaPlot version 11.0 (Systat software, Inc., San Jose, CA, United States[3]) (Wass, 2009).

RESULTS

AGXX® Inhibits the Growth of *S. aureus* 04-02981

To analyze the effect of Ag, and AGXX® on the growth of *S. aureus* 04-02981, disk diffusion tests with Ag, and AGXX® were performed in accordance with NCCLS-CLSI guidelines (Naas et al., 2006). The agar plates were monitored at 24 h intervals for 5 days to check if Ag or AGXX® exhibited an inhibitory effect on the pathogen, in the form of a zone of inhibition on the agar plate. The diameter of the inhibition zones was measured in 'cm.' The mean diameter of the inhibition zone was calculated to be 1.2 cm for AGXX® while no zone of inhibition was observed for Ag.

To verify the inhibitory effect of AGXX® on *S. aureus* 04-02981 as demonstrated in the agar diffusion tests, experiments in TSB medium were performed measuring the CFU/mL every hour for a period of 8 h, using the A: V ratio (metal mesh: medium volume) of 0.8, as described in Section "Materials and Methods." As observed in the disk diffusion assay, Ag did not show a significant inhibitory effect on the growth of *S. aureus* 04-02981 in liquid cultures. In contrast, AGXX® had a profound inhibitory effect on this strain. The OD_{600} of *S. aureus* 04-02981 in presence of AGXX® was very low, (OD_{600} AGXX® at t8 = 0.149) as compared to Ag (OD_{600} Ag at t8 = 3.086) and the control (OD_{600} Control at t8 = 3.173) (**Supplementary Table S2**). The CFU/mL of *S. aureus* 04-02981 grown in the batch culture with AGXX® increased from 2.77×10^6 in the 1st hour to 3.99×10^{10} in the 4th hour, but then decreased to 1.08×10^7 in the 8th hour. The colony counts of *S. aureus* 04-02981 + AGXX® (after 8 h of growth) were much lower than that of the same strain with Ag (1.27×10^{11}) or without metal amendment (1.73×10^{11}) (**Table 1**). These data confirm the antimicrobial effect of AGXX® on *S. aureus* 04-02981.

AGXX® Strongly Reduces Biofilm Formation of *S. aureus* 04-02981

The effect of AGXX®, and Ag on biofilm formation of *S. aureus* 04-02981 was analyzed using the Crystal Violet assay. *E. faecalis* 12030, a strong biofilm former served as a positive control (Huebner et al., 1999), and TSB as the negative control (**Figure 1**). **Figure 1A** shows the biofilm formation by *S. aureus* 04-02981, measured at 570 nm, **Figure 1B** shows the biofilm formation (OD_{570}) normalized to the bacterial growth (OD_{600}) to take the antimicrobial effect of AGXX® into account.

To determine the bacterial killing activity of AGXX® under these conditions (after 24 h of growth, prior to adding crystal violet), we measured the CFU per mL of the planktonic cultures and the biofilms in the presence as well as in absence of the two different metal sheets.

The following values were obtained for the biofilms: For *S. aureus* 04-02981 without metal sheet (control), $2.34 \times 10^9 \pm 8.49 \times 10^7$ CFU per mL, for the strain in presence of Ag, $2.13 \times 10^9 \pm 2.40 \times 10^8$, and in presence of AGXX®, $1.80 \times 10^4 \pm 1.41 \times 10^3$. When we measured the CFU per mL in the respective planktonic cultures, for the control, $2.55 \times 10^8 \pm 2.12 \times 10^7$, and for the strain in presence of Ag, $2.00 \times 10^8 \pm 1.41 \times 10^7$ CFU per mL were obtained. However, no colonies were observed in presence of AGXX®. Thus, we conclude that in contrast to Ag, all planktonic bacteria were killed by AGXX® and after exposure to AGXX®, only a drastically reduced number of bacteria (1.80×10^4 CFU per mL) survived in the biofilm in comparison to Ag (2.13×10^9 CFU per mL).

In summary, the biofilm formation measures normalized to the bacterial growth show that AGXX® reduced biofilm

[3]www.systatsoftware.com

TABLE 1 | Colony forming units (CFU)/mL of *Staphylococcus aureus* 04-02981 (without sheet = control), in the presence of AGXX® or Ag.

Sample	0 h	1 h	2 h
Control	$1.27 \times 10^6 \pm 2.0 \times 10^5$	$7.63 \times 10^7 \pm 1.5 \times 10^6$	$8.71 \times 10^9 \pm 1.0 \times 10^9$
Ag	$1.23 \times 10^6 \pm 4.7 \times 10^5$	$6.80 \times 10^6 \pm 1.1 \times 10^6$	$6.23 \times 10^9 \pm 1.2 \times 10^9$
AGXX®	$7.67 \times 10^5 \pm 4.1 \times 10^5$	$2.77 \times 10^6 \pm 1.4 \times 10^6$	$4.93 \times 10^7 \pm 3.4 \times 10^7$
	3 h	**4h**	**5 h**
Control	$1.11 \times 10^{10} \pm 1.1 \times 10^9$	$1.26 \times 10^{11} \pm 7.5 \times 10^9$	$\mathbf{1.73 \times 10^{11} \pm 3.3 \times 10^9}$
Ag	$1.02 \times 10^{10} \pm 1.2 \times 10^9$	$1.25 \times 10^{11} \pm 1.0 \times 10^{10}$	$\mathbf{1.27 \times 10^{11} \pm 6.2 \times 10^8}$
AGXX®	$2.06 \times 10^8 \pm 4.7 \times 10^7$	$3.99 \times 10^{10} \pm 1.1 \times 10^9$	$\mathbf{1.71 \times 10^8 \pm 1.6 \times 10^7}$
	6 h	**7 h**	**8 h**
Control	$1.66 \times 10^{11} \pm 5.7 \times 10^9$	$1.26 \times 10^{11} \pm 3.5 \times 10^9$	$\mathbf{1.17 \times 10^{11} \pm 4.4 \times 10^9}$
Ag	$1.22 \times 10^{10} \pm 4.6 \times 10^8$	$1.12 \times 10^{10} \pm 1.4 \times 10^9$	$\mathbf{1.02 \times 10^{10} \pm 1.4 \times 10^9}$
AGXX®	$1.99 \times 10^8 \pm 7.1 \times 10^8$	$1.34 \times 10^7 \pm 1.4 \times 10^6$	$\mathbf{1.08 \times 10^7 \pm 1.8 \times 10^6}$

The values for 5th hour and 8th hour are bolded because after t = 5h, the CFU values of MRSA + Ag decreased. And until t = 8h, the CFU values for all the three samples (MRSA, MRSA + Ag, and MRSA + AGXX) decreased.

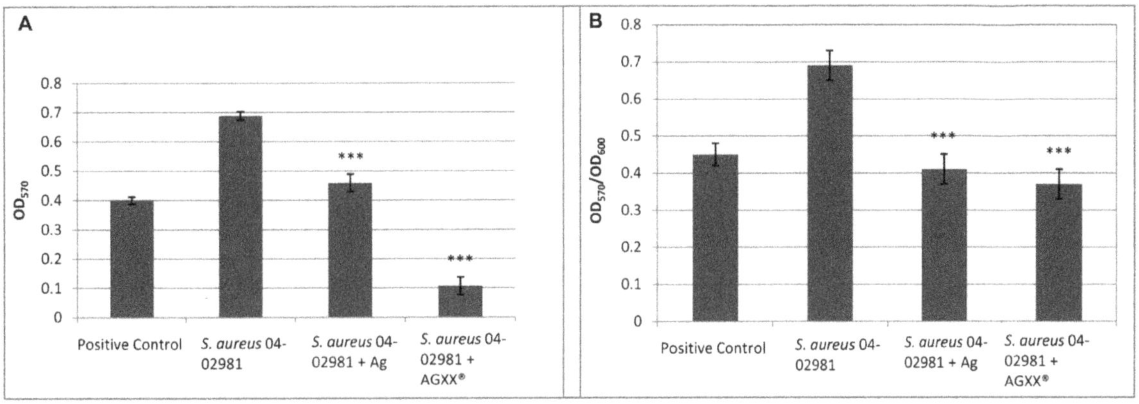

FIGURE 1 | AGXX® reduces biofilm formation of *S. aureus* 04-02981. Mean of five OD values of the biofilm assay with AGXX®, Ag, and without metal amendment with standard deviation are shown. **(A)** Shows the biofilm formation by *Staphylococcus aureus aureus* 04-02981, measured at 570 nm; **(B)** shows the biofilm formation (OD_{570}) normalized to the bacterial growth (OD_{600}) to take the antimicrobial effect of AGXX® into account. Asterisks indicate the *p*-values obtained from *t*-test using SigmaPlot (*****p* < 0.0001, ****p* < 0.001, ***p* < 0.01, **p* < 0.05) representing the statistical significance of the data.

formation of *S. aureus* 04-02981 by 46%, whereas the inhibitory effect of Ag on biofilm formation was less pronounced (41%).

The strong reduction of biofilm formation by AGXX® was confirmed by Hoechst 33342/propidium iodide staining of biofilms grown for 24 h in presence of AGXX®, Ag and without antimicrobial sheet (**Figure 2**). The inhibitory effect of Ag was also clearly visible, although it was less distinct.

AGXX® Strongly Induces Stress Response and Represses Pathogenesis in *S. aureus* 04-02981

The raw RNA sequence data obtained were aligned to the *S. aureus* 04-02981 genome. High sequencing depth was achieved as a mean value of ~12.4 million reads was obtained. The numbers of reads per sample ranged from ~8.4 million reads (Ag_24) to 175 million reads (Control_120) (**Supplementary Table S3** and **Supplementary Figure S1**). From the data, it is clear that the antimicrobial coating has a strong impact on the transcriptome of *S. aureus* 04-02981. In total, 2864 genes were differentially expressed in *S. aureus* 04-02981 on exposure to AGXX® and Ag (**Supplementary Table S4**). The number of differentially expressed genes in presence of AGXX® or Ag at different time-points is presented in **Figure 3**.

From **Figure 3A**, it can be seen that the number of differentially expressed genes at t24, t80, and t120 was quite similar. The maximum impact of AGXX® on the transcriptome of *S. aureus* 04-02981 was reached already after exposure for 24 min (723 genes up-regulated and 823 genes down-regulated) and remained nearly the same after exposure for 80 min (716 genes up- and 822 genes down-regulated), and 120 min (726 genes up- and 836 genes down-regulated). The lowest number of genes was differentially expressed at t6.

The differentially expressed genes were categorized as per Gene Ontology (GO) using the GSEA_Pro option in the RNA-Seq analysis section in the T-REx RNA-Seq analysis pipeline (de Jong et al., 2015). Several GOs were obtained via

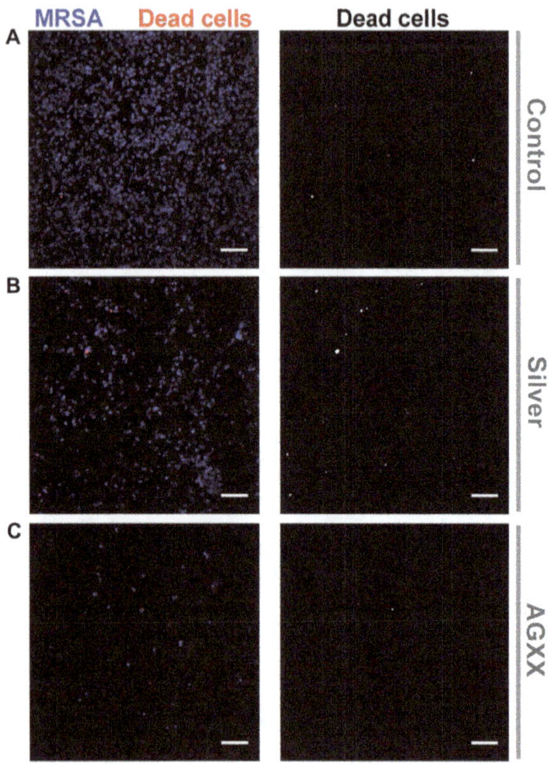

FIGURE 2 | Confocal images of biofilm formation by *S. aureus* 04-02981 (MRSA). The pathogen was grown on sterile cover slips for 24 h with the following conditions: **(A)** without antimicrobial sheet, **(B)** with a silver sheet or **(C)** with an AGXX® sheet. Biofilms were then stained with Hoechst 33342 (to mark out *S. aureus* 04-02981; blue) and propidium iodide (to identify dead cells; red) followed by acquisition of small *Z*-stacks (500 nm spacing) through the biofilms using a confocal microscope. Images show an average of Z-projection (average of 4–5 Z planes containing the biofilm) of the fluorescence signal through the biofilms with the propidium iodide staining shown alone in the images on the right (gray scale images). Scale bars = 10 μm.

GSEA_Pro, namely, oxidoreductase process, lipopolysaccharide synthesis, ATP binding, membrane transport, metabolism, metal binding, pathogenesis, transcription regulation, response to heat shock, iron-siderophore transporter activity, serine protease activity, etc. (**Supplementary Table S5**). In the GO "lipopolysaccharide synthesis," the *cap* genes mediating capsular polysaccharide synthesis (*cap*5A, *cap*A, and *cap*8C) were all down- regulated. Genes (*clp*B, *cts*R, *clp*C, and *groES*) involved in response to heat shock were up-regulated. Among the genes related to virulence (pathogenesis), 10 out of 11 genes were down-regulated, while only one gene was up-regulated at t120 (staphylokinase, a plasminogen activator). Among the responding transcriptional regulator genes, nine were up-regulated and 25 were down-regulated. **Figure 4** shows the differential expression of these GOs in *S. aureus* 04-02981 exposed to AGXX®.

hlgA (SA2981_RS09385) was the most differentially expressed gene associated with virulence, it was down-regulated at t24 (378 fold), at t80 (192 fold), and at t120 (16 fold). The protein encoded by *hlgA* functions as a two-component toxin along with leukocidins in the lysis of erythrocytes (Gouaux et al., 1997). Among the transcriptional regulators, the gene of the LysR family transcriptional regulator, *lysR* was the most significantly influenced one by AGXX®, being down-regulated about 4700 fold at t80, and about 11,000 fold at t120. One of the LysR family transcriptional regulators, HutR is involved in metabolic processes of *S. aureus* (Ibarra et al., 2013). AGXX® had the highest impact on the expression of *cap*A, of all the genes mediating capsular polysaccharide synthesis. *cap*A was down-regulated by 329 fold at t80. Among the most differentially expressed genes in response to heat shock was *clp*B. It is a member of the stress-induced multi-chaperone system and works with DnaK, DnaJ, and GrpE in the recovery of the cell from heat-shock damage (Frees et al., 2005). Among the genes in the GO families influenced by AGXX®, only those involved in enterotoxin (SA2981_RS09440), and staphylokinase production were also influenced by Ag, by -533 fold, and -2 fold, respectively, at t80 (**Supplementary Table S6**). In addition to the GO families, the effect of AGXX®, and Ag on the expression of operons in the pathogen was analyzed using the GSEA_Pro option on the T-REx pipeline. The results are presented in **Supplementary Tables S7**, **S8**, respectively.

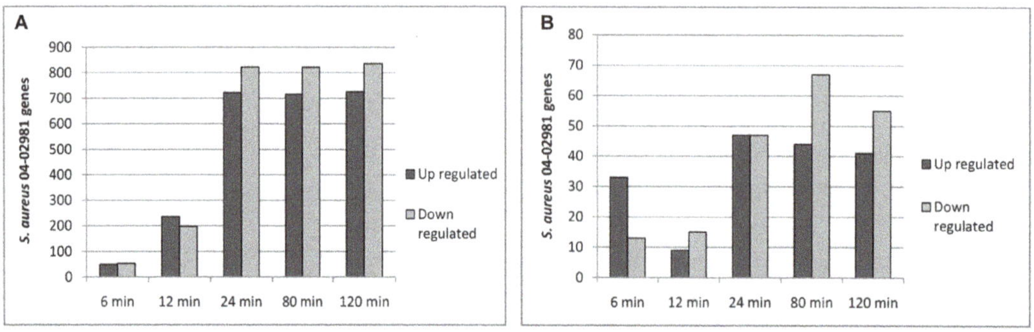

FIGURE 3 | (A,B) Show the number of differentially expressed genes in *S. aureus* 04-02981 on exposure to AGXX® **(A)** and Ag **(B)** for the indicated time-period, compared to control (*S. aureus* 04-02981 grown without a metal sheet).

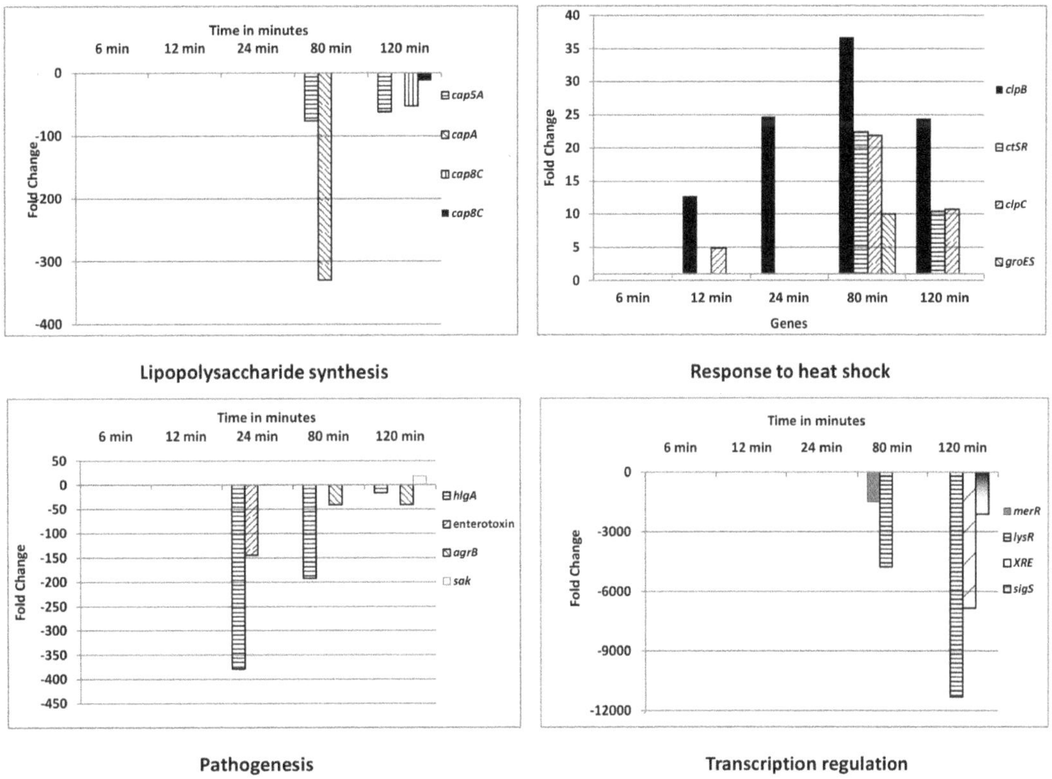

FIGURE 4 | Gene Ontology (GO) categories of differentially expressed genes of *S. aureus* 04-02981 exposed to AGXX® for different time-periods. Only significantly differentially expressed genes, which are likely relevant in the response of *S. aureus* to AGXX®, are presented. Details on the other genes are provided in the **Supplementary Table S5**, along with all of the other GOs, Gene IDs, and fold-change in gene expression, for each time-point. Capsular polysaccharide synthesis mediating *cap* genes are shown in 'Lipopolysaccharide synthesis.' Genes encoding chaperone ClpB (*clpB*), transcriptional regulator CtsR (*ctsR*), ATP-dependent Clp protease ATP-binding subunit ClpC (*clpC*), and chaperonin GroES (*groES*) are presented in the panel showing response to heat shock. In the pathogenesis panel, the genes for γ-hemolysin subunit A (*hlgA*), enterotoxin, accessory gene regulator subunit B (*agrB*), and staphylokinase (*sak*) are presented. Among the transcriptional regulator genes, that of the LysR family transcriptional regulator (*lysR*), MerR family transcriptional regulator (*merR*), XRE family transcriptional regulator (*XRE*) and RNA polymerase sigma factor SigS (*sigS*) are shown.

AGXX® Represses the Expression of Biofilm and Virulence-Associated Genes

We checked the effect of AGXX®, and Ag on the expression of genes associated with biofilm formation and virulence in *S. aureus* 04-02981. Many genes that are known to be crucial for biofilm formation and virulence were differentially expressed on exposure to AGXX® while Ag had an effect on just a few of them. The genes affected by AGXX® encode virulence factors, methicillin resistance, surface adhesins, capsular polysaccharide, two-component systems, and other biofilm- associated genes, as well as toxins (**Table 2**).

Upon exposure to AGXX®, the QS system genes *agrA*, *agrB*, *agrC*, and *agrD* of *S. aureus* 04-02981 were all down-regulated. Genes involved in the synthesis of capsular polysaccharide were also down-regulated. In general, the response of *S. aureus* 04-02981 to AGXX® was clearly visible after 24 min of exposure time. Genes encoding adhesins, *isdC*, *srtB*, and *sdrC* were also down-regulated. The *mecA* gene was down-regulated at t24. The up-regulation of genes inducing biofilm formation in *S. aureus*, such as *saeR* (2.3 fold at t120), *icaA* (36 fold at t24, 29 fold at t80 and 27 fold at t120), *icaB* (8 fold at t120) and *icaD* (55 fold at t12, and 6 fold at t120) was intriguing. The genes *icaB*, *icaA*, and *icaD* are involved in *ica*-dependent biofilm formation. In addition, other key genes associated with biofilm formation and virulence, such as, *codY*, *srrA*, *luxS*, and genes for toxins like leukocidins, enterotoxins, hemolysins, were all differentially expressed at least at one of the time-points (**Figure 5**). Description of all locus tags and Gene IDs shown to the right of the heatmap is given in **Table 2**.

In general, it was observed that AGXX® had a huge impact on the transcriptome of *S. aureus* 04-02981, in particular at the later time-points 24, 80, and 120 min. In contrast, the effect of Ag was much less pronounced as already visible in the growth kinetics and to a lesser extent in the biofilm assays. Although, quite a number of *S. aureus* 04-02981 genes were differentially expressed upon exposure to Ag, only very few belong to the group of biofilm or virulence-associated genes. Among those, which were significantly differentially expressed in the presence of Ag, were *fmtC*, which is associated with methicillin resistance (approximately 3 fold up-regulated at t80; in the presence of AGXX® it was 2 fold up-regulated at t24), transcriptional regulator *sarR* (approximately 3 fold down-regulated at t24;

TABLE 2 | Differential expression of biofilm, and virulence-associated genes in *S. aureus* 04-02981 on exposure to AGXX®.

Locus tag	Abbreviation	Description	6 min	12 min	24 min	80 min	120 min
SA2981_RS10640	*agrD*	Accessory gene regulator D			−8.9	−11.9	−17.5
SA2981_RS10645	**agrC*	Histidine kinase of the competence regulon ComD			−9.9	−7.7	−7
SA2981_RS10635	*agrB*	Accessory gene regulator B		−5.7	−18.3	−40.7	−40.8
SA2981_RS10650	*agrA*	Two-component system, LytR family, response regulator AgrA			−2.2		
SA2981_RS05970	PSM- β	Phenol-soluble modulin Beta				−10.5	−22.7
SA2981_RS05965	PSM- β	Phenol-soluble modulin Beta				−10.2	−12.4
SA2981_RS10825	*sigB*	RNA polymerase Sigma-B factor			−2.3	−4	−4.9
SA2981_RS07680	**srrA*	DNA-binding response regulator SrrA		−4.6	−9.1	−4.6	−5.3
SA2981_RS00190	*mecA*	mecA-Penicillin- binding Protein 2		−2.3	−5.5		
SA2981_RS12040	*sarR*	Transcriptional regulator SarR					
SA2981_RS06390	*codY*	GTP-sensing transcriptional pleiotropic repressor CodY			−2.3	2.4	3.2
SA2981_RS00550	*sarH1*	Staphylococcal accessory regulator A			−2.4		
SA2981_RS05325	*FmtA*	FmtA protein involved in methicillin resistance				−2.6	
SA2981_RS05940	*arcD*	Arginine/ornithine antiporter ArcD					2.4
SA2981_RS00770	*capF*	Capsular polysaccharide synthesis enzyme Cap8F			−4.8	−3.1	−2
SA2981_RS05275	*sspb*	Staphopain B precursor				−2.3	
SA2981_RS13390	*cidA*	Holin-like protein				2.8	8.2
SA2981_RS13925	*arcA*	Arginine deiminase	−2.6	−11.8	−114.7	−20	−4.6
SA2981_RS03620	*saeR*	two-component system, OmpR family, response regulator SaeR					2.3
SA2981_RS06960	*FmtC*	Protein involved in methicillin resistance/L-lysine modification of phosphatidylglycerol			2.3		
SA2981_RS05900	hemolysin II	Alpha-hemolysin precursor				3.8	3.7
SA2981_RS01335	*lrgA*	Antiholin-like protein		−2.6	−3.1		−3.7
SA2981_RS00745	**capA*	Capsular polysaccharide synthesis enzyme Cap5A			−34.9	−76.9	−61.9
SA2981_RS00750	*capB*	Tyrosine-protein kinase EpsD/capsular polysaccharide synthesis enzyme		2.4	−4.9	−29	−38.1
SA2981_RS00755	*capC*	protein-tyrosine phosphatase/capsular polysaccharide synthesis enzyme		2.8	−15	−20205	−52.9
SA2981_RS13940	*aur*	Zinc metalloproteinase precursor/aureolysin			−12.4		
SA2981_RS02875	**sdrC*	Serine-aspartate repeat-containing protein C			−13.4		−9.8
SA2981_RS02035	Exotoxin 6	Superantigen-like protein			−8.7	4	−12.6
SA2981_RS13920	*arcB*	Ornithine carbamoyltransferase			−16.3	−26.1	−8.6
SA2981_RS05715	*isdC*	NPQTN cell wall anchored protein IsdC				−5	−9.1
SA2981_RS05735	*srtB*	Sortase B				−8.3	−22.4
SA2981_RS14090	*icaD*	Polysaccharide intercellular adhesin (PIA) biosynthesis protein		54.7			5.6
SA2981_RS14085	*icaA*	Polysaccharide intercellular adhesin (PIA) biosynthesis *N*-glycosyltransferase			35.7	29.9	26.5
SA2981_RS14095	*icaB*	Polysaccharide intercellular adhesin (PIA) biosynthesis deacetylase				−102.4	7.8
SA2981_RS09385	**lukE*	Leukotoxin/leukocidin			−378.9	−192.2	

**Genes selected for validation via RT-qPCR.*

not differentially expressed in the presence of AGXX®), the gene of the holin-like protein CidA (approximately 4 fold down-regulated at t24; ∼2 and ∼8 fold up-regulated at t80 and t120, respectively, with AGXX®), the arginine deaminase gene *arcA* (approximately 6 fold down-regulated at t120 and 4.6 fold down-regulated with AGXX®), the hemolysin II gene (approximately 2 fold down-regulated at t24 and approximately 3 fold down-regulated at t120; ∼3.7 fold up-regulated with AGXX® at t80 and t120) and the gene of the antiholin-like protein lrgA (approximately 6 fold up-regulated at t6 with Ag, in the presence of AGXX®, it was ∼3- to 3.7 fold down-regulated at t24, t80, and t120).

Validation of RNA-Sequencing Data Using RT-qPCR

From the RNA-seq data, we observed that AGXX® affected genes encoding two-component systems, surface adhesins, capsular polysaccharides, and toxins. In total, five, highly

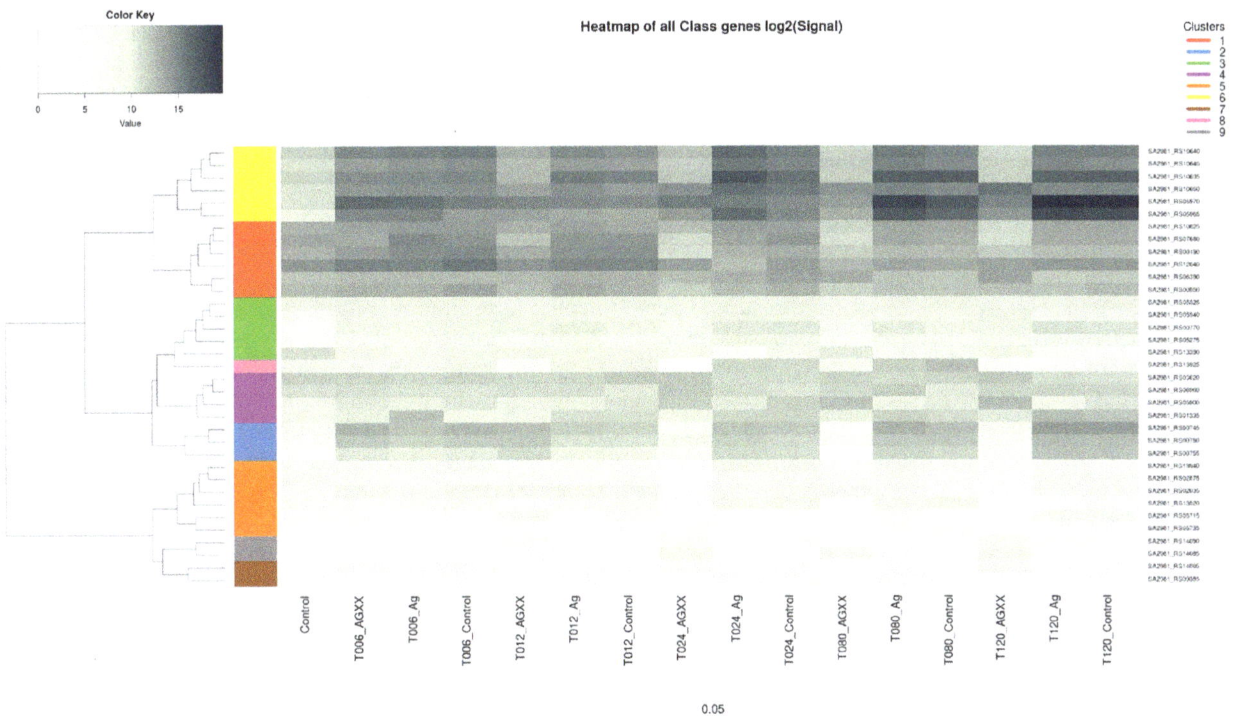

FIGURE 5 | Heatmap of differential expression of biofilm, and virulence-associated genes in *S. aureus* 04-02981. The genes are clustered as indicated by the dendrograms on the left side of the heatmap. Yellow represents genes *agrD*, **agrC*, *agrB*, *agrA*, and *PSM-β*, red represents genes *sigB*, **srrA*, *mecA*, *sarR*, *codY*, and *sarH1*. Green color is for genes *fmtA*, *arcD*, *capF*, *sspb*, *cidA*, pink represents *arcA* while purple is for *saeR*, *fmtC*, hemolysin II, and *lrgA* genes. Blue represents genes mediating capsular polysaccharide synthesis, namely, **capA*, *capB*, *capC*. *aur*, **sdrC*, exotoxin 6, *arcB*, *isdC*, and *srtB* are shown in orange. Gray represents *icaD*, and *icaA* and brown color represents *icaB*, and **lukE* genes. *Indicates genes selected for RT-qPCR.

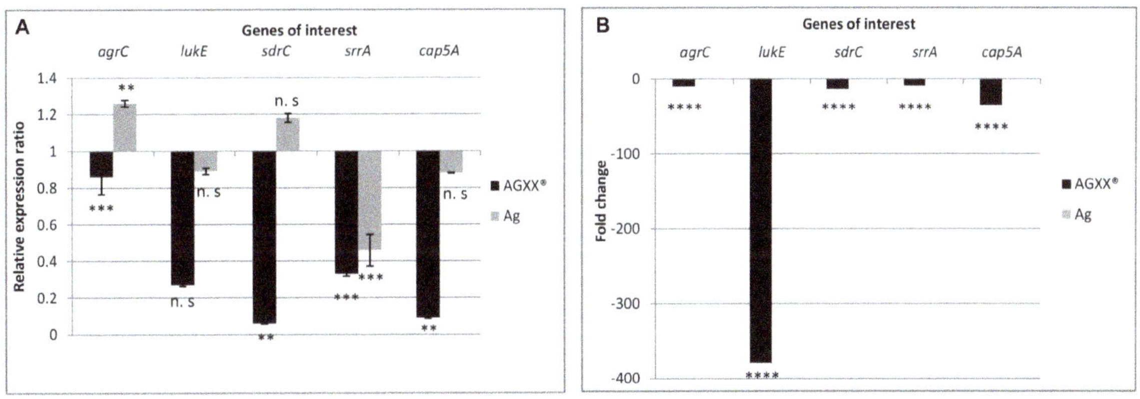

FIGURE 6 | Differential expression of *agrC*, *lukE*, *sdrC*, *srrA*, and *cap5A* in *S. aureus* 04-02981 on 24-min exposure to Ag or AGXX®, obtained via RT-qPCR **(A)**. Expression ratio of the genes of interest in *S. aureus* 04-02981 on exposure relative to control (untreated culture of *S. aureus* 04-02981) normalized to *gyrB*. **(B)** Shows differential expression of *agrC*, *lukE*, *sdrC*, *srrA*, and *cap5A* in *S. aureus* 04-02981 on 24-min exposure to Ag or AGXX®, obtained via RNA-seq as fold change. Error bars indicate standard deviation. Asterisks indicate *p*-values showing statistical significance. They were obtained from *t*-test using SigmaPlot 11.0 (****$p < 0.0001$, ***$p < 0.001$, **$p < 0.01$, *$p < 0.05$; n.s, not significant).

differentially expressed genes encoding these functions were selected to validate the RNA-seq derived transcriptional response of *S. aureus* 04-02981 to exposure to Ag or AGXX®. The validation experiment was performed on RNA extracted from *S. aureus* 04-02981 cultures exposed for 24, and 80 min to Ag or AGXX® since the selected genes were most differentially expressed at these time-points. The five selected genes were, *agrC*, and *srrA* which are part of the two-component systems AgrCA and SrrAB, respectively (Baker et al., 2010; Wu et al., 2015), *lukE* which encodes a toxin (Liu et al., 2016), *sdrC* specifying a surface adhesin (Barbu et al., 2014), and *cap5A* mediating the synthesis of capsular polysaccharides

(Qin et al., 2014). *gyrB* was used as the house-keeping gene (Smith et al., 2010; Cheung et al., 2011). **Figures 6, 7** show the results of these experiments.

After exposure to AGXX® for 24 min, all five genes were down-regulated both in RNA-seq analysis and in RT-qPCR studies as can be seen in **Table 2**, and **Figure 6**. However, after exposure to AGXX® for 80 min, *sdrC* was down-regulated in RT-qPCR assays but it was not differentially expressed in RNA-seq. All the other genes were down-regulated in both approaches as seen in **Table 2** and **Figure 7**, respectively. On exposure to AGXX® for 24 min, *sdrC* was the most down-regulated gene followed by *cap5A*, *lukE*, *srrA*, and *agrC*, whereas after 80 min, *agrC* was the most down-regulated gene followed by *srrA*, *lukE*, *cap5A*, and *sdrC*. On exposure to Ag for 24 min, *srrA* was the most down-regulated gene, whereas *agrC* was the most up-regulated gene, and after 80 min, *cap5A* was the most down-regulated gene while *sdrC* was the only up-regulated gene, as observed in the RT-qPCR experiments.

DISCUSSION

Multiple drug resistant, biofilm forming nosocomial pathogens such as MRSA pose a severe threat to public health demanding the development of novel antimicrobials as well as potent biofilm inhibitors. AGXX® is an effective antimicrobial that is active against many Gram-positive and Gram-negative bacteria (Guridi et al., 2015). AGXX® has been demonstrated to kill *S. aureus* 04-02981 as shown here by disk diffusion assay and growth kinetics experiments. In addition, AGXX® inhibited biofilm formation of *S. aureus* 04-02981 by ~46%. Moreover, for all time-points examined, the number of differentially expressed *S. aureus* 04-02981 genes was much higher upon exposure to AGXX® (in total 2391) than to Ag (317). For t120, the time-point showing the highest number of differentially expressed *S. aureus* 04-02981 genes, 1562 genes were differentially expressed in presence of AGXX®, while only 96 genes were affected by Ag.

Up-regulation of genes of Gene Ontology (GOs) groups "response to heat shock" and "oxidoreductases" involved in oxidative stress response, and down-regulation of genes of GOs "pathogenesis" and "lipopolysaccharide synthesis" involving genes mediating capsular polysaccharide synthesis important for biofilm formation, point to a role of AGXX® as an antimicrobial and potent biofilm inhibitor. Together with results of a recent study where we have shown that the QS system of *S. aureus* 04-02981, *agr* was completely repressed after 4 h of exposure to AGXX® (Probst et al., 2016), we propose that AGXX® acts as a potential biofilm inhibitor. In *S. aureus*, two main mechanisms of biofilm formation are known, namely *ica*-dependent biofilm formation, which involves the production of polysaccharide intercellular adhesin (PIA), and *ica*-independent biofilm formation (Kirmusaoglu, 2016). Here we show that, in the presence of AGXX®, *icaA*, *icaD* were up-regulated and *icaB* was down-regulated. *icaA* and *icaD* contribute to the production of PIA (polymer). *icaD* transfers PIA to the cell surface of the bacteria while *icaB* deacylates PIA by fixing PIA to the outer surface of the bacteria (Kirmusaoglu, 2016). In our study, intercellular adhesion biosynthesis *N*-deacetylase, *icaB* gene was down-regulated at t80 by ~100 fold. The structural development of exopolysaccharide-based biofilm requires deacetylation of PIA (Arciola et al., 2015). Since *icaB* was strongly down-regulated at t80, deacetylation of PIA probably does not occur which would obstruct the development of an exopolysaccharide-based biofilm. Fitzpatrick et al. (2005) showed that biofilm formation was unaffected in an *icaADBC* operon-deleted MRSA strain, while the same mutation in a methicillin sensitive strain of *S. aureus* (MSSA) impaired biofilm formation, suggesting strain-specificity in *ica*-dependent biofilm formation.

A two-component system associated with *ica*-dependent biofilm formation is SrrAB that acts as an autoregulator of biofilm formation. Deletion of *srrAB* inhibited *S. aureus* biofilm formation under oxic as well as microaerobic conditions (Wu et al., 2015). In our study, *srrA* was down-regulated 4 to 5 fold after 24, 80, and 120 min of exposure to AGXX®.

Global regulatory systems such as the *agr QS* system are among the best-studied factors involved in *ica*-(PIA) independent biofilm formation. Other proteins involved in such biofilms are SasG, SasC, Protein A, FnbB, FnbA, ATLA or ATLE, SdrG, SdrC, SdrD, biofilm associated protein (Bap) and lipoteichoic acid (Kirmusaoglu, 2016). We observed that two of these genes were down-regulated when AGXX® was present, namely *sdrD* and *sdrC*, *sdrC* was down-regulated 13- to 10 fold at t24 and t120, while *sdrD* was down-regulated 2 to 3 fold at t24 and t80. Moreover, the expression of lipoteichoic acid synthase, an enzyme responsible for the synthesis of lipoteichoic acid (Karatsa-Dodgson et al., 2010) was down-regulated approximately 4 fold after 24, 80, or 120 min of AGXX® presence. These data suggest that AGXX® might be working in an *ica*-independent manner to inhibit biofilm formation.

The *agr* locus contains five genes, *agrA*, *agrB*, *agrC*, *agrD*, and *hld*. On exposing *S. aureus* 04-02981 to AGXX®, only *hld* was not differentially expressed at any time-point, while all the other four genes were significantly down-regulated. The *agr* gene cluster regulates the expression of virulence factors such as phenol soluble modulins (PSMs), proteins that are closely associated with human skin and soft tissue infections (SSTIs) (Sully et al., 2014). "AgrD is a precursor peptide of autoinducer peptide (AIP)" (Quave and Horswill, 2014), AgrB is a membrane protease, which is involved in proteolytic processing and export of AgrD. It is also involved in AIP production (Njoroge and Sperandio, 2009; Quave and Horswill, 2014). AgrBD produce and secrete AIPs. AgrC, a sensor histidine kinase is activated when AIPs bind to AgrC. As a consequence, AgrC undergoes phosphorylation to activate AgrA, which is a DNA-binding response regulator (Njoroge and Sperandio, 2009). In our study, the *agrB* gene was the most down-regulated, at t80, and t120 (approximately 41 fold in both cases), while *agrA* was differentially expressed only at t24 (2 fold down-regulated). At t12, only *agrB* was differentially expressed, approximately 6 fold down-regulated. None of the *agr* genes was differentially expressed at t6. PSMs are staphylococcal toxins playing a role in acute infection (Kirmusaoglu, 2016); they are required for maturation and detachment of biofilm (Ma

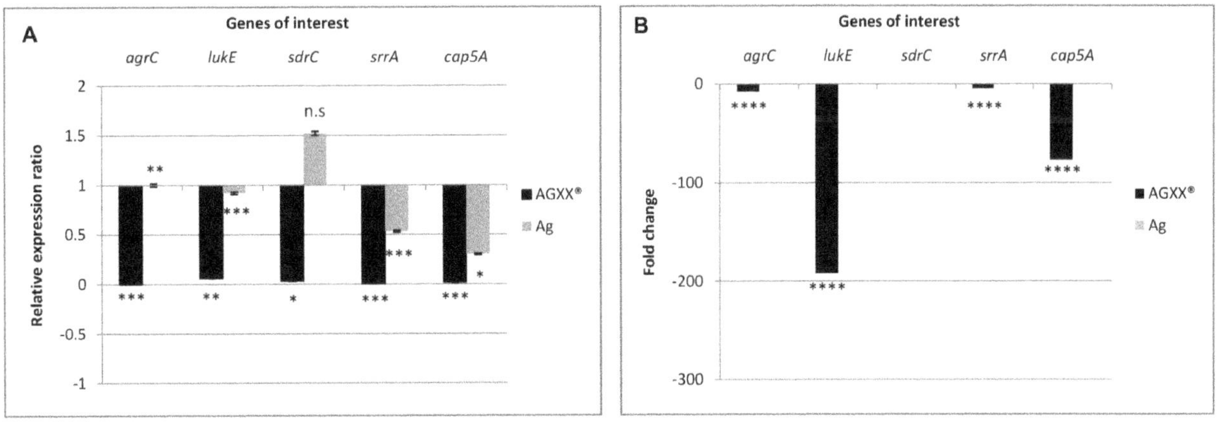

FIGURE 7 | Differential expression of *agrC*, *lukE*, *sdrC*, *srrA*, and *cap5A* in *S. aureus* 04-02981 on 80-min exposure to Ag or AGXX®, obtained via RT-qPCR. Expression ratio of the genes of interest in *S. aureus* 04-02981 on exposure relative to control (untreated culture of *S. aureus* 04-02981) normalized to *gyrB* are shown **(A)**. **(B)** Shows differential expression of *agrC*, *lukE*, *sdrC*, *srrA*, and *cap5A* in *S. aureus* 04-02981 on 80-min exposure to Ag or AGXX®, obtained via RNA-seq as fold change. Error bars indicate standard deviation. Asterisks indicate *p*-values showing statistical significance. They were obtained from *t*-test using SigmaPlot 11.0 (****$p < 0.0001$, ***$p < 0.001$, **$p < 0.01$, *$p < 0.05$; n.s, not significant).

et al., 2012). PSMs were also down-regulated in presence of AGXX® by ~10 fold at t80, and by 12 and 23 fold at t120. *agr* also regulates the expression of *sspB* which encodes a cysteine protease. *sspB* is positively associated with biofilm formation (Ma et al., 2012). It was down-regulated by 2.3 fold at t80. Inactivation of the alternative sigma factor SigB decreases biofilm formation in *S. aureus* (Ma et al., 2012). In presence of AGXX®, *sigB* was down-regulated 2–5 fold at the longer exposure times (t24, t80, and t120). In summary, down-regulation of all of the genes mentioned in this paragraph will likely reduce biofilm formation by *S. aureus*.

The two component systems, AgrCA and SaeRS influence biofilm formation in *S. aureus*, by the production of PSMs and by suppressing the synthesis of extracellular proteases, respectively (Baldry et al., 2016). The extracellular proteases degrade proteins that are important for biofilm formation (Baldry et al., 2016). In *S. aureus*, the *saeRS* system regulates the production of many virulence factors such as leukocidins, superantigens, proteases, surface proteins, and hemolysins (Liu et al., 2016). The gene for LukE, which enables *S. aureus* evasion from phagocytic cells by damaging the phagocytes was strongly down-regulated at t24 (379 fold) and t80 (192 fold). SplA is a serine protease, which is directly controlled by the *saeRS* system. *splA* was down-regulated 135 fold after 80 min of AGXX® presence. Mutations in genes for extracellular proteases (*spl*ABCDEF) in *S. aureus* SH1000 induced an increase in extracellular protease activity, which was associated with a reduction in biofilm formation (Chen et al., 2013). These facts taken together with *saeRS* not being differentially expressed at any time-point in the presence of AGXX®, except for a slight 2.3 fold up-regulation of *saeR* at t120, might suggest that *saeR* is not expressed in the mid exponential phase of growth of *S. aureus* 04-02981.

Capsular polysaccharides are also possible targets of the *saeRS* system (Liu et al., 2016). They play an important role in the virulence of the organism (Tuchscherr et al., 2010). The synthesis of capsular polysaccharides is mediated by the *cap5ABCFG* genes (Qin et al., 2014). Among these genes, only *cap*G was not differentially expressed, all other genes were significantly down-regulated, especially at t24, t80, and t120, suggesting a role of AGXX® in repression of virulence in *S. aureus* 04-02981.

Another QS system, which significantly influences biofilm formation and virulence in Staphylococci is the *luxS* system. *luxS* impacts biofilm formation in a similar way as *agr* does, but by regulating different factors. *luxS* negatively regulates biofilm formation via cell-cell interactions based on autoinducer 2 secretion (Xu et al., 2006). The gene was 2.9 fold up-regulated at t24 in the presence of AGXX®.

In addition, the genes *isdC*, *srtB*, *sdrC*, encoding adhesins, were all down-regulated in the pathogen exposed to AGXX®. Iron regulated surface determinant IsdC is necessary for the primary attachment of *S. aureus* to surfaces such as polystyrene, as well as for the accumulation phase of biofilm formation; as such, it induces biofilm formation (Missineo et al., 2014). IsdC is anchored to the cell wall by sortase B (Hammer and Skaar, 2011). Serine-aspartate repeat containing protein C precursor (SdrC) assists bacteria in adhering to surfaces and promotes biofilm formation (Barbu et al., 2014). In *S. aureus* 04-02981 exposed to AGXX®, *isdC* was down-regulated by 5 and 9 fold at t80 and t120, respectively. The sortase B gene *srtB* was also down-regulated in cells treated with AGXX®, at t80 (8 fold) and t120 (22 fold). *sdrC*, too, was down-regulated some 10 to 13 fold at t24 and t120. Thus, we suggest that AGXX® inhibits biofilm formation in *S. aureus* 04-02981, also by repressing the expression of adhesins.

Reverse transcription quantitative PCR assays were performed on RNA extracted from *S. aureus* 04-02981 cultures exposed to Ag or AGXX® for 24 min, and 80 min to validate the RNA-seq data. In RT-qPCR, on exposure to AGXX® for 24 min, *agrC*, *sdrC*, *srrA*, and *cap5A* were statistically significantly down-regulated, whereas the down-regulation of *lukE* was not statistically significant. In agreement with these data, the five genes were also significantly down-regulated in RNA-seq. By contrast, none of the five genes was significantly differentially expressed after 24 min in presence of Ag, as determined by RNA-seq, whereas RT-qPCR revealed a statistically significant down-regulation of *srrA* and a statistically significant up-regulation of *agrC*. The difference in expression of the other three genes *lukE, sdrC,* and *cap5A* was statistically not significant. When *S. aureus* 04-02981 was exposed to AGXX® for 80 min, all the five genes were down-regulated in RT-qPCR. The effect was statistically significant while in RNA-seq all genes were significantly down-regulated except *sdrC*. On exposure to Ag for 80 min, only *sdrC* was non-statistically significantly up-regulated. Thus, the trends in gene expression of *S. aureus* 04-02981 on exposure to AGXX® observed in RNA-seq and in RT-qPCR were similar.

In previous studies by others, differential gene expression of *S. aureus* in planktonic and biofilm mode has been examined. Resch et al. (2005) observed that in biofilms, genes encoding polysaccharide intercellular adhesin, and enzymes associated with cell envelope synthesis were significantly up-regulated (Resch et al., 2005). To combat biofilms, many metals have been tested for their capacity to inhibit bacterial biofilm formation. Specifically, silver nanoparticles have received much attention with respect to their antimicrobial nature. However, the minimum concentration of silver nanoparticles (AgNPs) required to eliminate biofilm formation is considered to have toxic effects on mammalian cells (Loo et al., 2016). They studied the effect of AgNPs and curcumin nanoparticles (Cur-NPs) on *S. aureus* and discovered that the combination of both nanoparticles was more effective than the individual AgNPs or Cur-NPs. Curcumin interferes with the QS system as was observed by the down-regulation of genes involved in QS, upon exposure to the substance (Loo et al., 2016). Ma et al. (2012) investigated the effect of two novel anti-virulence compounds on growth and biofilm formation of *S. aureus*. The compounds inhibited biofilm formation by repressing genes associated with biofilm formation such as *lrgA, sdrD, sspB, sigB, codY*, which were also down-regulated in our studies at least at one of the five time-points (Ma et al., 2012).

In summary, based on our findings, we conclude that AGXX® is an effective antimicrobial substance which might also act as a biofilm inhibitor based on our molecular data. The mechanism of inhibition is likely *ica*-independent without the production of PIA, by interfering with the QS system and by repressing genes associated with surface adhesin and lipopolysaccharide synthesis. In addition, the antimicrobial might also reduce pathogenesis of *S. aureus* 04-02981 by down-regulating the synthesis of toxins and virulence factors.

AUTHOR CONTRIBUTIONS

AV performed all the microbiological and molecular experiments, drafted the manuscript, and designed the figures. AdJ supervised and discussed bioinformatics analyses of RNA-seq, and prepared and deposited the RNA-seq data at NCBI. DW performed the confocal microscopy and analyzed the data. JK drafted part of the discussion and gave insightful suggestions on molecular biology of Gram-positive pathogens. EG designed the project and supervised all the experiments. All authors discussed and corrected the manuscript.

ACKNOWLEDGMENTS

We thank U. Landau and C. Meyer from Largentec GmbH, Berlin, for providing us with the antimicrobial AGXX® and for the helpful discussions and G. Werner and J. Bender from Robert Koch Institute, Wernigerode Branch, for the gift of *S. aureus* 04-02981.

REFERENCES

Arciola, C. R., Baldassarri, L., and Montanaro, L. (2001). Presence of *icaA* and *icaD* genes and slime production in a collection of staphylococcal strains from catheter-associated infections. *J. Clin. Microbiol.* 39, 2151–2156.

Arciola, C. R., Campoccia, D., Ravaioli, S., and Montanaro, L. (2015). Polysaccharide intercellular adhesin in biofilm: structural and regulatory aspects. *Front. Cell. Infect. Microbiol.* 5:7. doi: 10.3389/fcimb.2015.00007

Baker, J., Sitthisak, S., Sengupta, S., Johnson, M., Jayaswal, R. K., and Morrissey, J. A. (2010). Copper stress induces a global stress response in *Staphylococcus aureus* and represses *sae* and *agr* expression and biofilm formation. *Appl. Environ. Microbiol.* 76, 150–160. doi: 10.1128/AEM.02268-09

Baldry, M., Nielsen, A., Bojer, M. S., Zhao, Y., Friberg, C., Ifrah, D., et al. (2016). Norlichexanthone reduces virulence gene expression and biofilm formation in *Staphylococcus aureus*. *PLOS ONE* 11:e0168305. doi: 10.1371/journal.pone.0168305

Barbu, E. M., Mackenzie, C., Foster, T. J., and Höök, M. (2014). SdrC induces staphylococcal biofilm formation through a hemophilic interaction. *Mol. Microbiol.* 94, 172–185. doi: 10.1111/mmi.12750

Bouchard, A. (2011). AgXX Glass Microspheres. In Vitro Evaluation of Cytotoxicity by Neutral Red Assay Using MRC-5 Cell Line with a Direct Contact Procedure. Report 20100326STP. Dresden: APOGEPHA Arzneimittel GmbH.

Chen, C., Krishnan, V., Macon, K., Manne, K., and Schneewind, O. (2013). Secreted proteases control autolysin-mediated biofilm growth of *Staphylococcus aureus*. *J. Biol. Chem.* 288, 29440–29452. doi: 10.1074/jbc.M113.502039

Cheung, G. Y. C., Wang, R., Khan, B. A., Sturdevant, D. E., and Otto, M. (2011). Role of the accessory gene regulator *agr* in community- associated methicillin-resistant *Staphylococcus aureus* pathogenesis. *Infect. Immun.* 79, 1927–1935. doi: 10.1128/IAI.00046-11

Chopra, I. (2007). The increasing use of silver-based products as antimicrobial agents: a useful development or a cause for concern? *J. Antimicrob. Chemother.* 59, 587–590. doi: 10.1093/jac/dkm006

Clauss-Lendzian, E., Vaishampayan, A., de Jong, A., Landau, U., Meyer, C., Kok, J., et al. (2017). Stress response of a clinical *Enterococcus faecalis* isolate subjected to a novel antimicrobial surface coating. *Microbiol. Res.* doi: 10.1016/j.micres.2017.11.006

de Jong, A., van der Meulen, S., Kuipers, O. P., and Kok, J. (2015). T-REx: transcriptome analysis webserver for RNA-seq expression data. *BMC Genomics* 16:663. doi: 10.1186/s12864-015-1834-4

Ellis, M. W., Schlett, C. D., Millar, E. V., Crawford, K. B., Cui, T., Lanier, J. B., et al. (2014). Prevalence of nasal colonization and strain concordance in patients with community-associated *Staphylococcus aureus* skin and soft- tissue infections. *Infect. Control Hosp. Epidemiol.* 35, 1251–1256. doi: 10.1086/678060

Fitzpatrick, F., Humphreys, H., and O'Gara, J. P. (2005). Evidence for *icaADBC*-independent biofilm development mechanism in methicillin-resistant *Staphylococcus aureus* clinical isolates. *J. Clin. Microbiol.* 43, 1973–1976.

Frees, D., Chastanet, A., Qazi, S., Sorensen, K., Hill, P., Msadek, T., et al. (2005). Clp ATPases are required for stress tolerance, intracellular replication and biofilm formation in *Staphylococcus aureus*. *Mol. Microbiol.* 54, 1445–1462. doi: 10.1111/j.1365-2958.2004.04368.x

Gouaux, A., Hobaugh, M., and Song, L. (1997). α-hemolysin, γ-hemolysin, and leukocidin from *Staphylococcus aureus*: distant in sequence but similar in structure. *Protein Sci.* 6, 2631–2635. doi: 10.1002/pro.5560061216

Grass, G., Rensing, C., and Solioz, M. (2011). Metallic copper as an antimicrobial surface. *Appl. Environ. Microbiol.* 77, 1541–1547. doi: 10.1128/AEM.02766-10

Gupta, A., Matsui, K., Lo, J.-F., and Silver, S. (1999). Molecular basis for resistance to silver cations in *Salmonella*. *Nat. Med.* 5, 183–188. doi: 10.1038/5545

Guridi, A., Diederich, A. K., Aguila-Arcos, S., Garcia-Moreno, M., Blasi, R., Broszat, M., et al. (2015). New antimicrobial contact catalyst killing antibiotic resistant clinical and water borne pathogens. *Mater. Sci. Eng. C Mater. Biol. Appl.* 50, 1–11. doi: 10.1016/j.msec.2015.01.080

Hammer, N. D., and Skaar, E. P. (2011). Molecular mechanisms of *Staphylococcus aureus* iron acquisition. *Annu. Rev. Microbiol.* 65, 129–147. doi: 10.1146/annurev-micro-090110-102851

Huebner, J., Wang, Y., Krueger, W. A., Madoff, L. C., Martirosian, G., Boisot, S., et al. (1999). Isolation and chemical characterization of a capsular polysaccharide antigen shared by clinical isolates of *Enterococcus faecalis* and vancomycin-resistant *Enterococcus faecium*. *Infect. Immun.* 67, 1213–1219.

Ibarra, J. A., Pérez-Rueda, E., Carroll, R. K., and Shaw, L. N. (2013). Global analysis of transcriptional regulators in *Staphylococcus aureus*. *BMC Genomics* 14:126. doi: 10.1186/1471-2164-14-126

Karatsa-Dodgson, M., Woermann, M. E., and Gruendling, A. (2010). *In vitro* analysis of the *Staphylococcus aureus* lipoteichoic acid synthase enzyme using fluorescently labeled lipids. *J. Bacteriol.* 192, 5341–5349. doi: 10.1128/JB.00453-10

Kirmusaoglu, S. (2016). "Staphylococcal biofilms: pathogenicity, mechanism and regulation of biofilm formation by Quorum-Sensing system and antibiotic resistance mechanisms of biofilm-embedded microorganisms," in *Microbial Biofilms - Importance and Applications*, ed. D. Dhanasekaran (Rijeka: In Tech).

Kluytmans, J., van Belkum, A., and Verbrugh, H. (1997). Nasal carriage of *Staphylococcus aureus*: epidemiology, underlying mechanisms, and associated risks. *Clin. Microbiol. Rev.* 10, 505–520.

Landau, U. (2013). AGXX - Eine nachhaltige lösung für die Entkeimung wässriger Lösungen. *Galvanotechnik* 11, 2169–2184.

Landau, U., Meyer, C., and Grohmann, E. (2017a). AGXX – Beitrag der Oberflächentechnik zur Vermeidung von Biofilmen (Teil 1). *Galvanotechnik* 108, 885–890.

Landau, U., Meyer, C., and Grohmann, E. (2017b). AGXX – Beitrag der Oberflächentechnik zur Vermeidung von Biofilmen (Teil 2). *Galvanotechnik* 108, 1110–1121.

Landsdown, A. B. G. (2010). A pharmacological and toxicological profile of silver as an antimicrobial agent in medical devices. *Adv. Pharmacol. Sci.* 2010:910686. doi: 10.1155/2010/910686

Langmead, B., and Salzberg, S. (2012). Fast gapped-read alignment with Bowtie 2. *Nat. Methods* 9, 357–359. doi: 10.1038/nmeth.1923

Li, H., Handsaker, B., Wysoker, A., Fennell, T., Ruan, J., Homer, N., et al. (2009). The sequence alignment/map (SAM) format and SAMtools. *Bioinformatics* 25, 2078–2079. doi: 10.1093/bioinformatics/btp352

Lister, J. L., and Horswill, A. R. (2014). *Staphylococcus aureus* biofilms: recent developments in biofilm dispersal. *Front. Cell. Infect. Microbiol.* 4:178. doi: 10.3389/fcimb.2014.00178

Liu, Q., Yo, W.-S., and Bae, T. (2016). The SaeRS two-component system of *Staphylococcus aureus*. *Genes* 7:81. doi: 10.3390/genes7100081

Loo, C.-Y., Rohanizadeh, R., Young, P. M., Traini, D., Cavaliere, R., Whitchurch, C. B., et al. (2016). Combination of silver nanoparticles and curcumin nanoparticles for enhanced anti-biofilm activities. *J. Agric. Food Chem.* 64, 2513–2522. doi: 10.1021/acs.jafc.5b04559

Ma, Y., Xu, Y., Yestrepsky, B. D., Sorenson, R. J., Chen, M., Larsen, S. D., et al. (2012). Novel inhibitors of *Staphylococcus aureus* virulence gene expression and biofilm formation. *PLOS ONE* 7:e47255. doi: 10.1371/journal.pone.0047255

Maillard, J., and Hartemann, P. (2013). Silver as an antimicrobial: facts and gap in knowledge. *Crit. Rev. Microbiol.* 39, 373–383. doi: 10.3109/1040841X.2012.713323

Marathe, N. P., Nagarkar, S. S., Vaishampayan, A. A., Rasane, M. H., Samant, S. A., Dohe, V., et al. (2015). High prevalence of class 1 integrons in clinical isolates of methicillin-resistant *Staphylococcus aureus* from India. *Indian J. Med. Microbiol.* 33, 231–236. doi: 10.4103/0255-0857.154905

Missineo, A., Poto, D. A., Geoghegan, J. A., Rindi, S., Heilbronner, S., Gianotti, V., et al. (2014). IsdC from *Staphylococcus lugdunensis* induces biofilm formation under low-iron growth conditions. *Infect. Immun.* 82, 2448–2459. doi: 10.1128/IAI.01542-14

Naas, T., Coignard, B., Carbonne, A., Blanckaert, K., Bajolet, O., Bernet, C., et al. (2006). VEB-1 Extended-spectrum beta-lactamase-producing *Acinetobacter baumannii*, France. *Emerg. Infect. Dis.* 12, 1214–1222. doi: 10.3201/eid1208.051547

Njoroge, J., and Sperandio, V. (2009). Jamming bacterial communication: new approaches for the treatment of infectious diseases. *EMBO Mol. Med.* 1, 201–210. doi: 10.1002/emmm.200900032

Nuebel, U., Dordel, J., Kurt, K., Strommenger, B., Westh, H., Shukla, S. K., et al. (2010). A timescale for evolution, population expansion, and spatial spread of an emerging clone of methicillin resistant *Staphylococcus aureus*. *PLOS Pathog.* 6:e1000855. doi: 10.1371/journal.ppat.1000855

Nyenje, M. E., Green, E., and Ndip, R. N. (2013). Evaluation of the effect of different growth media and temperature on the suitability of biofilm formation by *Enterobacter cloacae* strains isolated from food samples in South Africa. *Molecules* 18, 9582–9593. doi: 10.3390/molecules18089582

Paniagua-Contreras, G., Sáinz- Espuñes, T., Monroy-Pérez, E., Rodríguez-Moctezuma, J. R., Arenas-Aranda, D., Negrete-Abascal, E., et al. (2012). Virulence markers in *Staphylococcus aureus* strains isolated from hemodialysis catheters of Mexican patients. *Adv. Microbiol.* 2, 476–487. doi: 10.4236/aim.2012.24061

Paraje, M. G. (2011). "Antimicrobial resistance in biofilms," in *Science against Microbial Pathogens: Communicating Current Research and Technological Advances*, Vol. 2, ed. A. Mendez-Vilas (Badajoz: Formatex Research Center), 736–744.

Probst, I., Vaishampayan, A., Kuechler, V., and Grohmann, E. (2016). Antimikrobielle Oberflächenbeschichtung tötet multiresistente Krankheitserreger. *Flug Reisemed.* 23, 14–17.

Qin, N., Tan, X., Jiao, Y., Liu, L., Zhao, W., Yang, S., et al. (2014). RNA-Seq-based transcriptome analysis of methicillin-resistant *Staphylococcus aureus* biofilm inhibition by ursolic acid and resveratrol. *Sci. Rep.* 4:5467. doi: 10.1038/srep05467

Quave, C. L., and Horswill, A. R. (2014). Flipping the switch: tools for detecting small molecule inhibitors of staphylococcal virulence. *Front. Microbiol.* 5:706. doi: 10.3389/fmicb.2014.00706

Resch, A., Rosenstein, R., Nerz, C., and Goetz, F. (2005). Differential gene expression profiling of *Staphylococcus aureus* cultivated under biofilm and planktonic conditions. *Appl. Environ. Microbiol.* 71, 2663–2676.

Sause, W. E., Buckley, P. T., Strohl, W. R., Lynch, A. S., and Torres, V. J. (2015). Antibody-based biologics and their promise to combat *Staphylococcus aureus* infections. *Trends Pharmacol. Sci.* 37, 231–241. doi: 10.1016/j.tips.2015.11.008

Schäberle, T. F., and Hack, I. M. (2014). Overcoming the current deadlock in antibiotic resistance. *Trends Microbiol.* 22, 165–167. doi: 10.1016/j.tim.2013.12.007

Schiwon, K., Arends, K., Rogowski, K. M., Fuerch, S., Prescha, K., Sakinc, T., et al. (2013). Comparison of antibiotic resistance, biofilm formation and conjugative transfer of *Staphylococcus* and *Enterococcus* isolates from International Space Station and Antarctic research station Concordia. *Microb. Ecol.* 65, 638–651. doi: 10.1007/s00248-013-0193-4

Smith, K., Gould, K. A., Gordon, R., Gemmell, C. G., Hinds, J., and Lang, S. (2010). Influence of tigecycline on expression of virulence factors in biofilm-associated cells of methicillin- resistant *Staphylococcus aureus*. *Antimicrob. Agents Chemother.* 54, 380–387. doi: 10.1128/AAC.00155-09

Warnes, S. L., and Keevil, C. W. (2013). Inactivation of norovirus on dry copper alloy surfaces. *PLOS ONE* 8:e75017. doi: 10.1371/journal.pone.0075017

Wass, Y. A. (2009). SigmaPlot 11: Now with total sigmaStat integration. *Sci. Comput.* 26, 21.

Wu, Y., Wu, Y., Zhu, T., Han, H., Liu, H., Xu, T., et al. (2015). *Staphylococcus epidermidis* SrrAB regulates bacterial growth and biofilm formation differently under oxic and microaerobic conditions. *J. Bacteriol.* 197, 459–476. doi: 10.1128/JB.02231-14

Xu, L., Li, H., Vuong, C., Vadyvaloo, V., Wang, J., Yao, Y., et al. (2006). Role of the *luxS* Quorum-Sensing system in biofilm formation and virulence of *Staphylococcus epidermidis*. *Infect. Immun.* 74, 488–496.

Yarwood, J. M., Bartels, D. J., Volper, E. M., and Greenberg, E. P. (2004). Quorum Sensing in *Staphylococcus aureus* biofilms. *J. Bacteriol.* 186, 1838–1850. doi: 10.1128/JB.186.6.1838-1850.2004

Sully, E. K., Malachowa, N., Elmore, B. O., Alexander, S. M., Femling, J. K., Gray, B. M., et al. (2014). Selective chemical inhibition of *agr* Quorum Sensing in *Staphylococcus aureus* promotes host defense with minimal impact on resistance. *PLOS Pathog.* 10:e1004174. doi: 10.1371/journal.ppat.1004174

Trapnell, C., Williams, B. A., Pertea, G., Mortazavi, A., Kwan, G., van Baren, M. J., et al. (2010). Transcript assembly and quantification by RNA-Seq reveals unannotated transcripts and isoform switching during cell differentiation. *Nat. Biotechnol.* 28, 511–515. doi: 10.1038/nbt.1621

Tuchscherr, L., Loeffler, B., Buzzola, F. R., and Sordelli, D. O. (2010). *Staphylococcus aureus* adaptation to the host and persistence: role of loss of capsular polysaccharide expression. *Future Microbiol.* 5, 1823–1832. doi: 10.2217/fmb.10.147

13

Phages in the Human Body

*Ferran Navarro[1] and Maite Muniesa[2]**

[1] Servei de Microbiologia, Hospital de la Santa Creu i Sant Pau, Institut d'Investigació Biomèdica Sant Pau, Barcelona, Spain, [2] Department of Microbiology, University of Barcelona, Barcelona, Spain

Bacteriophages, viruses that infect bacteria, have re-emerged as powerful regulators of bacterial populations in natural ecosystems. Phages invade the human body, just as they do other natural environments, to such an extent that they are the most numerous group in the human virome. This was only revealed in recent metagenomic studies, despite the fact that the presence of phages in the human body was reported decades ago. The influence of the presence of phages in humans has yet to be evaluated; but as in marine environments, a clear role in the regulation of bacterial populations could be envisaged, that might have an impact on human health. Moreover, phages are excellent vehicles of genetic transfer, and they contribute to the evolution of bacterial cells in the human body by spreading and acquiring DNA horizontally. The abundance of phages in the human body does not pass unnoticed and the immune system reacts to them, although it is not clear to what extent. Finally, the presence of phages in human samples, which most of the time is not considered, can influence and bias microbiological and molecular results; and, in view of the evidences, some studies suggest that more attention needs to be paid to their interference.

Keywords: bacteriophages, human biomes, homeostasis, metagenomics, diagnosis, virome

Edited by:
Manuel Espinosa,
Consejo Superior de Investigaciones Cientificas (CSIC), Spain

Reviewed by:
Guillem Prats,
Autonomous University of Barcelona, Spain
Steven P. T. Hooton,
University of Nottingham, UK
Radoslaw Pluta,
International Institute of Molecular and Cell Biology in Warsaw, Poland

***Correspondence:**
Maite Muniesa
mmuniesa@ub.edu

INTRODUCTION

Bacteriophages were discovered in the second decade of the 20th century (Twort, 1915; D'Herelle, 1917). It was initially suggested the idea they could be used to lyse pathogenic bacteria as a treatment of infectious diseases. However, the idea was rapidly abandoned in western countries due to the introduction of antibiotics. For decades, phages have been the most common model entities for the study of viruses and their replication cycles. Studies of certain model phages have contributed significantly to the advancement of molecular biology, for example in identifying the basis of genetic material, as the code of nucleotide triplets of individual amino acids (Crick et al., 1961), and the restriction enzymes (Dussoix and Arber, 1962). Moreover, the first sequenced genome was that of an *Escherichia coli* phage: ϕX174 (Sanger et al., 1977). For some years, the interest for phages was limited to ecological studies and proposals for their use as indicators of fecal pollution (IAWPRC Study Group on Health Related Water Microbiology, 1991; Jofre et al., 2016); in general, bacteriophages have deserved less interest in comparison to their bacterial hosts or to animal viruses.

Nevertheless, the remarkable estimated number of 10^{31} phages on the Earth (Suttle, 2005) is commonly used by researchers to highlight the importance of phages, which are believed to outnumber any other class of biological entity on the planet. Phages have recently re-emerged as powerful regulators of the bacterial populations in natural ecosystems (Fuhrman, 1999). Moreover, because of the appearance of resistances to different antimicrobial agents, their potential use as antimicrobials has been revisited (Reardon, 2014). Most significantly, recent metagenomes describe the abundance of viral sequences both outside and inside bacterial cells. This highlights their ubiquity as mobile genetic elements that contribute and affect bacterial evolution, causing the emergence of new bacterial pathogens, mobilizing genes outside the cells, and other functions. Interest in metagenomes includes the study of human microbiomes, where phages again appear as extremely abundant and diverse elements. Researchers are only now starting to suspect that phages actively contribute to the homeostasis of the bacterial flora (De Paepe et al., 2014). Because many studies focus on the role of the human symbiotic microbiota in our wellness, phages thus appear as contributing actors that are directly related with human health (Manrique et al., 2016), and therefore the interest in them is rising.

PHAGES AS A PART OF HUMAN AND ANIMAL MICROBIOTA

Many metagenomic analyses of human microbiomes show the abundance of phages, which is generally greater than that of eukaryotic viruses. This has been shown in metagenomic analysis of lung, vaginal, skin, oral or intestinal microbiota (Breitbart et al., 2003; Colomer-Lluch et al., 2011a; Minot et al., 2011; Oh et al., 2014; Virgin, 2014). More recently, infectious phages have been found in different clinical samples such as ascitic fluid and urine (Brown-Jaque et al., 2016). It was suggested that they could reach the peritoneal cavity after translocation from the intestine (Górski et al., 2006), where they are present (**Figure 1**) and abundant. They are also present in voided urine (Brown-Jaque et al., 2016), probably coming from the periurethral area. In animals, phages infecting *Bacteroides* were found in serum (Keller and Traub, 1974), confirming their presence in the blood stream. Translocation of phages from blood to mouse fetal tissues has also been demonstrated in pregnant mice (Srivastava et al., 2004a).

In the light of these results, and as a second level of study, some researchers have analyzed solely the virome fraction of these microbiomes. To do this, they have devised methods that allow discrimination of the viral fraction, while discarding bacterial and free DNA. Those studies have yielded some surprising results; many viral particles in fact carry sequences identified as bacterial DNA. Shared genetic content is observed when analyzing the phage and bacterial DNA fractions of the same sample (Breitbart et al., 2003; Minot et al., 2011; Colombo et al., 2016; Howe et al., 2016), including sequences belonging to CRISPR-Cas systems (Dutilh et al., 2014).

CRISPR-Cas systems constitute a immune system that protect bacteria against bacteriophages and foreign DNA (Mojica and Rodriguez-Valera, 2016), that has later been applied for genome engineering in bacteria and eukaryotes. The different activity of the CRISPR-Cas systems influences the allowance of bacterial cells to foreign DNA or their immunity to phage infection, and this can shape the evolution of human microbiomes. Besides the use of CRISPR-Cas systems in genome engineering, the analysis of CRISPR sequences from raw metagenomic data has revealed unidentified phages, as crAssphage phage, that is claimed to be present in the majority of human fecal microbiomes, although it has never been isolated (Dutilh et al., 2014).

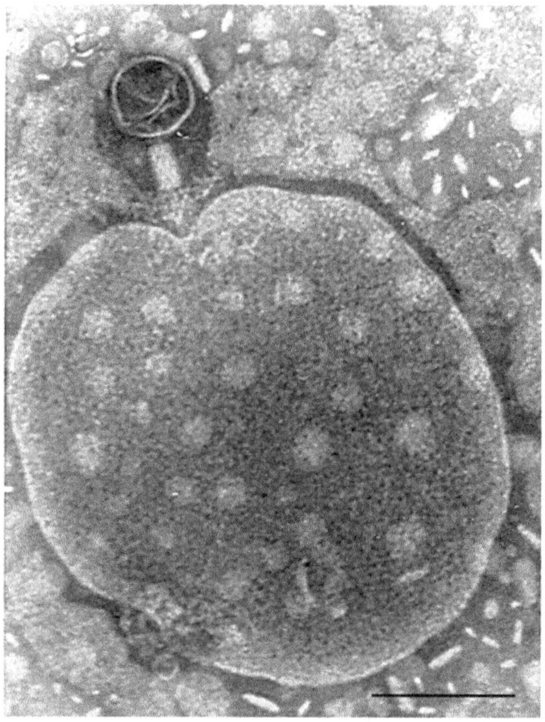

FIGURE 1 | Bacteriophage of *Myoviridae* morphology isolated from a fecal sample, attached to an unidentified particle. Bar 100 nm.

PHAGES AS MOBILE GENETIC ELEMENTS

Transduction, the process by which the DNA is mobilized between cell by a virus or viral vector was reported the last century (Zinder and Lederberg, 1952), although the rates of this mobilization has never been well defined. For this reason, the detection of an important proportion of bacterial DNA in phage particles observed in metagenomic analysis was indeed a surprise, and it initially prompted the belief that the methods for segregating phage and bacterial particles were not accurate enough, and either bacterial or free DNA contaminated the phage samples. However, the protocols have been optimized allowing specific extraction of packaged DNA. Another suspicion is that the bioinformatic analysis failed to identify phage DNA sequences correctly and they were mistaken for bacterial DNA. Nevertheless, subsequent repetitions and more accurate approaches have shown that despite some of these problems occurring, a relevant fraction of the virome is actually mobilizing bacterial DNA. This has led to the suspicion that bacterial cells use the numerous capsid genes that they possess, probably inherited from ancient prophage remnants, to build protein capsids that pack and spread their DNA content (Asadulghani et al., 2009; Lang et al., 2012; Penades et al., 2015).

The fact that phage capsids can mobilize bacterial DNA has multiple consequences, such as, for example, the fact that they can mobilize and transduce virulence genes (O'Brien et al., 1984; Griffiths et al., 2000; Allué-Guardia et al., 2011; Penades et al., 2015), antibiotic resistances (Muniesa et al., 2004; Colomer-Lluch et al., 2011b; Ross and Topp, 2015; Haaber et al., 2016) or genes related to fitness (Lindell et al., 2004; Müller et al., 2013) to new bacterial hosts. This causes horizontal genetic exchange and leads to the evolution of bacterial populations.

PHAGES AS REGULATORS OF POPULATIONS

Bacterial populations can change and evolve through acquisition of new genes transferred by phages, but also by predation and lysis caused by phages. Experimental evidence from chemostats and observations of phages/hosts in open systems has shown that for some bacterial species, populations of phages and hosts oscillate over time, following a "Red Queen/kill-the-winner" dynamics," which describes prey–predator variations (Rodriguez-Brito et al., 2010; Jover et al., 2013; Lim et al., 2015). However, phage–host dynamics can change in accordance with the homogeneity and structure of the environment, and also depending on the conditions that facilitate phage–cell encounters (De Paepe et al., 2014).

Changes or a total replacement of the microbiome by a fecal transplant in diseases without a well-defined etiological agent, such as inflammatory bowel diseases (Crohn's disease or ulcerative colitis), can result in different disease outcomes (Loh and Blaut, 2012; Moayyedi et al., 2015). Comparison of the viromes of individuals suffering from Crohn's disease and healthy relatives revealed differences in composition and variability (Pérez-Brocal et al., 2013; Wagner et al., 2013). Whether changes in the phagome of human biomes is a cause or a consequence of dysbiosis in such diseases has not yet been established. Considering that the phagome could influence bacterial populations, two options are plausible: changes in bacteria could cause variations in the distribution of phage groups; or changes in the phagome could be responsible of dysbacteriosis (Norman et al., 2015; Pérez-Brocal et al., 2015).

Similarly, phages have been detected in the metagenomes of sputum of patients suffering cystic fibrosis (Willner et al., 2009); and both the phage diversity and relative abundances were reported to be different from those of non-cystic fibrosis patients. It is hard, however, to conclude from these results what the cause of these differences is. Some variations in bacteria are caused by phages and those variations could be harmful to the patients. For example, mucoid isolates of *Pseudomonas fluorescens* are more virulent than their non-mucoid isogenic variants. This mucoid overproduction is a virulence factor contributing to more persistent infections in cystic fibrosis patients (Scanlan and Buckling, 2012). This phenotypic characteristic is favorably selected in the presence of phages, because it confers protection against phage infection. Accordingly, the mucoid isolates became resistant to the phages with the corresponding detrimental consequence for the patients (Scanlan and Buckling, 2012).

A different example of the regulation of human bacterial populations by phages is observed when we look at the competition between *Streptococcus pneumoniae* and *Staphylococcus aureus*. The former produces hydrogen peroxide; an agent that induces the bacterial SOS response and can induce temperate prophages. Meanwhile, the vast majority of *S. aureus* strains carry prophages that could be induced in the presence of the concentrations of H_2O_2 produced by *S. pneumoniae*. *S. pneumoniae* prophages, in turn, are not induced at these concentrations. The result is that *S. pneumoniae* prevails by killing *S. aureus* lysogenic strains via induction of prophages that cause the subsequent lysis of the cell (Selva et al., 2009).

Yet another example of how bacteriophages can impact the dynamics of bacterial populations has been observed in *Enterococcus faecalis* V583. This strain produces a composite phage ΦV1/7, derived from two distinct chromosomally encoded prophage elements. Prophage ΦV1 produces the capsids, while prophage ΦV7 is in charge of infection of susceptible hosts and V583 can produce infectious ΦV1/7. The induction of ΦV1/7 is highly enhanced by the availability of free amino acids in the medium. The strain producing ΦV1/7 has an advantage over other *E. faecalis* strains in the intestine, because these are lysed by ΦV1/7, while V583 is resistant to superinfection, enhancing the success of *E. faecalis* V583 during competitive growth (Duerkop et al., 2012).

INTERACTIONS WITH THE IMMUNE SYSTEM

It is not clear whether phages can easily be detected by the immune system, or whether they interact with it. Because the size of phage particles is usually bigger than eukaryotic viruses, activation of the immune system might occur as for other viruses. The desire to use phages to treat bacterial infections has led to explorations of the responses that phages might cause within the human immune system.

Very soon after the discovery of phages, it was observed that antibodies against bacteriophages in humans or animals were produced (Jerne, 1952, 1956); and it is easy to generate phage antiserums by immunization of humans or animals with phage lysates (Puig et al., 2001; Gorski et al., 2012; Bacon et al., 2017). The sera of non-immunized individuals (humans or animals) present antibodies against phages, although at low levels; the so-called "natural antibodies." For instance, antibodies against T4 phages are naturally present in human serum (Dabrowska et al., 2014) presumably as a consequence of the confirmed constant presence of phages in human biomes (Górski et al., 2006; Brown-Jaque et al., 2016). However, the origin of natural antibodies, generally of IgM class, with broad cross reactivity and low affinity, is not clear in the majority of cases.

The innate immune system, particularly by the components of the reticuloendothelial system (RES), could be a mechanism for removing phages that are circulating in the human body (Gorski et al., 2012). Certainly, this system was credited with the rapid removal of administered wild-type phage λ from the circulatory system in humans (Geier et al., 1973). Moreover, different phage λ mutants could induce different host responses. When using certain phage λ mutants that were capable of circumventing the RES immune response, these mutants prevailed for longer periods in the blood stream than the wild-type phage (Merril et al., 1996).

Data on anti-phage cellular responses are very scarce in comparison with data on phage-humoral responses. The only study we are aware of, evaluated the cellular response to MS2 phage that was intradermally administrated in guinea pigs. The presence of the phage produced erythema and induration, that are signs of cell-mediated immunity (Langbeheim et al., 1978). In contrast, another study showed that the permanence of phages in blood is the same when comparing immunocompetent mice or those deficient in T-cells, indicating no specific role of T-cell response in phage inactivation (Srivastava et al., 2004b).

When administered together with the host bacteria, some studies showed that phages seem to stimulate bacterial phagocytosis, and this is attributed to certain "opsonization" of the bacterial cells by phages. In addition, phages can remain active and infective when adsorbed onto the bacteria on intake by granulocytes. Therefore, some authors have suggested that during phagocytosis, phages continue lysing the phagocytosed bacteria, helping the activity of phagocytic cells. This process is limited in time and phages are no longer active after the completion of phagocytosis (Gorski et al., 2012). Despite these descriptions, there is no definitive evidence that phages activate phagocytosis by themselves, and some years ago, a contrary outcome was reported (Kantoch et al., 1958). In those studies, when used at very high doses (10^{10}/ml), phages inhibited phagocytosis of their host bacteria, and this inactivation was observed using either infectious or heat-inactivated phages (Kantoch et al., 1958). Inhibition was greater when using antibody-treated phages, and therefore the authors suggested that the immunocomplexes phage-antibody would be inactivating factors particularly active (Kantoch et al., 1958). Moreover, purified phages have anti-inflammatory effects via suppression of ROS (reactive oxygen species) production and inhibition of NF-κβ activity, affecting the production of cytokines [for a review, see (Gorski et al., 2012)]. Despite this evidence, it should be borne in mind that many experiments have been conducted with phage lysates, which on many occasions could contain remnants of bacteria lysed by the phages (e.g., lipopolysaccharide) or perhaps fragments of the host bacterial cell wall adhered to the phage tails. This makes it extremely difficult to determine the components truly responsible for the modulation of the immune response.

INTERFERENCE WITH CLINICAL DIAGNOSES

Assuming the relative occurrence and distribution of phages throughout the human body described above, coincident with

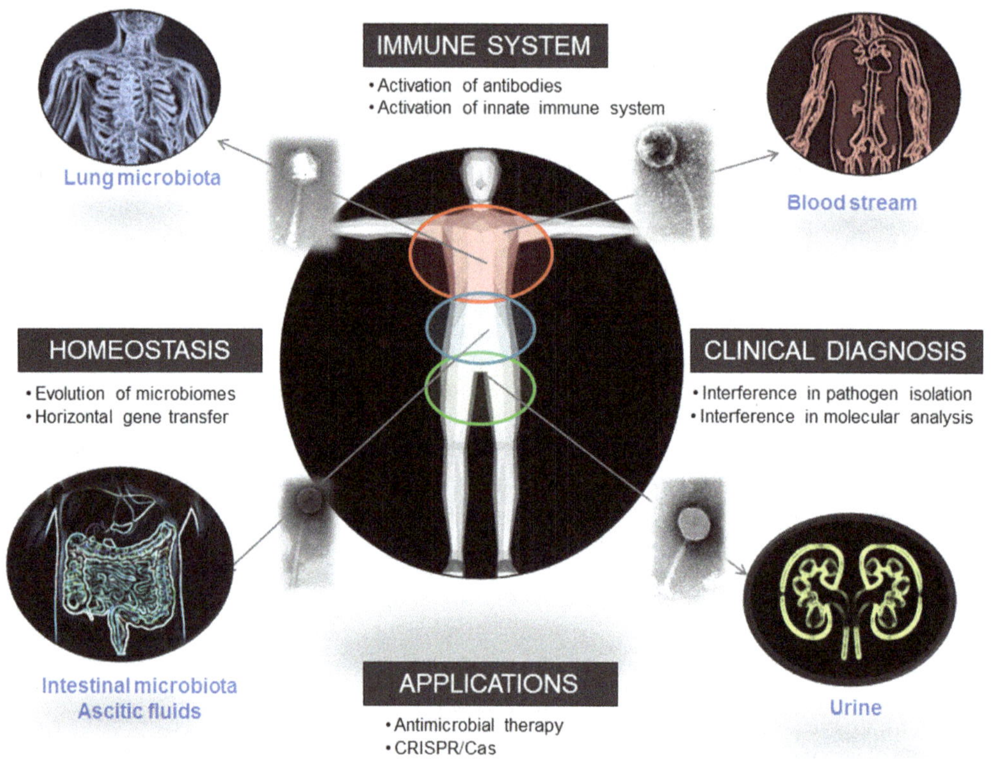

FIGURE 2 | Biomes where the presence of phages has been reported and direct and indirect ways in which phages influence human health. Pictures has been adapted from Pixabay.

the location of their bacterial hosts, and highly plausible translocation of phage particles to other areas of the body, some reports indicate that the neglected presence of phages in human samples could have an important influence by interfering in clinical practice (Brown-Jaque et al., 2016).

It has been shown that the presence of phages could interfere with many protocols intended to isolate bacteria by enrichment broth, since the phages in the sample destroy the bacterial cells during the enrichment procedure (Muniesa et al., 2005; Quiros et al., 2015). Phages might also interfere in clinical settings during bacterial isolation. As indicated above, there is evidence of a lack of, or reduced, bacterial isolation in clinical samples (ascitic fluid and urine) carrying a high titer of phages, because the lytic activity of the phages disturbs the isolation of the target bacteria (Brown-Jaque et al., 2016).

In addition, some results obtained using molecular methods when targeting some virulence genes present in pathogenic bacteria could be confusing. This is because some genes are located in temperate phages and DNA extraction methods do not distinguish between bacterial and phage DNA. This is the case for phages encoding the Shiga toxin gene, which can be detected in the absence of Shiga toxin-producing bacteria (Martínez-Castillo and Muniesa, 2014). Detection of certain bacterial groups by 16SrDNA qPCR or by genomic sequencing in mixed samples might also be confusing if the sample contains phages and the DNA in the phages is actually what is amplified in the absence of intact bacterial cells. This might be an explanation of the mismatch between the high number of gene copies of 16SrDNA obtained by qPCR amplification and the lack of bacterial isolation observed sometimes (Esparcia et al., 2011). Among others, one hypothesis could be that the positive results were due to amplification of bacterial DNA within phage particles or of bacterial DNA released in the sample after phage-mediated lysis.

USE OF PHAGES AGAINST HUMAN BACTERIAL PATHOGENS

The problems of fighting antibiotic resistance in bacteria are continually increasing and severely undermine our capacity to control bacterial infectious diseases. After the increased incidence of bacterial resistance to antibiotics over recent decades, phages have surfaced again as alternative or complementary therapies to control bacterial infections (Fischetti et al., 2006; Doyle and Erickson, 2012; Hertwig et al., 2013; Reardon, 2014; Schmelcher and Loessner, 2014; Górski et al., 2016).

CONCLUDING REMARKS

Phages, the most abundant entities on the planet, are also present in human biomes. This presence is known and recognized, but sometimes neglected; and it has a strong influence on the distribution and dynamics of different bacterial populations. Considering the influence of these populations in human health, as their reported ability to improve digestive health, it is clear that phages can be directly related with human well-being (**Figure 2**). The influence of phages in different mechanisms of our immune system suggests a long-term relationship that we are just starting to elucidate. Moreover, considering our interest in isolating and identifying bacterial pathogens, the presence of phages could certainly interfere with that analysis if it is not considered. A One Health multidisciplinary approach, not restricted to academic or clinical settings and not limited either to microbiological studies, is advisable to evaluate the real extent of and the role played by the phagome in human bodies.

AUTHOR CONTRIBUTIONS

All authors listed, have made substantial, direct and intellectual contribution to the work, and approved it for publication.

ACKNOWLEDGMENT

Authors thank Prof. P. Coll and Prof. J. Jofre for useful comments on the manuscript.

REFERENCES

Allué-Guardia, A., García-Aljaro, C., and Muniesa, M. (2011). Bacteriophage-encoding cytolethal distending toxin type V gene induced from nonclinical *Escherichia coli* isolates. *Infect. Immun.* 79, 3262–3272. doi: 10.1128/IAI.05071-11

Asadulghani, M., Ogura, Y., Ooka, T., Itoh, T., Sawaguchi, A., Iguchi, A., et al. (2009). The defective prophage pool of *Escherichia coli* O157: prophage-prophage interactions potentiate horizontal transfer of virulence determinants. *PLoS Pathog.* 5:e1000408. doi: 10.1371/journal.ppat.1000408

Bacon, E. J., Richmond, S. J., Wood, D. J., Stirling, P., Bevan, B. J., and Chalmers, W. S. (2017). Serological detection of phage infection in *Chlamydia psittaci* recovered from ducks. *Vet. Rec.* 119, 618–620.

Breitbart, M., Hewson, I., Felts, B., Mahaffy, J. M., Nulton, J., Salamon, P., et al. (2003). Metagenomic analyses of an uncultured viral community from human feces. *J. Bacteriol.* 185, 6220–6223. doi: 10.1128/JB.185.20.6220-6223.2003

Brown-Jaque, M., Muniesa, M., and Navarro, F. (2016). Bacteriophages in clinical samples can interfere with microbiological diagnostic tools. *Sci. Rep.* 6:33000. doi: 10.1038/srep33000

Colombo, S., Arioli, S., Guglielmetti, S., Lunelli, F., and Mora, D. (2016). Virome-associated antibiotic-resistance genes in an experimental aquaculture facility. *FEMS Microbiol. Ecol.* 92:fiw003. doi: 10.1093/femsec/fiw003

Colomer-Lluch, M., Imamovic, L., Jofre, J., and Muniesa, M. (2011a). Bacteriophages carrying antibiotic resistance genes in fecal waste from cattle, pigs, and poultry. *Antimicrob. Agents Chemother.* 55, 4908–4911. doi: 10.1128/AAC.00535-11

Colomer-Lluch, M., Jofre, J., and Muniesa, M. (2011b). Antibiotic resistance genes in the bacteriophage DNA fraction of environmental samples. *PLoS ONE* 6:e17549. doi: 10.1371/journal.pone.0017549

Dutilh, B. E., Cassman, N., McNair, K., Sanchez, S. E., Silva, G. G. Z., Boling, L., et al. (2014). A highly abundant bacteriophage discovered in the unknown sequences of human faecal metagenomes. *Nat. Commun.* 5, 4498. doi: 10.1038/ncomms5498

Crick, F. H., Barnett, L., Brenner, S., and Watts-Tobin, R. J. (1961). General nature of the genetic code for proteins. *Nature* 192, 1227–1232. doi: 10.1038/1921227a0

Dabrowska, K., Miernikiewicz, P., Piotrowicz, A., Hodyra, K., Owczarek, B., Lecion, D., et al. (2014). Immunogenicity studies of proteins forming the T4 phage head surface. *J. Virol.* 88, 12551–12557. doi: 10.1128/JVI.02043-14

De Paepe, M., Leclerc, M., Tinsley, C. R., and Petit, M.-A. (2014). Bacteriophages: an underestimated role in human and animal health? *Front. Cell. Infect. Microbiol.* 4:39. doi: 10.3389/fcimb.2014.00039

D'Herelle, F. (1917). sur un microbe invisible antagonist des bacilles disenterique. *C. R. Acad. Sci. Ser. D* 165, 373–375.

Doyle, M. P., and Erickson, M. C. (2012). Opportunities for mitigating pathogen contamination during on-farm food production. *Int. J. Food Microbiol.* 152, 54–74. doi: 10.1016/j.ijfoodmicro.2011.02.037

Duerkop, B. A., Clements, C. V., Rollins, D., Rodrigues, J. L. M., and Hooper, L. V. (2012). A composite bacteriophage alters colonization by an intestinal commensal bacterium. *Proc. Natl. Acad. Sci. U.S.A.* 109, 17621–17626. doi: 10.1073/pnas.1206136109

Dussoix, D., and Arber, W. (1962). Host specificity of DNA produced by *Escherichia coli*. II. Control over acceptance of DNA from infecting phage lambda. *J. Mol. Biol.* 5, 37–49. doi: 10.1016/S0022-2836(62)80059-X

Esparcia, O., Montemayor, M., Ginovart, G., Pomar, V., Soriano, G., Pericas, R., et al. (2011). Diagnostic accuracy of a 16S ribosomal DNA gene-based molecular technique (RT-PCR, microarray, and sequencing) for bacterial meningitis, early-onset neonatal sepsis, and spontaneous bacterial peritonitis. *Diagn. Microbiol. Infect. Dis.* 69, 153–160. doi: 10.1016/j.diagmicrobio.2010.10.022

Fischetti, V. A., Nelson, D., and Schuch, R. (2006). Reinventing phage therapy: are the parts greater than the sum? *Nat. Biotechnol.* 24, 1508–1511. doi: 10.1038/nbt1206-1508

Fuhrman, J. A. (1999). Marine viruses and their biogeochemical and ecological effects. *Nature* 399, 541–548. doi: 10.1038/21119

Geier, M. R., Trigg, M. E., and Merril, C. R. (1973). Fate of bacteriophage lambda in non-immune germ-free mice. *Nature* 246, 221–223. doi: 10.1038/246221a0

Gorski, A., Miedzybrodzk, R., Borysowski, J., Dabrowska, K., Wierzbicki, P., Ohams, M., et al. (2012). Chapter 2-"Phage as a modulator of immune responses: practical implications for phage therapy," in *Advances in Virus Research. Bacteriophages Part B*, eds M. Lobocka and W. Szybalski (Sant Diego, CA: Academic Press), 41–72.

Górski, A., Miêdzybrodzki, R., Weber-Dąbrowska, B., Fortuna, W., Letkiewicz, S., Rogóż, P., et al. (2016). Phage therapy: combating infections with potential for evolving from merely a treatment for complications to targeting diseases. *Front. Microbiol.* 7:1515. doi: 10.3389/fmicb.2016.01515

Górski, A., Wazna, E., Dabrowska, B.-W., Dabrowska, K., Switała-Jeleñ, K., and Miedzybrodzki, R. (2006). Bacteriophage translocation. *FEMS Immunol. Med. Microbiol.* 46, 313–319. doi: 10.1111/j.1574-695X.2006.00044.x

Griffiths, A. J., Miller, J. H., Suzuki, D. T., Lewontin, R. C., and Gelbart, W. M. (2000). "Gene transfer in bacteria and their viruses," in *An Introduction to Genetic Analysis*, 7th Edn, ed. W. Freeman (New York, NY: W. H. Freeman).

Haaber, J., Leisner, J. J., Cohn, M. T., Catalan-Moreno, A., Nielsen, J. B., Westh, H., et al. (2016). Bacterial viruses enable their host to acquire antibiotic resistance genes from neighbouring cells. *Nat. Commun.* 7:13333. doi: 10.1038/ncomms13333

Hertwig, S., Hammerl, J. A., Appel, B., and Alter, T. (2013). Post-harvest application of lytic bacteriophages for biocontrol of foodborne pathogens and spoilage bacteria. *Berl. Münch. Tierärztl. Wochenschr.* 126, 357–369.

Howe, A., Ringus, D. L., Williams, R. J., Choo, Z.-N., Greenwald, S. M., Owens, S. M., et al. (2016). Divergent responses of viral and bacterial communities in the gut microbiome to dietary disturbances in mice. *ISME J.* 10, 1217–1227. doi: 10.1038/ismej.2015.183

IAWPRC Study Group on Health Related Water Microbiology (1991). Bacteriophages as model viruses in water quality control. *Water Res* 25, 529–545. doi: 10.1016/0043-1354(91)90126-B

Jerne, N. K. (1952). Bacteriophage inactivation by antiphage serum diluted in distilled water. *Nature* 169, 117–118. doi: 10.1038/169117b0

Jerne, N. K. (1956). The presence in normal serum of specific antibody against bacteriophage T4 and its increase during the earliest stages of immunization. *J. Immunol.* 76, 209–216.

Jofre, J., Lucena, F., Blanch, A., and Muniesa, M. (2016). Coliphages as model organisms in the characterization and management of water resources. *Water* 8, 199. doi: 10.3390/w8050199

Jover, L. F., Cortez, M. H., and Weitz, J. S. (2013). Mechanisms of multi-strain coexistence in host-phage systems with nested infection networks. *J. Theor. Biol.* 332, 65–77. doi: 10.1016/j.jtbi.2013.04.011

Kantoch, M., Skurski, A., and Wieczorek, Z. (1958). In vitro blockade of bacterial phagocytosis of leukocytes by means of bacterial viruses. *Schweiz. Z. Pathol. Bakteriol.* 21, 1106–1119. doi: 10.1159/000160571

Keller, R., and Traub, N. (1974). The characterization of *Bacteroides fragilis* bacteriophage recovered from animal sera: observations on the nature of *Bacteroides* phage carrier cultures. *J. Gen. Virol.* 24, 179–189. doi: 10.1099/0022-1317-24-1-179

Lang, A. S., Zhaxybayeva, O., and Beatty, J. T. (2012). Gene transfer agents: phage-like elements of genetic exchange. *Nat. Rev. Microbiol.* 10, 472–482. doi: 10.1038/nrmicro2802

Langbeheim, H., Teitelbaum, D., and Arnon, R. (1978). Cellular immune response toward MS-2 phage and a synthetic fragment of its coat protein. *Cell. Immunol.* 38, 193–197. doi: 10.1016/0008-8749(78)90046-1

Lim, E. S., Zhou, Y., Zhao, G., Bauer, I. K., Droit, L., Ndao, I. M., et al. (2015). Early life dynamics of the human gut virome and bacterial microbiome in infants. *Nat. Med.* 21, 1228–1234. doi: 10.1038/nm.3950

Lindell, D., Sullivan, M. B., Johnson, Z. I., Tolonen, A. C., Rohwer, F., and Chisholm, S. W. (2004). Transfer of photosynthesis genes to and from *Prochlorococcus* viruses. *Proc. Natl. Acad. Sci. U.S.A.* 101, 11013–11018. doi: 10.1073/pnas.0401526101

Loh, G., and Blaut, M. (2012). Role of commensal gut bacteria in inflammatory bowel diseases. *Gut Microbes* 3, 544–555. doi: 10.4161/gmic.22156

Manrique, P., Bolduc, B., Walk, S. T., van der Oost, J., de Vos, W. M., and Young, M. J. (2016). Healthy human gut phageome. *Proc. Natl. Acad. Sci. U.S.A.* 113, 10400–10405. doi: 10.1073/pnas.1601060113

Martínez-Castillo, A., and Muniesa, M. (2014). Implications of free Shiga toxin-converting bacteriophages occurring outside bacteria for the evolution and the detection of Shiga toxin-producing *Escherichia coli*. *Front. Cell. Infect. Microbiol.* 4:46. doi: 10.3389/fcimb.2014.00046

Merril, C. R., Biswas, B., Carlton, R., Jensen, N. C., Creed, G. J., Zullo, S., et al. (1996). Long-circulating bacteriophage as antibacterial agents. *Proc. Natl. Acad. Sci. U.S.A.* 93, 3188–3192. doi: 10.1073/pnas.93.8.3188

Minot, S., Sinha, R., Chen, J., Li, H., Keilbaugh, S. A., Wu, G. D., et al. (2011). The human gut virome: inter-individual variation and dynamic response to diet. *Genome Res.* 21, 1616–1625. doi: 10.1101/gr.122705.111

Moayyedi, P., Surette, M. G., Kim, P. T., Libertucci, J., Wolfe, M., Onischi, C., et al. (2015). Fecal Microbiota transplantation induces remission in patients with active ulcerative colitis in a randomized controlled trial. *Gastroenterology* 149, 102–109.e6. doi: 10.1053/j.gastro.2015.04.001

Mojica, F. J. M., and Rodriguez-Valera, F. (2016). The discovery of CRISPR in archaea and bacteria. *FEBS J.* 283, 3162–3169. doi: 10.1111/febs.13766

Müller, M. G., Ing, J. Y., Cheng, M. K.-W., Flitter, B. A., and Moe, G. R. (2013). Identification of a phage-encoded Ig-binding protein from invasive *Neisseria meningitidis*. *J. Immunol.* 191, 3287–3296. doi: 10.4049/jimmunol.1301153

Muniesa, M., Blanch, A. R., Lucena, F., and Jofre, J. (2005). Bacteriophages may bias outcome of bacterial enrichment cultures. *Appl. Environ. Microbiol.* 71, 4269–4275. doi: 10.1128/AEM.71.8.4269-4275.2005

Muniesa, M., García, A., Miró, E., Mirelis, B., Prats, G., Jofre, J., et al. (2004). Bacteriophages and diffusion of beta-lactamase genes. *Emerg. Infect. Dis.* 10, 1134–1137. doi: 10.3201/eid1006.030472

Norman, J. M., Handley, S. A., Baldridge, M. T., Droit, L., Liu, C. Y., Keller, B. C., et al. (2015). Disease-specific alterations in the enteric virome in inflammatory bowel disease. *Cell* 160, 447–460. doi: 10.1016/j.cell.2015.01.002

O'Brien, A. D., Newland, J. W., Miller, S. F., Holmes, R. K., Smith, H. W., and Formal, S. B. (1984). Shiga-like toxin-converting phages from *Escherichia coli* strains that cause hemorrhagic colitis or infantile diarrhea. *Science* 226, 694–696. doi: 10.1126/science.6387911

Oh, J., Byrd, A. L., Deming, C., Conlan, S., Kong, H. H., and Segre, J. A. (2014). Biogeography and individuality shape function in the human skin metagenome. *Nature* 514, 59–64. doi: 10.1038/nature13786

Penades, J. R., Chen, J., Quiles-Puchalt, N., Carpena, N., and Novick, R. P. (2015). Bacteriophage-mediated spread of bacterial virulence genes. *Curr. Opin. Microbiol.* 23, 171–178. doi: 10.1016/j.mib.2014.11.019

Pérez-Brocal, V., García-López, R., Nos, P., Beltrán, B., Moret, I., and Moya, A. (2015). Metagenomic analysis of Crohn's disease patients identifies changes in the virome and microbiome related to disease status and therapy, and detects potential interactions and biomarkers. *Inflamm. Bowel Dis.* 21, 2515–2532. doi: 10.1097/MIB.0000000000000549

Pérez-Brocal, V., García-López, R., Vázquez-Castellanos, J. F., Nos, P., Beltrán, B., Latorre, A., et al. (2013). Study of the viral and microbial communities associated with Crohn's disease: a metagenomic approach. *Clin. Transl. Gastroenterol.* 4, e36. doi: 10.1038/ctg.2013.9

Puig, A., Araujo, R., Jofre, J., and Frias-Lopez, J. (2001). Identification of cell wall proteins of *Bacteroides fragilis* to which bacteriophage B40-8 binds specifically. *Microbiology* 147, 281–288. doi: 10.1099/00221287-147-2-281

Quiros, P., Martinez-Castillo, A., and Muniesa, M. (2015). Improving detection of Shiga toxin-producing *Escherichia coli* by molecular methods by reducing the interference of free Shiga toxin-encoding bacteriophages. *Appl. Environ. Microbiol.* 81, 415–421. doi: 10.1128/AEM.02941-14

Reardon, S. (2014). Phage therapy gets revitalized. *Nature* 510, 15–16. doi: 10.1038/510015a

Rodriguez-Brito, B., Li, L., Wegley, L., Furlan, M., Angly, F., Breitbart, M., et al. (2010). Viral and microbial community dynamics in four aquatic environments. *ISME J.* 4, 739–751. doi: 10.1038/ismej.2010.1

Ross, J., and Topp, E. (2015). Abundance of antibiotic resistance genes in bacteriophage following soil fertilization with dairy manure or municipal biosolids, and evidence for potential transduction. *Appl. Environ. Microbiol.* 81, 7905–7913. doi: 10.1128/AEM.02363-15

Sanger, F., Air, G. M., Barrell, B. G., Brown, N. L., Coulson, A. R., Fiddes, C. A., et al. (1977). Nucleotide sequence of bacteriophage phi X174 DNA. *Nature* 265, 687–695. doi: 10.1038/265687a0

Scanlan, P. D., and Buckling, A. (2012). Co-evolution with lytic phage selects for the mucoid phenotype of *Pseudomonas fluorescens* SBW25. *ISME J.* 6, 1148–1158. doi: 10.1038/ismej.2011.174

Schmelcher, M., and Loessner, M. J. (2014). Application of bacteriophages for detection of foodborne pathogens. *Bacteriophage* 4:e28137. doi: 10.4161/bact.28137

Selva, L., Viana, D., Regev-Yochay, G., Trzcinski, K., Corpa, J. M., Lasa, I., et al. (2009). Killing niche competitors by remote-control bacteriophage induction. *Proc. Natl. Acad. Sci. U.S.A.* 106, 1234–1238. doi: 10.1073/pnas.0809600106

Srivastava, A. S., Chauhan, D. P., and Carrier, E. (2004a). In utero detection of T7 phage after systemic administration to pregnant mice. *Biotechniques* 37, 81–83.

Srivastava, A. S., Kaido, T., and Carrier, E. (2004b). Immunological factors that affect the in vivo fate of T7 phage in the mouse. *J. Virol. Methods* 115, 99–104.

Suttle, C. A. (2005). Viruses in the sea. *Nature* 437, 356–361. doi: 10.1038/nature04160

Twort, F. W. (1915). An investigation on the nature of ultra-microscopic viruses. *Lancet* 186, 1241–1243. doi: 10.1016/S0140-6736(01)20383-3

Virgin, H. W. (2014). The virome in mammalian physiology and disease. *Cell* 157, 142–150. doi: 10.1016/j.cell.2014.02.032

Wagner, J., Maksimovic, J., Farries, G., Sim, W. H., Bishop, R. F., Cameron, D. J., et al. (2013). Bacteriophages in gut samples from pediatric Crohn's disease patients: metagenomic analysis using 454 pyrosequencing. *Inflamm. Bowel Dis.* 19, 1598–1608. doi: 10.1097/MIB.0b013e318292477c

Willner, D., Furlan, M., Haynes, M., Schmieder, R., Angly, F. E., Silva, J., et al. (2009). Metagenomic analysis of respiratory tract DNA viral communities in cystic fibrosis and non-cystic fibrosis individuals. *PLoS ONE* 4:e7370. doi: 10.1371/journal.pone.0007370

Zinder, N. D., and Lederberg, J. (1952). Genetic exchange in *Salmonella*. *J. Bacteriol.* 64, 679–699.

14

Double-Face Meets the Bacterial World: The Opportunistic Pathogen *Stenotrophomonas maltophilia*

Felipe Lira[1], Gabriele Berg[2] and José L. Martínez[1]*

[1] Centro Nacional de Biotecnología, Consejo Superior de Investigaciones Científicas, Madrid, Spain, [2] Institute of Environmental Biotechnology, Graz University of Technology, Graz, Austria

Edited by:
Chew Chieng Yeo,
Universiti Sultan Zainal Abidin, Malaysia

Reviewed by:
Radoslaw Pluta,
International Institute of Molecular and Cell Biology in Warsaw (IIMCB), Poland
Prabhu B. Patil,
Institute of Microbial Technology (CSIR), India
Gregory Anderson,
Indiana University – Purdue University Indianapolis, United States

***Correspondence:**
José L. Martínez
jlmtnez@cnb.csic.es

Most studies on bacterial virulence focus on the pathogen itself. However, it is important to recall that the in-host behavior and the virulence of bacterial pathogens constitute a complex situation that depends on both the microorganisms and the infected host. While healthy people (the community) is infected by classical pathogenic microorganisms, able to cope with the anti-infection defenses of the host, in the case of people with basal diseases, debilitated or immunodepressed, the range of pathogens able to cause infection is wider and includes the so-named opportunistic pathogens, which lack the inherent ability to cause disease in healthy hosts and rarely produce infections in the community. Some of the most relevant opportunistic pathogens, as *Stenotrophomonas maltophilia,* have an environmental origin and, in occasions, present interesting biotechnological properties. Consequently, it is important knowing whether *S. maltophilia* isolates recovered from infections constitute a specific phylogenetic branch that has evolved toward acquiring a virulent phenotype as it happens in the case of classical pathogens or rather, any member of this bacterial species is capable of producing infection and its pathogenic behavior is mainly a consequence of the host situation. To address this question, we analyzed a set of environmental and clinical *S. maltophilia* strains. Our results indicate that this opportunistic pathogen presents a large core genome and that the distribution of genes in general, and of known virulence determinants in particular, is similar among environmental and clinical isolates. The majority of genes not belonging to the *S. maltophilia* core genome are present in just one or two of the analyzed strains. This indicates that, more than speciation into different lineages (virulent and environmental), the evolution of *S. maltophilia* is based in the strain-specific acquisition of genes, likely involved in the adaptation of this bacterial species to different microniches. In addition, both environmental and clinical isolates present low susceptibility to several antimicrobials. Altogether our results support that *S. maltophilia* does not present a specific evolutionary branch toward virulence and most likely infection is mainly the consequence of the impaired anti-infective response of the infected patients.

Keywords: *Stenotrophomonas maltophilia*, opportunistic pathogens, comparative genomics, pangenome, core genome, antibiotic resistance

INTRODUCTION

Stenotrophomonas maltophilia is an opportunistic pathogen, with an environmental origin, which causes a variety of infections at hospitals (Brooke, 2012; Chang et al., 2015; Jeon et al., 2016; Brooke et al., 2017), particularly in those patients under previous therapy with broad-spectrum antibiotics (Chang et al., 2015), and in patients with underlying diseases as cystic fibrosis (Nicoletti et al., 2011; Pompilio et al., 2016; Esposito et al., 2017). *S. maltophilia* infections are difficult to treat because this pathogen displays low susceptibility to several antimicrobials (Sanchez et al., 2009; Sanchez, 2015). As the consequence of this situation and likely also because *S. maltophilia* infects mainly severely debilitated individuals, the mortality of patients suffering *S. maltophilia* infections is high (Jeon et al., 2016). Consequently, understanding the underlying features by which this pathogen can traverse different ecological allocations, from its natural habitat toward infecting humans, may help in the development of strategies to improve the treatment of infections due to this microorganism.

Besides its clinical relevance, different *S. maltophilia* strains exert an extraordinary range of activities with biotechnological relevance (Mukherjee and Roy, 2016), such as bioremediation (Dungan et al., 2003; Berg and Martinez, 2015), degradation of toxic compounds (Lee et al., 2002), biosynthesis (Jakobi et al., 1996; Nangia et al., 2009; Zonaro et al., 2015) and biological control in agriculture (Dunne et al., 2000; Alavi et al., 2013), among others.

Given these two aspects of *S. maltophilia*, it is highly relevant to determine whether infective and environmental (non-clinical) *S. maltophilia* isolates constitute different evolutionary branches in this species as it has been shown in the case of the *Burkholderia cepacia* complex (Chiarini et al., 2006) or if, by contrary, any strain can infect a compromised human host, as it has been described for *Pseudomonas aeruginosa* (Alonso et al., 1999; Morales et al., 2004; Wiehlmann et al., 2007). This is particularly relevant in order to evaluate the risks for human health associated to the use of *S. maltophilia* for biotechnological purposes, mainly for non-confined applications, as agriculture.

Different works, based in classical Multi-Locus Sequence Typing (MLST), *in silico* MLST and whole genome analyses, have been published to address the phylogenetic structure of this species and of others belonging to the same complex (Rocco et al., 2009; Adamek et al., 2014; Gherardi et al., 2015; Youenou et al., 2015; Esposito et al., 2017; Ochoa-Sanchez and Vinuesa, 2017). Nevertheless, it is still unclear whether or not clinical isolates are predominant in any of these branches. In addition, studies on the potential correlation between the presence in the genome of virulence determinants and antibiotic resistance with the origin of the strains (clinical or environmental) are extremely limited, despite the relevance of these features for the nosocomial infections produced by *S. maltophilia*.

In order to address whether or not clinical and environmental isolates belong to different phylogenetic branches in *S. maltophilia*, in the present work we have sequenced 20 *S. maltophilia* isolates (10 from clinical environments and 10 from environmental samples). Four complete genomes sequences were also included in the study as references, two clinical strains *S. maltophilia* K279a (Crossman et al., 2008) and D457 (Lira et al., 2012) and two environmental isolates, *S. maltophilia* R551-3 (Lucas et al., 2008) and JV3 (Lucas et al., 2011). In addition, in the present work we present the phenotypic analysis of the studied isolates in order to determine whether or not clinical isolates are more resistant to antibiotics than environmental ones, information that cannot be obtained from the simple inspection of *S. maltophilia* available genomes.

MATERIALS AND METHODS

DNA Extraction and Genome Sequencing of 20 New Strains of *Stenotrophomonas maltophilia*

The complete DNAs of 20 isolates of *S. maltophilia* (**Table 1**) were extracted using the GENOME DNA Kit (MP Biomedicals LLC, Illkrich, France). Whole-genome sequencing was performed at the facility of the Madrid Science Park (Madrid, Spain), using Illumina MiSeq technology (Illumina, San Diego, CA, United States) from DNA libraries with insertion sizes between 700 and 800 bp, to generate paired-end reads with 260–300 bp length.

Quality Control and Sequence Assembling

Quality score of the sequences of all strains was checked using FastQC v.0.11.2, to identify adapters and contaminant sequences remaining after sequencing. Contaminant sequences were removed using the AlienTrimmer v.0.4.0 software (Criscuolo and Brisse, 2013) and a customized database of adapters adding the contaminant sequences recognized by FastQC. Sequence trimming and filtering were performed by PRINSEQ-Lite (Schmieder and Edwards, 2011) to filter the sequences by length and quality score (Phred $\geq$ 22, minimum read length = 90 bp). Each set of reads was submitted to *de novo* assembling using the Spades v.3.9 assembler (Bankevich et al., 2012) in a local server (24 cores and 512Gb RAM). After assembling, contigs with a minimal 5.000 bp length were selected. The synteny of the generated contigs was ordered using Mauve aligner (Darling et al., 2004) and two reference genomes, the model strains *S. maltophilia* D457 (Lira et al., 2012) and *S. maltophilia* K279 (Crossman et al., 2008). Both genomes were chosen because they were the largest complete genomes available. Contigs alignment did not presented divergences with respect to the reference genomes synteny.

Open Reading Frames Detection, Gene Prediction and Annotation

For the prediction and annotation of the Open Read Frames (ORFs) from each set of contigs, we used two approaches: (a) In a first step the ORFs were predicted using Prodigal v2.6.1

(Hyatt et al., 2010), avoiding truncated genes The parameters were set to predict genes containing both start and stop codons. This approach allowed the elimination of fragmented genes located at the edges of the contigs. Predicted ORFs were annotated performing a local alignment with BLASTp (Camacho et al., 2009) against the NCBI non-redundant database setting the expected value (*e-value*) of 10^{-10}. In a second step, all contigs were submitted to the NCBI Prokaryotic Genome Annotation Pipeline (PGAP). Divergences between the local annotation and the PAGP were checked and curated manually.

Comparative Genomics

Twenty draft genomes of *S. maltophilia* obtained in this study, and the complete genomes of four strains were used to estimate the preliminary core genome and pangenome sizes of *S. maltophilia.* The complete genomes of two clinical strains: D457 (NC_017671.1) and K279a (NC_010943.1) and of two environmental strains: R551-3 (NC_011071.1) and JV3 (NC_015947.1) were also used for the analysis. The accession numbers of draft genomes of the 20 strains of *S. maltophilia* analyzed in this study are: clinical strains: E729 (NERH00000000), E759 (NERG00000000), E999 (NERF00000000), G51 (NERE00000000), E301 (NERD 00000000), D388 (NERC00000000), E861 (NERB00000000), C357 (NERA00000000), E539 (NEQZ00000000), E824 (NEQY00000000); environmental strains: NS26 (NEQO00 000000), EP13 (NEQX00000000), EA22 (NEQW00000000), EA1 (NEQV00000000), PS5 (NEQU00000000), EA23 (NEQT00000000), EP20 (NEQS00000000), EP5 (NEQR0 0000000), EA21 (NEQQ00000000), EA63 (NEQP00000000) (**Table 1**).

The pangenome and the core genome of the sequenced strains were analyzed using the script GET_HOMOLOGUES v.07112016 (Contreras-Moreira and Vinuesa, 2013). Clusters of homologous gene families were generated using the COGtriangles algorithm. To form clusters and estimate the core genome and pangenome of *S. maltophilia*, coverage and identity thresholds of 90% and of 95%, respectively were used.

The complete Coding DNA Sequence (CDS) composition and the clusters generated for all strains were used to perform a comparative analysis and to calculate the genome similarity distance to determine the relationship of clinical and environmental isolates. Clustered genes were used to compile the corresponding pangenome matrix using the script compare_cluster.pl with default settings, embedded in the GET_HOMOLOGUES software package. The clusters formed were classified considering the distribution of ortholog genes through the strains. The core genome contains those genes belonging to all strains, the soft-core genome the genes present in, at least, 95% of the strains, the shell genome the genes present in less than 95% and more than 10% of the genomes and the cloud genome the genes present in less than 10% of the genomes (Koonin and Wolf, 2008; Kaas et al., 2012).

TABLE 1 | Accession numbers of the genomes and origins of the *S. maltophilia* isolates of used in this study.

	Strains	Accession number	Origin	Reference
Clinical strains	E729	NERH00000000	Urine	Alonso and Martinez, 2001
	E759	NERG00000000	Sputum	Alonso and Martinez, 2001
	E999	NERF00000000	Respiratory secretion	Alonso and Martinez, 2001
	G51	NERE00000000	Blood	Alonso and Martinez, 2001
	E301	NERD00000000	Urine	Alonso and Martinez, 2001
	D388	NERC00000000	Urine	Alonso and Martinez, 2001
	E861	NERB00000000	Sputum	Alonso and Martinez, 2001
	C357	NERA00000000	Urine	Alonso and Martinez, 2001
	E539	NEQZ00000000	Pus from a wound	Alonso and Martinez, 2001
	E824	NEQY00000000	Blood	Alonso and Martinez, 2001
	K279a*	NC_010943.1	Blood	Crossman et al., 2008
	D457*	NC_017671.1	Respiratory secretion	Lira et al., 2012
Environmental strains	NS26	NEQO00000000	Dune soil	De Boer et al., 2001;Ribbeck-Busch et al., 2005
	EP13	NEQX00000000	Rhizosphere of oilseed rape	Minkwitz and Berg, 2001
	EA22	NEQW00000000	Sewage	Minkwitz and Berg, 2001
	EA1	NEQV00000000	Brackish water	Minkwitz and Berg, 2001
	PS5	NEQU00000000	Rhizosphere of oilseed rape.	Berg et al., 1996
	EA23	NEQT00000000	Eye-care solution	Suckstorff and Berg, 2003
	EP20	NEQS00000000	Rhizosphere of potato	Minkwitz and Berg, 2001
	EP5	NEQR00000000	Rhizosphere of *Brassica napus* L.	Minkwitz and Berg, 2001
	EA21	NEQQ00000000	Sewage	Minkwitz and Berg, 2001
	EA63	NEQP00000000	Sewage	Gabrielle Berg's lab collection
	R551-3*	NC_011071.1	Endosphere	Lucas et al., 2008
	JV3*	NC_015947.1	Rhizosphere	Lucas et al., 2011

**Complete genomes of S. maltophilia.*

In Silico Multi-Locus Sequences Typing and Polymorphic Sites in the Core Genome

In silico MLST analysis (Larsen et al., 2012) was performed using the web server of the Centre for Genomic Epidemiology[1]. The alleles from each strain were identified individually and their nucleotides sequences were further concatenated (separated by 10 Ns) to perform a Multiple Sequence Alignment (MSA) using ClustalW2. A phylogenetic tree based in this alignment was calculated using the same software based on the similarity distance between concatenated sequences.

The identification of polymorphic sites was performed using Snippy[2] using *S. maltophilia* K279a as reference strain (Accession number: NC_010943). Polymorphic sites in genes shared by all strains formed the core of Single Nucleotide Polymorphisms SNPs, that was used to perform a MSA. A phylogenetic tree from the derived information was constructed by using the maximum likelihood method.

Genomic Composition and Comparative Genomics

Putative functional similarities and differences between the clinical and the environmental strains were estimated by a subsystem classification using the RAST server[3] (Aziz et al., 2008) and the coding sequences from each genome were classified according to their protein families (*FIGfams*). All strains were compared by the presence/absence of 20 subsystems and 35 functional roles included in the category of "Virulence, Disease and Defense". A local database containing a set of specific genes, described as responsible for the virulence phenotype of *S. maltophilia* (Adamek et al., 2014) was used to retrieve similar genes from the studied strains. Hierarchical clustering was performed in R functions (Langfelder and Horvath, 2012). For this purpose, each one of the resulting tables containing the information about presence/absence of these genes was converted into a square similarity matrix to measure the distance between strains (R function 'dist'), clustered based on the matrix data (R function 'hclust') and plotted as heatmap (R function 'heatmap.2').

Quorum-Sensing Signals

It has been described that the alleles of the quorum-sensing system (QS) *rpfF* gene, *rpf*F1 (GenBank: KJ149475) and *rpf*F2 (GenBank: KJ149552), are markers of two different phylogenetic branches, each one presenting differences in terms of virulence (Huedo et al., 2014). To address whether or not the presence of a specific *rpfF* allele could be linked to clinical strains, the 108 N-terminal residues of RpfF, which has been proposed

[1] www.cbs.dtu.dk/services/MLST

[2] https://github.com/tseemann/snippy

[3] http://rast.nmpdr.org/

TABLE 2 | Overall characteristics of the genomes analyzed in the current article.

	Strains	Bases	Contigs	Largest contig	GC%	Predicted genes
Clinical strains	K279a*	4,851,126	1	4851126	66.3	4354
	D457*	4,769,156	1	4769156	66.8	4254
	E861	4,658,203	31	653242	66.4	4191
	D388	4,659,986	30	740754	66.4	4190
	E539	4,555,541	18	1731480	66.5	4057
	C357	4,810,581	17	954550	66.2	4310
	E824	5,041,912	14	1834293	65.9	4502
	E729	5,005,550	12	1548184	66.6	4540
	E999	4,414,069	11	1140634	66.7	3879
	G51	4,852,740	8	2066793	66.1	4368
	E301	4,428,328	5	3885998	66.8	3965
	E759	4,546,405	4	2470865	66.5	4083
Environmental strains	R551-3*	4,573,969	1	4573969	66.3	4023
	JV3*	4,544,477	1	4544477	66.9	4040
	EA23	4,752,304	29	642831	66.4	4283
	EP13	4,755,757	27	744273	66.4	4281
	NS26	4,689,165	18	1729723	66.2	4152
	EP20	4,625,290	16	2060034	66.1	4087
	EA1	4,752,176	16	918730	66.6	4234
	EA22	4,759,594	10	1721891	66.2	4265
	EA63	4,885,042	10	1847362	66	4390
	EA21	4,732,256	9	1707011	66.2	4246
	PS5	4,600,476	7	2135136	66.4	4076
	EP5	4,600,182	7	2134905	66.4	4075

**Complete genomes of S. maltophilia.*

to be used as markers for distinguishing the two RpfF variants (Huedo et al., 2014) were aligned using ClustalW2 (Larkin et al., 2007). A phylogenetic tree derived from this information was established using JalView v.2 (Waterhouse et al., 2009)

Antibiotic Susceptibility

Minimum Inhibitory Concentrations (MICs) were determined in Mueller Hinton agar medium using MIC Test strips (Liofilchem) of the following antibiotics Trimethoprim/Sulfamethoxazole (SXT); Tigecyclin (TGC); Ceftazidime (CAZ); Cefepime (PM); Gentamicin (CN); Gatifloxacin (GAT); Colistin (CS); Chloramphenicol (CL); Imipenem (IMI); Ertapenem (ETP); Moxifloxacin (MXF); Nalidixic Acid (NA).

RESULTS AND DISCUSSION

Genome Assembling and Annotation of Clinical and Environmental Strains of *Stenotrophomonas maltophilia*

Although the number of sequenced genomes of the opportunistic pathogen *S. maltophilia* has increased since the first genome was published, specific analyses on the core genome and pangenome (Esposito et al., 2017) as well as on the genomic relationships of clinical and environmental isolates of this species are scarce. In addition, the quality (in terms of number of contigs) of the different available draft genomes is diverse, which makes their comparison difficult in occasions. Finally, clear information on the origin of the isolates (clinical or environmental) is not always available, making the use of these sequences difficult for the purposes of this work. Consequently, to analyze whether clinical and environmental isolates present different genomic features or, by contrary they do not form two different phylogenetic branches, we decided to sequence and analyze twenty isolates of *S. maltophilia* for which the origin has been well established (10 clinical and ten environmental). The assembling of all strains generated a total of 94 Mbp comprising 299 contigs. The genome length average of the sequenced strains was 4.7 Mb and their average GC% content 66.36% (**Table 2**). These data were similar to those of the available *S. maltophilia* complete genomes from strains D457, K279a, R551-3 and JV3, whose genome length and GC% content are, in average, 4.6 Mb and 66.57%, respectively. All contigs were submitted to the Prokaryotes Genome Annotation Pipeline (PGAP) (Tatusova et al., 2014) from NCBI, retrieving an average of 4206 CDS/strain (min = 3879; max = 4540) (**Table 2**). A presence/absence matrix was generated and used for the phylogenetic clustering of the different isolates based in the CDS composition of their genomes. As shown in **Figure 1**, and although branch D comprised just strains isolated from the rhizosphere, the other branches included both clinical and environmental isolates. This fact indicates that, at least in a whole view, there is not a clear differentiation in the CDS composition between the genomes of clinical and environmental *S. maltophilia* isolates.

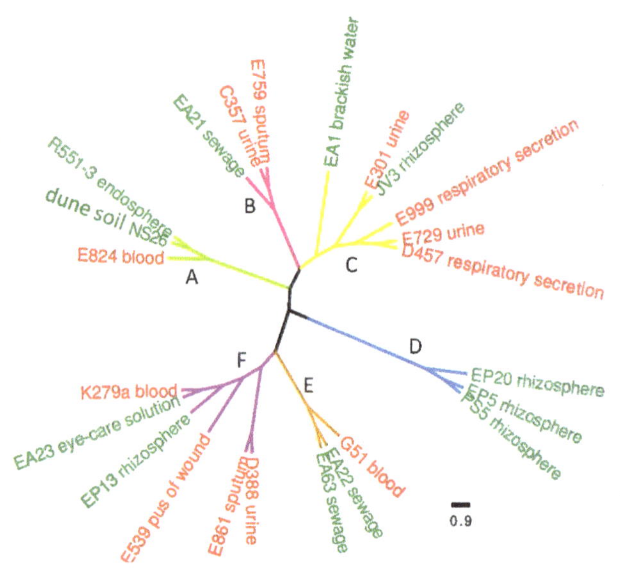

FIGURE 1 | Genetic similarity of clinical and environmental *S. maltophilia* isolates. The complete CDS composition of all strains was used to generate a presence/absence matrix. Based in this matrix, a similarity plot was generated. Red clinical isolates, green environmental strains. As shown, most clusters present both clinical and environmental strains. The horizontal bar shows the genomic similarity distance based on the presence/absence of CDS in each genome.

Effect of the Origin of *S. maltophilia* Isolates in Their Pangenome and Core Genome

The pangenome and the core genome of *S. maltophilia* were calculated using the draft genomes of the 20 sequenced strains as well as the four full genomes used as references in our work. The number of total genes was plotted as a function of the number of genomes added to the analysis. As shown in **Figure 2A**, an asymptotical increase in the number of genes with respect to the number of analyzed strains was detected. In agreement with previous information (Yu et al., 2016), this indicates that *S. maltophilia* has an open pangenome based on the analysis of the 24 genomes examined. To estimate the core genome, the number of genes shared by all stains was plotted as a function of the number of *S. maltophilia* genomes sequentially added to the analysis (**Figure 2B**). The core genome was estimated in 2762 genes, corresponding to 38% of the pangenome of *S. maltophilia* (**Figure 2B**). To estimate the tendency of the core genome two approaches were performed. Following the approach and terminology of Tettelin et al. (2005), the *S. maltophilia* core genome presents a 'relative constancy' after several genomes are included in the analysis (red line in **Figure 2B**), whereas the predictions using the approach of Willenbrock et al. (2007) is that the incorporation of novel genomes should produce a decay in the number of genes that compose the core genome of *S. maltophilia* (blue line in **Figure 2B**). Compositional analysis retrieved a pangenome composed by 7108 orthologous groups, although this number should likely increase when more genomes

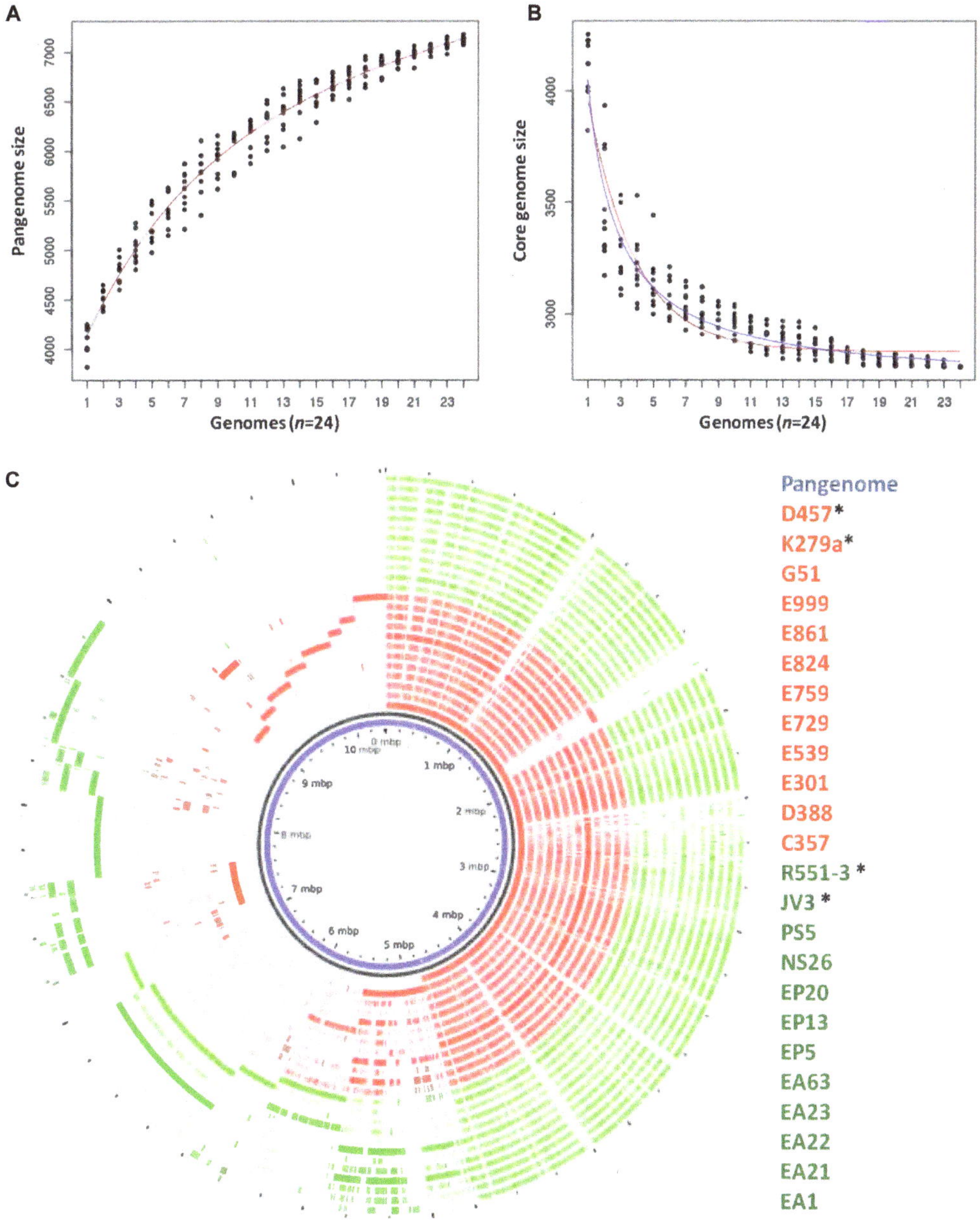

FIGURE 2 | Comparison of the genomes of environmental and clinical *S. maltophilia* isolates. The pangenome and the core genome of *S. maltophilia* were calculated by random sampling of the 24 genomes: **(A)** The number of total genes plotted as a function of the number of genomes added to the analysis is presented. As shown, *S. maltophilia* has an open pangenome. **(B)** The curve shows the number of genes shared by all trains as a function of the number of genomes of *S. maltophilia* added sequentially. Red and blue lines were plotted as an estimation of the tendency of the core genome. Red line indicates that the core genome of *S. maltophilia* should maintain, following the terminology and the estimation rules of Tettelin and collaborators (Tettelin et al., 2005), a 'relative constancy' after several genomes are included in the analysis. Blue line indicates that, following the approach of Willenbrock and collaborators, the incorporation of novel genomes might produce a decay in the number of genes that compose the core genome of *S. maltophilia* (Willenbrock et al., 2007). **(C)** Representation of the pangenome obtained by analyzing 24 genomes of *S. maltophilia* isolates. Each circle represents the contribution of each genome to the composition analysis. Genes shared by several strains are clustered at the right side of the circle and strain-specific genes are clustered at its left side. The list of strains displays their names from the inner to the outer circle. *Complete genomes used in this study. Red: clinical isolates. Green: environmental strains.

are analyzed (**Figure 2C**). It is important to notice that, since draft genomes are analyzed, the lack of genes in one specific strain may be the consequence of its presence at the edge of one contig, in which case will be annotated as a truncated gene, although this putative truncation will be the consequence of the method of analysis, not of a real absence. Consequently, the "soft-core genome" (Kaas et al., 2012) was also analyzed. By using this approach, we estimated the number of orthologous genes

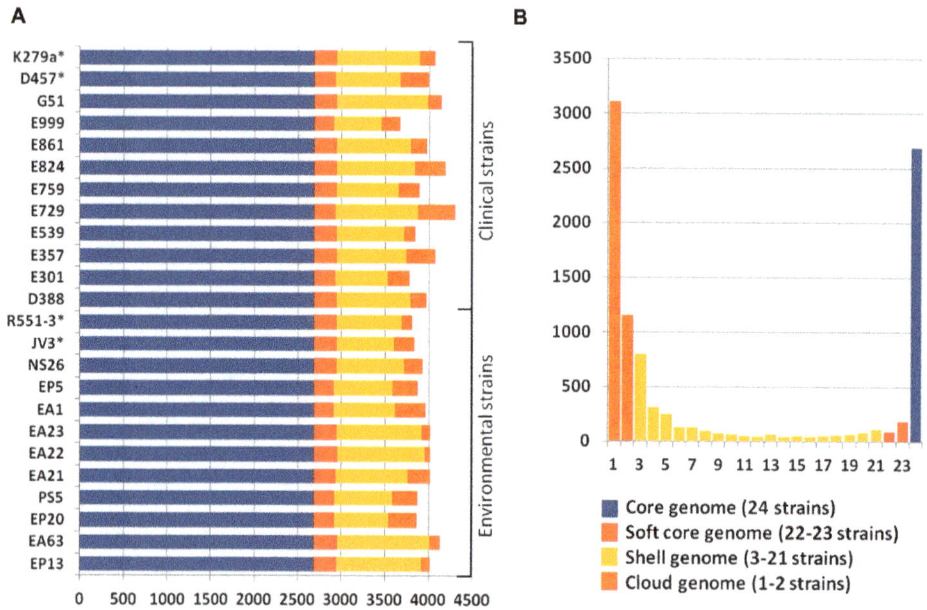

FIGURE 3 | Estimation of *S. maltophilia* core, soft-core, shell and cloud genome. **(A)** Distribution of the predicted CDSs along the 24 *S. maltophilia* genomes representing the core genome (blue bars), the soft-core genome (red bars), the shell genome (yellow bars), and the cloud genome (orange bars) of each of the strains is shown. **(B)** Representation of the overall distribution of genes contributing to the core, the soft, the shell and the cloud genomes. As shown, most genes not belonging to the core/soft-core genome are present in just one or two strains, suggesting that most of the *S. maltophilia* pangenome is strain specific, and does not depend on the environmental or clinical origin of the isolate.

shared by ~90% of the organisms included in the comparative analysis. Applying the soft-core genome concept, the number of orthologous clusters increased to 3045. When the 24 genomes were analyzed independently, we estimated that the size of the core genome for each *S. maltophilia* isolate comprised around 59.11% (minimum 54.6%; maximum 64%) of the CDS from each genome. Further, the analysis of the pangenome shows that most of the genes carried by *S. maltophilia* and not belonging to the core genome are strain-specific, suggesting specific adaptations for each isolate more than a common pattern of speciation of some members of the population toward virulence. Indeed, among those genes not belonging to the soft-core genome, and shared by 3–21 strains (dubbed as the 'shell genome'), just 1337 gene clusters, from the total of 7108 orthologous genes present in the pangenome, were found, indicating that the vast majority of *S. maltophilia* genes, not belonging to its core genome, are strain-specific (**Figure 3**). The speciation of bacterial pathogens usually involve the acquisition by horizontal gene transfer (HGT) of virulence genes, followed by the loss of other genes and the selection of mutations that allow the fine tuning of the metabolism (Martinez, 2013), a process very well studied in the case of *Yersinia* (Achtman et al., 1999; Wren, 2003; Achtman et al., 2004; Zhou and Yang, 2009). HGT is the consequence of either transformation or either the acquisition of mobile elements. Once these mobile elements are acquired, they can be fixed or spread to other hosts, a situation highly relevant in the case of antibiotic resistance (Martinez et al., 2009, 2017). Despite that the presence of several genes in the cloud genome of *S. maltophilia* suggests that this process has largely contributed to the diversification of this pathogen, clear information on its mobilome has not been published. Indeed, only three whole sequenced *S. maltophilia* plasmids are available, which makes difficult to estimate the role of these mobile elements in the evolution of *S. maltophilia*.

In addition to the finding that there are several strain-specific genes, is important to recall that, as **Figure 2C** shows, a differential distribution of genes, not belonging to the core genome, was not found when environmental and clinical isolates of *S. maltophilia* were compared. This result further suggests that there is not a specific phylogenetic branch, deriving from the acquisition of a specific set of virulence determinants by the clinical *S. maltophilia* isolates, which can drive the speciation of this microorganism toward virulence.

It is important to highlight that several articles analyzing bacterial core genomes make use of draft genomes in which genes at the edges of contigs are interrupted, which introduce some noise in the analysis that can produce an underestimation of the size of core genomes. Hence to avoid such noise, and since generation of complete genomes is by far more expensive than draft genomes, we propose using the soft-core genome as the right estimator of the number of genes that are common to all members of a given bacterial species.

In Silico Multi-Locus Sequences Typing (MSLT) and Core Genome SNPs

Phylogenetic branches do not depend just on the presence/absence of genes, but in the fixation of specific mutations that can also provide differentiation of clinical and

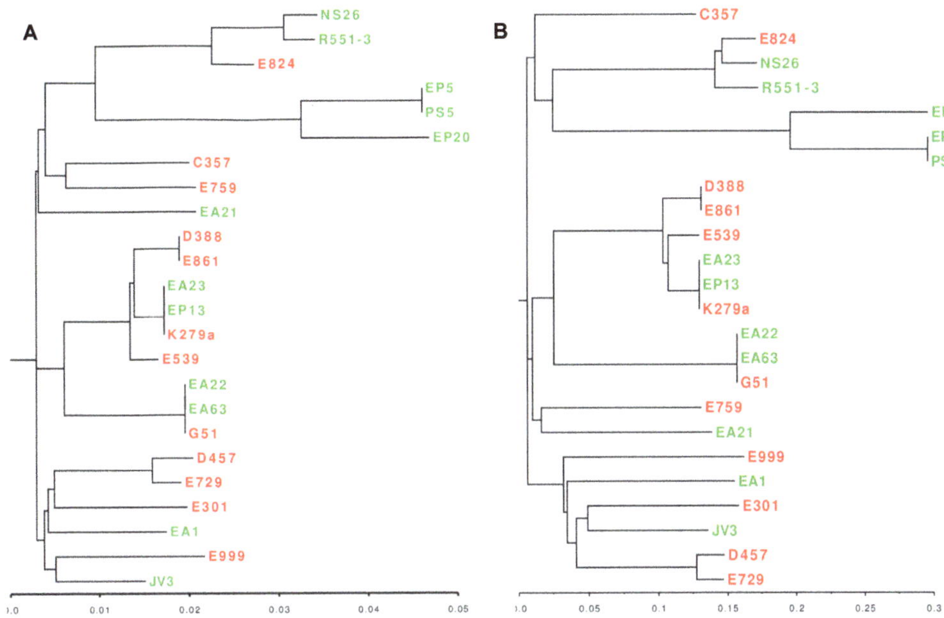

FIGURE 4 | Phylogenetic distribution of clinical and *environmental S. maltophilia* isolates. **(A)** Phylogenetic dendrogram based on the *in* vitro MLST analysis of seven concatenated genes (*atpD, guaA, nuoD, recA, gapA, mutM, ppsA*) **(B)** Phylogenetic dendrogram based on the alignment of SNPs present in the core genome of *S. maltophilia.* Red: clinical isolates. Green: environmental strains. As shown, both analysis grouped the strains in three major clusters, each one containing clinical and environmental strains.

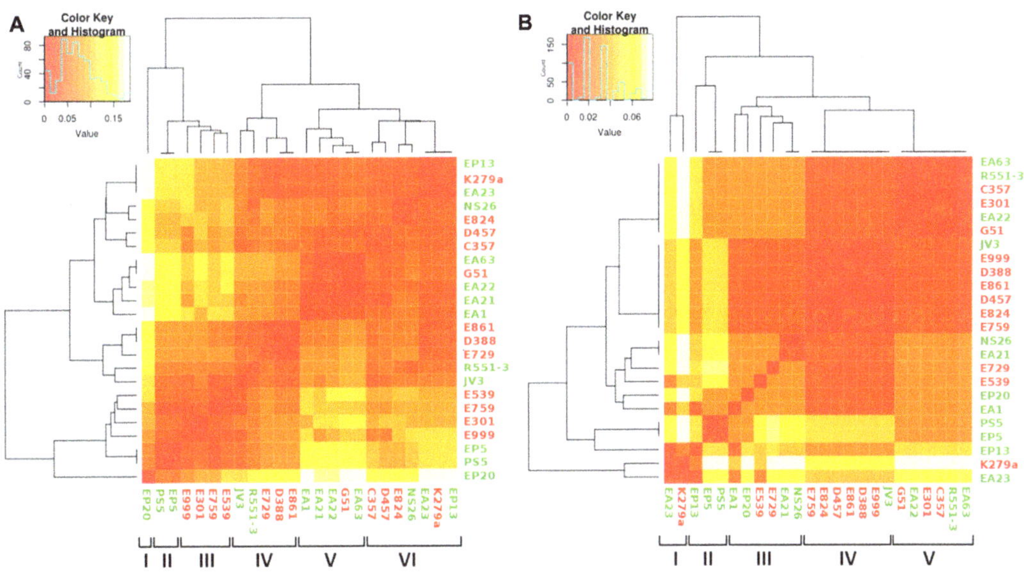

FIGURE 5 | Analysis of genome composition of clinical and environmental *S. maltophilia* isolates based on functional categories. Color key represents the scale of similarities from red (high) to yellow (low), and counts axis is the number of observed pairs (x, y) that fall into each binary event (presence/absence of shared functional categories for each isolate) represented by the histograms (blue lines). Green, environmental isolates; red clinical isolates. **(A)** Heatmap showing the clustering of clinical and environmental *S. maltophilia* isolates based in the genes with functional roles classified at FIGfam within the subsystem 'Virulence, Disease and Defense' in the RAST server. The clustering, based on a presence/absence matrix, revealed that most clusters contain both clinical and environmental strains. **(B)** Heatmap showing the clustering of clinical and environmental *S. maltophilia* strains based in the presence/absence of a specific set of virulence determinants described in Adamek et al. (2014). As shown, most branches contain both environmental and clinical isolates.

environmental isolates in different phylogenetic branches. To address this possibility, we performed two types of complementary analysis, namely *in silico* MLST and study of the core genome SNPs. Seven genes were used as markers for the MLST analysis: *atpD*, *gapA*, *guaA*, *mutM*, *nuoD*, *ppsA* and *recA*. The phylogenetic tree based on the alignment of

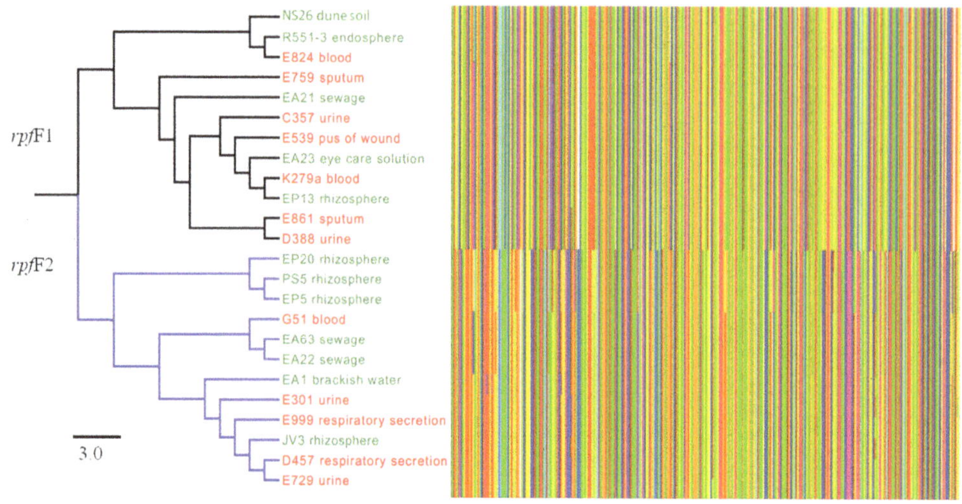

FIGURE 6 | Comparative analysis of the amino acid sequences of the two variants of RpfF of clinical and environmental *S. maltophilia* strains. The Figure shows the alignment of the 108 N-terminal residues of RpfF, which has been proposed to be used as markers for distinguishing the two RpfF variants (Huedo et al., 2014). Each amino acid is represented by a bar with different color. As shown, a clear separation between the two variants of RpfF are detected, but there is not a clear distinction between clinical and environmental isolates. Clinical strains are represented in red and environmental strains in green. The bar shows the sequence similarity distance.

these genes consisted of three major groups, each one of them comprised by clinical and environmental strains (**Figure 4A**). All SNPs were identified using *S. maltophilia* K279a as reference and phylogenetic dendrogram based on the core SPNs alignment was consistent with the topology and branches of the MSLT-based tree (**Figure 4B**). The data combining the genotypic profiling provided by the MSLT and the evaluation of the core SNPs of the 24 strains presented in this study revealed that *S. maltophilia* is a diverse complex, forming an interlaced taxon, sharing the same attributes between clinical and environmental strains without preference with respect to their origin.

Functional-Based Comparison between Clinical and Environmental Strains of *S. maltophilia*

Even though we did not find a clear distinction between the genome sizes and their CDS composition of clinical and environmental *S. maltophilia* strains, it is still possible that some functional categories, particularly those dealing with virulence are enriched as a function of the habitat (clinical or environmental) from which these strains have been isolated. Consequently, the 20 sequenced genomes and the four complete genomes used as reference were analyzed according to the functional groups of the CDS present in each of the genomes to further explore the relationship between habitat and genome composition. The presence of genes classified into the FIGfam subsystem 'Virulence, Disease and Defense' (Overbeek et al., 2005) was analyzed in all strains (Supplementary Table S1). From this information, a customized set of genes, containing only genes that were not present in all the isolates (Supplementary Table S2) was used to create a presence/absence matrix with roles not shared by all strains. Calculation of the average distance of strains and further clustering indicated the formation of six hierarchical clusters (**Figure 5A**). From the six clusters, only clusters II and III were formed exclusively by strains isolated from the same habitat. Environmental strains isolated from the rhizosphere, EP5 and PS5, composed the cluster II and both lacked some functional roles attributed to copper resistance. Cluster III was constituted by the clinical isolates E999, E301, E759 and E539 isolated from respiratory secretion, urine, sputum and pus, respectively. These isolates did not present five functional roles responsible for copper resistance. Other three clusters presented both clinical and environmental isolates in their composition, and EP20 did not group with other strains. Cluster IV, despite its composition including clinical and environmental isolates, presented a subdivision in those branches, creating two distinct sub-clusters composed by two environmental strains isolated from plants, JV3 and R551-3. Clinical isolates E861 and E729 were obtained from patients presenting urinary infection and strain D388 obtained from blood sample. The isolates grouped in cluster V present different origins and were characterized by the presence of genes related with Hg resistance (Supplementary Table S2). The presence of these genes suggest that these strains are able to inactivate Hg toxic forms into less toxic compounds. Finally, cluster VI was characterized by the diversity of organisms isolates from different habitats.

Taking into consideration that most clusters contain clinical and environmental isolates, and that the observed differences do not involve the presence of specific virulence genes in the clinical isolates, our results reinforce the notion that there are not clear distinctions between clinical and environmental *S. maltophilia* strains, even when the analysis is based in the distribution of functional categories.

It is worth mentioning, however, that virulence can be due to the presence of a small subset of genes and global analysis would not be sufficient to distinguish the presence or absence of such genes. Consequently, we screened for the presence of a set of genes that has been described as markers for *S. maltophilia* virulence (Adamek et al., 2014) (Supplementary Table S3). By using this dataset, the presence of five clusters was shown with four of them mixing clinical and environmental strains from different origins (**Figure 5B**). Only cluster II contained exclusively environmental strains (three), all obtained from plant rhizosphere. These results reinforced the idea that the genomic composition is not sufficient to establish a clear separation between clinical and environmental strains of *S. maltophilia*. Cluster I grouped the isolates EA23 and K279a that presents genes encoding filamentous hemaglutinins, which are important for adhesion and spread of bacteria through the respiratory tract (Colombi et al., 2006;Crossman et al., 2008). Despite it was not grouped in the same cluster, the isolate EP13 presented as well filamentous hemaglutinins genes. Isolates EP5 and PS5 were clustered in the same branch, in agreement with their complete CDS composition. Seven isolates did not present five functional roles responsible for copper resistance: four of them, E759, E999, E301 and E539, were clinical strains that shared the same cluster when the analysis was performed using the classification based in functional roles (**Figure 5A**). Otherwise, they did not share the same cluster when analyzed using the set of virulence factors (**Figure 5B**). The same happened with the environmental isolates PS5 and EP5 that shared the same clusters in both types of analysis. Altogether, the phylogenetical relationship of all strains analyzed in this study, calculated in base of their CDS composition and the clustering in orthologous groups, demonstrated that clinical and environmental strains did not form two independent evolutionary lineages. These results support the idea that clinical and environmental isolates are closely related and the pathogenic behavior does not depend on the acquisition of a specific set of virulence genes.

Quorum-Sensing Signals

The quorum-sensing system (QS) is responsible for the synchronization of particular bacterial behaviors on a population scale. In the case of *S. maltophilia* this process is relevant for *S. maltophilia* virulence and for its interaction with plants (Alavi et al., 2013, 2014), and depends on the Diffusible Signal Factor QS (DSF-QS), which has been identified as the fatty acid *cis*-11-methyl-2-dodecenoic acid (Fouhy et al., 2007).

In a previous study the existence of two different alleles for the *rpfF* gene, that is essential for the synthesis of DSF has been described. Each of the alleles defined a branch presenting a different virulence behavior (Huedo et al., 2014). It is then still possible that environmental and clinical isolates could present a differential virulence based in the presence/absence of a specific *rpfF* allele. Since these variants are markers of two different

TABLE 3 | Minimal Inhibitory Concentrations (MICs) of 20 new sequenced strains and of the model strain *S. maltophilia* D457.

	Strains	SXT	TGC	CAZ*	PM*	CN	GAT	CS	CL*	IMI*	ETP*	MXF	NA
Clinical strains	E729	0.75	3	1	4	2	1	4	24	>32	>32	0.75	8
	E759	1	2	>256	>256	1	128	6	32	>32	>32	0.25	4
	E999	0.38	2	4	12	24	0.25	4	16	>32	>32	0.13	8
	G51	0.38	0.75	>256	48	2	0.06	24	3	>32	>32	0.09	6
	E301	0.19	0.75	1.5	6	0.38	0.13	24	4	>32	>32	0.19	6
	D388	0.25	0.75	96	64	24	0.13	24	6	>32	>32	0.06	3
	E861	0.38	0.5	128	64	16	0.13	24	8	>32	>32	0.13	4
	C357	0.75	6	192	96	32	3	48	128	>32	>32	3	48
	E539	0.5	0.75	256	64	1	0.13	256	8	>32	>32	0.13	6
	E824	0.64	0.09	12	16	0.5	0.05	3	8	>32	>32	0.06	4
Environmental strains	NS26	0.19	0.75	16	64	2	0.13	12	6	>32	>32	0.13	3
	EP13	0.09	0.19	16	32	4	96	12	16	>32	>32	0.13	0.8
	EA22	0.5	0.75	32	48	2	0.13	32	6	>32	>32	0.03	3
	EA1	0.5	0.19	32	48	4	0.02	8	16	>32	>32	0.03	2
	PS5	0.02	0.05	64	>256	0.09	0.02	16	>256	0.06	0.2	0.03	8
	EA23	0.19	0.25	24	64	4	0.19	8	6	>32	>32	0.09	2
	EP20	0.09	0.19	16	32	6	0.09	48	8	>32	>32	0.03	2
	EP5	0.02	0.75	24	32	4	96	8	6	>32	>32	0.03	2
	EA21	0.19	0.5	>256	64	1	0.09	16	16	>32	>32	0.06	3
	EA63	0.13	0.38	32	48	6	0.06	48	8	>32	>32	0.06	6
	D457	0.13	1.5	1.5	16	6	0.5	32	12	32	32	0.25	8
	MIC50	0.25	0.75	32	48	2	0.13	16	8	32	32	0.09	4
	MIC90	0.75	2	256	96	24	96	48	32	32	32	0.25	8

*Values of the MIC50 (Minimum Inhibitory Concentration required to inhibit the growth of 50% of the isolates) and MIC90 (Minimum Inhibitory Concentration required to inhibit the growth of 90% of the isolates). Antibiotics: SXT, Trimethoprim/Sulfamethoxazole; TGC, Tigecyclin; CAZ, Ceftazidime; PM, Cefepime; CN, Gentamicin; GAT, Gatifloxacin; CS, Colistin; CL, Chloramphenicol; IM, Imipenem; ETP, Ertapenem; MXF, Moxifloxacin; NA, Nalidixic Acid. *At least one strain presented a MIC above the highest antibiotic concentration in the strip test.*

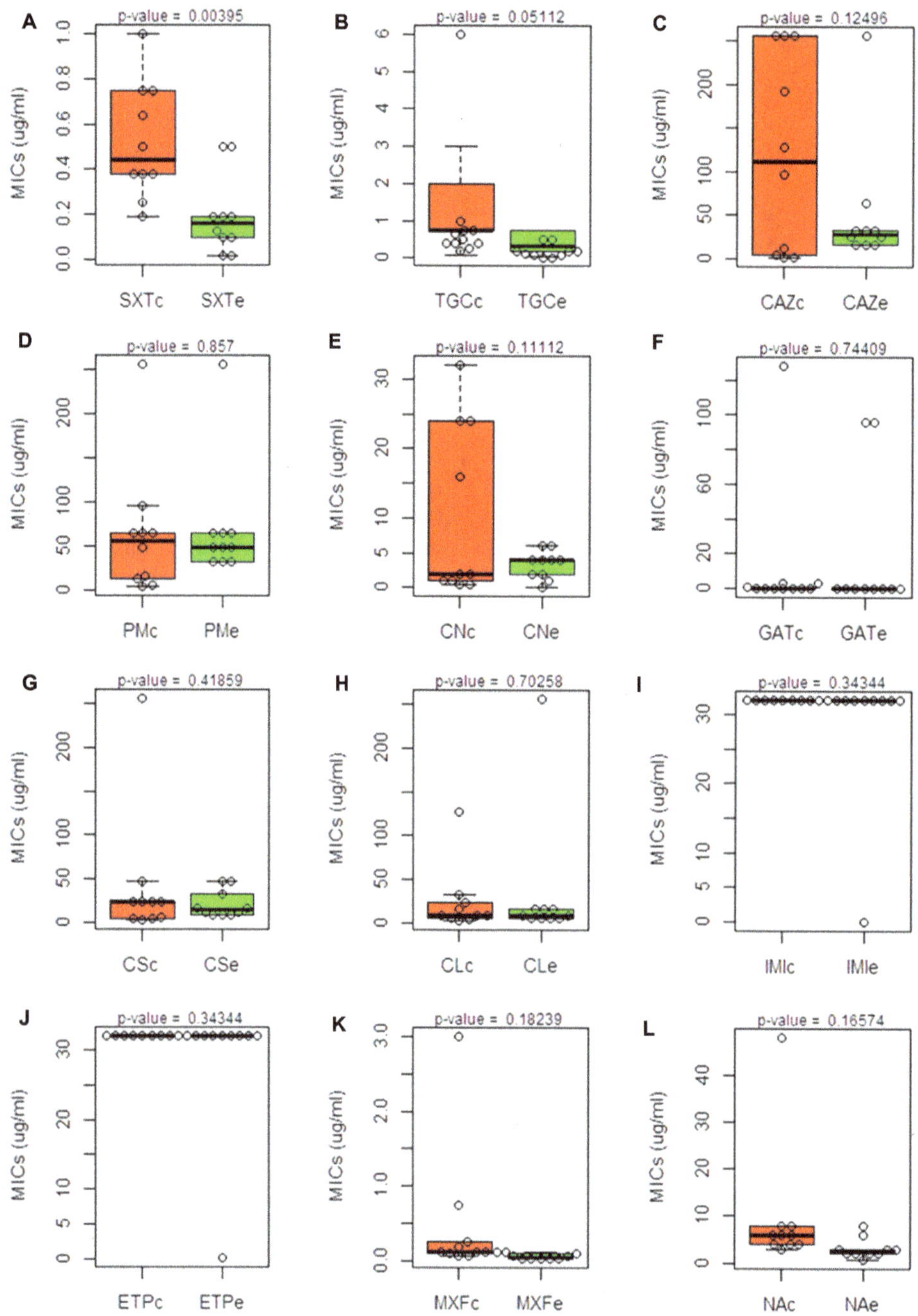

FIGURE 7 | Comparison of the susceptibility to antibiotics of clinical and environmental *S. maltophilia* isolates. Boxplot charts representing the Minimal Inhibitory Concentrations (MICs) for all clinical and environmental isolates obtained using antibiogram strip-tests of 12 antibiotics from different families: (SXT, Trimethoprim/Sulfamethoxazole; TGC, Tigecyclin; CAZ, Ceftazidime; PM, Cefepime; CN, Gentamicin; GAT, Gatifloxacin; CS, Colistin; CL, Chloramphenicol; IMI, Imipenem; ETP, Ertapenem; MXF, Moxifloxacin; NA, Nalidixic Acid). The median and the quartiles for the MIC values in each group are shown. Clinical isolates are represented in red box plots; environmental isolates are represented in green box plots. Statistical significance of the results was estimated by using the *t*-Student test. A significant difference (p-value < 0.05) was found just in the case of Trimethoprim/Sulfamethoxazole. Trimethoprim/Sulfamethoxazole (SXT); Tigecyclin (TGC); Ceftazidime (CAZ); Cefepime (PM); Gentamicin (CN); Gatifloxacin (GAT); Colistin (CS); Chloramphenicol (CL); Imipenem (IMI); Ertapenem (ETP); Moxifloxacin (MXF); Nalidixic Acid (NA). Each panel, from **(A–L)** represents the MICs of one antibiotic.

phylogenetic branches, each one presenting differences in terms of virulence, we analyzed their presence in the 24 studied genomes. Using the available sequences for *rpf*F1 and *rpf*F2, a direct search was performed for the corresponding DNA region of *rpf*F in all 24 genomes. All 24 strains of *S. maltophilia* harbor this gene with different length and variable residues along the

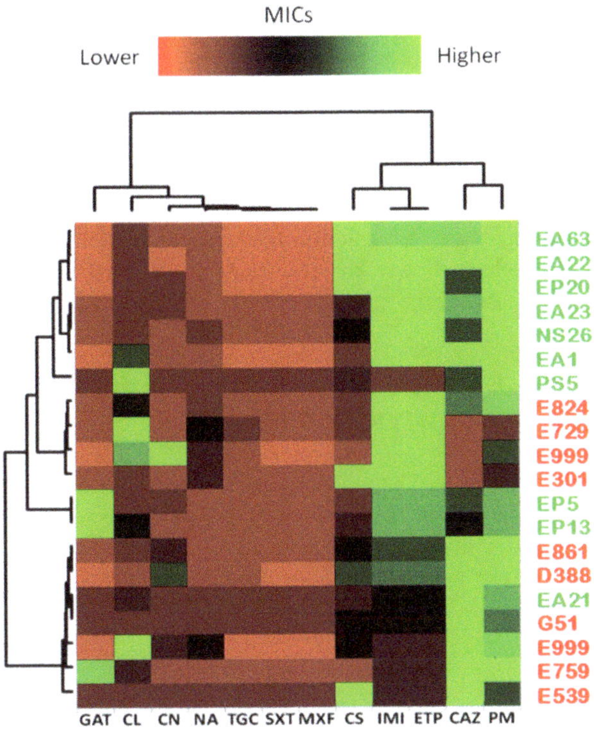

FIGURE 8 | Normalized susceptibility to antibiotics of clinical and environmental *S. maltophilia* isolates. Clustering of strains was obtained by normalizing the MICs from each independent isolate to the MIC50 (minimal concentration at which 50 % of the strains are susceptible) of all strains, and expressed in logarithmic scale ($\log_2$). Trimethoprim/Sulfamethoxazole (SXT); Tigecyclin (TGC); Ceftazidime (CAZ); Cefepime (PM); Gentamicin (CN); Gatifloxacin (GAT); Colistin (CS); Chloramphenicol (CL); Imipenem (IMI); Ertapenem (ETP); Moxifloxacin (MXF); Nalidixic Acid (NA). Green plots represent MICs higher than MIC50 while red plots represent MICs lower than MIC50. As shown, there is not a common trend toward multi-resistance among clinical isolates when compared with environmental strains.

sequence. In agreement with previous results (Huedo et al., 2014), the studied isolates are distributed into two distinct groups, each one presenting a different *rpf*F variant. Each group comprised 12 strains (**Figure 6**); however, there was not a clear difference in the distribution of clinical and environmental isolates between both groups. The cluster with the RpfF1 variant, which displays detectable DSF production (Huedo et al., 2014), comprised four environmental strains and eight clinical isolates, while the cluster containing the RpfF2 allele, with no significant effect on virulence-related phenotypes, presented seven environmental and five clinical strains.

Antibiotic Susceptibility of Clinical and Environmental Isolates of *S. maltophilia*

The Minimal Inhibitory Concentrations (MIC) of 12 different antibiotics, belonging to a wide range of structural families and presenting different targets, were established for the 20 isolates. The strain D457, which has been used in several studies on antibiotic resistance in *S. maltophilia* (Alonso and Martinez, 1997) was included as a control. The results (**Table 3**) were plotted in quartiles. As shown in **Figure 7**, the clinical isolates, as a group, present a trend toward higher levels of resistance than the environmental ones.

The environmental strain PS5 was the only isolate susceptible to IMI and ETP, while the other strains grew over the maximum value of this strip-test (>32 μg/ml), a feature that fits with previous information showing that *S. maltophilia* is resistant to these antibiotics (Howe et al., 1997). Notably, the same isolate, PS5, presented the highest level of resistance to CL (>256 μg/ml), followed by the strain C357 (MIC 128 μg/ml). For the other strains, the values ranged between 3 and 32 μg/ml. This may suggest that all *S. maltophilia* isolates, independently from their origin, present similar chances to acquire resistance to this antibiotic.

Although all isolates displayed low susceptibility to the tested antibiotics, when we analyze just the antibiotic concentration ranges where the values did not exceed the maximum concentration of the strip tests, the clinical strains presented overall higher MIC values for the antibiotics SXT, TGC, GAT, MXF and NA when compared with the environmental isolates (**Figure 7**). Nevertheless, this difference was statistically significant only in the case of STX. Therefore, despite there seems to be a trend toward lower MIC values in the environmental isolates, and in agreement with other studies (Berg et al., 1999), the multiple-antibiotic-resistance pattern of both clinical and environmental strains does not present significant differences and might be explained by the intrinsic resistome linked to the core genome of this species.

The previous analysis was based in the independent analysis of each of the antibiotics in the full population. To analyze a different aspect of the problem: the susceptibility to several antibiotics in each independent isolate, further comparisons of the clinical and environmental isolates were performed normalizing the obtained MICs by the MIC50 of all strains (**Figure 8**). Normalization of the MICs by the MIC50 of the 20 isolates grouped the environmental strains NS26, EA22, EA1, PS5, EA23, EP20 and EA63 in one branch presenting overall less susceptibility to carbapenems, imipenem and ertapenem, and the cefalosporins, ceftazidime and cefepime, than the other strains, indicating that, at least for some antibiotics, environmental isolates can present higher levels of resistance than clinical strains. Previous publications have shown that both clinical and environmental *S. maltophilia* isolates are highly resistant to antibiotics (Berg et al., 1999). Our results confirm this issue: the MICs of most antibiotics are high in all isolates as compared with other bacterial pathogens, and in occasions environmental strains are even less susceptible than clinical isolates.

CONCLUSION

When looking to the structure of bacterial species presenting infective and non-infective habitats, three situations can be foreseen. Either the species present specific virulence branches, as it happens in the case of *Escherichia coli*, either all isolates can produce an infection in healthy and sick people as in the case of *Yersinia pestis*, either all isolates can produce infection, but only in people with a previous basal disease, as it has been described in the case of *P. aeruginosa*. The consequences in terms of preventing infections by each one of these species would be different. For the first type of microorganisms, surveillance must be taken at the clonal level: some clones constitute a risk while some others are not dangerous. For the second, each member of the species must be considered as a risk for human health. In the third case, the risk is not mainly due to the organism itself, which does not infect the community, but to the situation of the potential host to be infected. Our results indicate that *S. maltophilia* belongs to the third category; all strains are likely equivalent in their capability of infecting humans, but only patients presenting severe underlying diseases including cystic fibrosis would be infected by this pathogen. Given the high biotechnological potential of *S. maltophilia*, both for confined and non-confined applications, there are concerns on the risk that this use may have for human health. Our results indicate that this concern applies just for people with underlying diseases and not for the community and, given that *S. maltophilia* is an environmental ubiquitous and cosmopolitan organism, its use in the habitats that this bacterium regularly colonizes will likely produce just an incremental risk of acquiring infections, even in the case of patients presenting underlying diseases.

A final reflection concerns the distribution of *S. maltophilia* pangenome. Most genes not belonging to the core genome are present in just one or a few strains. Together with the finding that *S. maltophilia* presents an open genome, this suggests that *S. maltophilia* can likely colonize a full range of microniches and, for such colonization, each member of this bacterial species is capable of acquiring a specific set of genes through HGT.

AUTHOR CONTRIBUTIONS

FL performed the experiments and bioinformatic analysis presented in the work. All authors contributed to the design and interpretation of the results, as well as to writing the article, and approved it for publication.

REFERENCES

Achtman, M., Morelli, G., Zhu, P., Wirth, T., Diehl, I., Kusecek, B., et al. (2004). Microevolution and history of the plague bacillus, *Yersinia pestis*. *Proc. Natl. Acad. Sci. U.S.A.* 101, 17837–17842. doi: 10.1073/pnas.0408026101

Achtman, M., Zurth, K., Morelli, G., Torrea, G., Guiyoule, A., and Carniel, E. (1999). *Yersinia pestis*, the cause of plague, is a recently emerged clone of *Yersinia pseudotuberculosis*. *Proc. Natl. Acad. Sci. U.S.A.* 96, 14043–14048. doi: 10.1073/pnas.96.24.14043

Adamek, M., Linke, B., and Schwartz, T. (2014). Virulence genes in clinical and environmental *Stenotrophomonas maltophilia* isolates: a genome sequencing and gene expression approach. *Microb. Pathog.* 67-68, 20–30. doi: 10.1016/j.micpath.2014.02.001

Alavi, P., Muller, H., Cardinale, M., Zachow, C., Sanchez, M. B., Martinez, J. L., et al. (2013). The DSF quorum sensing system controls the positive influence of *Stenotrophomonas maltophilia* on plants. *PLOS ONE* 8:e67103. doi: 10.1371/journal.pone.0067103

Alavi, P., Starcher, M. R., Thallinger, G. G., Zachow, C., Muller, H., and Berg, G. (2014). *Stenotrophomonas* comparative genomics reveals genes and functions that differentiate beneficial and pathogenic bacteria. *BMC Genomics* 15:482. doi: 10.1186/1471-2164-15-482

Alonso, A., and Martinez, J. L. (1997). Multiple antibiotic resistance in *Stenotrophomonas maltophilia*. *Antimicrob. Agents Chemother.* 41, 1140–1142.

Alonso, A., and Martinez, J. L. (2001). Expression of multidrug efflux pump SmeDEF by clinical isolates of *Stenotrophomonas maltophilia*. *Antimicrob. Agents Chemother.* 45, 1879–1881. doi: 10.1128/AAC.45.6.1879-1881.2001

Alonso, A., Rojo, F., and Martinez, J. L. (1999). Environmental and clinical isolates of *Pseudomonas aeruginosa* show pathogenic and biodegradative properties irrespective of their origin. *Environ. Microbiol.* 1, 421–430. doi: 10.1046/j.1462-2920.1999.00052.x

Aziz, R. K., Bartels, D., Best, A. A., Dejongh, M., Disz, T., Edwards, R. A., et al. (2008). The RAST Server: rapid annotations using subsystems technology. *BMC Genomics* 9:75. doi: 10.1186/1471-2164-9-75

Bankevich, A., Nurk, S., Antipov, D., Gurevich, A. A., Dvorkin, M., Kulikov, A. S., et al. (2012). SPAdes: a new genome assembly algorithm and its applications to single-cell sequencing. *J. Comput. Biol.* 19, 455–477. doi: 10.1089/cmb.2012.0021

Berg, G., Marten, P., and Ballin, G. (1996). *Stenotrophomonas maltophilia* in the rhizosphere of oilseed rape occurrence, characterization and interaction with phytopathogenic fungi. *Microbiol. Res.* 151, 18–27. doi: 10.1016/S0944-5013(96)80051-6

Berg, G., and Martinez, J. L. (2015). Friends or foes: can we make a distinction between beneficial and harmful strains of the *Stenotrophomonas maltophilia* complex? *Front. Microbiol.* 6:241. doi: 10.3389/fmicb.2015.00241

Berg, G., Roskot, N., and Smalla, K. (1999). Genotypic and phenotypic relationships between clinical and environmental isolates of *Stenotrophomonas maltophilia*. *J. Clin. Microbiol.* 37, 3594–3600.

Brooke, J. S. (2012). *Stenotrophomonas maltophilia*: an emerging global opportunistic pathogen. *Clin. Microbiol. Rev.* 25, 2–41. doi: 10.1128/CMR.00019-11

Brooke, J. S., Di Bonaventura, G., Berg, G., and Martinez, J. L. (2017). Editorial: a multidisciplinary look at *Stenotrophomonas maltophilia*: an emerging multi-drug-resistant global opportunistic pathogen. *Front. Microbiol.* 8:1511. doi: 10.3389/fmicb.2017.01511

Camacho, C., Coulouris, G., Avagyan, V., Ma, N., Papadopoulos, J., Bealer, K., et al. (2009). BLAST+: architecture and applications. *BMC Bioinformatics* 10:421. doi: 10.1186/1471-2105-10-421

Chang, Y. T., Lin, C. Y., Chen, Y. H., and Hsueh, P. R. (2015). Update on infections caused by *Stenotrophomonas maltophilia* with particular attention to resistance mechanisms and therapeutic options. *Front. Microbiol.* 6:893. doi: 10.3389/fmicb.2015.00893

Chiarini, L., Bevivino, A., Dalmastri, C., Tabacchioni, S., and Visca, P. (2006). *Burkholderia cepacia* complex species: health hazards and biotechnological potential. *Trends Microbiol.* 14, 277–286. doi: 10.1016/j.tim.2006.04.006

Colombi, D., Oliveira, M. L., Campos, I. B., Monedero, V., Perez-Martinez, G., and Ho, P. L. (2006). Haemagglutination induced by *Bordetella pertussis* filamentous haemagglutinin adhesin (FHA) is inhibited by antibodies produced against FHA(430-873) fragment expressed in *Lactobacillus casei*. *Curr. Microbiol.* 53, 462–466. doi: 10.1007/s00284-005-0388-0

Contreras-Moreira, B., and Vinuesa, P. (2013). GET_HOMOLOGUES, a versatile software package for scalable and robust microbial pangenome analysis. *Appl. Environ. Microbiol.* 79, 7696–7701. doi: 10.1128/AEM.02411-13

Criscuolo, A., and Brisse, S. (2013). AlienTrimmer: a tool to quickly and accurately trim off multiple short contaminant sequences from high-throughput sequencing reads. *Genomics* 102, 500–506. doi: 10.1016/j.ygeno.2013.07.011

Crossman, L. C., Gould, V. C., Dow, J. M., Vernikos, G. S., Okazaki, A., Sebaihia, M., et al. (2008). The complete genome, comparative and functional analysis of *Stenotrophomonas maltophilia* reveals an organism heavily shielded by drug resistance determinants. *Genome Biol.* 9:R74. doi: 10.1186/gb-2008-9-4-r74

Darling, A. C., Mau, B., Blattner, F. R., and Perna, N. T. (2004). Mauve: multiple alignment of conserved genomic sequence with rearrangements. *Genome Res.* 14, 1394–1403. doi: 10.1101/gr.2289704

De Boer, W., Klein Gunnewiek, P. J., Kowalchuk, G. A., and Van Veen, J. A. (2001). Growth of chitinolytic dune soil beta-subclass *Proteobacteria* in response to invading fungal hyphae. *Appl. Environ. Microbiol.* 67, 3358–3362. doi: 10.1128/AEM.67.8.3358-3362.2001

Dungan, R. S., Yates, S. R., and Frankenberger, W. T. Jr. (2003). Transformations of selenate and selenite by *Stenotrophomonas maltophilia* isolated from a seleniferous agricultural drainage pond sediment. *Environ. Microbiol.* 5, 287–295. doi: 10.1046/j.1462-2920.2003.00410.x

Dunne, C., Moenne-Loccoz, Y., De Bruijn, F. J., and O'gara, F. (2000). Overproduction of an inducible extracellular serine protease improves biological control of *Pythium ultimum* by *Stenotrophomonas maltophilia* strain W81. *Microbiology* 146(Pt 8), 2069–2078. doi: 10.1099/00221287-146-8-2069

Esposito, A., Pompilio, A., Bettua, C., Crocetta, V., Giacobazzi, E., Fiscarelli, E., et al. (2017). Evolution of *Stenotrophomonas maltophilia* in cystic fibrosis lung over chronic infection: a genomic and phenotypic population study. *Front. Microbiol.* 8:1590. doi: 10.3389/fmicb.2017.01590

Fouhy, Y., Scanlon, K., Schouest, K., Spillane, C., Crossman, L., Avison, M. B., et al. (2007). Diffusible signal factor-dependent cell-cell signaling and virulence in the nosocomial pathogen *Stenotrophomonas maltophilia*. *J. Bacteriol.* 189, 4964–4968. doi: 10.1128/JB.00310-07

Gherardi, G., Creti, R., Pompilio, A., and Di Bonaventura, G. (2015). An overview of various typing methods for clinical epidemiology of the emerging pathogen *Stenotrophomonas maltophilia*. *Diagn. Microbiol. Infect. Dis.* 81, 219–226. doi: 10.1016/j.diagmicrobio.2014.11.005

Howe, R. A., Wilson, M. P., Walsh, T. R., and Millar, M. R. (1997). Susceptibility testing of *Stenotrophomonas maltophilia* to carbapenems. *J. Antimicrob. Chemother.* 40, 13–17. doi: 10.1093/jac/40.1.13

Huedo, P., Yero, D., Martinez-Servat, S., Estibariz, I., Planell, R., Martinez, P., et al. (2014). Two different rpf clusters distributed among a population of *Stenotrophomonas maltophilia* clinical strains display differential diffusible signal factor production and virulence regulation. *J. Bacteriol.* 196, 2431–2442. doi: 10.1128/JB.01540-14

Hyatt, D., Chen, G. L., Locascio, P. F., Land, M. L., Larimer, F. W., and Hauser, L. J. (2010). Prodigal: prokaryotic gene recognition and translation initiation site identification. *BMC Bioinformatics* 11:119. doi: 10.1186/1471-2105-11-119

Jakobi, M., Winkelmann, G., Kaiser, D., Kempler, C., Jung, G., Berg, G., et al. (1996). Maltophilin: a new antifungal compound produced by *Stenotrophomonas maltophilia* R3089. *J. Antibiot.* 49, 1101–1104. doi: 10.7164/antibiotics.49.1101

Jeon, Y. D., Jeong, W. Y., Kim, M. H., Jung, I. Y., Ahn, M. Y., Ann, H. W., et al. (2016). Risk factors for mortality in patients with *Stenotrophomonas maltophilia* bacteremia. *Medicine* 95:e4375. doi: 10.1097/MD.0000000000004375

Kaas, R. S., Friis, C., Ussery, D. W., and Aarestrup, F. M. (2012). Estimating variation within the genes and inferring the phylogeny of 186 sequenced diverse *Escherichia coli* genomes. *BMC Genomics* 13:577. doi: 10.1186/1471-2164-13-577

Koonin, E. V., and Wolf, Y. I. (2008). Genomics of bacteria and archaea: the emerging dynamic view of the prokaryotic world. *Nucleic Acids Res.* 36, 6688–6719. doi: 10.1093/nar/gkn668

Langfelder, P., and Horvath, S. (2012). Fast R functions for robust correlations and hierarchical clustering. *J. Stat. Softw.* 46, 1–17. doi: 10.18637/jss.v046.i11

Larkin, M. A., Blackshields, G., Brown, N. P., Chenna, R., Mcgettigan, P. A., Mcwilliam, H., et al. (2007). Clustal W and Clustal X version 2.0. *Bioinformatics* 23, 2947–2948. doi: 10.1093/bioinformatics/btm404

Larsen, M. V., Cosentino, S., Rasmussen, S., Friis, C., Hasman, H., Marvig, R. L., et al. (2012). Multilocus sequence typing of total-genome-sequenced bacteria. *J. Clin. Microbiol.* 50, 1355–1361. doi: 10.1128/JCM.06094-11

Lee, E. Y., Jun, Y. S., Cho, K. S., and Ryu, H. W. (2002). Degradation characteristics of toluene, benzene, ethylbenzene, and xylene by *Stenotrophomonas maltophilia* T3-c. *J. Air Waste Manag. Assoc.* 52, 400–406. doi: 10.1080/10473289.2002.10470796

Lira, F., Hernandez, A., Belda, E., Sanchez, M. B., Moya, A., Silva, F. J., et al. (2012). Whole-genome sequence of *Stenotrophomonas maltophilia* D457, a clinical isolate and a model strain. *J. Bacteriol.* 194, 3563–3564. doi: 10.1128/JB.00602-12

Lucas, S., Copeland, A., Lapidus, A., Glavina Del Rio, T., Dalin E., and Tice, H. (2008). Complete sequence of *Stenotrophomonas maltophilia* R551-3. *EMBL/GenBank/DDBJ Databases.*

Lucas, S., Han, J., Lapidus, A., Cheng, J.-F., Goodwin, L., Pitluck, S., et al. (2011). Complete sequence of *Stenotrophomonas maltophilia* JV3. *EMBL/GenBank/DDBJ Databases.*

Martinez, J. L. (2013). Bacterial pathogens: from natural ecosystems to human hosts. *Environ. Microbiol.* 15, 325–333. doi: 10.1111/j.1462-2920.2012.02837.x

Martinez, J. L., Coque, T. M., Lanza, V. F., De La Cruz, F., and Baquero, F. (2017). Genomic and metagenomic technologies to explore the antibiotic resistance mobilome. *Ann. N. Y. Acad. Sci.* 1388, 26–41. doi: 10.1111/nyas.13282

Martinez, J. L., Fajardo, A., Garmendia, L., Hernandez, A., Linares, J. F., Martinez-Solano, L., et al. (2009). A global view of antibiotic resistance. *FEMS Microbiol. Rev.* 33, 44–65. doi: 10.1111/j.1574-6976.2008.00142.x

Minkwitz, A., and Berg, G. (2001). Comparison of antifungal activities and 16S ribosomal DNA sequences of clinical and environmental isolates of *Stenotrophomonas maltophilia*. *J. Clin. Microbiol.* 39, 139–145. doi: 10.1128/JCM.39.1.139-145.2001

Morales, G., Wiehlmann, L., Gudowius, P., Van Delden, C., Tummler, B., Martinez, J. L., et al. (2004). Structure of *Pseudomonas aeruginosa* populations analyzed by single nucleotide polymorphism and pulsed-field gel electrophoresis genotyping. *J. Bacteriol.* 186, 4228–4237. doi: 10.1128/JB.186.13.4228-4237.2004

Mukherjee, P., and Roy, P. (2016). Genomic potential of *Stenotrophomonas maltophilia* in bioremediation with an assessment of its multifaceted role in our environment. *Front. Microbiol.* 7:967. doi: 10.3389/fmicb.2016.00967

Nangia, Y., Wangoo, N., Goyal, N., Shekhawat, G., and Suri, C. R. (2009). A novel bacterial isolate *Stenotrophomonas maltophilia* as living factory for synthesis of gold nanoparticles. *Microb. Cell Fact.* 8:39. doi: 10.1186/1475-2859-8-39

Nicoletti, M., Iacobino, A., Prosseda, G., Fiscarelli, E., Zarrilli, R., De Carolis, E., et al. (2011). *Stenotrophomonas maltophilia* strains from cystic fibrosis patients: genomic variability and molecular characterization of some virulence determinants. *Int. J. Med. Microbiol.* 301, 34–43. doi: 10.1016/j.ijmm.2010.07.003

Ochoa-Sanchez, L. E., and Vinuesa, P. (2017). Evolutionary genetic analysis uncovers multiple species with distinct habitat preferences and antibiotic resistance phenotypes in the *Stenotrophomonas maltophilia* complex. *Front. Microbiol.* 8:1548. doi: 10.3389/fmicb.2017.01548

Overbeek, R., Begley, T., Butler, R. M., Choudhuri, J. V., Chuang, H. Y., Cohoon, M., et al. (2005). The subsystems approach to genome annotation and its use in the project to annotate 1000 genomes. *Nucleic Acids Res.* 33, 5691–5702. doi: 10.1093/nar/gki866

Pompilio, A., Crocetta, V., Ghosh, D., Chakrabarti, M., Gherardi, G., Vitali, L. A., et al. (2016). *Stenotrophomonas maltophilia* phenotypic and genotypic diversity during a 10-year colonization in the lungs of a cystic fibrosis patient. *Front. Microbiol.* 7:1551. doi: 10.3389/fmicb.2016.01551

Ribbeck-Busch, K., Roder, A., Hasse, D., De Boer, W., Martinez, J. L., Hagemann, M., et al. (2005). A molecular biological protocol to distinguish potentially human pathogenic *Stenotrophomonas maltophilia* from plant-associated *Stenotrophomonas rhizophila*. *Environ. Microbiol.* 7, 1853–1858. doi: 10.1111/j.1462-2920.2005.00928.x

Rocco, F., De Gregorio, E., Colonna, B., and Di Nocera, P. P. (2009). *Stenotrophomonas maltophilia* genomes: a start-up comparison. *Int. J. Med. Microbiol.* 299, 535–546. doi: 10.1016/j.ijmm.2009.05.004

Sanchez, M. B. (2015). Antibiotic resistance in the opportunistic pathogen *Stenotrophomonas maltophilia*. *Front. Microbiol.* 6:658. doi: 10.3389/fmicb.2015.00658

Sanchez, M. B., Hernandez, A., and Martinez, J. L. (2009). *Stenotrophomonas maltophilia* drug resistance. *Future Microbiol.* 4, 655–660. doi: 10.2217/fmb.09.45

Schmieder, R., and Edwards, R. (2011). Quality control and preprocessing of metagenomic datasets. *Bioinformatics* 27, 863–864. doi: 10.1093/bioinformatics/btr026

Suckstorff, I., and Berg, G. (2003). Evidence for dose-dependent effects on plant growth by *Stenotrophomonas* strains from different origins. *J. Appl. Microbiol.* 95, 656–663. doi: 10.1046/j.1365-2672.2003.02021.x

Tatusova, T., Ciufo, S., Fedorov, B., O'neill, K., and Tolstoy, I. (2014). RefSeq microbial genomes database: new representation and annotation strategy. *Nucleic Acids Res.* 42, D553–D559. doi: 10.1093/nar/gkt1274

Tettelin, H., Masignani, V., Cieslewicz, M. J., Donati, C., Medini, D., Ward, N. L., et al. (2005). Genome analysis of multiple pathogenic isolates of *Streptococcus agalactiae*: implications for the microbial "pan-genome". *Proc. Natl. Acad. Sci. U.S.A.* 102, 13950–13955. doi: 10.1073/pnas.0506758102

Waterhouse, A. M., Procter, J. B., Martin, D. M., Clamp, M., and Barton, G. J. (2009). Jalview version 2–a multiple sequence alignment editor and analysis workbench. *Bioinformatics* 25, 1189–1191. doi: 10.1093/bioinformatics/btp033

Wiehlmann, L., Wagner, G., Cramer, N., Siebert, B., Gudowius, P., Morales, G., et al. (2007). Population structure of *Pseudomonas aeruginosa*. *Proc. Natl. Acad. Sci. U.S.A.* 104, 8101–8106. doi: 10.1073/pnas.0609213104

Willenbrock, H., Hallin, P. F., Wassenaar, T. M., and Ussery, D. W. (2007). Characterization of probiotic *Escherichia coli* isolates with a novel pan-genome microarray. *Genome Biol.* 8:R267. doi: 10.1186/gb-2007-8-12-r267

Wren, B. W. (2003). The yersiniae–a model genus to study the rapid evolution of bacterial pathogens. *Nat. Rev. Microbiol.* 1, 55–64. doi: 10.1038/nrmicro730

Youenou, B., Favre-Bonte, S., Bodilis, J., Brothier, E., Dubost, A., Muller, D., et al. (2015). Comparative genomics of environmental and clinical *Stenotrophomonas maltophilia* strains with different antibiotic resistance profiles. *Genome Biol. Evol.* 7, 2484–2505. doi: 10.1093/gbe/evv161

Yu, D., Yin, Z., Li, B., Jin, Y., Ren, H., Zhou, J., et al. (2016). Gene flow, recombination, and positive selection in *Stenotrophomonas maltophilia*: mechanisms underlying the diversity of the widespread opportunistic pathogen. *Genome* 59, 1063–1075. doi: 10.1139/gen-2016-0073

Zhou, D., and Yang, R. (2009). Molecular Darwinian evolution of virulence in *Yersinia pestis*. *Infect. Immun.* 77, 2242–2250. doi: 10.1128/IAI.01477-08

Zonaro, E., Lampis, S., Turner, R. J., Qazi, S. J., and Vallini, G. (2015). Biogenic selenium and tellurium nanoparticles synthesized by environmental microbial isolates efficaciously inhibit bacterial planktonic cultures and biofilms. *Front. Microbiol.* 6:584. doi: 10.3389/fmicb.2015.00584

15

Fluorescence Imaging of *Streptococcus pneumoniae* with the *Helix pomatia* agglutinin (HPA) as a Potential, Rapid Diagnostic Tool

Mirian Domenech [1, 2] *and Ernesto García* [1, 2*]

[1] *Departamento de Microbiología Molecular y Biología de las Infecciones, Centro de Investigaciones Biológicas, Consejo Superior de Investigaciones Científicas, Madrid, Spain,* [2] *Centro de Investigación Biomédica en Red de Enfermedades Respiratorias, Madrid, Spain*

Edited by:
Chew Chieng Yeo, Universiti Sultan Zainal Abidin, Malaysia

Reviewed by:
Manuel Espinosa, Centro de Investigaciones Biológicas (CSIC), Spain
Darío García De Viedma, Hospital General Universitario Gregorio Marañón, Spain
Analia Rial, Facultad de Medicina, Universidad de la Republica de Uruguay, Uruguay
Natalia Munoz Wolf, Trinity College, Dublin, Ireland

***Correspondence:**
Ernesto García
e.garcia@cib.csic.es

Streptococcus pneumoniae is a common human pathogen and a major causal agent of life-threatening infections that can either be respiratory or non-respiratory. It is well known that the *Helix pomatia* (edible snail) agglutinin (HPA) lectin shows specificity for terminal αGalNAc residues present, among other locations, in the Forssman pentasaccharide (αGalNAc1→3βGalNAc1→3αGal1→4βGal1→4βGlc). Based on experiments involving choline-independent mutants and different growth conditions, we propose here that HPA recognizes the αGalNAc terminal residues of the cell wall teichoic and lipoteichoic acids of *S. pneumoniae*. In addition, experimental evidence showing that pneumococci can be specifically labeled with HPA when growing as planktonic cultures as well as in mixed biofilms of *S. pneumoniae* and *Haemophilus influenzae* has been obtained. It should be underlined that pneumococci were HPA-labeled despite of the presence of a capsule. Although some non-pneumococcal species also bind the agglutinin, HPA-binding combined with fluorescence microscopy constitutes a suitable tool for identifying *S. pneumoniae* and, if used in conjunction with Gram staining and/or other suitable technique like antigen detection, it may potentially facilitate a fast and accurate diagnosis of pneumococcal infections.

Keywords: *Streptococcus pneumoniae*, Forssman antigen, binding lectins, teichoic acids, fluorescence microscopy

INTRODUCTION

Streptococcus pneumoniae, the pneumococcus, is a leading human pathogen and one of the foremost etiologic agents of invasive diseases such as bacteremic community-acquired pneumonia (CAP), bacteremia, and meningitis, mainly in children, the elderly, and immunocompromised patients. Indeed, the pneumococcus is one of the major causes of non-invasive diseases such as non-bacteremic CAP, acute otitis media, sinusitis and conjunctivitis. In 2015, CAP accounted for 16% of all deaths of children under 5 years old globally and 920,000 deaths globally in children of all ages (World Health Organization, 2016). *S. pneumoniae* is the commonest bacterial causative agent across all ages, accounting for 30–40% of CAP cases (Haq et al., 2017). Other bacterial causes of CAP include *Streptococcus pyogenes* (group A streptococci; GAS), and *Streptococcus agalactiae* (group B streptococci; GBS) in infants. *Staphylococcus aureus* is associated with round pneumonia, a well-defined round area of consolidation visible on chest radiographs. Despite the current vaccination programs, *Haemophilus influenzae* remains prevalent in several developing countries (Hajjeh et al., 2013) and *Mycoplasma pneumoniae* accounts for up to a third of all cases and is a common cause of atypical CAP.

Since mortality rates are highest during the first days of admission (Ewig et al., 2009), early diagnosis and treatment may have a crucial role in curing the patient or in reducing their morbidity and mortality, particularly in the era of antibiotic resistance (Vila et al., 2017). Rapid identification techniques are even more important in bacterial meningitis patients, since delayed initiation of antibiotic treatment is strongly associated with death and poor outcome; as a consequence, it has been recently recommended that antibiotic treatment in these patients should be started as soon as possible, and the time period from entering the hospital to initiation of antibiotic treatment should not exceed 1 h (van de Beek et al., 2016). In every case, the gold standard diagnostic method is still culture. Unfortunately, this is time consuming (24–72 h) and includes the inoculation of appropriate media, subculturing, and phenotype-based characterization via biochemical testing, along with antibiotic susceptibility testing. Currently, bacteriological diagnosis is making progress in molecular biology techniques such as PCR, matrix-assisted laser desorption/ionization time-of-flight mass spectrometry (MALDI-TOF MS), and whole genome sequencing (Clark, 2015; Pai et al., 2015; McGill et al., 2016; Torres et al., 2016). These techniques have been developed to decrease the time for the initiation of an optimal antimicrobial therapy for improving clinical outcomes. Unfortunately, although microbiological diagnosis is very important for a correct clinical management of the disease—particularly because spreading of antibiotic multiresistance is an increasing global concern—, microbiological identification is lacking in many instances approaching 50% in CAP cases (Cilloniz et al., 2016).

Most bacteria in nature exist in multispecies communities known as biofilms. Multispecies biofilms are structured and spatially defined communities where species interact both intra- and interspecifically (Røder et al., 2016). Imaging techniques are useful for identifying multiple species, which provides information on their spatial organization. Fluorescence *in situ* hybridization (FISH) and many advanced versions of the FISH technique have been implemented for different purposes; these are well-established means of visualizing and identifying microbial groups or species in natural and artificially created multispecies biofilms (Amann and Fuchs, 2008; Liu et al., 2016; Røder et al., 2016). Although FISH is typically time-consuming and destructive to the samples, it is however possible to visualize different species simultaneously (Schimak et al., 2016). Fluorescence imaging of individual species can also be achieved by genetically marking the species with genes coding for different fluorescent or bioluminescent proteins (Kjos et al., 2015). Unfortunately, not all bacteria can be fluorescently tagged, particularly those from natural samples.

Polymicrobial biofilms are abundant in clinical diseases such as acute otitis media, a significant public health problem worldwide (Monasta et al., 2012), particularly among children (Ahmed et al., 2014). Acute otitis media is preceded by the nasopharyngeal carriage of bacterial pathogens like *S. pneumoniae*, non-typeable (NT) *H. influenzae*, and *Moraxella catarrhalis* (Ngo et al., 2016). Moreover, the concurrent carriage of these pathogens is a predictor of clinical pneumonia (Chochua et al., 2016). When growing in an *in vitro* mixed biofilm, *S. pneumoniae* and NT *H. influenzae* cells appear to accomplish a strong inter-population cooperation, i.e., metabolic interdependence or mutualism (Momeni et al., 2013), as indicated by the finding that pneumococci were intermixed with NT *H. influenzae* cells throughout the biofilm (Domenech and García, 2017). This was determined using the *Helix pomatia* (edible snail) agglutinin (HPA) that unambiguously stained *S. pneumoniae* cells in the mixed biofilms. The binding preference of HPA has been reported to be the Forssman pentasaccharide (αGalNAc1→3βGalNAc1→3αGal1→4βGal1→4βGlc) >blood group A substance (αGalNAc1→3[αFuc1→2]Gal] >Tn antigen (αGalNAc-Ser/Thr) >GalNAc >GlcNAc, confirming its specificity for terminal αGalNAc residues (Wu and Sugii, 1991; Cooling, 2015). The Forssman antigen (FA) can be defined as a substance that provokes the appearance of sheep red blood cell hemolytic antibodies when injected into rabbits; it is a glycolipid with the structure GalNAcα1→3GalNAcβ1→3Galα1→4Galβ1→4Glcβ1→1Cer (Siddiqui and Hakomori, 1971). Originally found in the tissues of different animals (although not humans; Yamamoto et al., 2012), FA was subsequently discovered in some bacteria, including *S. pneumoniae* (Jenkin, 1963). It has been shown that the Forssman cross-reactive material of *S. pneumoniae* is the type IV, ribitol phosphate-containing, membrane anchored lipoteichoic acid (pnLTA) (Briles and Tomasz, 1973; Gisch et al., 2013). The non-lipid terminus of pnLTA consists of 6-*O*-*P*Cho-α-D-Gal*p*NAc-(1→3)-6-*O*-*P*Cho-β-D-Gal*p*NAc (Seo et al., 2008); this disaccharide represents a structural feature that is able to partly explain the FA properties of pnLTA (Gisch et al., 2013). In addition and unlike other bacterial species, pnLTA and the *S. pneumoniae* peptidoglycan-bound teichoic acid (pnWTA) have identical chain structures (for a recent review, see reference by Gisch et al. (2015a). Based on this information it can be assumed that HPA binds to both WTA and LTA of *S. pneumoniae*.

In the present study we report that HPA binds to the teichoic acids (TA) of encapsulated and non-encapsulated pneumococcal cells grown either planktonically or forming biofilms. In addition to *S. pneumoniae*, HPA labeling of other bacterial species, e.g., some *S. aureus* strains, has also been observed. We propose that, in combination with other widespread rapid techniques, labeling with HPA in biological fluids may represent a helpful technique for the fast and accurate diagnosis of pneumococcal diseases.

MATERIALS AND METHODS

Bacteria and Growth Conditions

The bacteria used in this study are listed in **Table 1**. Streptococci, staphylococci, enterococci, and *Pseudomonas aeruginosa* were grown in Todd-Hewitt broth supplemented with 0.5% yeast extract (THY). For planktonic growth, the NT *H. influenzae* strain 54997 was incubated in brain heart infusion (BHI)

TABLE 1 | Bacterial strains used in this study.

Strain	Description	Source or References[a]	HPA labeling[b]
S. pneumoniae			
D39	Serotype 2; clinical isolate	Lanie et al., 2007	+
P007	Serotype 3; laboratory transformant	Domenech et al., 2009	+
R6	Non-encapsulated D39 derivative	Hoskins et al., 2001	+
MNZ67	Non-encapsulated; clinical isolate	Park et al., 2012	+
1001	Serotype 3; clinical isolate	J. Liñares[c]	+
2951/13	Serotype 12F; clinical isolate	Domenech et al., 2015	+
Tupelo_VT	Serotype 14; clinical isolate	Moscoso et al., 2010	+
1064	Serotype 19F; clinical isolate	A. Fenoll[d]	+
1178/17	Serotype 22F; clinical isolate	L. Vicioso[d]	+
2971/13	Serotype 23B; clinical isolate	Domenech et al., 2015	+
JY2190	Choline-independent Rx1 mutant; non-encapsulated	Yother et al., 1998	+
P501	Choline-independent R6 mutant; non-encapsulated	González et al., 2008	+
Streptococcus pseudopneumoniae			
CCUG 49455	Type strain	CCUG	–
Streptococcus mitis			
NCTC 12261	Type strain	NCTC	–
SK137	Biovar 1; strain with choline-containing teichoic acids	Bergström et al., 2000	+
SK598	Biovar 1; strain with ethanolamine-containing teichoic acids	Bergström et al., 2003	–
Streptococcus oralis			
NCTC 11427	Type strain	Kilpper-Bälz et al., 1985	–
Streptococcus pyogenes			
CECT 985	Type strain; Lancefield group A	CECT	–
Streptococcus agalactiae			
CECT 183	Type strain; Lancefield group B	CECT	–
Streptococcus dysgalactiae subsp. *equisimilis*			
CECT 926	ATCC 9542; Lancefield group C	CECT	+
Streptococcus suis			
298	Serotype 9; Lancefield group D[e]	A. I. Vela[f]	+/–
357	Serotype 2; Lancefield group D	A. I. Vela	+/–
Staphylococcus aureus			
CECT 86	Type strain	Shiroma et al., 2015; CECT	+
15981	Clinical isolate; good biofilm former	Valle et al., 2003	+
Staphylococcus epidermidis			
CECT 231	Good biofilm former	Domenech et al., 2011; CECT	–
Enterococcus faecalis			
CECT 481	Type strain	CECT	–
Haemophilus influenzae			
54997	Nontypeable	García-Cobos et al., 2014	–
Pseudomonas aeruginosa			
PAO1	ATCC 15692	ATCC	–

[a]*ATCC, American Type Culture Collection; CECT, Colección Española de Cultivos Tipo; CCUG, Culture Collection, University of Göteborg; NCTC, National Collection of Type Cultures.*
[b]*+, positive; –, negative; +/–, slightly positive.*
[c]*Instituto de Investigación Biomédica de Bellvitge (IDIBEL); Barcelona (Spain).*
[d]*Laboratorio de Referencia de Neumococos; Centro Nacional de Microbiología (CNM-ISCIII); Majadahonda (Madrid; Spain).*
[e]*Although early studies reported that some strains of* S. suis *contained streptococcal group D antigen, more recent results indicated that the group R and group D antigens were similar and crossreacted. To date, the species belonging to the* Streptococcus bovis *group constitute the non-enterococcal group D streptococci (Dekker and Lau, 2016).*
[f]*Facultad de Veterinaria; Universidad Complutense de Madrid; Madrid (Spain).*

supplemented with haemin and NAD (15 μg/ml each) (sBHI). In some experiments, *S. pneumoniae* was grown in C medium (Lacks and Hotchkiss, 1960) supplemented with 0.08% yeast extract (Difco Laboratories; C+Y medium) or a chemically defined medium (Cden) supplemented (or not) with either choline (5 μg/ml) (Cden-choline) or ethanolamine (40 μg/ml) (Cden-EA) (Tomasz, 1968). Cells were incubated at 37°C without shaking. Bacterial growth was monitored by measuring the absorbance at 595 nm (A_{595}).

Formation of mixed biofilms of *S. pneumoniae* and NT *H. influenzae* was performed as described elsewhere (Domenech and García, 2017). Briefly, cultures of *S. pneumoniae* strain R6 and *H. influenzae* strain 54997 were grown to mid-exponential phase in C+Y medium supplemented with haemin and NAD (15 μg/ml each) [s(C+Y)], diluted to $\approx 5 \times 10^6$ colony-forming units (cfu)/ml and mixed in an 1:1 proportion. Two milliliter of the mixtures were then distributed into the wells of a glass-bottomed dish (WillCo-dish, WillCo Wells B. V., The Netherlands) and incubated for 6 h at 37°C under 5% CO_2. For species biofilm formation, 2 ml of the individual cultures (5×10^6 cfu/ml each) were independently inoculated as indicated above for mixed biofilms.

All studies which involved the handling of virulent bacteria, whole blood, or blood derivatives were undertaken at the biosafety level II laboratory of Centro de Investigaciones Biológicas. It should be mentioned that, according to the supplier (Innovative Research), the whole human blood used had been tested by FDA-approved methods for human immunodeficiency virus RNA, antibodies to immunodeficiency virus, antibodies to hepatitis C virus, hepatitis C virus RNA, hepatitis B virus, hepatitis B surface antigen, and syphilis.

Staining with HPA Lectin

Exponentially growing cells of various bacterial species were centrifuged, washed and suspended in phosphate-buffered saline (PBS). After incubation for 15 min at room temperature in the dark with HPA lectin conjugated to Alexa Fluor-488 (2.5–25 μg/ml), cells were centrifuged again and resuspended in PBS. Bacteria were observed under a Leica DM4000B fluorescence microscope equipped with an L5 filter (bandpass 480/40), and viewed under a Leica HCX PL FLUOTAR 40×/0.75 objective or an HC PL APO 63×/1.40–0.60 oil objective. In early experiments, bacteria were also diluted into fetal bovine serum (from Sigma-Aldrich), or into defibrinated sheep blood (from Biomedics or Oxoid). Afterwards, experiments were also carried out with groups A and O citrated human whole blood (from Innovative Research). It should be underlined that since HPA labeling does not require the presence of divalent cations (Kobayashi et al., 2014), blood treated to prevent its coagulation (e.g., citrate- or EDTA-treated) can be used.

For biofilm observation, the culture medium was removed and the biofilm rinsed with sterile water to remove non-adherent bacteria. Staining was performed with HPA and SYTO 59 and biofilms were gently rinsed with PBS. Observations were made using a Leica TCS-SP2-AOBS-UV confocal laser scanning microscope (CLSM) equipped with an argon ion laser. Images were analyzed using LCS software from Leica. Projections were obtained in the planes *x–y* (individual scans at 0.5 μm intervals) and *x–z* (images at 5 μm intervals).

RESULTS

HPA Binding by Different Bacteria

Pneumococci were clearly labeled with HPA when this was present in concentrations of 2.5–25 μg/ml (**Figure 1**). Closer examination showed that HPA labeling was not uniform across the pneumococcal surface; reduced fluorescence was noticed in the equatorial zone of growth, the place where new cell wall material is incorporated (Gisch et al., 2015a). The ability to bind HPA was not exclusive to pneumococci; the surfaces of other bacteria too were labeled with the lectin, i.e., *Streptococcus mitis* SK137 and some strains of *Streptococcus dysgalactiae* subsp. *equisimilis*, *Streptococcus suis* and *S. aureus* (**Figure 2**, **Table 1**). Notably, HPA was unable to label the type strains ([T]) of *Streptococcus pseudopneumoniae, S. mitis,* or *Streptococcus oralis*, the three closest relatives of the pneumococcus. Notably, cells of other relevant pathogens, either Gram-positive (GAS, GBS, *Enterococcus faecalis* and *Staphylococcus epidermidis*) or Gram-negative (*H. influenzae* and *P. aeruginosa*), did not bind the lectin (**Table 1**). Besides, and in agreement with previous results (Domenech and García, 2017), *S. pneumoniae* (but not NT *H. influenzae*) were also labeled with HPA when growing as mixed biofilms (**Figure 3**).

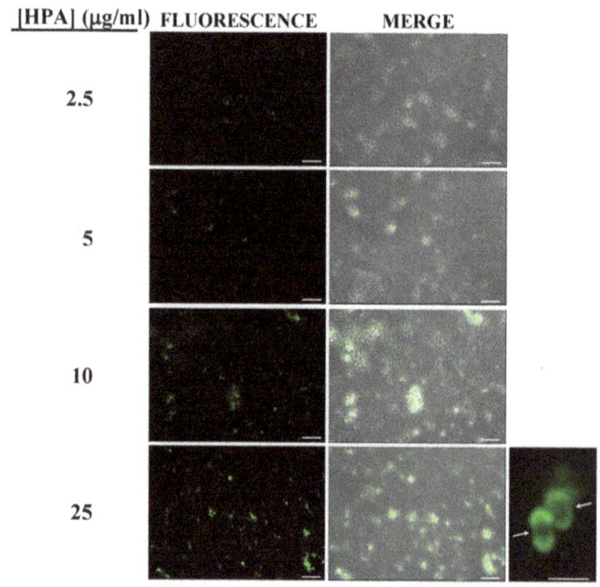

FIGURE 1 | Fluorescent labeling of the non-encapsulated *S. pneumoniae* strain R6 with HPA. Exponentially growing cultures of *S. pneumoniae* R6 in C+Y medium were incubated with the indicated concentrations of the lectin and observed for fluorescence (HCX PL FLUOTAR 40×/0.75 objective). Merges of fluorescence and phase-contrast images are also shown; bar = 25 μm. Enlarged view of two diplococci showing reduced fluorescence at the equatorial zone of growth (indicated by arrows; 63× objective). Bar = 2 μm.

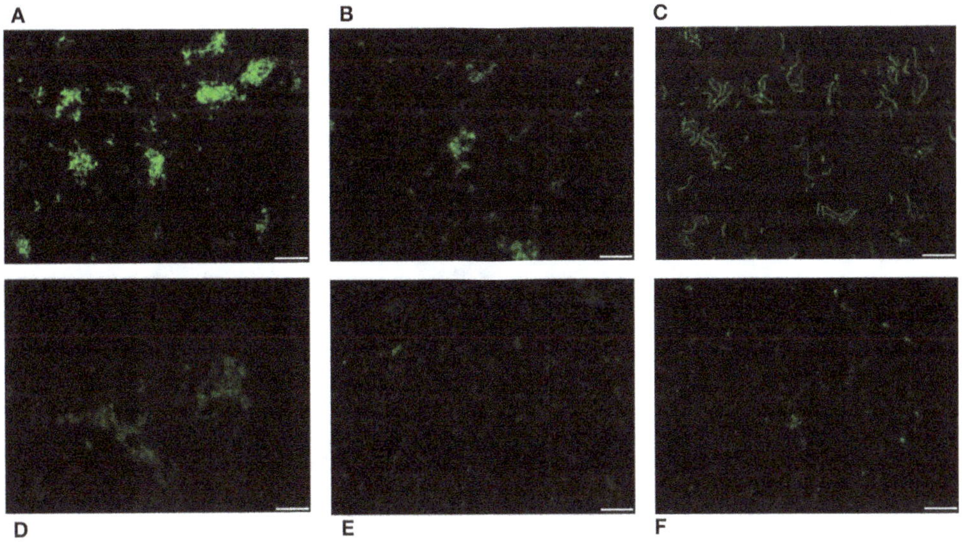

FIGURE 2 | HPA labeling of various streptococci in C+Y medium. First row: *S. pneumoniae* strains R6 (non-encapsulated) **(A)**, D39 (serotype 2) **(B)**, and P007 (serotype 3) **(C)**. Second row: *S. suis* 298 **(D)**, *S. dysgalactiae* subsp. *equisimilis* CECT 926 **(E),** and *S. aureus* type strain **(F)**. Bar = 25 μm.

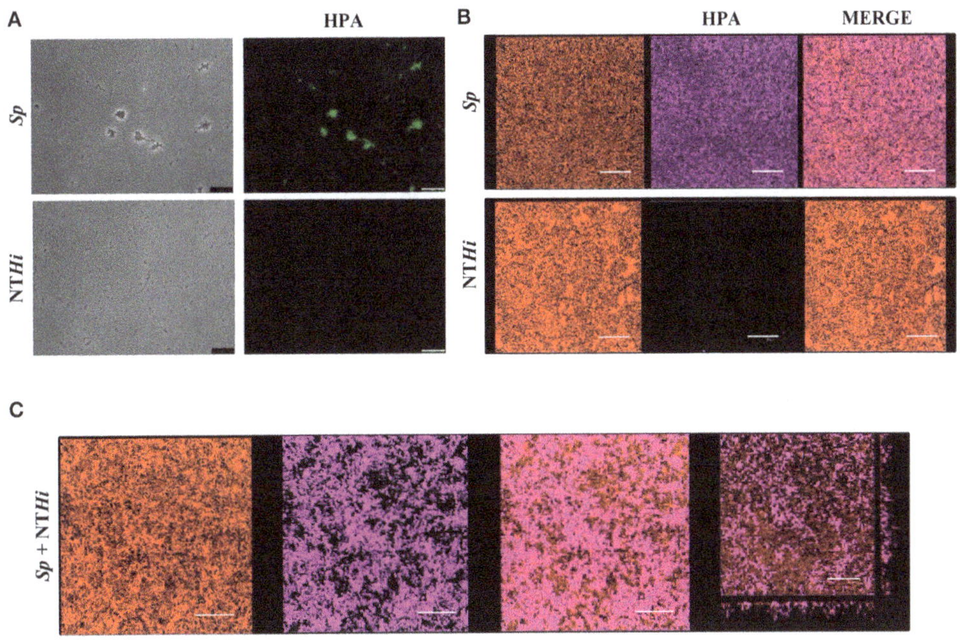

FIGURE 3 | *S. pneumoniae* labeling with HPA in mixed biofilms. **(A)** Fluorescent labeling of the *S. pneumoniae* (*Sp*) strain R6 and the non-typeable *H. influenzae* (NT*Hi*) 54997 with HPA. Biofilms formed by *Sp* R6 or NT*Hi* 54997 **(B)**, or both pathogens **(C)** were stained with a combination of SYTO 59 (red) and HPA conjugated to Alexa Fluor-488 (pink). An orthogonal projection of a CLSM image showing a representative region of the *x*–*y* plane over the depth of the biofilm in both *x*–*z* and *y*–*z* dimensions of the mixed biofilm is also shown at the bottom right part of the figure. Planktonic and biofilms cultures were incubated with HPA (25 μg/ml). Bar = 25 μm.

The presence of the capsule does not appear to hinder HPA binding to the *S. pneumoniae* surface since strains D39 (serotype 2) and P007 (a heavily encapsulated serotype 3 transformant) were efficiently labeled (**Figures 2B,C**). Cells of *S. pneumoniae* D39 were also positive for HPA binding when diluted to 2.5 × 10^6 cfu/ml into defibrinated sheep blood or into fetal bovine serum, although excess fluorescence on the erythrocyte surface may partly hinder a distinct pneumococcal identification (**Figure 4**). As HPA also binds to the human blood group A antigen (Matsui et al., 2001), that interference, which is due to the fact that sheep erythrocytes harbor FA on their surface (see above), should disappear using group non-A

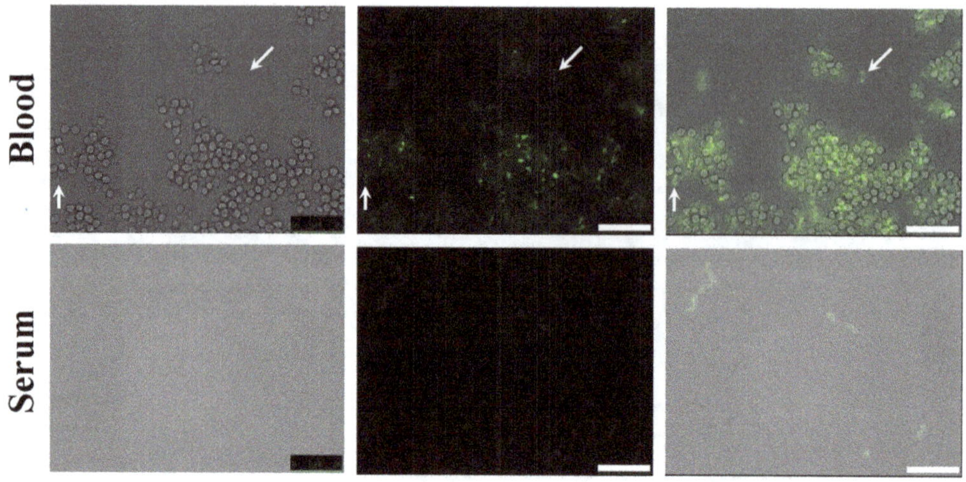

FIGURE 4 | HPA labeling of *S. pneumoniae* in non-human blood and serum. The type 2 encapsulated *S. pneumoniae* strain D39 was diluted to 2.5×10^6 cfu/ml into sheep blood or fetal bovine serum (63× objective). Cultures were incubated with HPA (25 μg/ml). Arrows point to pneumococcal cells. Bar = 25 μm.

(e.g., group O) human whole blood. In addition, it should be mentioned that up to 98 pneumococcal capsular polysaccharides differing in sugar composition and linkages have been described to date (Geno et al., 2017). Besides, invasive pneumococcal disease and pneumonia rates have decreased in most countries following the introduction of conjugate pneumococcal vaccines (PCVs). However, after PCV implementation, current data show that more non-vaccine serotypes increased in frequency than decreased, which is consistent with vaccine-induced replacement. Clinical pneumococcal isolates of six different serotypes—including three emerging serotypes that are currently among the predominant non-PCV13 serotypes worldwide, i.e., serotypes 12F, 22F, and 23B; Balsells et al., 2017)—were diluted to about 2.5×10^6 cfu/ml into whole human blood, incubated with HPA, and observed under the microscope. Moreover, due to the increasing clinical importance of non-encapsulated *S. pneumoniae* (Keller et al., 2016), a representative strain (MNZ67) was also investigated. All the pneumococcal isolates tested bound the lectin and, as expected, HPA-labeled pneumococci were particularly noticeable using group O human blood (**Figure 5**). It is worth mentioning that any possible interference caused by an excess fluorescence on the surface of group A human erythrocytes could be virtually abolished by partly sedimenting the blood cells at low speed (1,000 × g; 1 min; room temperature) before HPA addition (**Figure 5A**, bottom right).

HPA Labeling of Pneumococcal Teichoic Acids

With the possible exception of *S. pseudopneumoniae* (González et al., 2008), the nutritional requirement of pneumococci for the amino alcohol choline (as a component of pnWTA and pnLTA) appears to be an exclusive trait (Rane and Subbarow, 1940). Nevertheless, several choline-independent pneumococcal mutants have been characterized in the last years, and it has been suggested that the absence of choline incorporation might affect the structure of TAs as well as the composition of the cell wall. Actually, it has been shown that pnLTA and pnWTA isolated from these mutants were free of phosphocholine and other phosphorylated aminoalcohols (Yother et al., 1998). However, and as deduced from the *in vivo* cell labeling with HPA of two double *tacF* mutants that form long chains of cells (and are autolysis-defective) when grown in media lacking any amino alcohol (strains JY2190 and P501), the absence of choline residues in TAs does not appear to modify HPA binding by *S. pneumoniae* (**Figure 6**).

It is well known that pneumococci growing in Cden medium containing EA instead of choline form long chains, do not autolyze, and are non-transformable (Tomasz, 1968). In the present work, the EA-grown cells were unable to bind HPA (**Figures 7A,D**). The same was seen when *S. pneumoniae* R6 was incubated in C+Y supplemented with 2% choline chloride (data not shown). It has been reported that upon the addition of choline (5 μg/ml) to EA-grown pneumococci, these cells revert to the normal phenotype (Tomasz et al., 1975). As expected, EA-grown cells became entirely HPA-labeled 180 min (≈4 generations) after shifting to a choline-containing medium (**Figures 7B,E**). As an alternative and complementary model, we examined the *S. mitis* strain SK598, which is unique in that its WTA and LTA contain EA instead of choline even when incubated in a choline-containing medium (Bergström et al., 2003). As observed for EA-grown *S. pneumoniae* cells, *S. mitis* SK598 was unable to bind HPA (**Figures 7C,F**).

DISCUSSION

In a previous work we have shown that, in contrast to *S. pneumoniae*, NT *H. influenzae* strains did not bind HPA; thus fluorescent staining of *S. pneumoniae* with the HPA lectin revealed that pneumococci were evenly distributed throughout the *in vitro* biofilm and interspersed with NT *H. influenzae*

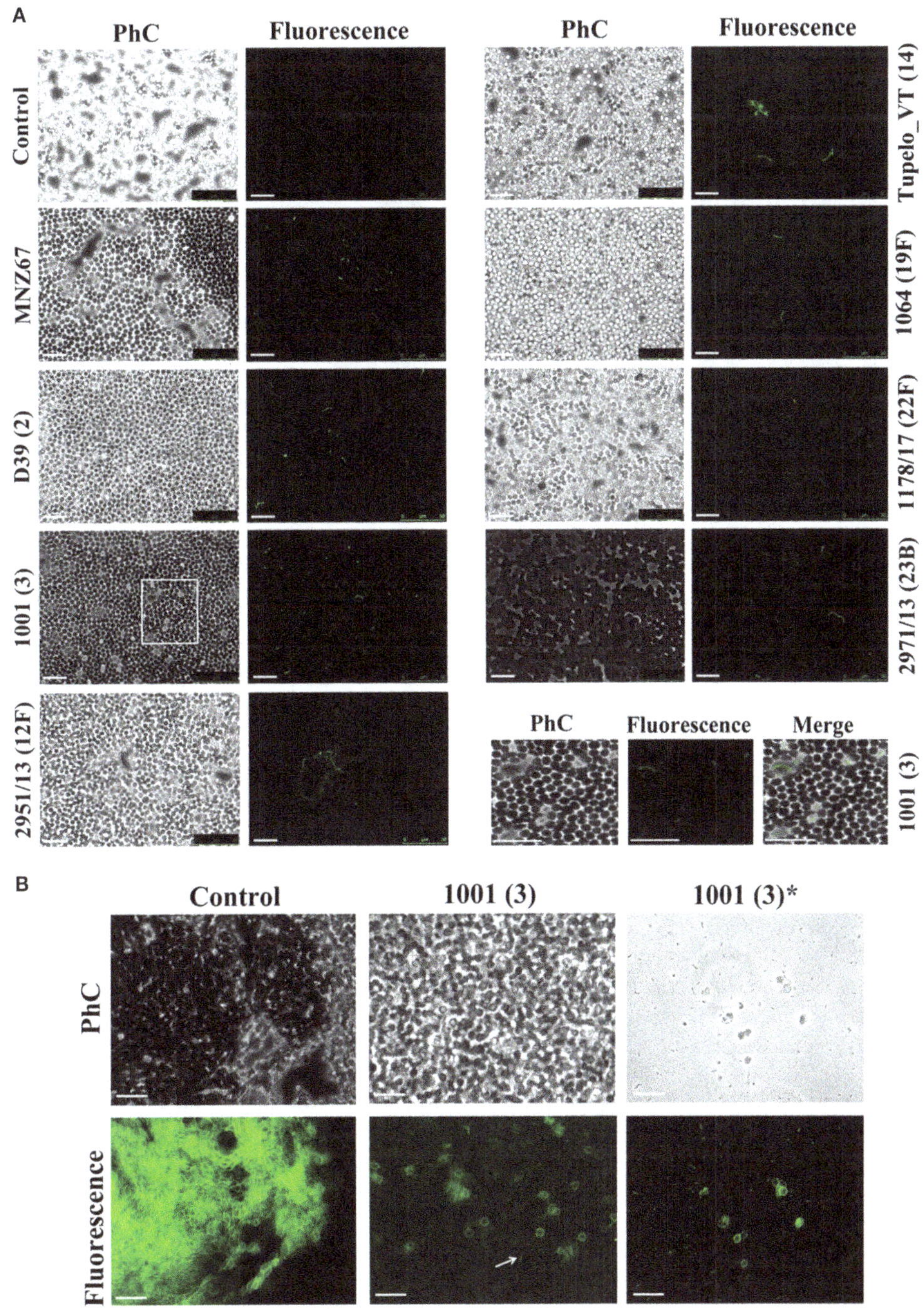

FIGURE 5 | HPA labeling of clinical pneumococcal isolates in human blood. The *S. pneumoniae* strains were diluted to 2.5×10^6 cfu/ml into group O **(A)** and A **(B)** human blood (40× objective). Serotypes are indicated in parentheses. Strain MNZ67 is a non-encapsulated clinical isolate. In **(A)**, the images shown at the bottom right are enlarged visions of the area marked with a rectangle in strain 1001. Control denotes non-infected blood samples. In **(B)**, those labeled with an asterisk correspond to views of the supernatant of infected group A blood that had been centrifuged (1,000 × g; 1 min; room temperature) before HPA labeling. Samples were incubated with HPA (25 μg/ml). The arrow points to a pneumococcal diplococcus. Phc, phase-contrast micrograph; Bar = 25 μm.

(Domenech and García, 2017). We have shown here that HPA recognition of the *S. pneumoniae* surface does not require the presence of choline residues in TAs. This observation is in agreement with the results of Gisch et al. who recently reported that the presence/absence of phosphorylcholine in the FA terminus of pnLTA to have no effect on detection by an anti-Forssman antibody (Gisch et al., 2013). This is relevant since the number of repeating units and the phosphocholine content per repeat (mono-substituted or di-substituted) for pneumococcal TAs slightly vary among strains (Gisch et al., 2015a). It is interesting, however, that, in the present work, EA-grown cells or *S. pneumoniae* R6 incubated in C+Y supplemented with 2% choline chloride did not bind HPA. Incubation in high choline concentrations inhibits cell separation, leaving pneumococci to grow in long chains of cells, just like EA-grown cells (Briese and Hakenbeck, 1984; Giudicelli and Tomasz, 1984). This phenotype is, at least partly, the result of the inhibition of LytB—a member of the family of choline-binding proteins (CBPs)—and the release of this and other CBPs to the medium (López and García, 2004). Both processes would be expected to occur in choline-independent strains growing in the absence of any amino alcohol (see above), although when this was performed in the present work HPA-labeling was unaffected. Interestingly, in an early study, Briles and Tomasz reported the yield of heterophile (sheep hemolytic) antibodies elicited by pneumococci to be at least 10-fold greater in choline- than in EA-containing media (Briles and Tomasz, 1975). Moreover, these authors reported that pneumococci growing in C+Y medium (choline-containing) elicit antibodies which bind poorly to EA-grown bacteria, whereas the latter elicit antibodies which bind well to choline-grown cells. The reasons for these unexpected results are still unclear, although the possibility of the existence of a choline-dependent regulatory pathway for the synthesis of pneumococcal TAs warrants future research.

To our knowledge, the concentration of free choline (or EA) in human lungs has not been reported so far. However, the normal levels of free choline and EA in human fluids are quite similar, e.g., about 1 µM each in serum (Forteschi et al., 2016; Dereziński et al., 2017) and 2 µM for choline or 9–15 µM for EA in cerebrospinal fluid (CSF) (Kruse et al., 1985; Frölich et al., 1998; Ogawa et al., 2015). These data together with the early findings that choline is an effective inhibitor of the cellular incorporation of EA: addition of as little as 0.1 µg/ml choline to a culture growing in the presence of 40 µg/ml EA immediately inhibited further EA incorporation, and that choline was incorporated by such cultures without any detectable lag (Tomasz, 1968), strongly suggest that EA-grown cells (or pneumococci grown in high choline concentrations) are not expected to be found in nature, although they represent important model systems for *in vitro* studies.

HPA labeling allows the recognition of *S. pneumoniae* cells among a variety of other species. However, it is clear that HPA specificity is not restricted to pneumococci (**Table 1**). Interestingly, neither *S. pseudopneumoniae*^{T}, *S. mitis*^{T} nor *S. oralis*^{T} binds the lectin; this is of note since, although these three species are very closely related to *S. pneumoniae*, monoclonal antibodies directed against the backbone and the phosphocholine residues of TAs react only with some strains of these three species (Kilian et al., 2008). The binding of HPA to these bacteria could, therefore, be mostly strain-specific. For example, in contrast to that observed for *S. mitis*^{T}, *S. mitis* SK137 was susceptible to HPA labeling (**Table 1**). This was not unexpected since this particular strain has choline-containing TAs with a carbohydrate backbone identical to that of pnWTA/pnLTA, which forms the Lancefield group O antigen (Bergström et al., 2000). Quite unexpectedly, the SK137 strain only showed an average 67.1% nucleotide similarity to the

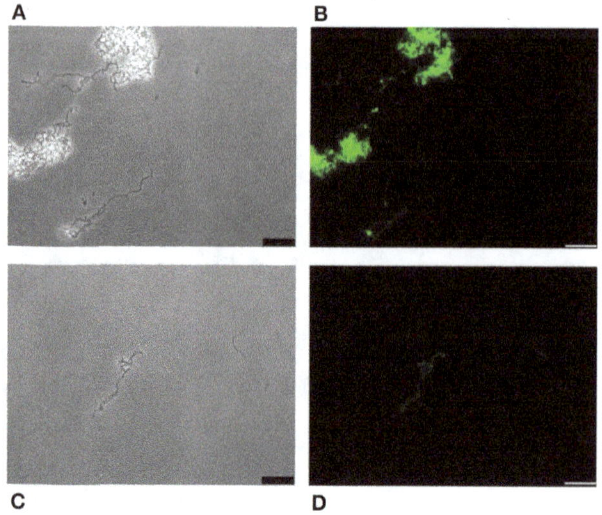

FIGURE 6 | HPA labeling of two choline-independent *S. pneumoniae* strains. Pneumococcal strains JY2190 **(A,B)** and P501 **(C,D)** were incubated in a chemically-defined medium (Cden) lacking any amino alcohol, and labeled with HPA (25 µg/ml). Bar = 25 µm.

S. mitis^{T} in a DNA–DNA hybridization assay (Kilian et al., 2008), slightly below the 70% level typically expected for two strains of the same species (Wayne et al., 1987). Whether strain SK137 represents a distinct species is, however, debatable, according to recent taxonomic proposals (Tindall et al., 2010).

The type strain of *S. oralis* (NCTC 11427)—the LTA structure of which is unknown—did not bind HPA (**Table 1**). However, a recent study has revealed that, in contrast to pnLTA, in which the structural element αGalNAc1→3βGalNAc1→ is present (Gisch et al., 2013), only a βGalNAc1→ moiety is detectable in the *S. oralis* Uo5 LTA repeating unit (Gisch et al., 2015b). Assuming an identical LTA structure for *S. oralis*^{T}, the lack of the αGalNAc1→ residue at the non-reducing end would fit in with the absence of HPA labeling. It should be underlined that at least three biochemical variants of choline-containing TAs may occur in *S. oralis* and some *S. mitis* strains, according to recent results (Denapaite et al., 2016).

Among the few other bacterial species tested, Lancefield group C and D streptococci were positive for HPA labeling. These results were expected in view of previous reports of *Streptococcus* belonging to groups C (Coligan et al., 1977; Sørensen and Henrichsen, 1987; Köhler and Nagai, 1989) and D (Kurl et al., 1989). From a diagnostic perspective, however, it is important to underline that non-pneumococcal, viridans (i.e., α-hemolytic) streptococci are usually considered as commensals and seldom cause CAP. It is well known that viridans streptococci produce a range of invasive disease in humans (e.g., infective endocarditis) and are also emerging as a cause of bloodstream infections, but mainly in immunocompromised patients (Doern and Burnham, 2010). *S. dysgalactiae* subsp. *equisimilis*, which belongs to the β-hemolytic group C and G pyogenic group of streptococci, also binds HPA and is also currently considered as an emergent

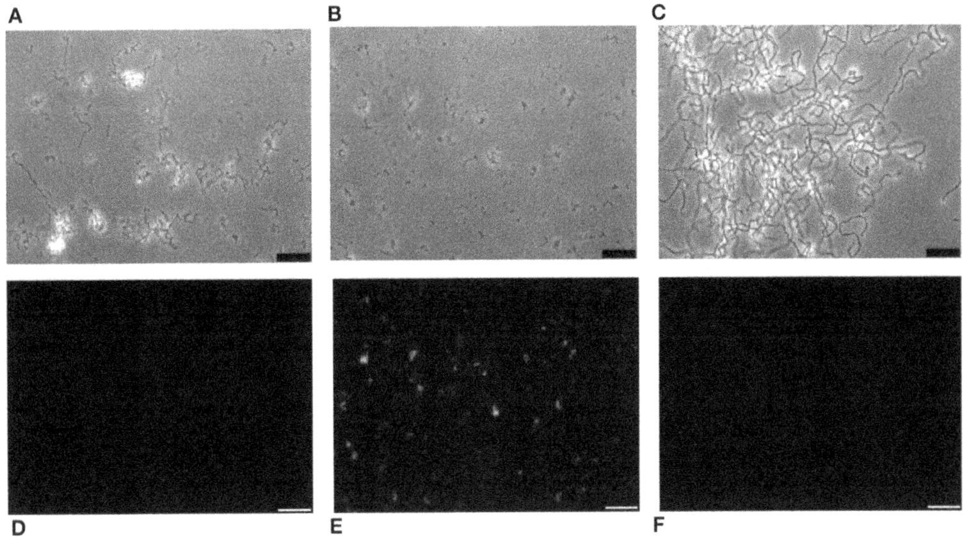

FIGURE 7 | Lack of HPA labeling in ethanolamine-containing streptococci. *S. pneumoniae* R6 cells were incubated for several generations in Cden medium containing ethanolamine (Cden-EA) **(A,D)**. A portion of the culture then received choline chloride (5 μg/ml) and incubation proceeded at 37°C for 3 h **(B,E)**. The *S. mitis* SK598 strain was grown in THY medium **(C,F)**. The three cultures were incubated with HPA (25 μg/ml) **(D–F)**. Bar = 25 μm.

human pathogen. Nevertheless, and as mentioned above for viridans streptococci, it is a frequent cause of invasive disease only in patients having underlying conditions (Broyles et al., 2009).

S. aureus is an important opportunistic pathogen that persistently colonizes about 20% of the human population and is intermittently associated with the remainder. This organism is one of the most frequent and important human pathogens and is implicated in a range of infections, including superficial skin infections, abscesses, and food poisoning as well as life-threatening invasive diseases (Tong et al., 2015). Although, *S. aureus* possesses surface carbohydrates that might be recognized by HPA (Krivan et al., 1988; Payne et al., 1992), the molecular basis for HPA binding is not well understood. It has been shown, however, that certain methicillin-susceptible *S. aureus* isolates—in particular those belonging to the atypical sequence type (ST) 395 lineage (e.g., strain PS187; Winstel et al., 2013)—produce a glycerol TA modified with αGalNAc (Winstel et al., 2014; Lee et al., 2015) that may be responsible for HPA binding. It should be noted, however, that the chemical structures of WTA and LTA of *S. aureus* are similar but not identical (Xia et al., 2010; Brown et al., 2013), that the ST of the type strain of *S. aureus* (ST8) is only distantly related to ST395, and that strain PS187 appears to be more closely related to several coagulase-negative staphylococcal species than to other *S aureus* isolates (Winstel et al., 2013); the ST of *S. aureus* 15981 is unknown. It has been demonstrated that staphylococci can be accurately differentiated from streptococci in Gram-stained preparations (cocci in clusters and diplococci or short chains, respectively; Agger and Maki, 1978), and that *S. aureus* is more uncommon than the latter in CAP and acute meningitis cases. However, our results warrant additional studies to determine whether HPA labeling is common among *S. aureus* isolates, particularly in methicillin-resistant isolates that represent a global health care problem (Tong et al., 2015).

In agreement with a previous report (Domenech and García, 2017), NT *H. influenzae* strains did not bind HPA; identical results are shown here for *P. aeruginosa* PAO1. Together with *S. pneumoniae*, both Gram-negative species are common in biofilms formed during acute otitis media and live-threatening, chronic respiratory diseases such as chronic obstructive pulmonary disease or cystic fibrosis (Blasi et al., 2016). The use of HPA as *S. pneumoniae*-specific lectin for fluorescence imaging should provide a powerful tool for future research on these and other relevant human pathogens forming multispecies biofilms.

HPA labeling combined with Gram staining and/or antigen detection may also constitute an appropriate combination for the rapid diagnosis of CAP and, perhaps, other conditions such as bacterial meningitis (McGill et al., 2016). This may facilitate the rapid implementation of an appropriate antibiotic regime, which is conditional on the age of the patient and the regional rate of decreased susceptibility of *S. pneumoniae* to β-lactam antibiotics (van de Beek et al., 2016). As a proof of concept, cultures of pneumococci, *H. influenzae* and GBS were mixed either with sheep blood or fetal bovine serum. Also using human whole blood, only pneumococci (either encapsulated or non-encapsulated) were clearly identified by HPA labeling when used at a relatively low bacterial concentration (2.5×10^6 cfu/ml). This concentration is lower than that frequently found in the blood of CAP patients (between 7×10^7 and 8×10^8 cfu/ml; Gadsby et al., 2016) or in the CSF of children with confirmed pneumococcal meningitis (median bacterial load $\approx 5 \times 10^7$ DNA copies/ml) at the time of admittance (Roine et al., 2009). If required, bacteria could be concentrated by centrifugation before or after staining.

There are a number of limitations to our study: (1) although the HPA labeling technique does not require a very specialized personnel, fluorescence microscopy may be unavailable in many laboratories of developing countries. To circumvent this problem, biotinilated HPA, and alkaline phosphatase- or ferritin-conjugated HPA could be employed. (2) Only seven different bacterial genera—including 15 streptococcal and 3 staphylococcal strains—, were tested. It should be noted, however, that the bacteria tested here include some of the microorganisms most frequently causing CAP and other severe diseases. (3) Although pneumococci and staphylococci are morphologically different, the finding that two *S. aureus* strains also bind HPA may sometimes represent a diagnostic drawback and deserves further research. It should be noted, however, that, for example, severe sepsis—a serious complication of CAP—is caused by *S. pneumoniae* about 100 times more frequently than by *S. aureus*, whereas methicillin-resistant *S. aureus* is an important cause of antimicrobial-resistant hospital-acquired infections worldwide and remains a public health priority in Europe (Montull et al., 2016). (4) This represents an *in vitro* study and an appropriate evaluation of the benefits of HPA labeling for diagnostic purposes should be performed directly with clinical samples, e.g., sputum, brochoalveolar fluid, blood, and/or CSF.

AUTHOR CONTRIBUTIONS

MD and EG conceived and designed the experiments. MD performed the experiments. MD and EG analyzed the data and wrote the paper.

ACKNOWLEDGMENTS

The authors thank M. H. Nam (University of Alabama at Birmingham, USA) for kindly providing strain MNZ67, M. Moscoso, J. Yuste, and P. García for helpful comments and for critically reading the manuscript, A. Burton for correcting the English version, and E. Cano and S. Ruiz for skillful technical assistance. This work was supported by a grant from the Ministerio de Economía y Competitividad (MINECO) (SAF2012-39444-C02-01). CIBER de Enfermedades Respiratorias (CIBERES) is an initiative of the Instituto de Salud Carlos III.

REFERENCES

Agger, W. A., and Maki, D. G. (1978). Efficacy of direct Gram stain in differentiating staphylococci from streptococci in blood cultures positive for gram-positive cocci. *J. Clin. Microbiol.* 7, 111–113.

Ahmed, S., Shapiro, N. L., and Bhattacharyya, N. (2014). Incremental health care utilization and costs for acute otitis media in children. *Laryngoscope* 124, 301–305. doi: 10.1002/lary.24190

Amann, R., and Fuchs, B. M. (2008). Single-cell identification in microbial communities by improved fluorescence *in situ* hybridization techniques. *Nat. Rev. Micro.* 6, 339–348. doi: 10.1038/nrmicro1888

Balsells, E., Guillot, L., Nair, H., and Kyaw, M. H. (2017). Serotype distribution of *Streptococcus pneumoniae* causing invasive disease in children in the post-PCV era: a systematic review and meta-analysis. *PLoS ONE* 12:e0177113. doi: 10.1371/journal.pone.0177113

Bergström, N., Jansson, P. -E., Kilian, M., and Sørensen, U. B. S. (2003). A unique variant of streptococcal group O-antigen (C-polysaccharide) that lacks phosphocholine. *Eur. J. Biochem.* 270, 2157–2162. doi: 10.1046/j.1432-1033.2003.03569.x

Bergström, N., Jansson, P. E., Kilian, M., and Sørensen, U. B. S. (2000). Structures of two cell wall-associated polysaccharides of a *Streptococcus mitis* biovar 1 strain. A unique teichoic acid-like polysaccharide and the group O antigen which is a C-polysaccharide in common with pneumococci. *Eur. J. Biochem.* 267, 7147–7157. doi: 10.1046/j.1432-1327.2000.01821.x-i2

Blasi, F., Page, C., Rossolini, G. M., Pallecchi, L., Matera, M. G., Rogliani, P., et al. (2016). The effect of N-acetylcysteine on biofilms: implications for the treatment of respiratory tract infections. *Respir. Med.* 117, 190–197. doi: 10.1016/j.rmed.2016.06.015

Briese, T., and Hakenbeck, R. (1984). "Influence of lipoteichoic acid and choline on the autolytic enzyme activity of Streptococcus pneumoniae," in *Microbial Cell Wall Synthesis and Autolysis*, ed C. Nombela (Amsterdam: Elsevier Science Publishers), 201–206.

Briles, E. B., and Tomasz, A. (1973). Pneumococcal Forssman antigen. A choline-containing LTA. *J. Biol. Chem.* 248, 6394–6397.

Briles, E. B., and Tomasz, A. (1975). Physiological studies on the pneumococcal Forssman antigen: a choline-containing lipoteichoic acid. *J. Gen. Microbiol.* 86, 267–274. doi: 10.1099/00221287-86-2-267

Brown, S., Santa Maria, J. P. Jr., and Walker, S. (2013). Wall teichoic acids of gram-positive bacteria. *Annu. Rev. Microbiol.* 67, 313–336. doi: 10.1146/annurev-micro-092412-155620

Broyles, L. N., Van Beneden, C., Beall, B., Facklam, R., Shewmaker, P. L., Malpiedi, P., et al. (2009). Population-based study of invasive disease due to β-hemolytic streptococci of groups other than A and B. *Clin. Infect. Dis.* 48, 706–712. doi: 10.1086/597035

Chochua, S., D'Acremont, V., Hanke, C., Alfa, D., Shak, J., Kilowoko, M., et al. (2016). Increased nasopharyngeal density and concurrent carriage of *Streptococcus pneumoniae*, *Haemophilus influenzae*, and *Moraxella catarrhalis* are associated with pneumonia in febrile children. *PLoS ONE* 11:e0167725. doi: 10.1371/journal.pone.0167725

Cilloniz, C., Martin-Loeches, I., Garcia-Vidal, C., San Jose, A., and Torres, A. (2016). Microbial etiology of pneumonia: epidemiology, diagnosis and resistance patterns. *Int. J. Mol. Sci.* 17:2120. doi: 10.3390/ijms17122120

Clark, J. E. (2015). Determining the microbiological cause of a chest infection. *Arch. Dis. Child.* 100, 193–197. doi: 10.1136/archdischild-2013-305742

Coligan, J. E., Fraser, B. A., and Kindt, T. J. (1977). A disaccharide hapten from streptococcal group C carbohydrate that cross-reacts with the Forssman glycolipid. *J. Immunol.* 118, 6–11.

Cooling, L. (2015). Blood groups in infection and host susceptibility. *Clin. Microbiol. Rev.* 28, 801–870. doi: 10.1128/CMR.00109-14

Dekker, J. P., and Lau, A. F. (2016). An update on the *Streptococcus bovis* group: classification, identification, and disease associations. *J. Clin. Microbiol.* 54, 1694–1699. doi: 10.1128/JCM.02977-15

Denapaite, D., Rieger, M., Köndgen, S., Brückner, R., Ochigava, I., Kappeler, P., et al. (2016). Highly variable *Streptococcus oralis* strains are common among viridans streptococci isolated from primates. *mSphere* 1:e00041-15. doi: 10.1128/mSphere.00041-15

Dereziński, P., Klupczynska, A., Sawicki, W., Pałka, J. A., and Kokot, Z. J. (2017). Amino acid profiles of serum and urine in search for prostate cancer biomarkers: a pilot study. *Int. J. Med. Sci.* 14, 1–12. doi: 10.7150/ijms.15783

Doern, C. D., and Burnham, C.-A. D. (2010). It's not easy being green: the viridans group streptococci, with a focus on pediatric clinical manifestations. *J. Clin. Microbiol.* 48, 3829–3835. doi: 10.1128/JCM.01563-10

Domenech, M., and García, E. (2017). *N*-Acetyl-L-cysteine and cysteamine: new strategies against mixed biofilms of non-encapsulated *Streptococcus pneumoniae* and non-typeable *Haemophilus influenzae*. *Antimicrob. Agents Chemother.* 61:e01992-16. doi: 10.1128/AAC.01992-16

Domenech, M., Damián, D., Ardanuy, C., Liñares, J., Fenoll, A., and García, E. (2015). Emerging, non-PCV13 serotypes 11A and 35B of *Streptococcus pneumoniae* show high potential for biofilm formation *in vitro*. *PLoS ONE* 10:e0125636. doi: 10.1371/journal.pone.0125636

Domenech, M., García, E., and Moscoso, M. (2009). Versatility of the capsular genes during biofilm formation by *Streptococcus pneumoniae*. *Environ. Microbiol.* 11, 2542–2555. doi: 10.1111/j.1462-2920.2009.01979.x

Domenech, M., García, E., and Moscoso, M. (2011). *In vitro* destruction of *Streptococcus pneumoniae* biofilms with bacterial and phage peptidoglycan hydrolases. *Antimicrob. Agents Chemother.* 55, 4144–4148. doi: 10.1128/AAC.00492-11

Ewig, S., Birkner, N., Strauss, R., Schaefer, E., Pauletzki, J., Bischoff, H., et al. (2009). New perspectives on community-acquired pneumonia in 388 406 patients. Results from a nationwide mandatory performance measurement programme in healthcare quality. *Thorax* 64, 1062–1069. doi: 10.1136/thx.2008.109785

Forteschi, M., Zinellu, A., Assaretti, S., Mangoni, A. A., Pintus, G., Carru, C., et al. (2016). An isotope dilution capillary electrophoresis/tandem mass spectrometry (CE-MS/MS) method for the simultaneous measurement of choline, betaine, and dimethylglycine concentrations in human plasma. *Anal. Bioanal. Chem.* 408, 7505–7512. doi: 10.1007/s00216-016-9848-6

Frölich, L., Dirr, A., Götz, M. E., Gsell, W., Reichmann, H., Riederer, P., et al. (1998). Acetylcholine in human CSF: methodological considerations and levels in dementia of Alzheimer type. *J. Neural Transm.* 105, 961–973. doi: 10.1007/s007020050105

Gadsby, N. J., Russell, C. D., McHugh, M. P., Mark, H., Conway Morris, A., Laurenson, I. F., et al. (2016). Comprehensive molecular testing for respiratory pathogens in community-acquired pneumonia. *Clin. Infect. Dis.* 62, 817–823. doi: 10.1093/cid/civ1214

García-Cobos, S., Moscoso, M., Pumarola, F., Arroyo, M., Lara, N., Pérez-Vázquez, M., et al. (2014). Frequent carriage of resistance mechanisms to β-lactams and biofilm formation in *Haemophilus influenzae* causing treatment failure and recurrent otitis media in young children. *J. Antimicrob. Chemother.* 69, 2394–2399. doi: 10.1093/jac/dku158

Geno, K. A., Saad, J. S., and Nahm, M. H. (2017). Discovery of novel pneumococcal serotype 35D, a natural WciG-deficient variant of serotype 35B. *J. Clin. Microbiol.* 55, 1416–1425. doi: 10.1128/JCM.00054-17

Gisch, N., Kohler, T., Ulmer, A. J., Müthing, J., Pribyl, T., Fischer, K., et al. (2013). Structural reevaluation of *Streptococcus pneumoniae* lipoteichoic acid and new insights into its immunostimulatory potency. *J. Biol. Chem.* 288, 15654–15667. doi: 10.1074/jbc.M112.446963

Gisch, N., Peters, K., Zähringer, U., and Vollmer, W. (2015a). "The pneumococcal cell wall," in *Streptococcus pneumoniae: Molecular Mechanisms of Host-Pathogen Interactions*, eds J. Brown, S. Hammerschimdt, and C. Orihuela (San Diego, CA: Elsevier, Inc.), 145–167.

Gisch, N., Schwudke, D., Thomsen, S., Heb, N., Hakenbeck, R., and Denapaite, D. (2015b). Lipoteichoic acid of *Streptococcus oralis* Uo5: a novel biochemical structure comprising an unusual phosphorylcholine substitution pattern compared to *Streptococcus pneumoniae*. *Sci. Rep.* 5:16718. doi: 10.1038/srep16718

Giudicelli, S., and Tomasz, A. (1984). "Inhibition of the *in vitro* and *in vivo* activity of the pneumococcal autolytic enzyme by choline and phosphorylcholine," in *Microbial Cell Wall Synthesis and Autolysis*, ed C. Nombela (Amsterdam: Elsevier Science Publishers), 207–212.

González, A., Llull, D., Morales, M., García, P., and García, E. (2008). Mutations in the *tacF* gene of clinical strains and laboratory transformants of *Streptococcus pneumoniae*: impact on choline auxotrophy and growth rate. *J. Bacteriol.* 190, 4129–4138. doi: 10.1128/JB.01991-07

Hajjeh, R., Mulholland, K., Schuchat, A., and Santosham, M. (2013). Progress towards demonstrating the impact of *Haemophilus influenzae* type b conjugate vaccines globally. *J. Pediatr.* 163, S1–S3. doi: 10.1016/j.jpeds.2013.03.022

Haq, I. J., Battersby, A. C., Eastham, K., and McKean, M. (2017). Community acquired pneumonia in children. *BMJ* 356:j686. doi: 10.1136/bmj.j686

Hoskins, J., Alborn, W. E. Jr., Arnold, J., Blaszczak, L. C., Burgett, S., DeHoff, B. S., et al. (2001). Genome of the bacterium *Streptococcus pneumoniae* strain R6. *J. Bacteriol.* 183, 5709–5717. doi: 10.1128/JB.183.19.5709-5717.2001

Jenkin, C. R. (1963). Heterophile antigens and their significance in the host–parasite relationship. *Adv. Immunol.* 3, 351–376. doi: 10.1016/S0065-2776(08)60816-0

Keller, L. E., Robinson, D. A., and McDaniel, L. S. (2016). Nonencapsulated *Streptococcus pneumoniae*: emergence and pathogenesis. *mBio* 7:e01792-15. doi: 10.1128/mBio.01792-15

Kilian, M., Poulsen, K., Blomqvist, T., Håvarstein, L. S., Bek-Thomsen, M., Tettelin, H., et al. (2008). Evolution of *Streptococcus pneumoniae* and its close commensal relatives. *PLoS ONE* 3:e2683. doi: 10.1371/journal.pone.0002683

Kilpper-Bälz, R., Wenzig, P., and Schleifer, K. H. (1985). Molecular relationships and classification of some viridans streptococci as *Streptococcus oralis* and emended description of *Streptococcus oralis* (Bridge and Sneath 1982). *Int. J. Syst. Bacteriol.* 35, 482–488. doi: 10.1099/00207713-35-4-482

Kjos, M., Aprianto, R., Fernandes, V. E., Andrew, P. W., van Strijp, J. A. G., Nijland, R., et al. (2015). Bright fluorescent *Streptococcus pneumoniae* for live-cell imaging of host-pathogen interactions. *J. Bacteriol.* 197, 807–818. doi: 10.1128/JB.02221-14

Kobayashi, Y., Tateno, H., Ogawa, H., Yamamoto, K., and Hirabayashi, J. (2014). Comprehensive list of lectins: origins, natures, and carbohydrate specificities. *Methods Mol. Biol.* 1200, 555–577. doi: 10.1007/978-1-4939-1292-6_45

Köhler, W., and Nagai, T. (1989). Reactions of the lectin anti-A_{HP} from the edible snail *Helix pomatia* with *N*-acetyl-D-galactosamine of streptococci. *Kitasato Arch. Exp. Med.* 62, 107–113.

Krivan, H. C., Roberts, D. D., and Ginsburg, V. (1988). Many pulmonary pathogenic bacteria bind specifically to the carbohydrate sequence GalNAcβ1-4Gal found in some glycolipids. *Proc. Natl. Acad. Sci. U.S.A.* 85, 6157–6161. doi: 10.1073/pnas.85.16.6157

Kruse, T., Reiber, H., and Neuhoff, V. (1985). Amino acid transport across the human blood-CSF barrier. An evaluation graph for amino acid concentrations in cerebrospinal fluid. *J. Neurol. Sci.* 70, 129–138. doi: 10.1016/0022-510X(85)90082-6

Kurl, D. N., Haataja, S., and Finne, J. (1989). Hemagglutination activities of group B, C, D, and G streptococci: demonstration of novel sugar-specific cell-binding activities in *Streptococcus suis*. *Infect. Immun.* 57, 384–389.

Lacks, S., and Hotchkiss, R. D. (1960). A study of the genetic material determining an enzyme activity in Pneumococcus. *Biochim. Biophys. Acta* 39, 508–518. doi: 10.1016/0006-3002(60)90205-5

Lanie, J. A., Ng, W. -L., Kazmierczak, K. M., Andrzejewski, T. M., Davidsen, T. M., Wayne, K. J., et al. (2007). Genome sequence of Avery's virulent serotype 2 strain D39 of *Streptococcus pneumoniae* and comparison with that of unencapsulated laboratory strain R6. *J. Bacteriol.* 189, 38–51. doi: 10.1128/JB.01148-06

Lee, J. -H., Kim, N. -H., Winstel, V., Kurokawa, K., Larsen, J., An, J. -H., et al. (2015). Surface glycopolymers are crucial for *in vitro* anti-wall teichoic acid IgG-mediated complement activation and opsonophagocytosis of *Staphylococcus aureus*. *Infect. Immun.* 83, 4247–4255. doi: 10.1128/IAI.00767-15

Liu, W., Røder, H. L., Madsen, J. S., Bjarnsholt, T., Sørensen, S. J., and Burmølle, M. (2016). Interspecific bacterial interactions are reflected in multispecies biofilm spatial organization. *Front. Microbiol.* 7:1366. doi: 10.3389/fmicb.2016.01366

López, R., and García, E. (2004). Recent trends on the molecular biology of pneumococcal capsules, lytic enzymes, and bacteriophage. *FEMS Microbiol. Rev.* 28, 553–580. doi: 10.1016/j.femsre.2004.05.002

Matsui, T., Hamako, J., Ozeki, Y., and Titani, K. (2001). Comparative study of blood group-recognizing lectins toward ABO blood group antigens on neoglycoproteins, glycoproteins and complex-type oligosaccharides. *Biochim. Biophys. Acta* 1525, 50–57. doi: 10.1016/S0304-4165(00)00170-7

McGill, F., Heyderman, R. S., Panagiotou, S., Tunkel, A. R., and Solomon, T. (2016). Acute bacterial meningitis in adults. *Lancet* 388, 3036–3047. doi: 10.1016/S0140-6736(16)30654-7

Momeni, B., Brileya, K. A., Fields, M. W., and Shou, W. (2013). Strong inter-population cooperation leads to partner intermixing in microbial communities. *eLife* 2:e00230. doi: 10.7554/eLife.00230

Monasta, L., Ronfani, L., Marchetti, F., Montico, M., Vecchi Brumatti, L., Bavcar, A., et al. (2012). Burden of disease caused by otitis media: systematic review and global estimates. *PLoS ONE* 7:e36226. doi: 10.1371/journal.pone.0036226

16

DNA Delivery and Genomic Integration into Mammalian Target Cells through Type IV A and B Secretion Systems of Human Pathogens

Dolores L. Guzmán-Herrador[1], Samuel Steiner[2], Anabel Alperi[1], Coral González-Prieto[1†], Craig R. Roy[2] and Matxalen Llosa[1]*

[1] Departamento de Biología Molecular, Universidad de Cantabria (UC), Instituto de Biomedicina y Biotecnología de Cantabria (IBBTEC, UC-CSIC-SODERCAN), Santander, Spain, [2] Department of Microbial Pathogenesis, Boyer Center for Molecular Medicine, Yale University School of Medicine, New Haven, CT, United States

Edited by:
Manuel Espinosa,
Centro de Investigaciones Biológicas (CSIC), Spain

Reviewed by:
Elisabeth Grohmann,
Beuth University of Applied Sciences, Germany
Jose Angel Ruiz-Masó,
Centro de Investigaciones Biológicas (CSIC), Spain

***Correspondence:**
Matxalen Llosa
llosam@unican.es

We explore the potential of bacterial secretion systems as tools for genomic modification of human cells. We previously showed that foreign DNA can be introduced into human cells through the Type IV A secretion system of the human pathogen *Bartonella henselae*. Moreover, the DNA is delivered covalently attached to the conjugative relaxase TrwC, which promotes its integration into the recipient genome. In this work, we report that this tool can be adapted to other target cells by using different relaxases and secretion systems. The promiscuous relaxase MobA from plasmid RSF1010 can be used to deliver DNA into human cells with higher efficiency than TrwC. MobA also promotes DNA integration, albeit at lower rates than TrwC. Notably, we report that DNA transfer to human cells can also take place through the Type IV secretion system of two intracellular human pathogens, *Legionella pneumophila* and *Coxiella burnetii*, which code for a distantly related Dot/Icm Type IV B secretion system. This suggests that DNA transfer could be an intrinsic ability of this family of secretion systems, expanding the range of target human cells. Further analysis of the DNA transfer process showed that recruitment of MobA by Dot/Icm was dependent on the IcmSW chaperone, which may explain the higher DNA transfer rates obtained. Finally, we observed that the presence of MobA negatively affected the intracellular replication of *C. burnetii*, suggesting an interference with Dot/Icm translocation of virulence factors.

Keywords: protein secretion, bacterial conjugation, *Legionella pneumophila*, *Coxiella burnetii*, *Bartonella henselae*, conjugative relaxase, intracellular pathogen, gene therapy

INTRODUCTION

Bacterial Type IV secretion systems (T4SS) selectively deliver macromolecules to other cells or to the extracellular media. An outstanding feature of these secretion systems is their ability to secrete both, protein and DNA molecules, a particularity that distinguishes them from other types of secretion systems. In addition, the secreted substrates can be delivered to either prokaryotic or eukaryotic cells. This plasticity allows T4SS to be involved in bacterial processes as diverse as horizontal DNA transfer or virulence (Christie, 2016).

Bacterial Type IV secretion systems are multiprotein complexes formed by different constitutive elements: a core complex spanning both bacterial membranes, which forms the transport conduit; a pilus-like appendage, whose function as a transport channel is still under debate; a series of cytoplasmic ATPases, which energize the transport process; and elements necessary to recruit and present the substrates to the translocation machine, including chaperones that are variable for each system (Zechner et al., 2012). Within the family of T4SS, two sub-families were described based on sequence homologies: The Type IV A-IV B secretion systems (T4ASS and T4BSS, respectively). The formers are homologous to the prototypical VirB T4SS of *Agrobacterium tumefaciens* and have been characterized extensively, both functionally and structurally (Chandran Darbari and Waksman, 2015). Members of this family form part of conjugative systems of plasmids such as R388 or RP4; others are encoded in the genomes of human pathogens such as *Bartonella henselae (Bh), Brucella melitensis* or *Helicobacter pylori* among others, and their main role is to inject virulence factors to the target human cell. Similarly, T4BSS members are encoded in conjugative plasmids such as F, and in the chromosomes of human pathogens such as *Legionella pneumophila (Lp)* and *Coxiella burnetii (Cb).* Research on T4BSS structure and function lags behind T4ASS; however, extensive work has been done regarding the role of T4BSS-delivered effectors within human cell (Hubber and Roy, 2010; Rolando and Buchrieser, 2014; Personnic et al., 2016).

As aforementioned, a distinctive feature of T4SS is their ability to secrete DNA molecules. This is the main molecular function of T4SS belonging to the conjugative machinery of self-transmissible plasmids (Cabezon et al., 2015). In order to secrete DNA, at least two components are essential in addition to the T4SS machinery: an origin of transfer (*oriT*), which is the DNA sequence required in *cis* on a DNA molecule to be transferred, and a conjugative relaxase, which cuts the DNA strand to be transferred at the *oriT*. Many plasmids also encode for accessory nicking proteins, which assist the DNA processing by the relaxase. The DNA is transferred as a single strand covalently attached to the relaxase, which itself is the substrate of the T4SS; the nucleoprotein complex enters the recipient cell, where the relaxase catalyzes the recircularization of the transferred DNA strand (Garcillan-Barcia et al., 2007; Gonzalez-Perez et al., 2007).

Notably, some conjugative relaxases have the ability to catalyze site-specific recombination between two copies of *oriT*. This phenomenon was first described for the R388 relaxase TrwC (Llosa et al., 1994). TrwC acts as a site-specific recombinase on supercoiled substrates containing minimal target sequences (Cesar et al., 2006). This ability is shared by some, but not all, conjugative relaxases, and it is unclear why. MobA, the relaxase of the mobilizable plasmid RSF1010 (virtually identical to plasmid R1162), is able to catalyze *oriT*-*oriT* recombination on single-stranded substrates but not on supercoiled plasmid substrates (Meyer, 1989). TrwC can also catalyze the integration of the transferred DNA molecule into a target sequence present in the recipient bacterium (Draper et al., 2005); moreover, the protein can catalyze integration into DNA sequences present in the human genome that resemble its natural target, the *oriT* (Agundez et al., 2012), opening the possibility that this relaxase could work as a site-specific integrase in human cells (Gonzalez-Prieto et al., 2013). Recently, we have shown that the relaxase TrwC is active in a human cell after delivery by the T4SS of *Bartonella henselae*, where it can promote the integration of foreign DNA into the human genome, although without site-specificity (Gonzalez-Prieto et al., 2017). The integration rate of the foreign DNA introduced by TrwC was about 100 times higher compared to when it was introduced by the Mob relaxase from *Bartonella* cryptic plasmid pRGB1, or by transfection.

Gene therapy strategies combine methods to introduce DNA into specific human cell types and to promote DNA integration in the human genome for stable expression. Bacteria have previously been used as vectors for DNA delivery into mammalian cells; the process, known as bactofection, is based on the engulfment of bacteria by an eukaryotic cell, which causes bacterial lysis and DNA release (Celec and Gardlik, 2017). We have previously shown that DNA of any origin and length can be introduced into specific human cell types using *B. henselae* as a delivery agent (Fernandez-Gonzalez et al., 2011). In contrast to bactofection, in this case the DNA is secreted by the living bacterium. *B. henselae* encodes a T4ASS named VirB/D4, which translocates effector proteins to the infected human cell, contributing to its virulence (Saenz et al., 2007). We showed that the VirB/D4 T4SS is also capable of translocating relaxase-DNA complexes via a process resembling bacterial conjugation. DNA transfer was dependent on the conjugative elements required to process the DNA in the donor bacterium, which in this case were derived from the conjugative plasmid R388. No DNA transfer occurred in the absence of the relaxase TrwC, and it was severely impaired in the absence of the conjugative coupling protein TrwB. In a parallel work, Schroder et al. (2011) similarly showed DNA transfer through the *B. henselae* VirB/D4 using the Mob relaxase of a natural plasmid of *Bartonella*; in this case, it was necessary to fuse the known T4 recruiting signal (the BID domain) to the relaxase in order to attain efficient DNA transfer. This discovery had interesting biological implications, opening the possibility that pathogens naturally send DNA to their host cell, and potential biotechnological applications, constituting a new way of DNA delivery to specific human cells (Llosa et al., 2012).

In this work, we asked whether this DNA delivery system could be extended to T4SS from other human pathogens targeting different cell types. We infect cultured mammalian cell lines with *B. henselae, L. pneumophila*, or *C. burnetii*, all containing mobilizable plasmids with markers for eukaryotic selection and encoding different conjugative relaxases. We report that DNA can be delivered to human cells through the T4BSS of *L. pneumophila* and *C. burnetii*, which belong to a distant family of T4SS. This suggests that DNA transfer may be an intrinsic feature of T4SS. DNA transfer and integration rates depend on the relaxase used. All these elements could add to the development of useful tools for *in vivo* genetic modification of human cells. In addition, DNA is a trackable substrate which could be used to study the T4 secretion process in the mammalian host.

MATERIALS AND METHODS

Bacterial Strains and Growth Conditions

Bacterial strains used in this work are listed in **Table 1**. *Escherichia coli* (*Ec*) strains DH5α and D1210 were used for DNA manipulations. *B. henselae* strain RSE247, *L. pneumophila* serogroup 1 strain Lp01 (*hsdR, rpsL*; Berger and Isberg, 1993), and *C. burnetii* strain RSA439 Nine Mile phase II (NMII), or derivatives from these strains as indicated, were used for infection of cultured cells.

Escherichia coli strains were grown at 37°C in Luria-Bertani broth, supplemented with agar for growth on plates. *B. henselae* was grown on Columbia blood agar (CBA) plates at 37°C under a 5% CO_2 atmosphere. *L. pneumophila* strains were grown on charcoal yeast extract (CYE) plates [1% yeast extract, 1% *N*-(2-acetamido)-2-aminoethanesulfonic acid (ACES; pH 6.9), 3.3 mM L-cysteine, 0.33 mM $Fe(NO_3)_3$, 1.5% Bacto agar, 0.2% activated charcoal] at 37°C, supplemented with 100 μg/ml thymidine if required. *C. burnetii* was grown axenically in liquid acidified citrate cysteine medium 2 (ACCM-2) for 6 days or on ACCM-2 agarose for >8 days at 37°C, 5% CO_2, and 2.5% O_2 as previously described (Omsland et al., 2011).

For plasmid selection, antibiotics were added at the following final concentrations: ampicillin (Ap), 100 μg/ml; kanamycin monosulfate (Km), 20 μg/ml (*L. pneumophila*), 50 μg/ml (*E. coli, B. henselae*) or 375 μg/ml (*C. burnetii*); streptomycin (Sm), 300 μg/ml (*E. coli*) or 100 μg/ml (*B. henselae, L. pneumophila*); gentamicin sulfate (Gm), 10 μg/ml (*E. coli, B. henselae*) or 5 μg/ml (*L. pneumophila*); chloramphenicol (Cm), 25 μg/ml (*E. coli*) or 3 μg/ml (*C. burnetii*).

Plasmids and Plasmid Constructions

Bacterial plasmids are listed in **Table 2**. Oligonucleotides used for plasmid constructions are listed in **Table 3**. Plasmids pAA58, pLG03, pLG04, pMTX808, pMTX821, and pMTX822 were constructed by the isothermal assembly method (Gibson et al., 2009) using the HiFi assembly cloning kit (New England Biolabs). Plasmids pLG05 and pLG06 were constructed by standard restriction cloning techniques (Sambrook and Russell, 2001).

pAA58 was generated by assembling the eGFP eukaryotic expression cassette from pHP161 into the PstI sites of RSF1010K, which was itself amplified in two overlapping PCR fragments. To generate pLG03, pLG04, pLG05, and pLG06, the hygromycin resistance cassette from pMTX708 was amplified and assembled into the SgsI site of pMTX808 and pAA58, or into the ClaI site of pMTX821 and pMTX822, respectively. pMTX808 was constructed by insertion of an ampicillin resistance cassette (amplified from pJB-KAN) into the *mobA* gene of pAA58. The cassette was inserted at the unique BstZ17I site which lies at nt 320 of *mobA*, leaving unaffected the downstream *mobB* and *repB* ORFs which overlap *mobA*. pMTX821 and pMTX822 were generated by insertion of a kanamycin resistance cassette from pJB-KAN into the gentamicin resistance cassette of pHP159 and pHP181, respectively.

Plasmids were routinely introduced in all strains by electroporation. The protocol for *C. burnetii* electroporation was previously described (Newton et al., 2014); electroporation was carried out with a Bio-Rad GenePulser Xcell (settings: 1.8 kV, 500 Ω, 25 μF). To make competent *L. pneumophila* cells, bacteria were collected from 48 h-patches grown on CYE plates, resuspended in 1 ml ice-cold sterile ddH_2O, and centrifuged for 2 min in Eppendorf tubes. The washing step was repeated three times. The pellet was resuspended in 1 ml ice-cold sterile glycerol, pelleted for 5 min and resuspended in 1 ml ice-cold sterile glycerol, from which 100 μl aliquots were either frozen at −80°C or used for transformation. Electroporation was

TABLE 1 | Bacterial strains used in this work.

Name	Relevant genotype	Description/comments	Reference
Escherichia coli			
D1210	*recA hspR hsdM rpsl lacI^q*	Sm^R, LacI^q constitutive expression	Sadler et al., 1980
DH5α T1 phage resistant	*F- φ80lacZΔM15 Δ(lacZYA-argF)U169 recA1 endA1 hsdR17(rk−, mk+) phoA supE44 λ-thi-1 gyrA96 relA1 tonA*	Nx^R, T1 phage resistant strain	Killmann et al., 1996
Bartonella henselae			
RSE247	Sm^R	Sm^R spontaneous mutant of ATCC 49882	Schmid et al., 2004
Legionella pneumophila			
Lp02	Lp01 *thyA*	Spontaneous thymidine auxotroph	Berger and Isberg, 1993
Lp03	Lp02 *dotA*	Spontaneous *dotA* mutant	Berger and Isberg, 1993
CR503	Lp01 *ΔicmSΔicmW*		Coers et al., 2000
Coxiella burnetii			
RSA439	Wild type	Plaque-purified Nine Mile phase II (NMII) clone 4	Williams et al., 1981
RSA439 *dotA*::Tn	*dotA*::TnA7	Transposon insertion mutant in *dotA* (CBU_1648), Cm^R, mCherry	Newton et al., 2014
RSA439 intergenic::Tn	intergenic::TnA7	Transposon insertion mutant between *hemD* and CBU_2078, Cm^R, mCherry; shows intracellular replication comparable to wild type	Newton et al., 2014

TABLE 2 | Plasmids used in this work.

Relaxase	Other conjugative elements	Plasmid	Selection markers[1]	Description	Reference
Mob-BID	pBGR *oriT*	pRS130	Km^{R} Neo^{R}	pBGR::*mob:BID+gfp+neo*	(Schroder et al., 2011)
MobA	RSF1010 *oriT mobB mobC*	RSF1010K	Km^{R}	RSF1010 *ΔSm KmR*	(Lessl et al., 1993)
MobA	RSF1010 *oriT mobB mobC*	pAA58	Km^{R}	RSF1010K::*egfp*	This work
MobA	RSF1010 *oriT mobB mobC*	pLG04	Km^{R} Hyg^{R}	pAA58::*hyg*	This work
TrwC	R388 *oriT trwA trwB*	pHP159	Gm^{R}	pBBR6::*oriT trwABC+egfp*	(Fernandez-Gonzalez et al., 2011)
TrwC	R388 *oriT trwA trwB*	pHP161	Gm^{R}	pBBR6::*oriT trwABC+egfp*	(Fernandez-Gonzalez et al., 2011)
TrwC	R388 *oriT trwA trwB*	pMTX821	Km^{R}	pHP159::*Km ΔGm*	This work
TrwC	R388 *oriT trwA trwB*	pCOR31	Gm^{R} Neo^{R}	pHP159::*neo*	(Gonzalez-Prieto et al., 2017)
TrwC	R388 *oriT trwA trwB*	pLG05	Km^{R} Hyg^{R}	pMTX821::*hyg*	This work
TrwC-RalF	R388 *oriT trwA trwB*	pAA12	Gm^{R}	pHP159::*trwC-RalF TS*	(Alperi et al., 2013)
–	RSF1010 *oriT mobB mobC*	pMTX808	Km^{R} Ap^{R}	pAA58::*Ap MobA-*	This work
–	RSF1010 *oriT mobB mobC*	pLG03	Km^{R} Ap^{R} Hyg^{R}	pMTX808::*hyg*	This work
–	R388 *oriT trwA trwB*	pHP181	Gm^{R}	pBBR6::*oriT trwAB+egfp*	(Fernandez-Gonzalez et al., 2011)
–	R388 *oriT trwA trwB*	pMTX822	Km^{R}	pHP181::*Km ΔGm*	This work
–	R388 *oriT trwA trwB*	pCOR35	Gm^{R} Neo^{R}	pHP181::*neo*	(Gonzalez-Prieto et al., 2017)
–	R388 *oriT trwA trwB*	pLG06	Km^{R}Hyg^{R}	pMTX822::*hyg*	This work
nr[2]	nr[2]	pMTX708	Ap^{R} Hyg^{R}	pTRE2hyg::*Ptac-oriT*	(Gonzalez-Prieto et al., 2017)
nr[2]	nr[2]	pJB-KAN	Km^{R} Ap^{R}	Cloning vector	(Omsland et al., 2011)

[1]*R, resistance to Ampicillin (Ap), Gentamycin (Gm), Kanamycin (Km), Hygromycin (Hyg) or Neomycin (Neo).* [2]*nr, not relevant.*

TABLE 3 | Oligonucleotides used for plasmid constructions.

Plasmid constructed (IA/RC)[1]	Oligonucleotide sequence (5′ to 3′)[2]	Amplified fragment
pLG03, pLG04 (IA)	TCCAGATGTATGCTCTTCTGCTCGGCGCGCC**TTTCGTCTCGAGGCAGTG**	Hyg^{R} cassette
	TGCGATGATAAGCTGTCAAACAGGCGCGCC**GTCAGTTAGGGTGTGGAAAG**	
pLG05, pLG06 (RC)	CCAAACATCGAT**GTCAGTTAGGGTGTGGAAAG**	Hyg^{R} cassette
	CCAAACATCGAT**CTTTCGTCTCGAGGCAGTG**	
pAA58 (IA)	**AGCTTGCCGCCGCCGCAG**	RSF1010K
	GGTCTATTGCCTCCCGGTATTCCTGT	
	CGCCCAGATCATCGACTTACAGGAATAC	
	GAGCAGAAGAGCATACATCTGGAAGC	
	GCCGCTTTCCTGGCTTTGCTTCCAGATGTATGCTCTTCTGCTCGGCGCGCC**TGTTTGACAGCTTATCATCGCAG**	eGFP cassette
	GTGCGGATGAAGTCAGCTCCACCTGCGGCGGCGGCAAGCTCCTGCAGG**CCCCGACACCCGCCAACAC**	
pMTX808 (IA)	GCACCTGACCGGTGCCGAGCGCCTGCCGTATTG**AGAAGGCCATCCTGACGGA**	Ap^{R} cassette
	TCGCCGCCACCGGCATGGATGGCCAGCGTA**TTACCAATGCTTAATCAGTGAG**	
pMTX821, pMTX822 (IA)	AGTATGGGCATCATTCGCACATGAA**GGCGATTCGCCGCTTTC**	Km^{R} cassette
	GGTGGCGGTACTTGGGTCGAT**TTATCAGAAGAACTCGTCAAG**	Km^{R} cassette

[1]*IA, isothermal assembly; RC, restriction cloning.* [2]*Nucleotides annealing to the PCR template are shown in bold, and restriction sites used for cloning are underlined.*

carried out adding 500 ng DNA and transferring the mixture to a cooled Bio-Rad 0.2-cm cuvette for electroshock with a Bio-Rad GenePulser Xcell set at 2.0 kV, 25 μF, and 200 Ω. After electroporation, 1 ml of AYE broth [1% yeast extract, 1% ACES pH 6.9, 3.3 mM L-cysteine, 0.33 mM $Fe(NO_3)_3$] was added, supplemented with thymidine when required, and the mixture was transferred to a 10 ml tube for incubation for 6 h at 37°C with orbital shaking. The cells were then plated on CYE supplemented with the appropriate antibiotics.

For *B. henselae,* a plate grown for 2 to 3 days was harvested with a sterile cotton swab and resuspended in 950 μl of LB. The suspension was centrifuged at 4,000 rpm for 5 min at 4°C, and the pellet was washed in 950 μl of ice-cold 10% glycerol (three times); 40 μl of these competent cells was transferred to a cooled tube, and 3 μl of DNA (300 ng/μl) was added. The mixture was incubated on ice for 15 min and transferred to a cooled Bio-Rad 0.2-cm cuvette for electroshock with a Bio-Rad Pulse controller II at 2.5 kV/cm, 25 μF, and 200 Ω. After electroporation, 1 ml of SB broth (RPMI 1640 plus L-glutamine, 42 mM HEPES, 1% sodium pyruvate, 5% heat-inactivated fetal calf serum, and 5% sheep blood lysate) was added, and the mixture was transferred to an Eppendorf tube for incubation for 3.5 h at 37°C under 5% CO_2 conditions with slow shaking. The cells were then centrifuged at 4,000 rpm for 4 min at room temperature. The pellet was resuspended in 40 μl SB broth and plated on CBA supplemented with the appropriate antibiotics.

Cell Lines and Cell Culture Conditions

The cell lines used for bacterial infections are listed in **Table 4**. EA.hy926 and HeLa cell lines were routinely grown in Dulbecco's modified Eagle medium (DMEM; Lonza or Gibco), and Chinese Hamster Ovary (CHO) cells were maintained in minimal essential medium MEMα (Gibco); both media were supplemented with 10% heat inactivated fetal bovine serum (FBS; Lonza or Sigma). Cells were incubated at 37°C under 5% CO_2.

Infections

Bartonella henselae strains containing the appropriate plasmids were grown on CBA plates for 3 to 4 days. Human cells were seeded 1 day before infection. For routine infections, cells were seeded in 6-well plates (80,000 cells per well) in 3 ml of medium. When the purpose of the infection was to select human cells that had stably acquired the plasmid transferred from *B. henselae*, infections were performed in 10-cm tissue culture dishes seeded with 450,000 cells in 12 ml of medium. The day of infection, DMEM was replaced by M199 medium (Gibco) supplemented with 10% FBS and appropriate antibiotics to select for the *B. henselae* strains to be added. The bacteria were recovered from the CBA plate and resuspended in 1 ml of PBS. The number of bacteria was calculated considering that an OD_{600} of 1 corresponds to 10^9 bacteria/ml (Kirby and Nekorchuk, 2002). Bacteria were added to the human cells to get a multiplicity of infection (MOI) of 400 bacteria per host cell. The dishes or plates were incubated for 72 h at 37°C under 5% CO_2.

Coxiella burnetii strains containing the appropriate plasmids were grown for 6 days in liquid cultures. 25,000–50,000 HeLa 229 cells were seeded in DMEM 5% FBS into 24-well plates 6–8 h before they were infected at a MOI of 500, unless specified otherwise. Bacteria were quantified measuring genome equivalents (GE) as previously described (Newton et al., 2014). Infections were incubated for 96 h at 37°C under 5% CO_2. Wells for quantification of intracellular replication were washed once with PBS at approximately 15 h post infection (hpi) before the addition of fresh DMEM 5% FBS. Wells for flow cytometry experiments were not washed.

Legionella pneumophila strains containing the appropriate plasmids were harvested from a heavy patch (after 48 h growth on CYE plates), and used to infect CHO FcγRII cells, stably expressing the receptor FcγRII. This receptor allows *L. pneumophila* opsonized with anti-*Legionella* antibodies to be internalized efficiently by non-phagocytic cells (Arasaki and Roy, 2010). FcγRII cells were grown to near confluency in 24-well dishes. Bacteria were opsonized with rabbit anti-*Legionella* antibody diluted 1/1000 for 20 min at room temperature with shaking. Bacteria were then added to the cells at an estimated MOI of 10. The cells were centrifuged 5 min at 1000 rpm and incubated for 1 h, washed three times with PBS (Gibco) and incubated in fresh media for 24 h at 37°C under 5% CO_2.

TABLE 4 | Mammalian cell lines used in this work.

Name	Description	Reference
CHO FcγRII	Chinese hamster ovary cells producing the FcγRII protein	Joiner et al., 1990
EA.hy926	Fusion cell line of human umbilical vein endothelial cells (HUVEC) and adenocarcinomic human alveolar basal epithelial cells (A549)	ATCC CRL-2922
HeLa	Human epithelial cells of cervix adenocarcinoma	ATCC CCL-2
HeLa 229	Human epithelial cells of cervix adenocarcinoma	ATCC CCL-2.1

Detection of GFP Positive Cells by Flow Cytometry

At the indicated hours post infection (hpi) indicated for each bacteria, infected cells were washed with PBS, trypsinized, and analyzed by flow cytometry using a Cytomics FC500 flow cytometer (Beckman Coulter) for *B. henselae* infections, or a BD Accuri C6 flow cytometer (BD Biosciences) for *L. pneumophila* and *C. burnetii* infections. Data were analyzed using the software for each cytometer and FlowJo (Tree Star, Inc.) software. Singlet cells were gated based on SSC-H/FSC-H and GFP positive cells (detected in the FL1-H channel) were gated based on uninfected control cells. The gate was set to approximately 0.05% GFP^+ cells in the uninfected control sample.

Fluorescence Microscopy

At the indicated hpi, wells with infected cells were washed with PBS and the plates were placed directly on a Nikon Eclipse TE2000-S inverted fluorescence microscope with a 10× objective lens. Digital images were acquired with a microscope camera (Photometrics CoolSNAP EZ) controlled by SlideBook^{TM} (Intelligent Imaging Innovations).

Detection of Stable Integrants

At 72 hpi, either 500 μg/ml G418 disulfate salt (Sigma-Aldrich) or 300 μg/ml Hygromycin B (Invitrogen), as appropriate, were added to HeLa cells infected with *B. henselae*, and selection was maintained for 4 to 5 weeks. Resistant colonies on the plates were counted.

In order to calculate the integration rate, integration experiments were always performed in parallel with infections to measure GFP positive cells by flow cytometry. The resulting percentage of GFP positive cells was extrapolated to the number of cells in the 10-cm plate used to detect integrants, and the number of resistant colonies was divided by the inferred number of GFP positive cells.

Determination of Genome Equivalents (GE)

Quantification of *C. burnetii* intracellular replication was performed as described in Newton et al. (2014). Briefly, infected HeLa cells were lysed in ddH_2O at specific time points post infection. Total genomic DNA was extracted using the Illustra Bacteria GenomicPrep Mini Spin Kit (GE Healthcare) and GE were quantified by qPCR using *dotA*-specific primers (GCGCAATACGCTCAATCACA,

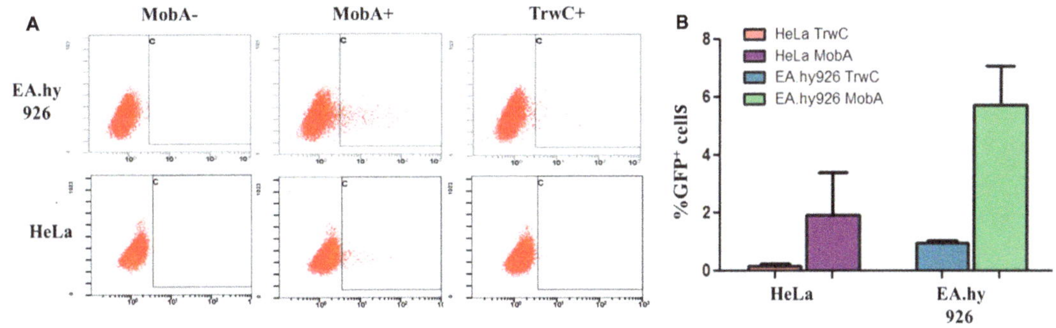

FIGURE 1 | Transient expression of transferred DNA in HeLa and EA.hy926 cells. Pools of cells obtained at 3 days post infection with *B. henselae* were analyzed by flow cytometry using uninfected cells as control. **(A)** Representative plots (cell granularity versus GFP intensity). The square marks the population considered as positive. The relaxase present in each experiment is indicated on top of the panels. **(B)** Mean values of the percentage of GFP positive cells from 2 to 4 independent experiments, after subtracting the background values of the negative controls with no relaxase. The relaxase present in each experiment is indicated.

CCATGGCCCCAATTCTCTT). The generation of this short PCR product is not affected by the presence of a transposon in the *dotA*::Tn mutant strain.

RESULTS

The conjugative relaxase TrwC can be translocated through the T4SS VirB/D4 of *B. henselae* to human cells, where it promotes the integration of the transferred DNA into the recipient genome (Gonzalez-Prieto et al., 2017). In this work, we wanted to test whether this is a unique feature of TrwC and VirB/D4, or other systems can also be combined to deliver and integrate DNA into human cells.

To test DNA transfer mediated by the relaxase MobA of the mobilizable plasmid RSF1010, we constructed a derivative carrying an eukaryotic eGFP expression cassette to detect gene expression from the human cell nucleus. An insertion of an ampicillin resistance cassette in *mobA* served as a negative control. The insertion is located in the 5′ region of the ORF, thus not affecting the expression of the ORFs *mobB* and especially *repB*', which encodes a DNA primase required for plasmid replication. We observed that this *mobA*$^{-}$ construct had a higher copy number than the parental plasmid, as judged from the amount of DNA extracted from parallel cultures (data not shown). This phenomenon has previously been reported, and attributed to the repressor role of MobA/RepB in replication (Frey et al., 1992).

These plasmids (pAA58 and pMTX808; **Table 2**) were introduced in *B. henselae*, and the resulting strains were used to infect both EA.hy926 and HeLa human cell lines. The former is derived from HUVEC cells, which are the natural target of *B. henselae in vivo*; however, HeLa cells can also be infected by *B. henselae* with lower efficiency, and we showed that TrwC-mediated DNA transfer takes place to HeLa cells as well (Gonzalez-Prieto et al., 2017). *B. henselae* carrying plasmids coding for either MobA or TrwC, or relaxase mutants as negative controls, were used for infections. To assess transfer of the plasmid DNA to the human cells, flow cytometry was used to quantify the expression of the eGFP cassette per cell, thus allowing the determination of the percentage of GFP positive cells. The results are shown in **Figure 1** and **Table 5**, top 8 rows. We observed DNA transfer when the plasmids encoded a functional relaxase, and background levels in the absence of a relaxase. DNA transfer rates were notably higher when using MobA as the leading relaxase compared to TrwC.

In order to measure genomic integration of the transferred DNA, we constructed plasmid derivatives encoding antibiotic resistance cassettes (see **Table 2**). The plasmids containing R388 conjugative elements carried a neomycin gene; however, this was not used in these experiment because of the presence of a kanamycin resistance gene in the RSF1010K backbone, which could lead to recombination between both cassettes. Instead, a hygromycin resistance cassette was inserted. In order to avoid an effect caused by the different antibiotic selections applied, we also constructed Hygromycin-resistant derivatives encoding TrwC (**Table 2**), and we found that TrwC-mediated integration rate did not vary when the selection applied was hygromycin B or Geneticin (data not shown).

HeLa cells were used as target cells to measure DNA integration, because in contrast to EA.hy926 cells HeLa cells show enhanced survival during the 4–5 weeks of antibiotic selection required to measure resistant colonies (Gonzalez-Prieto et al., 2017). The cells were infected with *B. henselae* carrying the different plasmids. A plasmid derived from the cryptic *Bartonella* plasmid pBGR1 was also assayed for comparison, since it has been reported that its relaxase mediates DNA transfer but does not promote integration of the transferred DNA (Gonzalez-Prieto et al., 2017). After applying the antibiotic selection, resistant colonies were counted, and integration rates were calculated dividing this number by the number of GFP positive cells determined in parallel infection experiments (see Materials and Methods for details). The results (**Figure 2**) indicate that the integration rate for the MobA constructs was approximately one-log higher than in case of Mob-BID, which suggest that MobA promotes integration of the transferred DNA. It can also be observed that TrwC has a stronger effect on integration than MobA (approximately five-fold higher DNA integration).

TABLE 5 | Rates of DNA transfer to mammalian cells through T4ASS and T4BSS.

Donor bacteria (genotype)	T4SS	Transfer system	Relaxase	Infected cells	GFP^{+} mammalian cells[1]	
					Flow cyt %	Scope
Bh RSE247 (wt)	Functional	RSF1010	MobA	EA.hy926	5.72 ± 1.37	nq[2]
Bh RSE247 (wt)	Functional	RSF1010	–	EA.hy926	0.29 ± 0.07	nq[2]
Bh RSE247 (wt)	Functional	R388	TrwC	EA.hy926	1.00 ± 0.09	nq[2]
Bh RSE247 (wt)	Functional	R388	–	EA.hy926	0.14 ± 0.19	nq[2]
Bh RSE247 (wt)	Functional	RSF1010	MobA	HeLa	2.00 ± 1.48	nq[2]
Bh RSE247 (wt)	Functional	RSF1010	–	HeLa	0.07 ± 0.05	nq[2]
Bh RSE247 (wt)	Functional	R388	TrwC	HeLa	0.20 ± 0.03	nq[2]
Bh RSE247 (wt)	Functional	R388	–	HeLa	0.04 ± 0.06	nq[2]
Lp Lp02 (wt)	Functional	RSF1010	MobA	CHO FcγRII	0.35 ± 0.12	nq[2]
Lp Lp03 *(dotA)*	No transport	RSF1010	MobA	CHO FcγRII	0.03 ± 0.05	$<5 \times 10^{-6}$
Lp Lp02 (wt)	Functional	RSF1010	–	CHO FcγRII	0.00 ± 0.00	$<5 \times 10^{-6}$
Lp CR503 *(icmS icmW)*	No chaperone	RSF1010	MobA	CHO FcγRII	0.00 ± 0.00	$<5 \times 10^{-6}$
Lp Lp02 (wt)	Functional	R388	TrwC	CHO FcγRII	0.00 ± 0.00	$<5 \times 10^{-6}$
Lp Lp02 (wt)	Functional	R388	TrwC-RalF	CHO FcγRII	0.00 ± 0.00	1×10^{-5}
Lp Lp03 *(dotA)*	No transport	R388	TrwC-RalF	CHO FcγRII	nq[2]	$<5 \times 10^{-6}$
Lp Lp02 (wt)	Functional	R388	–	CHO FcγRII	nq[2]	$<5 \times 10^{-6}$
Lp CR503 *(icmS icmW)*	No chaperone	R388	TrwC-RalF	CHO FcγRII	nq[2]	2×10^{-5}
Cb intergenic::Tn (wt)	Functional	RSF1010	MobA	HeLa	0.56 ± 0.53	nq[2]
Cb dotA::Tn	No transport	RSF1010	MobA	HeLa	0.04 ± 0.02	nq[2]
Cb intergenic::Tn (wt)	Functional	RSF1010	–	HeLa	0.10[3] ± 0.04	$<5 \times 10^{-6}$

[1]DNA transfer is measured as the ratio of mammalian recipient cells expressing GFP. Data from flow cytometry (left column) show the percentage of GFP positive cells (mean ± SD of two to eight independent assays). Infected cells were also screened visually under the microscope (right column). Positive cells were counted and divided by the total number of cells per well (estimated as 200.000). The screen was performed at least twice for each condition. [2]nq, not quantified. [3]Due to higher background (see text for details).

Earlier studies reported Dot/Icm-dependent conjugative DNA transfer of RSF1010 (Vogel et al., 1998), implying that MobA can mediate the translocation of an attached DNA substrate through the T4BSS Dot/Icm of *L. pneumophila*. Thus, we asked whether the Dot/Icm T4SS could also promote DNA transfer to mammalian cells upon infection by *L. pneumophila*. In addition to testing MobA-mediated transfer, we tested DNA transfer mediated by TrwC and TrwC-RalF, a fusion protein carrying the C-terminal 20 residues of the *L. pneumophila* Dot/Icm substrate RalF, that has been shown to be sufficient for translocation (Nagai et al., 2005). In contrast to the infection experiments done with *B. henselae*, for infections with *L. pneumophila* a MOI of 10 was used and DNA transfer was monitored at 24 hpi. As shown in **Figure 3A** and **Table 5**, we detected GFP positive cells after infection by a mechanism dependent on the Dot/Icm T4BSS and the relaxase MobA. Thus, we show for the first time that DNA transfer can occur through a T4BSS into mammalian cells. Using the same flow cytometry assay, we did not detect GFP positive cells above the background when the mobilizable plasmids encoded the relaxase TrwC or TrwC-RalF. However, inspection of the infected cells by fluorescence microscopy did reveal a small number of positive cells that expressed GFP uniformly and strongly after infection with *L. pneumophila* producing TrwC-RalF (**Figure 3B**). Positive cells were not observed in the negative controls or with TrwC-encoding plasmids.

The rate of DNA transfer was highly dependent on the conjugative DNA processing system used. This could be due to different relaxase recruitment efficiencies. The Dot/Icm T4BSS recruits a subset of its substrates through a chaperone complex formed by IcmS and IcmW (Cambronne and Roy, 2007). To determine if recruitment of the relaxases was dependent on this complex, a *L. pneumophila* Δ*icmS* Δ*icmW* mutant strain was used in infection experiments carrying plasmids which encode either MobA or TrwC-RalF. The results (**Table 5** and **Figure 3B**) indicate that the absence of IcmSW did not affect DNA transfer mediated by TrwC-RalF, while DNA transfer mediated by MobA was abolished in the absence of IcmSW.

The Dot/Icm T4BSS of *L. pneumophila* is closely related to that of *C. burnetii*, and several reports have shown that both can recruit the same effector proteins and cross-complement *icmSW* mutants (Zamboni et al., 2003; Zusman et al., 2003; Carey et al., 2011). Thus, we decided to test MobA-mediated DNA transfer through the Dot/Icm T4BSS of *C. burnetii*. HeLa cells were infected with *C. burnetii* strains harboring the plasmids with and without MobA at a MOI of 500, and GFP expression was investigated at 4 days post infection. The results are shown in **Table 5**, and **Figure 3C** shows representative plots. Similar to what was observed with *L. pneumophila*, GFP positive cells were only detected when the Dot/Icm T4BSS and the MobA relaxase were present.

Performing these experiments, we observed a difference in the background fluorescence intensity of HeLa cells depending

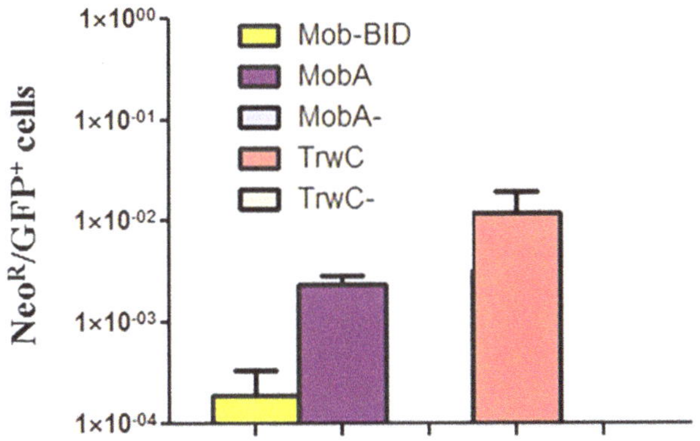

FIGURE 2 | DNA integration rate for different relaxases. HeLa cells were infected with *B. henselae* containing the plasmids encoding the indicated relaxases (or the negative controls lacking the relaxase) and at 3 days post infection, cells were either analyzed by flow cytometry (to determine transient expression of eGFP), or subjected to antibiotic selection (to determine permanent expression of the antibiotic resistance gene). The graph shows the ratio between the number of antibiotic-resistant cells and the number of GFP positive cells.

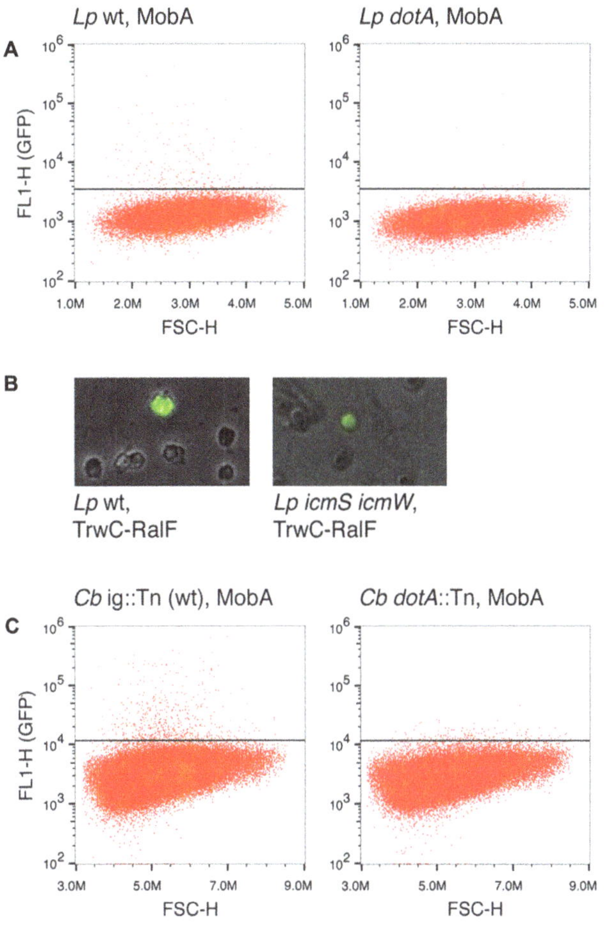

FIGURE 3 | DNA transfer through the Dot/Icm T4BSS. **(A)** Representative flow cytometry plots for quantification of GFP positive CHO FcγRII cells after infection with *L. pneumophila* strains Lp02 (wild type, left panel) or Lp03 (*dotA* mutant, right panel) harboring plasmid pAA58, which encodes MobA. **(B)** Fluorescence microscope images showing GFP positive CHO FcγRII cells infected with *L. pneumophila* strains Lp02 (wild type, left panel) or CR503 (Δ*icmS*Δ*icmW* mutant, right panel) harboring plasmid pAA12, which encodes TrwC-RalF. **(C)** Representative flow cytometry plots for quantification of GFP positive HeLa cells after infection with *C. burnetii* strains RSA439 intergenic::Tn (wild type, left panel) or RSA439 *dotA*::Tn (*dotA* mutant, right panel) harboring plasmid pAA58, which encodes MobA. ig, intergenic.

on the bacterial strain used for infection. A representative flow cytometry histogram is shown in **Figure 4A**. The background GFP fluorescence peak shifts toward a higher intensity when HeLa cells were infected with wild type *C. burnetii* or wild type *C. burnetii* harboring the plasmid with the *mobA* mutation, but not when cells were infected with wild type *C. burnetii* carrying the plasmid with the intact *mobA* gene. This higher fluorescence did not correspond to DNA transfer, since we did not detect any proper GFP positive cells by flow cytometry or using microscopy, but it contributed to a minimal raise in the background frequencies observed when infecting with a *mobA*$^-$ strain (see **Table 5**). However, the difference in background fluorescence may be attributed to a different amount of intracellular bacteria per cell. To test this hypothesis, HeLa cells were infected at a MOI of 50 and the number of intracellular *C. burnetii* was determined by measuring GE at two time points post infection. The results are shown in **Figure 4B**. A strain carrying the *mobA*-deficient plasmid replicates nearly as efficiently as a strain with no plasmid. In contrast, the same strain carrying a plasmid that encodes a functional MobA protein was severely impaired in intracellular replication. A *dotA* mutant that fails to replicate intracellularly due to the absence of a functional T4SS was used as a control in this assay.

DISCUSSION

In our previous reports, we showed that the conjugative relaxase TrwC can be translocated to human cells through the T4SS VirB/D4 of *B. henselae* (Fernandez-Gonzalez et al., 2011), and also that it promotes integration of the transferred DNA into the recipient genome (Gonzalez-Prieto et al., 2017). Whether these abilities were unique for TrwC and VirB/D4 remained to be tested. In this work, we report that different relaxases and T4SS can be used to transfer DNA to human cells and to promote DNA integration. In other words, relaxases and T4SS from various bacterial species can be combined to create tools intended to genetically modify specific human target cells in a permanent way, thus generating enormous biotechnological potential.

Firstly, we compared the ability of different relaxases to transfer DNA to mammalian cells and to promote DNA integration into the recipient genome when translocated by the same T4SS, VirB/D4. Human cells were infected with *B. henselae* carrying derivatives of the mobilizable plasmid RSF1010, encoding the relaxase MobA; with constructs containing the conjugative processing elements of the self-transferable plasmid R388, which encodes the relaxase TrwC; or with derivatives

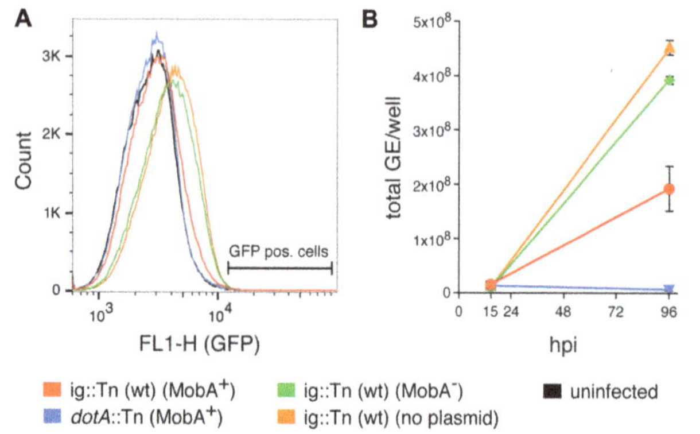

FIGURE 4 | The presence of MobA interferes with intracellular replication of *C. burnetii* in HeLa cells. **(A)** Flow cytometry histogram analysis of HeLa cells infected with the indicated *C. burnetii* strains at 4 days post infection. The location of GFP positive cells (see **Figure 3C** and Materials and Methods) is indicated. **(B)** Quantification of intracellular replication of *C. burnetii*. Total genome equivalents (GE)/well are shown at two time points post infection for infections with the indicated strains at a MOI of 50. The bottom strain legend applies for **(A,B)**. The presence or absence of plasmid-encoded MobA is indicated in brackets. ig, intergenic. hpi, hours post infection.

of *B. henselae* cryptic plasmid pBGR1, coding for the relaxase Mob fused to the BID signal for efficient recruitment by VirB/D4 (Schroder et al., 2011). When the three plasmids are compared in terms of DNA transfer and integration rates (**Figures 1**, **2**), we find that these vary significantly, with RSF1010 being the most efficiently transferred, while TrwC is the relaxase showing higher integration rates. The rate of DNA transfer is probably proportional to the efficiency with which the relaxase is recruited to the T4SS machinery; this assumption comes from previous works showing that the relaxase Mob itself could transfer DNA to human cells with barely detectable frequency, but when a recruitment secretion signal was fused to its C-terminal end, it transferred DNA to similar frequencies than TrwC (Schroder et al., 2011). In addition, in case of R388, a deletion of the conjugative coupling protein, a component believed to play a key role in the recruitment of the conjugative substrate, caused DNA transfer rates to drop 10-fold (Fernandez-Gonzalez et al., 2011). The relaxase MobA belongs to a mobilizable plasmid which hijacks the T4SS of co-residing conjugative plasmids, so it can be translocated through various T4SS; thus, it is plausible that the requirements for MobA recruitment are less stringent. In fact, the C-terminal 48 residues of MobA were shown to direct translocation of a Cre fusion through the VirB T4SS of *A. tumefaciens* into plant cells (Vergunst et al., 2005). Now, we show that MobA can also be translocated through a T4ASS into mammalian cells.

The ability to enhance integration of the transferred DNA into the recipient cell genome must reside in an intrinsic property of the relaxase, which is the only protein entering the recipient cell covalently attached to the transferred DNA strand. We report here that the promiscuous relaxase MobA also promotes DNA integration, resulting in resistant colonies with about 10-fold higher frequency than Mob-BID, which does not promote integration above background levels obtained by DNA transfection (Gonzalez-Prieto et al., 2017), but roughly five-fold lower frequency than TrwC. These differences observed among relaxases could be due to differential nuclear targeting, catalytic activity, or binding affinity to its target, which could protect the DNA ends, thus favoring integration by host-mediated mechanisms, as previously suggested (Gonzalez-Prieto et al., 2017). Subcellular localization of TrwC and MobA in human cells showed no preferential nuclear localization for either relaxase (Silby et al., 2007; Agundez et al., 2011). It is noteworthy that TrwC catalyzes site-specific recombination on supercoiled DNA substrates (Cesar et al., 2006), while MobA was shown to catalyze site-specific recombination between two *oriT* copies when the substrate was single-stranded (Meyer, 1989), and other relaxases do not catalyze this reaction at all. Although the integration pattern in the human genome is random (Gonzalez-Prieto et al., 2017), site-specific recombination ability could play a role in strand-transfer reactions when the nucleoprotein complex is directed to a nicked DNA strand by the host repair machinery.

MobA can be translocated by the T4BSS of *L. pneumophila*, alone or bound to DNA, into recipient bacteria (Vogel et al., 1998; Luo and Isberg, 2004). These results prompted us to test its translocation by T4BSS into mammalian cells. Our results (**Figure 3**) show for the first time that DNA transfer to human cells can also be accomplished through the Dot/Icm T4BSS of *L. pneumophila* and *C. burnetii*, only remotely related to T4ASS. Thus, it is reasonable to assume that DNA translocation may be an intrinsic ability of T4SS. An important difference between both Dot/Icm systems is the temporal pattern of secretion: while *L. pneumophila* has been shown to secrete effectors as internalization into host cells is initiated (Nagai et al., 2005) in case of *C. burnetii* effector translocation is initiated when the pathogen has reached an acidified lysosomal compartment (Newton et al., 2013); thus, DNA transfer in *C. burnetii* must occur from within the *Coxiella*-containing vacuole.

DNA transfer was dependent on the presence of the Dot/Icm T4SS and a functional relaxase, as expected for a *bona fide* conjugation-like DNA transfer process. The wide differences in DNA transfer rates depending on the relaxase (MobA, TrwC, or TrwC-RalF, including the translocation signal of the natural T4SS substrate RalF) and on the presence/absence of the chaperones IcmSW (see **Table 5**) support the concept that relaxase recruitment is the main driver of DNA transfer.

During the course of performing *C. burnetii* infection experiments, we noticed an inhibition of *C. burnetii* intracellular replication caused by the presence of RSF1010 derivatives carrying a functional MobA relaxase while isogenic strains with a *mobA* mutation did not affect growth (**Figure 4**). Similarly, RSF1010 conjugation was shown to inhibit intracellular replication and virulence of *L. pneumophila* (Segal and Shuman, 1998), probably by MobA interference with effector secretion by Dot/Icm. This result should be taken into account when using vectors based on RSF1010, which are the more commonly used by both *L. pneumophila* and *C. burnetii*.

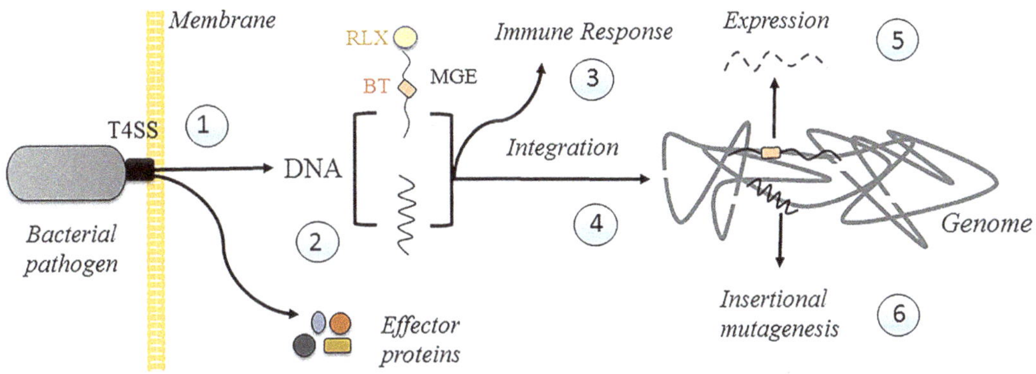

FIGURE 5 | Possible fates of translocated DNA in a human cell. See text for details. Random DNA is represented by the jagged line. Mobile genetic element (MGE) is represented by the wavy line. RLX, relaxase (small yellow sphere). BT, beneficial trait (small orange box).

Finally, an attractive question that remains open is the possible biological role, if any, of DNA transfer to mammalian cells by bacterial pathogens harboring a T4SS. Is the DNA transfer ability an evolutionary remnant of the conjugative T4SS from which the T4SS involved in virulence probably have evolved? Or is it an ability which the pathogens have evolved to use to their own benefit, in the same way as *A. tumefaciens* uses it to subvert its eukaryotic host cell?

In support of the first possibility, it is relevant to point out that in spite of many attempts, no T4 protein, protein domain or amino acid residue has been identified to date, which is specifically involved in DNA transfer. All analyzed mutants in T4 components, even in the conjugative coupling protein ATPase, affected DNA and protein translocation to the same extent, leading to the suggestion that relaxase and DNA translocation may have the same molecular requirements (de Paz et al., 2010; Larrea et al., 2017). Thus, the ability to transfer DNA could not be lost in a T4SS even if it evolved to only secrete proteins. However, the potential of DNA transfer for long-term subversion of the host cells makes it attractive to think that pathogens may utilize such a process for their own profit. **Figure 5** illustrates the possible fates of secreted DNA in a human cell. A pathogen translocates effector proteins and DNA through its T4SS once in contact with the membrane (1 in **Figure 5**), whether it is from within a vacuolar compartment, as in case of *C. burnetii*, or from the outside. The secreted DNA could either be random DNA, as proposed for *H. pylori* (Varga et al., 2016), or a specifically recruited mobile genetic element (MGE), in which case a dedicated transfer system would attach a relaxase to its end (2). The cytoplasmic DNA could elicit an immune response (3), as proposed for *H. pylori* (Varga et al., 2016), which could be used by the pathogen for its own benefit. DNA could also get integrated into the host cell genome (4) by the host repair/recombination systems, and/or by the covalently attached conjugative relaxase. Integration will lead to the stable expression of the encoded information (5), including any beneficial traits that the pathogen may have evolved to encode in MGE for that purpose. Finally, random integration has an inherent risk of insertional mutagenesis (6), which could lead to increased growth of the host cell, thereby promoting the extension of the niche of the pathogen.

In this context, it has to be stressed that human pathogens contain many poorly characterized MGE, which could be substrates for DNA transfer (in addition to the possibility of sporadic transfer of visiting promiscuous plasmids, such as RSF1010). As examples from the pathogens used in this study, the pBRG1 cryptic plasmid of *B. henselae* can be recruited by VirB/D4 and translocated to human cells (Schroder et al., 2011); conjugative transfer of chromosomal DNA has been reported for *L. pneumophila* (Miyamoto et al., 2003), and its genome includes several genomic islands; and notably, a cryptic plasmid in *C. burnetii* is enriched in important effector genes (Voth et al., 2011); it is tempting to speculate that this plasmid may be transferred to the host cell.

AUTHOR CONTRIBUTIONS

All authors contributed to the conception and design of the work, data acquisition and/or analysis. All authors contributed to drafting, revising, and final approval of the work. All authors agree to be accountable for all aspects of the work.

ACKNOWLEDGMENT

ML wishes to thank the Roy lab, and especially David Chetrit and Stephanie Shames, for their support with *Legionella* lab protocols.

REFERENCES

Agundez, L., Gonzalez-Prieto, C., Machon, C., and Llosa, M. (2012). Site-specific integration of foreign DNA into minimal bacterial and human target sequences mediated by a conjugative relaxase. *PLoS ONE* 7:e31047. doi: 10.1371/journal.pone.0031047

Agundez, L., Machon, C., Cesar, C. E., Rosa-Garrido, M., Delgado, M. D., and Llosa, M. (2011). Nuclear targeting of a bacterial integrase that mediates site-specific recombination between bacterial and human target sequences. *Appl. Environ. Microbiol.* 77, 201–210. doi: 10.1128/AEM.01371-10

Alperi, A., Larrea, D., Fernandez-Gonzalez, E., Dehio, C., Zechner, E. L., and Llosa, M. (2013). A translocation motif in relaxase TrwC specifically affects recruitment by its conjugative type IV secretion system. *J. Bacteriol.* 195, 4999–5006. doi: 10.1128/JB.00367-13

Arasaki, K., and Roy, C. R. (2010). *Legionella pneumophila* promotes functional interactions between plasma membrane syntaxins and Sec22b. *Traffic* 11, 587–600. doi: 10.1111/j.1600-0854.2010.01050.x

Berger, K. H., and Isberg, R. R. (1993). Two distinct defects in intracellular growth complemented by a single genetic locus in *Legionella pneumophila*. *Mol. Microbiol.* 7, 7–19. doi: 10.1111/j.1365-2958.1993.tb01092.x

Cabezon, E., Ripoll-Rozada, J., Pena, A., de la Cruz, F., and Arechaga, I. (2015). Towards an integrated model of bacterial conjugation. *FEMS Microbiol. Rev.* 39, 81–95. doi: 10.1111/1574-6976.12085

Cambronne, E. D., and Roy, C. R. (2007). The *Legionella pneumophila* IcmSW complex interacts with multiple Dot/Icm effectors to facilitate type IV translocation. *PLoS Pathog.* 3:e188. doi: 10.1371/journal.ppat.0030188

Carey, K. L., Newton, H. J., Luhrmann, A., and Roy, C. R. (2011). The *Coxiella burnetii* Dot/Icm system delivers a unique repertoire of type IV effectors into host cells and is required for intracellular replication. *PLoS Pathog.* 7:e1002056. doi: 10.1371/journal.ppat.1002056

Celec, P., and Gardlik, R. (2017). Gene therapy using bacterial vectors. *Front. Biosci.* 22:81–95. doi: 10.2741/4473

Cesar, C. E., Machon, C., de la Cruz, F., and Llosa, M. (2006). A new domain of conjugative relaxase TrwC responsible for efficient oriT-specific recombination on minimal target sequences. *Mol. Microbiol.* 62, 984–996. doi: 10.1111/j.1365-2958.2006.05437.x

Chandran Darbari, V., and Waksman, G. (2015). Structural biology of bacterial type IV secretion systems. *Annu. Rev. Biochem.* 84, 603–629. doi: 10.1146/annurev-biochem-062911-102821

Christie, P. J. (2016). The mosaic type IV secretion systems. *EcoSal Plus* 7, doi: 10.1128/ecosalplus.ESP-0020-2015

Coers, J., Kagan, J. C., Matthews, M., Nagai, H., Zuckman, D. M., and Roy, C. R. (2000). Identification of Icm protein complexes that play distinct roles in the biogenesis of an organelle permissive for *Legionella pneumophila* intracellular growth. *Mol. Microbiol.* 38, 719–736. doi: 10.1046/j.1365-2958.2000.02176.x

de Paz, H. D., Larrea, D., Zunzunegui, S., Dehio, C., de la Cruz, F., and Llosa, M. (2010). Functional dissection of the conjugative coupling protein TrwB. *J. Bacteriol.* 192, 2655–2669. doi: 10.1128/JB.01692-09

Draper, O., Cesar, C. E., Machon, C., de la Cruz, F., and Llosa, M. (2005). Site-specific recombinase and integrase activities of a conjugative relaxase in recipient cells. *Proc. Natl. Acad. Sci. U.S.A.* 102, 16385–16390. doi: 10.1073/pnas.0506081102

Fernandez-Gonzalez, E., de Paz, H. D., Alperi, A., Agundez, L., Faustmann, M., Sangari, F. J., et al. (2011). Transfer of R388 derivatives by a pathogenesis-associated type IV secretion system into both bacteria and human cells. *J. Bacteriol.* 193, 6257–6265. doi: 10.1128/JB.05905-11

Frey, J., Bagdasarian, M. M., and Bagdasarian, M. (1992). Replication and copy number control of the broad-host-range plasmid RSF1010. *Gene* 113, 101–106. doi: 10.1016/0378-1119(92)90675-F

Garcillan-Barcia, M. P., Jurado, P., Gonzalez-Perez, B., Moncalian, G., Fernandez, L. A., and de la Cruz, F. (2007). Conjugative transfer can be inhibited by blocking relaxase activity within recipient cells with intrabodies. *Mol. Microbiol.* 63, 404–416. doi: 10.1111/j.1365-2958.2006.05523.x

Gibson, D. G., Young, L., Chuang, R. Y., Venter, J. C., Hutchison, C. A. III, and Smith, H. O. (2009). Enzymatic assembly of DNA molecules up to several hundred kilobases. *Nat. Methods* 6, 343–345. doi: 10.1038/nmeth.1318

Gonzalez-Perez, B., Lucas, M., Cooke, L. A., Vyle, J. S., de la Cruz, F., and Moncalian, G. (2007). Analysis of DNA processing reactions in bacterial conjugation by using suicide oligonucleotides. *EMBO J.* 26, 3847–3857. doi: 10.1038/sj.emboj.7601806

Gonzalez-Prieto, C., Agundez, L., Linden, R. M., and Llosa, M. (2013). HUH site-specific recombinases for targeted modification of the human genome. *Trends Biotechnol.* 31, 305–312. doi: 10.1016/j.tibtech.2013.02.002

Gonzalez-Prieto, C., Gabriel, R., Dehio, C., Schmidt, M., and Llosa, M. (2017). The conjugative relaxase TrwC promotes integration of foreign DNA in the human genome. *Appl. Environ. Microbiol.* 83:e00207-17. doi: 10.1128/AEM.00207-17

Hubber, A., and Roy, C. R. (2010). Modulation of host cell function by *Legionella pneumophila* type IV effectors. *Annu. Rev. Cell Dev. Biol.* 26, 261–283. doi: 10.1146/annurev-cellbio-100109-104034

Joiner, K. A., Fuhrman, S. A., Miettinen, H. M., Kasper, L. H., and Mellman, I. (1990). *Toxoplasma gondii*: fusion competence of parasitophorous vacuoles in Fc receptor-transfected fibroblasts. *Science* 249, 641–646. doi: 10.1126/science.2200126

Killmann, H., Benz, R., and Braun, V. (1996). Properties of the FhuA channel in the *Escherichia coli* outer membrane after deletion of FhuA portions within and outside the predicted gating loop. *J. Bacteriol.* 178, 6913–6920. doi: 10.1128/jb.178.23.6913-6920.1996

Kirby, J. E., and Nekorchuk, D. M. (2002). *Bartonella*-associated endothelial proliferation depends on inhibition of apoptosis. *Proc. Natl. Acad. Sci. U.S.A.* 99, 4656–4661. doi: 10.1073/pnas.072292699

Larrea, D., de Paz, H. D., Matilla, I., Guzman-Herrador, D. L., Lasso, G., de la Cruz, F., et al. (2017). Substrate translocation involves specific lysine residues of the central channel of the conjugative coupling protein TrwB. *Mol. Genet. Genomics* doi: 10.1007/s00438-017-1331-3 [Epub ahead of print].

Lessl, M., Balzer, D., Weyrauch, K., and Lanka, E. (1993). The mating pair formation system of plasmid RP4 defined by RSF1010 mobilization and donor-specific phage propagation. *J. Bacteriol.* 175, 6415–6425. doi: 10.1128/jb.175.20.6415-6425.1993

Llosa, M., Bolland, S., Grandoso, G., and de la Cruz, F. (1994). Conjugation-independent, site-specific recombination at the oriT of the IncW plasmid R388 mediated by TrwC. *J. Bacteriol.* 176, 3210–3217. doi: 10.1128/jb.176.11.3210-3217.1994

Llosa, M., Schroder, G., and Dehio, C. (2012). New perspectives into bacterial DNA transfer to human cells. *Trends Microbiol.* 20, 355–359. doi: 10.1016/j.tim.2012.05.008

Luo, Z. Q., and Isberg, R. R. (2004). Multiple substrates of the *Legionella pneumophila* Dot/Icm system identified by interbacterial protein transfer. *Proc. Natl. Acad. Sci. U.S.A.* 101, 841–846. doi: 10.1073/pnas.0304916101

Meyer, R. (1989). Site-specific recombination at oriT of plasmid R1162 in the absence of conjugative transfer. *J. Bacteriol.* 171, 799–806. doi: 10.1128/jb.171.2.799-806.1989

Miyamoto, H., Yoshida, S., Taniguchi, H., and Shuman, H. A. (2003). Virulence conversion of *Legionella pneumophila* by conjugal transfer of chromosomal DNA. *J. Bacteriol.* 185, 6712–6718. doi: 10.1128/JB.185.22.6712-6718.2003

Nagai, H., Cambronne, E. D., Kagan, J. C., Amor, J. C., Kahn, R. A., and Roy, C. R. (2005). A C-terminal translocation signal required for Dot/Icm-dependent delivery of the *Legionella* RalF protein to host cells. *Proc. Natl. Acad. Sci. U.S.A.* 102, 826–831. doi: 10.1073/pnas.0406239101

Newton, H. J., Kohler, L. J., McDonough, J. A., Temoche-Diaz, M., Crabill, E., Hartland, E. L., et al. (2014). A screen of *Coxiella burnetii* mutants reveals important roles for Dot/Icm effectors and host autophagy in vacuole biogenesis. *PLoS Pathog.* 10:e1004286. doi: 10.1371/journal.ppat.1004286

Newton, H. J., McDonough, J. A., and Roy, C. R. (2013). Effector protein translocation by the *Coxiella burnetii* Dot/Icm type IV secretion system requires endocytic maturation of the pathogen-occupied vacuole. *PLoS ONE* 8:e54566. doi: 10.1371/journal.pone.0054566

Omsland, A., Beare, P. A., Hill, J., Cockrell, D. C., Howe, D., Hansen, B., et al. (2011). Isolation from animal tissue and genetic transformation of *Coxiella burnetii* are facilitated by an improved axenic growth medium. *Appl. Environ. Microbiol.* 77, 3720–3725. doi: 10.1128/AEM.02826-10

Personnic, N., Barlocher, K., Finsel, I., and Hilbi, H. (2016). Subversion of retrograde trafficking by translocated pathogen effectors. *Trends Microbiol.* 24, 450–462. doi: 10.1016/j.tim.2016.02.003

Rolando, M., and Buchrieser, C. (2014). *Legionella pneumophila* type IV effectors hijack the transcription and translation machinery of the host cell. *Trends Cell Biol.* 24, 771–778. doi: 10.1016/j.tcb.2014.06.002

Sadler, J. R., Tecklenburg, M., and Betz, J. L. (1980). Plasmids containing many tandem copies of a synthetic lactose operator. *Gene* 8, 279–300. doi: 10.1016/0378-1119(80)90005-0

Saenz, H. L., Engel, P., Stoeckli, M. C., Lanz, C., Raddatz, G., Vayssier-Taussat, M., et al. (2007). Genomic analysis of *Bartonella* identifies type IV secretion systems as host adaptability factors. *Nat. Genet.* 39, 1469–1476. doi: 10.1038/ng.2007.38

Sambrook, J., and Russell, D. W. (2001). *Molecular Cloning : A Laboratory Manual.* Cold Spring Harbor, NY: Cold Spring Harbor Laboratory Press.

Schmid, M. C., Schulein, R., Dehio, M., Denecker, G., Carena, I., and Dehio, C. (2004). The VirB type IV secretion system of *Bartonella henselae* mediates invasion, proinflammatory activation and antiapoptotic protection of endothelial cells. *Mol. Microbiol.* 52, 81–92. doi: 10.1111/j.1365-2958.2003.03964.x

Schroder, G., Schuelein, R., Quebatte, M., and Dehio, C. (2011). Conjugative DNA transfer into human cells by the VirB/VirD4 type IV secretion system of the bacterial pathogen *Bartonella henselae. Proc. Natl. Acad. Sci. U.S.A.* 108, 14643–14648. doi: 10.1073/pnas.1019074108

Segal, G., and Shuman, H. A. (1998). Intracellular multiplication and human macrophage killing by *Legionella pneumophila* are inhibited by conjugal components of IncQ plasmid RSF1010. *Mol. Microbiol.* 30, 197–208. doi: 10.1046/j.1365-2958.1998.01054.x

Silby, M. W., Ferguson, G. C., Billington, C., and Heinemann, J. A. (2007). Localization of the plasmid-encoded proteins TraI and MobA in eukaryotic cells. *Plasmid* 57, 118–130. doi: 10.1016/j.plasmid.2006.08.006

Varga, M. G., Shaffer, C. L., Sierra, J. C., Suarez, G., Piazuelo, M. B., Whitaker, M. E., et al. (2016). Pathogenic *Helicobacter pylori* strains translocate DNA and activate TLR9 via the cancer-associated cag type IV secretion system. *Oncogene* 35, 6262–6269. doi: 10.1038/onc.2016.158

Vergunst, A. C., van Lier, M. C., den Dulk-Ras, A., Stuve, T. A., Ouwehand, A., and Hooykaas, P. J. (2005). Positive charge is an important feature of the C-terminal transport signal of the VirB/D4-translocated proteins of *Agrobacterium. Proc. Natl. Acad. Sci. U.S.A.* 102, 832–837. doi: 10.1073/pnas.0406241102

Vogel, J. P., Andrews, H. L., Wong, S. K., and Isberg, R. R. (1998). Conjugative transfer by the virulence system of *Legionella pneumophila. Science* 279, 873–876. doi: 10.1126/science.279.5352.873

Voth, D. E., Beare, P. A., Howe, D., Sharma, U. M., Samoilis, G., Cockrell, D. C., et al. (2011). The *Coxiella burnetii* cryptic plasmid is enriched in genes encoding type IV secretion system substrates. *J. Bacteriol.* 193, 1493–1503. doi: 10.1128/JB.01359-10

Williams, J. C., Peacock, M. G., and McCaul, T. F. (1981). Immunological and biological characterization of *Coxiella burnetii*, phases I and II, separated from host components. *Infect. Immun.* 32, 840–851.

Zamboni, D. S., McGrath, S., Rabinovitch, M., and Roy, C. R. (2003). *Coxiella burnetii* express type IV secretion system proteins that function similarly to components of the *Legionella pneumophila* Dot/Icm system. *Mol. Microbiol.* 49, 965–976. doi: 10.1046/j.1365-2958.2003.03626.x

Zechner, E. L., Lang, S., and Schildbach, J. F. (2012). Assembly and mechanisms of bacterial type IV secretion machines. *Philos. Trans. R. Soc. Lond. B Biol. Sci.* 367, 1073–1087. doi: 10.1098/rstb.2011.0207

Zusman, T., Yerushalmi, G., and Segal, G. (2003). Functional similarities between the icm/dot pathogenesis systems of *Coxiella burnetii* and *Legionella pneumophila. Infect. Immun.* 71, 3714–3723. doi: 10.1128/IAI.71.7.3714-3723.2003

17

Secondary Bacterial Infections Associated with Influenza Pandemics

Denise E. Morris[1], David W. Cleary[1] and Stuart C. Clarke[1,2,3]**

[1] *Infectious Disease Epidemiology Group, Academic Unit of Clinical and Experimental Sciences, Faculty of Medicine, Institute for Life Sciences, University of Southampton, University Hospital Southampton Foundation NHS Trust, Southampton, United Kingdom,* [2] *Global Health Research Institute, University of Southampton, Southampton, United Kingdom,* [3] *NIHR Southampton Respiratory Biomedical Research Unit, Southampton, United Kingdom*

Lower and upper respiratory infections are the fourth highest cause of global mortality (Lozano et al., 2012). Epidemic and pandemic outbreaks of respiratory infection are a major medical concern, often causing considerable disease and a high death toll, typically over a relatively short period of time. Influenza is a major cause of epidemic and pandemic infection. Bacterial co/secondary infection further increases morbidity and mortality of influenza infection, with *Streptococcus pneumoniae*, *Haemophilus influenzae,* and *Staphylococcus aureus* reported as the most common causes. With increased antibiotic resistance and vaccine evasion it is important to monitor the epidemiology of pathogens in circulation to inform clinical treatment and development, particularly in the setting of an influenza epidemic/pandemic.

Keywords: influenza, *Streptococcus pneumoniae*, *Haemophilus influenzae*, *Staphylococcus aureus*, pandemic

Edited by:
Chew Chieng Yeo,
Universiti Sultan Zainal Abidin, Malaysia

Reviewed by:
Jeanette Teo,
National University Hospital, Singapore
Siomar De Castro Soares,
Universidade Federal do Triângulo Mineiro, Brazil
Chanwit Tribuddharat,
Siriraj Hospital, Thailand

****Correspondence:***
Stuart C. Clarke
s.c.clarke@soton.ac.uk
Denise E. Morris
d.e.morris@soton.ac.uk

INTRODUCTION

From the Plague of Athens to the present day, infectious disease has beset mankind throughout history. Medical and socio-economic advances have substantially reduced this burden, the eradication of smallpox in 1979 (World Health Organization, 2017) and the remarkable successes against polio and parasitic Guinea worm disease being three examples of an extensive list. Respiratory tract infections, however, continue to be a major cause of morbidity and mortality worldwide (Lozano et al., 2012; Morse et al., 2012; Zumla et al., 2014). When combined, lower and upper respiratory infections are the fourth highest cause of global mortality (Lozano et al., 2012). Epi- and pandemic outbreaks of respiratory infection are a major medical concern, often causing considerable disease and a high death toll, typically over a relatively short period of time. The unpredictable nature of these outbreaks, in terms of their etiology and the reservoirs from which they emerge, the constant emergence of new antigenic variants by mutation, combined with transmission within potentially immunologically naïve populations facilitates the characteristic high proficiency of spread (Morse et al., 2012).

It is well established that both animals and humans can act as reservoirs of infection within which pathogens may adapt and evolve. Examples include *Coxiella burnetii* which typically causes Q fever in cattle, sheep and goats but can also infect humans (Eldin et al., 2017), the plague causing *Yersinia pestis*, infamously transmitted to humans by rats via a flea vector (Yang et al., 2016b), human immunodeficiency virus (HIV) which originated in non-human primates before spreading into the human population (Rupp et al., 2016) and of course the most common example, influenza, which circulates within and between swine, avian and human hosts (amongst others). This cross-species flow can lead to adaptations that result in an increased pathogenicity to susceptible hosts, creating the potential for localized outbreaks or global spread (Murphy, 1998;

Karesh et al., 2012; Morse et al., 2012). Important evolutionary modifications can occur during the timespan of an individual infection, permitting new and evolved strains of pathogens to emerge at an increased rate (Karesh et al., 2012). The evolution of pathogens (particularly zoonotic pathogens which account for 60% of human infectious diseases), and development of pandemics and epidemics, can be described in ecological principles whereby changing environmental pressures or opportunities drive a pathogen to exploit new niches or hosts to survive and thrive. This evolution is influenced by a range of anthropogenic factors, which include population expansion, changing land use and habitat destruction, selective pressures of increased antimicrobial usage, vaccination, global trade and travel (Daszak, 2012; Karesh et al., 2012; Morse et al., 2012).

Pandemics are generally viral in cause. This is thought to be due to their high mutation rate, which is particularly true for RNA viruses such as influenza where high nucleotide substitution and poor proof reading leads to the accumulation of errors in newly synthesized RNA strands. Influenza can also undergo re-assortment during mixed infection. These factors can result in divergence of surface antigens, such as haemagglutinin (HA) and neuraminidase (NA), producing strains not recognized by the human immune system and not covered by extant vaccines (Holland et al., 1982; Webster et al., 1992; Chen and Holmes, 2006; Hampson and Mackenzie, 2006; Jones et al., 2008; Taubenberger and Morens, 2008; Dormitzer et al., 2011; Morse et al., 2012). For instance, influenza A is now known to have 18 subtypes of HA and 11 subtypes of NA (Li et al., 2012; Tong et al., 2012; Wu et al., 2014). This high mutation rate and the emergence of new strains can also make vaccine development and policy difficult to plan and carry out. Due to viral antigenic shift, yearly influenza vaccines are required so the population is sufficiently protected by the vaccine, however, vaccine composition is determined ~8 months in advance of administration. This lag may allow new strains to emerge or for antigenic drift to result in a poor match between vaccine and the circulating strain of influenza. Furthermore as seen in the 2009 influenza pandemic, governments and public health departments face considerable difficulties in the production and distribution of vaccines when faced with sudden or unexpected outbreaks of newly emerged strains (Houser and Subbarao, 2015).

A common complication of respiratory viral disease can be secondary bacterial infection. Noting this association is important as it has clear implications for global health, principally because bacterial co/secondary infection is known lead to increased morbidity (Smith and McCullers, 2014). Co/secondary bacterial infection, as the name suggests, is a bacterial infection that occurs during or after an infection from another pathogen, commonly viruses. A number of viral infections (including infection from influenza virus, respiratory syncytial virus, parainfluenza virus and human metapneumovirus) can be complicated by co/secondary infection by a variety of bacteria including *Streptococcus pneumoniae, Haemophilus influenzae,* and *Staphylococcus aureus.* This association leads to an increased severity of disease and sequela such as pneumonia (Smith and McCullers, 2014). In this review we dwell on influenza pandemics since the late 1800's, focussing on the associations and complications that arise from secondary bacterial infections.

INFLUENZA

Influenza viruses are important zoonotic pathogens as they are highly contagious and one of the most prevalent causes of respiratory infection. Worldwide annual epidemics reportedly cause up to five million cases of severe illness, which result in 250,000–500,000 deaths per year. The majority of deaths caused by influenza occur in young children and people over 65 (World Health Organization, 2016). Reports suggest that each year up to 20% of the United States population may be infected by influenza (Sullivan et al., 1993; Biggerstaff et al., 2014). The virus spreads easily from person to person via aerosol droplets (Hilleman, 2002; Taubenberger and Morens, 2008) and replicates in the upper and lower respiratory tract (Taubenberger and Morens, 2008). Commonly, in non-tropical regions, annual influenza epidemics occur during late autumn and winter. Although less frequent, tropical regions too suffer influenza epidemics, these generally coinciding with the rainy season (Cox and Subbarao, 2000; Biggerstaff et al., 2014).

There are three types of influenza virus, types A, B, and C, each differing in host range and pathogenicity (Taubenberger and Morens, 2008). Type A has been isolated from humans, avian, swine, horses, mink, dogs, seals, and ferrets (Jakeman et al., 1994; Taubenberger and Morens, 2008; Parrish et al., 2015), whilst type B has been isolated from humans, seals (Osterhaus et al., 2000) and ferrets (Jakeman et al., 1994), and type C from humans (Matsuzaki et al., 2002), swine and dogs (Youzbashi et al., 1996). Influenza A and B virions contain several structural antigens and three antigenic surface proteins; HA, NA, and M2/BM2 ion channels (Webster et al., 1992; Hampson and Mackenzie, 2006; Racaniello, 2009; Dormitzer et al., 2011). Influenza virus C only expresses one antigenic surface protein, haemagglutinin-esterase-fusion (HEF), and thus stimulates a lesser immune reaction than types A or B (Taubenberger and Morens, 2008; Racaniello, 2009). Influenza A is the fastest to evolve, at a rate 2–3 times faster than B, whilst C is the slowest (Yamashita et al., 1988). Antigenic drift allows the influenza virus to escape immunity acquired through previous exposure or vaccination; thus influenza A causes more epidemics and pandemics than either influenza B or C (Hampson and Mackenzie, 2006; Taubenberger and Morens, 2008). Whilst influenza B causes periodic/yearly epidemics but not pandemics, influenza C viruses only cause relatively infrequent mild respiratory problems (Taubenberger and Morens, 2008). Throughout the past 300 years there have been 12 pandemics caused by influenza A; the most infamous being the 1918 'Spanish flu' pandemic (Taubenberger and Morens, 2008). In the years between 1933 and 1957 there were nine influenza A (H1N1) epidemics and five influenza B epidemics. The worst of all these epidemics was the 1935–1936 influenza B epidemic that resulted in at least 55,000 deaths.

This was closely followed by the 1943–1944 influenza A (H1N1) epidemic which caused 53,000 deaths (Glezen, 1996). Evidently, although influenza B doesn't cause pandemics, it is still a cause for concern.

During an infection influenza virions attach to and enter host epithelial cells by the binding of viral HA to sialic acid on the host cell which instigates endocytosis and the movement of the virion into the cell within an endosome. The virus then uses/hijacks the host cells 'machinery' to replicate and transcribe viral RNA and produce more viral components (Samji, 2009). Progeny virions bud from the host cell, using the host cell membrane as a viral envelope, and go on to infect neighboring host cells (Nayak et al., 2009). As influenza infection develops the virus causes cell damage and death within the host's airways and up-regulates the production of toxins, causing further destruction. Influenza cytotoxins for example causes necrosis of host cells (Conenello and Palese, 2007; Iverson et al., 2011). Influenza infection, particularly pandemic influenza infection, is known to generate an increased inflammation response within the host, as the body works to rapidly deliver immune cells to the site of infection. This inflammation is a response to the expression of cytokines and chemokines (de Jong et al., 2006; Kash et al., 2006; Kobasa et al., 2007; Rock and Kono, 2008). Virally induced decreased mucociliary activity, the dysfunction of immune cells and the reduction of phagocytosis reduces clearance of the virus from the host airways and the host's ability to fight the virus (Brundage, 2006; Wu et al., 2011; Cauley and Vella, 2015). In an attempt to limit and control infection, the host immune system kills infected host cells. It does this in several ways, including; the production of perforin by Natural Killer (NK) cells which creates lesions/pores in cell membranes resulting in the induction of apoptosis, apoptosis from tumour necrosis factor (TNF) and FasL and the production of reactive oxygen species from macrophages and neutrophils causing oxidation of cellular lipids, proteins and DNA resulting in cell dysfunction and death (Topham and Hewitt, 2009; Kash et al., 2014; Kash and Taubenberger, 2015). Of course viral infection and/or interference with host processes can cause and direct the pathway of cell death, as is the case for necrosis. Host cell death, whether apoptosis, necrosis or pyroptosis, impacts on the severity and outcome of influenza disease in a variety of ways. Virally induced death of immune cells assist in the evasion of host defenses and hinders the clearance of the virus promoting the development of infection. Studies have shown a 90% reduction of alveolar macrophages in mice within a week of influenza infection, and evidence of necrosis in the remaining macrophages (Robinson et al., 2015). Necrosis and pyroptosis are pro-inflammatory due to their role in the release of cytokines. These cell death pathways allow for the rapid release of intracellular contents, including any viral components, from the infected host cell promoting host inflammatory responses and the formation of a cytokine storm which causes host tissue damage (Cundell et al., 1995; Rock and Kono, 2008; Lamkanfi and Dixit, 2010; Cauley and Vella, 2015). Furthermore infection with some influenza subtypes, for instance H1N1 and H5N1, typically result in lymphopenia, a state of abnormally low levels of lymphocytes, which has been associated with higher viral load. de Jong et al. (2006) found influenza infection caused lower levels of cytotoxic T cell lymphocytes, which would therefore negatively affect acquired immunity (de Jong et al., 2006; Cunha et al., 2009). Where lymphopenia occurs, studies have shown a corresponding increase in macrophages. Supporting the evidence for the increase in macrophages is the significant increase in IP-10 (a chemokine secreted in response to gamma interferon (IFNγ) which activates macrophages), Interleukin-8 (IL-8, a chemokine which is produced by macrophages), IL-6 (in this case, a pro-inflammatory cytokine secreted by macrophages), and MCP-1 (a chemokine that recruits monocytes, a type of leukocyte that can differentiate into macrophages) (de Jong et al., 2006; Kobasa et al., 2007).

INFLUENZA PANDEMICS SINCE THE LATE 1800's

Influenza pandemics, generally characterized by the emergence of a novel influenza A against which little or no immunity exists within the global populace, are a cause of high mortality and morbidity and are a major financial burden (Glezen, 1996). Since the 1800's these pandemics have arisen from a number of countries, spreading across the globe (**Figure 1**). Detailed below and in **Table 1** we have sought to describe some of the most significant influenza pandemics since the late 1800's to highlight the potential impact of influenza with respect to associations with bacterial infection.

BACTERIAL CO-INFECTION AND SECONDARY INFECTIONS

Laboratory, clinical and epidemiological research has made it abundantly clear that bacterial co/secondary infection can significantly increase the morbidity and mortality of viral infections (Gupta et al., 2008). Up to 75% of those infected with influenza that go on to acquire pneumonia, are confirmed to have bacterial co-infection (Zambon, 2001). Bacterial co/secondary infection of influenza infection appears to occur frequently. Studies have shown that up to 65% of laboratory confirmed cases of influenza infection exhibited bacterial co/secondary infection, although Klein et al. (2016) state that in the majority of the research included in their meta-analysis this figure ranged between 11 and 35%. In the setting of an influenza epidemic or pandemic bacterial co/secondary infection can have devastating consequences, particularly in at-risk groups such as the immunocompromised/immunosuppressed. Immunosuppression is associated with more severe morbidity and a much higher risk of mortality from co/secondary bacterial infection (Rice et al., 2012). During the 2009 Swine influenza pandemic, there was an increase in hospital pneumonia cases as a result of secondary bacterial pneumonia, which was identified in 29–55% of mortalities (Centers for Disease Control and Prevention, 2009; Gill et al., 2010; Weinberger et al., 2012).

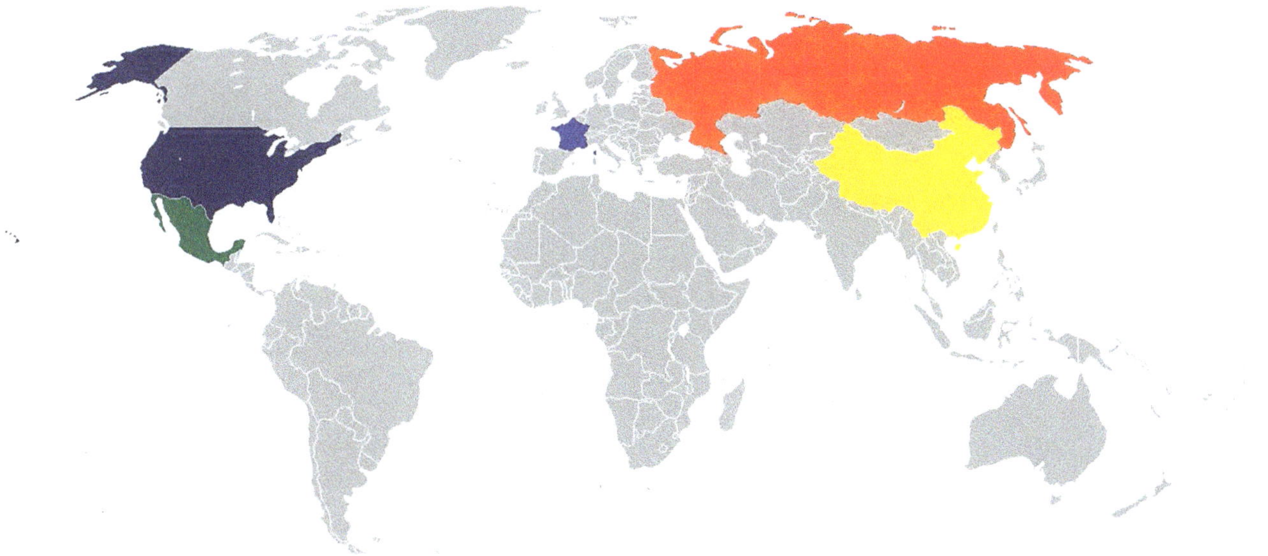

FIGURE 1 | World Map showing countries confirmed and suspected of being the origin of influenza pandemics. Blue – The origin of the 1918 Spanish is still unclear, although various papers suggest the United States (New York) or France as the origin; yellow – China the origin of the 1957 Asian flu pandemic; Hong Kong, the origin of the 1968 Hong Kong pandemic; red – Russia, the origin of the 1889 and 1977 Russian flu pandemics; green – Mexico, the origin of the 2009 Swine flu pandemic.

PATHOBIONTS ASSOCIATED WITH CO/SECONDARY BACTERIAL INFECTION

The upper respiratory tract has been shown to host a diverse microbiota, within which a number of bacterial pathobionts may be found, i.e., those bacterial species that can be pathogenic yet also harmlessly carried (Hooper et al., 2012; Cauley and Vella, 2015). *Legionella pneumophila* (Rizzo et al., 2010), *Streptococcus pyogenes* (Chertow and Memoli, 2013), *Neisseria meningitidis*, *Moraxella catarrhalis*, *S. pneumoniae*, *H. influenzae*, *S. aureus* (Dela Cruz and Wunderink, 2017), *Pseudomonas aeruginosa* as well as a number of other *Streptococcus* and *Staphylococcus* spp. (Yang et al., 2016a) have all been associated with co-infection of influenza. However, *S. pneumoniae*, *H. influenzae*, and *S. aureus* are the most commonly reported bacteria associated with co/secondary infections during influenza pandemics since the late 1800's.

Streptococcus pneumoniae

Streptococcus pneumoniae is the most common bacteria found in viral secondary bacterial infections, and is particularly associated with causing high mortality and morbidity during influenza epidemics and pandemics (Brundage, 2006; Joseph et al., 2013). *S. pneumoniae* is a Gram-positive diplococci and is the most common cause of community-acquired pneumonia and invasive disease, i.e., sepsis and meningitis worldwide, as well as less severe acute disease such as otitis media (Bridy-Pappas et al., 2005; McCullers et al., 2010). *S. pneumoniae* is grouped into >97 immunologically distinctive serotypes based on a polysaccharide capsule (Bentley et al., 2006; Park et al., 2007; Jin et al., 2009; Calix and Nahm, 2010; Calix et al., 2012). A burden to public health in its' own right, the WHO has reported that diseases caused by *S. pneumoniae* resulted in approximately 826,000 deaths in 2000 alone (Pittet and Posfay-Barbe, 2012). A more recent study shows that there are 4 million cases of disease caused by *S. pneumoniae* and 22,000 deaths annually in the United States (Huang et al., 2011). The current public health impact of *S. pneumoniae* infection is reduced by vaccine policies, with, for example, PCV-13 and PPV-23 being used for children and adults, respectively, in the United Kingdom (Pittet and Posfay-Barbe, 2012).

Many studies have shown that influenza infection facilitates the acquisition, colonization and development of disease from *S. pneumoniae* in people of all ages (Shrestha et al., 2013; Grijalva et al., 2014; Siegel et al., 2014). This is partly due to *S. pneumoniae's* ability to catabolise sialic acid which is released from host cells and mucus by influenza's NA. Influenza infection also results in increased mucus production, further increasing the amount of metabolite available for *S. pneumoniae*. The NA produced by *S. pneumoniae* also assists in the release of sialic acid (Siegel et al., 2014). Mouse models support the concept that influenza facilitates the development of disease from *S. pneumoniae*; they have provided evidence that influenza infection enhances secondary *S. pneumoniae* pneumonia (McCullers and Rehg, 2002; McCullers and Bartmess, 2003). Wu et al. (2011), showed that co-infection of a virus and a bacterium can either occur from mixed viral bacterial infection, or from a viral infection being sequentially followed by a bacterial infection. Sequential bacterial infection normally occurs within a 7-day period of the viral infection. Influenza infections and successive *S. pneumoniae* infections result is a time and dose

TABLE 1 | Details of significant influenza pandemics since the late 1800's.

Name of pandemic	Year	Strain	Disease burden	Additional information
Russian influenza pandemic	1889	A(H2N2)	132,000 deaths in England, Wales, and Ireland alone (Smith, 1995).	The 1889 'Russian Flu' as the name suggests started in Russia and spread across Europe reaching North America in 1890. In only 4 months the infection had spread across all of Europe and the United States. The United Kingdom encountered four waves of disease and it is thought that at least one third of the adult population in England and Ireland suffered at least one bout of disease (Smith, 1995; Valleron et al., 2010).
Spanish influenza pandemic	1918	A(H1N1)	Caused 40–50 million deaths worldwide. India alone suffered 7 million deaths (Potter, 2001; Hilleman, 2002; Brundage, 2006; Michaelis et al., 2009).	Considered the most devastating influenza pandemic ever recorded, infecting 50% of the world's population. The origin of this pandemic is unclear as it appeared in North America, Asia, and Europe at roughly the same time (Taubenberger et al., 2001; Hilleman, 2002). Reports of disease and mortality were initially suppressed in many countries, included the United Kingdom, France and the United States, to ensure wartime efforts and morale weren't negatively affected. In Spain the press were able to print freely, meaning the first publicized cases were reported from Spain facilitating the nickname 'the Spanish flu' (Johnson, 2016; Peckham, 2016). In contrast to its name, it has been suggested that the pandemic started in France/mainland Europe and that it reached Spain from France (Reid et al., 2001; Trilla et al., 2008) although more recent papers suggest New York as the origin due to evidence of a pre-pandemic wave of the H1N1 virus (Olson et al., 2005). What is remarkable is how far the pandemic spread; the pandemic reached as far as the Alaskan wilderness to remote Pacific islands (Burnet and Clark, 1942; Taubenberger et al., 2001). The pandemic experienced a couple of waves; the first of which was relatively mild. The second wave, however, was far more lethal (Hilleman, 2002). The first outbreaks were reported in military camps as males responded to the call for troops in the spring and summer of 1918. A period of dormancy was then recorded toward the end of summer in America, but this was short lived as transmission picked up as schools reopened in September after the summer holidays (Glezen, 1996).
Asian influenza pandemic	1957–1958	A(H2N2)	Although global death toll estimates vary [between 1.5 million (Gatherer, 2009) and 2–4 million (Michaelis et al., 2009)], the death toll in the United States is accurately reported to have been 69, 800 (Klimov et al., 1999; Hilleman, 2002).	The pandemic affected 40–50% of people worldwide (Potter, 2001), however, resulted in lot less mortality than the previous pandemic. This Asian influenza pandemic started in March 1957 in Southern China, where pigs, ducks, and humans live together closely. It reached Hong Kong in April, and then spread to Singapore, Taiwan, and Japan (Fukumi, 1959; Potter, 2001; Hilleman, 2002). The pandemic reached India, Australia, and Indonesia by May, Pakistan, Europe, North America, and the Middle East by June, South Africa, South America, New Zealand, and the Pacific Islands by July, and Central, West and East Africa, Eastern Europe, and the Caribbean by August (Dunn, 1958; Payne, 1958; Potter, 2001).
Hong Kong influenza pandemic	1968–1969	A(H3N2)	1–2 million people died worldwide (Michaelis et al., 2009). Overall 33,800 people died in the United States (Klimov et al., 1999) and England and Wales saw a 55% increase in respiratory deaths in 1969 (Tillett et al., 1983).	The 1968 Hong Kong pandemic started in July 1968 in Hong Kong and spread to the Southern hemisphere by June 1969 (Biggerstaff et al., 2014). The H3N2 virus was isolated and identified too late in the pandemic for vaccine intervention (Nakajima et al., 1978; Hilleman, 2002) so it was fortunate that in most countries, apart from the United States, the disease was mild (Cockburn et al., 1969). There are several proposed reasons for the reduced mortality of this compared to the Asian Flu. Firstly the N2 was seen in the Asian Flu so may have contributed some cross-reactive immunity to this H3N2 strain (Glezen, 1996). Although antibodies to NA do not prevent infection, they help to reduce the amount of newly formed virus released from infected cells (Couch et al., 1974; Glezen, 1996). Secondly, during the initial wave of this pandemic, the number of cases started to grow exponentially in December, at this point the school Christmas holidays began; it has been speculated that this removed an important susceptible population (Glezen, 1996).
Russian Flu influenza pandemic	1977–1978	A(H1N1)	Approximately 700,000 deaths globally (Michaelis et al., 2009).	This pandemic was caused by a reappearance of H1N1, identical to that of the Spanish flu virus (Michaelis et al., 2009). The disease mainly affected those born after the late 1950's, so those who had not been exposed to the pandemic H1N1 strain that had circulated previously (Hilleman, 2002).
Swine influenza pandemic	2009	A(H1N1)	By the end of the pandemic it is thought that there were 284,000 deaths worldwide (Chertow and Memoli, 2013).	In early 2009, an influenza A H1N1 virus outbreak was initially identified in Mexico and then the United States (Michaelis et al., 2009). In June 2009 the WHO declared the outbreak a pandemic. Within 4 weeks the outbreak had spread to 41 countries, resulting in 11,034 confirmed cases and 85 deaths (Michaelis et al., 2009; Wang and Palese, 2009). Disease/symptoms were generally mild (Peiris et al., 2009) however, complications of the disease did result in hospitalization, particularly in at risk groups (Wang and Palese, 2009).

(Continued)

TABLE 1 | Continued

Name of pandemic	Year	Strain	Disease burden	Additional information
				Unlike other pandemics and yearly epidemics, during this pandemic it was predominantly children and young adults that were affected, particularly those aged 12–22 (Gill et al., 2010). Overall this pandemic was relatively mild. It is thought that morbidity and mortality rates were reduced due to three main factors. Firstly, the quick responses of various governments in terms of school closures helped reduce the spread of the virus. Thousands of schools were shut worldwide, including the United States and Mexico. Japan alone closed almost 2000 schools (Wang and Palese, 2009; Jackson et al., 2014). Secondly, influenza A H1N1 strains have been circulating amongst the human population for decades, therefore prior exposure could have provided some degree of immunity against the 2009 pandemic strain. Lastly an important pathogenicity factor, PB1-F2, was not present making the strain milder than those present in previous pandemics (Wang and Palese, 2009).

dependent change in the host dendritic cells which produces enhanced inflammation. Berendt et al. (1975) inoculated squirrel monkeys with either influenza A, *S. pneumoniae* or influenza A and *S. pneumoniae.* Influenza alone caused minor illness such as mild tracheitis, with symptoms such as sneezing, coughing and fever (although some did develop bronchopneumonia) and had a 100% survival rate. *S. pneumoniae* again caused minor illness with a 100% survival rate. Co-infection of influenza A with *S. pneumoniae* resulted in severe morbidity with a 75% death rate within 40 h, clear evidence of the consequences of co/secondary bacterial infection (Berendt et al., 1975). These findings are reflected in several other studies, with some even showing that co-infection may assist in the spreading of *S. pneumoniae* infection to the lower respiratory tract (Takase et al., 1999; Seki et al., 2004).

An additional mouse model of infection provided comparable results whilst comparing the effect of different *S. pneumoniae* serotypes on co-infection (Sharma-Chawla et al., 2016). More cases of pneumonia and bacteraemia were observed in mice infected with both influenza A and *S. pneumoniae* than in mice infected with these pathogens individually. This was the case for all *S. pneumoniae* serotypes tested. More virulent pneumococcal serotypes caused a greater burden of disease in both the co-infected mice and those infected with *S. pneumoniae* alone. The highly invasive pneumococcal serotype 4 caused pneumonia in 58% of mice and bacteraemia in 21% in a single infection model. When co-infecting with influenza these figures increased to 100 and 90% for pneumonia and bacteraemia, respectively. Mortality rates increased from 0% for individual infection to 79% during co-infection. In comparison, individual infection with a carrier strain (of lower invasive potential) of serotype 19F, caused pneumonia in 91% of cases and bacteraemia in 0%. When co-infecting with influenza and 19F these figures increased to 100 and 33%. Mortality rose from 0% during individual infection to 63% during co-infection (Sharma-Chawla et al., 2016).

Pneumococcal vaccination has shown to ameliorate the risk of secondary bacterial pneumonia. During a vaccine efficacy study, the incidence of pneumonia in those with influenza reduced by 45% in groups vaccinated against *S. pneumoniae* (Madhi et al., 2007). However, whilst vaccine implementation has successfully reduced pneumococcal disease in a number of countries, lower levels of vaccine implementation in low and middle income countries coupled with fractional serotype coverage and increasing levels of antibiotic resistance, means the specter of influenza pandemic associated *S. pneumoniae* secondary infection remains a significant risk to global health.

Haemophilus influenzae

Haemophilus influenzae is another bacteria commonly found to co/secondarily infect viral infection, and has been associated with the complication of disease during influenza pandemics (Abrahams et al., 1919; Spooner et al., 1919; Brundage, 2006; Palacios et al., 2009). It is a Gram-negative fastidious coccobacillus. Typeable strains have a polysaccharide capsule and are categorized into six serotypes (A–F). *H. influenzae* serotype B was a major cause of invasive disease (Wenger, 1998; Murphy, 2003; Chin et al., 2005; Brouwer et al., 2010) although widespread implementation of the Hib vaccine has significantly reduced the burden of disease (Rosenstein and Perkins, 2000). Those *H. influenzae* that lack a capsule, denoted non-typeable *H. influenzae* (NTHi), remain a significant cause of bacterial meningitis, otitis media and exacerbations of chronic lung disease such as COPD worldwide (Langereis and de Jonge, 2015).

Various studies have shown the impact when *H. influenzae* co/secondarily infects with influenza, and some suggest a level of synergism. The effect of influenza and *H. influenzae* co-infection verses individual infection of both pathogens is tellingly different; Shope found that co-infection resulted in severe disease or death when on their own *H. influenzae* and influenza only caused mild infection or disease (Shope, 1931). More recently, Lee E.H. et al. (2010) undertook a similar study which provided comparable results and evidence that influenza and *H. influenzae* co-infection produces more epithelial cell destruction than single infection with either pathogen (Lee L.N. et al., 2010). Furthermore, they found individual infection caused mild bronchiolitis within 4 days of initial infection, from which the host lung was able to recover. Conversely, co-infection caused bronchial necrosis, bronchial inflammation and bronchitis within the same time period or less, and led to further complication such as epithelial erosion (Lee L.N. et al., 2010). It is now commonly accepted that co-infection results in more severe morbidity and poorer clinical outcome than infection of influenza *or H. influenzae* alone.

Further support of the impact of co-infection comes from Michaels et al. (1977), who dosed two groups of rats intranasally with *H. influenzae* with the intention of giving them meningitis. One group of rats were naive and the other had previously been dosed with influenza. In both groups ~50% of the rats acquired meningitis, however, the naïve rats required a 100-fold larger dose of *H. influenzae* (Michaels et al., 1977).

As is the case for many bacterial and viral co-infections, mortality from *H. influenzae* and influenza co-infection is highly dependent on the timing of the introduction of the secondary microbe as well as density of bacterial colonization. Studies have shown that when influenza virus and *H. influenzae* are introduced at the same time there is no synergistic relationship. When *H. influenzae* is introduced 7 or more days after influenza there is again no synergistic relationship; however, high lethality is exhibited when *H. influenzae* and influenza are introduced 3 or 4 days apart (Lee L.N. et al., 2010).

Staphylococcus aureus

Staphylococcus aureus is a Gram-positive cocci that has been found to complicate influenza infection; increasingly so in more recent years/pandemics (Hers et al., 1958; Palacios et al., 2009). *S. aureus* is transiently carried in the nose of 30% of the population, whilst 20% of the population have persistent nasal colonization (Wertheim et al., 2005). Like *H. influenzae* and *S. pneumoniae*, *S. aureus* is an opportunistic pathogen and a major cause of bacteraemia (Wertheim et al., 2005; Tong et al., 2015). It is also a common cause of pneumonia (Kollef et al., 2005); specifically necrotising pneumonia that is caused by community acquired Methicillin-resistant *Staphylococcus aureus* (MRSA) and has a 30% mortality rate (Murray et al., 2010). Necrotising pneumonia is highly associated with either the presence of Panton-Valentine leukocidin (PVL) or prior/co influenza infection (DeLeo and Musser, 2010). MRSA is a particularly problematic pathogen and concern for public health as it can be hard to treat due to its multidrug-resistant properties (Wu et al., 2005; Eom et al., 2014; Fishovitz et al., 2014).

Influenza infection has been shown to increase the adherence of *S. aureus* (as well as *H. influenzae* and *S. pneumoniae*) to host pharyngeal cells (Fainstein et al., 1980). In addition to this, mouse models have highlighted increased morbidity and mortality in mice that are pre-infected with influenza before they are exposed to *S. aureus* vs. those just exposed to *S. aureus*. Increased lung damage and bacterial density has also been shown (DeLeo and Musser, 2010; Lee M.H. et al., 2010; Iverson et al., 2011). Lee M.H. et al. (2010) showed that mice infected with low doses of influenza, low doses of *S. aureus* and high doses of *S. aureus* were able to survive. Those infected with high doses of influenza died within 4–7 days; however, all mice infected with a high dose of influenza and then a high dose *S. aureus* died within 2 days of bacterial exposure, showing how death can be accelerated by co-infection. When mice were infected with a low dose of influenza and then a high dose *S. aureus* they died at 7 days. The fact that the mice survived low influenza infection on its own, but could not survive co-infection with *S. aureus* shows the lethality of such co/secondary bacterial infection (Lee M.H. et al., 2010).

In an act of synergism, *S. aureus* infection may actually assist influenza infection by increasing the infectivity of influenza; when the virion is being moved into the host cell within an endosome the low pH in the endosome causes a conformational change to the HA [$HA_{(0)}$] allowing it to be cleaved by host proteases into two subunits [$HA_{(1)}$ and $HA_{(2)}$]. This cleaving 'activates' the HA, mediating fusion between the virus and endosome membrane, ready for the opening of the M2 ion channel so the vRNP (viral ribonucleoproteins) can be released into the host cell where the viral RNA is replicated and transcribed. Several strains of *S. aureus* produce proteases that cleaves influenza HA; the more protease that is available, the more HA can be cleaved meaning more vRNP can get into host cells meaning overall more progeny virions (Tashiro et al., 1987; Steinhauer, 1999; Samji, 2009). This aspect contributes to the increased severity of disease caused by co-infection verses individual influenza infection. And although not all strains of *S. aureus* produce proteases that cleave influenza HA, the proteases they do produce indirectly enhance morbidity by causing host inflammatory responses which result in the production of host enzymes that are capable of cleaving HA (Tashiro et al., 1987).

HISTORICAL EVIDENCE OF CO/SECONDARY BACTERIAL INFECTION DURING MAJOR INFLUENZA PANDEMICS

1918 Spanish Influenza Pandemic

The 1918 influenza pandemic was a result of influenza strain A (H1N1). It is considered the most devastating influenza pandemic ever recorded, infecting 50% of the world's population and resulting in approximately 40–50 million deaths worldwide. India alone suffered 7 million deaths (Potter, 2001; Hilleman, 2002; Brundage, 2006; Michaelis et al., 2009). The main groups of individuals affected by this pandemic were those aged 20–40 years old, in addition to infants and those over 65. Ordinarily only young children and the elderly are the age groups most at risk from influenza, showing how distinctive pandemic strains can be (Potter, 2001). It is suggested that war time efforts meant that influenza easily spread through military camps, allowing the 20–40 years old age range to be more at risk than usual.

There are many published examples of co/secondary bacterial infections during the 1918 influenza pandemic, and pneumonia as a consequence of bacterial infection is estimated to have occurred in up to 95% of deaths during this pandemic (Morens et al., 2008). A majority of those deaths due to secondary *S. pneumoniae* infection (Brundage and Shanks, 2008; Morens et al., 2008). Many of the examples that detail co/secondary bacterial infection come from outbreaks within army camps. Within a 1-month period in 1918 at the military Camp Devens, a quarter of all troops were diagnosed with influenza. Of those infected, 17% developed pneumonia, of which 35% of cases were fatal. Out of 37 autopsies performed, 43% were positive for pure growth of *H. influenzae* in at least one lobe of the lung. Blood culture revealed 65% had *S. pneumoniae*, 2.5% had *H. influenzae* and 1.3% had *S. aureus* (Spooner et al., 1919; Brundage, 2006). This pattern of invasive bacterial co/secondary infection has also been documented for several other camps during the same year, including Camp Logan. Here 2,487 influenza-associated hospitalizations were recorded, 17% acquired pneumonia with 4% of these cases being fatal. Post-mortems found *S. pneumoniae* in the lungs of 44% and heart blood of 33% (Hall et al., 1918; Brundage, 2006). At Camp Jackson, 17% of influenza cases progressed to pneumonia with a further 31% of pneumonia cases proving fatal. Autopsies found *S. pneumoniae* to be the bacterial co-infection most associated with pneumonia, however, 155 of 312 lung cultures were positive for *S. aureus* (Michael, 1942; Brundage, 2006). At Camp Custer, 21% of influenza cases progressed to pneumonia, of which 28% died. Sputum cultures proved the presence of *S. pneumoniae* in 26% of cases. Further investigation found 28% of lung and blood cultures were positive for *S. pneumoniae,* again acting as supporting evidence of the invasive potential of such co-infections (Blanton and Irons, 1918; Brundage, 2006). Camp Fremont experienced 2418 hospitalizations, 17% had pneumonia of which 36% were fatal. Nasopharyngeal and sputum samples from 158 pneumonia cases found *S. pneumoniae* in 41% of cases, *H. influenzae* in 38%, and other *Streptococcus* spp. in 29% (Brem et al., 1918).

Further lung tissue from fatalities of this pandemic were re-examined in 1919; *S. pyogenes longus* was found in 36% of cases, *S. pneumoniae* in 29% of cases and *H. influenzae* in 25% (Abrahams et al., 1919; Brundage, 2006). Additional post-mortems of lung tissue suggest that at least 90% of samples showed evidence of bacterial infection (Oxford et al., 2002; Morens et al., 2008; Chien et al., 2009). Overall 95% of deaths were due to co/secondary bacterial pneumonia (Opie et al., 1921; Morens et al., 2008).

Co-infection had also been reported as an issue prior to the official start of the pandemic. Influenza with secondary bacterial infection of *S. pneumoniae* (and other *Streptococcus* sp.), *H. influenzae* and/or *Staphylococcus* sp. was associated with major outbreaks of purulent bronchitis in 1916 and 1917 (Brundage, 2006; Joseph et al., 2013). Indeed in 1916–1917 British, Australian, Canadian, and American armed forces in England and France experienced an epidemic of purulent bronchitis. Out of 20 tested sputum specimens from a British army camp based in north France, 90% presented with *H. influenzae*, 65% presented with *S. pneumoniae*, 25% with other *Streptococcus* spp. and 15% with *Staphylococcus* spp. Out of the specimens positive for *H. influenzae*, many exhibited simultaneous *H. influenzae* and *S. pneumoniae* co-infection; with *H. influenzae* identified as the primary bacterial infector. *S. pneumoniae* infection first presented with low virulence, however, pathogenesis soon worsened, it has been suggested, as result of the symbiotic growth with *H. influenzae* (Brundage, 2006; Dennis Shanks et al., 2012). Of course it is known that there is a positive association between the colonization of *H. influenzae* and *S. pneumoniae*, and colonization is a prerequisite for disease, so the presence of such co-infection fits with current knowledge (Jacoby et al., 2007; Abdullahi et al., 2008).

1957 Asian Influenza Pandemic

This pandemic affected 40–50% of people worldwide. The cause was influenza strain A (H2N2) (Potter, 2001). Although global death toll estimates vary [between 1.5 million (Gatherer, 2009) and 2–4 million (Michaelis et al., 2009)], the death toll in the United States is accurately reported to have been 69, 800 (Klimov et al., 1999; Hilleman, 2002). Post-mortem cultures show evidence of bacterial infection in up to 80% of all severe and fatal cases (Hers et al., 1958; Morens et al., 2008; Gill et al., 2010).

During this pandemic the United States, and many other countries, experienced an increase in hospitalization rates. A majority were due to pneumonia, predominantly caused by *S. pneumoniae, H. influenzae,* and *S. aureus* infection (Petersdorf et al., 1959). There are similar documented reports from the Netherlands; of the 148 deaths presumed to be from the Asian pandemic influenza strain that were examined fully, 75% presented with bacterial pneumonia of which 15% were positive for *S. pneumoniae* and 59% were positive for *S. aureus* (Hers et al., 1958).

Robertson et al. (1958) unveiled similar findings when investigating the hospitalization of 140 people suffering pneumonia at Sheffield City General Hospital in 1957. A majority showed evidence of influenza A infection; 27% of those had co/secondary infection of *S. aureus* (which had a 47% death rate), 15% *S. pneumoniae* and 4% *H. influenzae*, although this is likely to be an underestimation as many patients had already started taking antibiotics (Robertson et al., 1958).

1968–1969 Hong Kong Influenza Pandemic

Worldwide 1–2 million people died during this pandemic which was caused by the influenza strain A(H3N2) (Michaelis et al., 2009). Although this is a lower death toll than engendered in previous pandemics, it is still an awfully high number of deaths. Overall 33,800 people died in the United States (Klimov et al., 1999) and the pandemic cost 3.9 billion dollars (Hilleman, 2002). In 1969, England and Wales saw a 55% increase in respiratory deaths, of which co/secondary bacterial infection was shown to be a major contributor (Tillett et al., 1983).

Staphylococcal pneumonia in particular was a major source of complication to influenza infection. A hospital in Atlanta suffered a threefold increase in cases of Staphylococcal pneumonia during this pandemic. Staphylococcal infection caused 26% of pneumonia cases during this period, and a high correlation was recognized between influenza infection and bacterial pneumonia (Schwarzmann et al., 1971). In addition, out of 79 cases of fatal influenza with pneumonia complications, 16% had bacterial co-infection with *S. pneumoniae* (6%), *S. pyogenes* (5%), and *S. aureus* (1%) being the main causes. More than one of these bacteria were present in 4% of cases (Schwarzmann et al., 1971; Metersky et al., 2012).

Another health care facility in the United States, the Mayo Clinic in Minnesota, also found *S. aureus* to be a major cause of complication. Of 129 adults diagnosed with pandemic influenza, pneumonia was established in 16%, of which 40% of these cases (6% of all 129 influenza cases) were fatal. *S. aureus* or *P. aeruginosa* bacterial infection was present in 75% of all fatal cases, indicating bacterial co/secondary infection was a major determinant of severe disease and death (Lindsay et al., 1970).

In previous pandemics *S. pneumoniae* has been proposed as the major contributor of mortality and morbidity, however, during this 1968–1969 Hong Kong and the 1957 Asian influenza pandemic *S. aureus* clearly had a larger impact. This is possibly a reflection of increased antibiotic use and increased antibiotic resistance.

2009 Swine Influenza Pandemic

Within 4 weeks this outbreak of influenza A(H1N1) had spread to 41 countries resulting in 11,034 confirmed cases and 85 deaths (Michaelis et al., 2009; Wang and Palese, 2009). By the end of the pandemic it is thought that there were 284,000 deaths worldwide, with Mexico and the United States being most severely affected (Chertow and Memoli, 2013). Unlike other pandemics and yearly epidemics, during this pandemic it was predominantly children and young adults that were affected, particularly those aged 12–22 (Gill et al., 2010). Influenza A (H1N1) strains have been circulating amongst the human population for many years therefore this prior exposure could have provided many adults with some degree of immunity against the 2009 pandemic strain, particularly older groups who were more likely exposed during previous pandemics.

Surveillance by the New York City Department of Health and Mental Hygiene has shown that during the 2009 Swine Flu Pandemic almost 30% of the first 47 deaths showed invasive bacterial disease. *S. pneumoniae* was the most common causative agent identified (followed by *S. pyogenes*) (Lee E.H. et al., 2010). In the United Kingdom, of the 457 fatalities 68 were autopsied. Of these, 41% were shown to have complications associated with secondary bacterial infection, most commonly (25% of cases) due to *S. pneumoniae* (Lucas, 2010).

Further studies in the United States have reviewed 77 deaths during the period of May–August 2009 and found bacterial co-infection in almost 30% of cases; 46% of which were with *S. pneumoniae*, 9% with *S. aureus* and 1% with *H. influenzae* (Centers for Disease Control and Prevention, 2009). Studies based in Argentina produced similar evidence for the presence of bacterial infection, showing this wasn't just a localized trend. Palacios et al. (2009) examined nasopharyngeal swab samples from almost 200 cases of pandemic influenza. *H. influenzae* was found in 52%, *S. pneumoniae* was found in 31% and *S. aureus* in 18% of samples. Although not the most common bacteria found, *S. pneumoniae* was the most strongly associated with severe disease (Palacios et al., 2009).

Additional research in pediatric intensive care units in the United States, investigated 838 critically ill children who were infected with pandemic influenza. Within 72 h of admission to the intensive care unit 33% exhibited bacterial co-infection; in 26% of these cases *S. aureus* was identified as the cause (48% of which were MRSA), 5.5% were positive for *S. pneumoniae* and 5% were positive for *H. influenzae.* Bacteraemia was observed in 5% of admissions, for which *S. aureus* was the main cause (Randolph et al., 2011). This study highlights how quickly co/secondary bacterial infection can become invasive particularly in at risk groups such as young children or the elderly. A point of concern is that almost half of the *S. aureus* were MRSA, and therefore inherently resistant to multiple antibiotics.

In another study of vulnerable and critically ill children in a pediatric intensive care unit in the United States, 51% of those with influenza infection had bacterial co/secondary infection. Of these 35% presented with *S. aureus*, 18% *P. aeruginosa*, 18% *M. catarrhalis*, 9% NTHI, 6% *S. pneumoniae* and 6% Group A Streptococcus. Those with *S. aureus* showed more severe morbidity and were more likely to develop disseminated intravascular coagulation which leads to a compromised blood flow within body tissue and therefore tissue damage (Nguyen et al., 2012).

In a retrospective study of 50 patients who were infected during pandemic influenza, 28% showed co/secondary bacterial infection (Dhanoa et al., 2011). *Mycoplasma pneumoniae* was found in 10%, making it the most common co/secondary infecting bacteria. This was followed by *S. aureus* found in 6%, *K. pneumoniae* and *S. pneumoniae* found in 4% and *M. catarrhalis*, *P. aeruginosa*, *S. pyogenes*, and *Streptococcus agalactiae* found in 2% of these patients (Dhanoa et al., 2011).

Moraxella catarrhalis is a bacteria of increasing importance being now acknowledged as the third most common cause of otitis media (OM), after *S. pneumoniae* and *H. influenzae* (Bluestone, 1986; Faden et al., 1994; Kilpi et al., 2001; Dupont et al., 2010) and the second most common cause of exacerbations in COPD, accounting for up to 4 million exacerbations per year in the United States alone (Murphy et al., 2005). *M. catarrhalis* is a cause of pneumonia (Berg and Bartley, 1987; Hager et al., 1987; Marchant, 1990; Verduin et al., 2002) and invasive disease such as bacteraemia (Ioannidis et al., 1995) and meningitis (Newing and Christie, 1947), with bacteraemia being a common complication of pneumonia, particularly in adults (Collazos et al., 1992; Ioannidis et al., 1995). Although this review has focused on *S. pneumoniae, H. influenzae,* and *S. aureus*, it has cited other bacteria seen as a source of co-infection during the various pandemics described. In early influenza pandemics such as the 1918 Spanish pandemic, *M. catarrhalis* rarely appears to be a noted cause of co-infection. However, in the 2009 pandemic it is seen in up to 18% of cases (Nguyen et al., 2012). We have therefore considered the importance of this. Data produced toward the end of the 1970's and throughout the 1980's demonstrated *M. catarrhalis*' potential to cause disease, however, before this *M. catarrhalis* was considered a non-pathogenic harmless commensal (McNeely et al., 1976; Johnson et al., 1981; Onofrio et al., 1981; McLeod et al., 1983; Feder and Garibaldi, 1984; Hager et al., 1987; Catlin, 1990). Therefore there are two possibilities to consider; perhaps *M. catarrhalis* wasn't present in early pandemics as a cause of co-infection and has become more of an issue in recent years; possibly as a result of vaccines, i.e., Hib and PCV, reducing the disease burden of other bacteria such as *S. pneumonia* and *H. influenzae*. Alternatively, we must consider that as *M. catarrhalis* was not considered a pathogen it was therefore missed or not commented upon prior to the 1980's. Retrospective studies may be able to address this. For example, autopsies from the 1918 pandemic were reviewed and it was found that *S. pneumoniae* was the most common co-infector, followed by *S hemolytic, S. aureus,* and *H. influenzae*. 'Other bacteria' were also highlighted within which *M. catarrhalis* was grouped (Morens et al., 2008).

Another point of consideration are changes of methodology. Pre-1983 laboratories would only undertake bacterial culture, however, in 2009 more sensitive methodology, i.e., PCR were available and commonly used in laboratories worldwide. The use of sensitive methods such as PCR, may have increased the likelihood of *M. catarrhalis* being detected, and as a known respiratory pathogen it would have been tested for, where as previously it may not have been. Alternatively maybe PCR detects bacteria that may have been out grown/not shown on a culture plate?

In contrast to *S. pneumoniae* and *H. influenzae* little research has been undertaken looking at influenza and *M. catarrhalis* co-infection and the dynamics and mechanisms of such infection. This is therefore an area worthy of future research. *M. catarrhalis* has been highlighted as a frequent source of co-infection for influenza since the early 1980's (Klein et al., 2016). In the setting of a pandemic it may therefore have a major public health impact.

FACTORS AFFECTING THE SEVERITY OF BACTERIAL CO/SECONDARY INFECTION

As discussed above, co/secondary bacterial infection can result in a deterioration of clinical condition with more severe disease. The severity of co/secondary infection depends on multiple factors such as the strain of virus and serotype/strain of bacteria, the lag between viral infection and bacterial exposure and density of bacterial colonization (Lee L.N. et al., 2010; McCullers et al., 2010; Smith and McCullers, 2014).

Virally Enhanced Colonization and Attachment of Bacteria

It has become clear that influenza, as well as other upper respiratory tract viral infections, leads not only to a greater risk of infection from bacterial pathobionts but also an increased likelihood that an individual may become colonized with bacteria such as *S. pneumoniae*, *H. influenzae* and *S. aureus* (Hament et al., 1999; Hilleman, 2002). Plotkowski et al. (1986) found enhanced colonization and adherence of *S. pneumoniae* to the tracheal cells of mice when they were infected with influenza (Plotkowski et al., 1986). Other studies have intranasally inoculated ferrets with influenza, finding prior viral infection increases colonization and adherence of *S. aureus* (Sanford and Ramsay, 1987). Furthermore, poor disease outcome has been linked to lost lung repair function and loss of basal epithelial cells, including alveolar epithelial cells; which is associated with increased bacterial attachment and apoptosis (Kash et al., 2011). Wadowsky et al. (1995) conducted a study in which adult subjects were inoculated with influenza and then screened for bacterial colonization. After 6 days 15% of the subjects were heavily colonized by *S. pneumoniae* (Wadowsky et al., 1995). Additionally, the effect of viral prevention methods further supports the idea of viruses predisposing a host to secondary bacterial infection (Peltola and McCullers, 2004; Lee et al., 2008). Studies have shown that influenza vaccination can reduce the occurrence of bacterial pneumonia (Fedson et al., 1993; Nichol et al., 1998).

Viral Factors Implicated in Severity of Infection

Research shows that influenza A is the type most commonly associated with co/secondary bacterial infection and subtypes with NA2 traditionally result in more severe infection (Peltola et al., 2005). Although reported less, influenza B has also been associated with severe bacterial co/secondary infection (Finelli et al., 2008; Aebi et al., 2010). Various factors are known to impact the severity of viral infection, which in turn increases the likelihood of bacterial co/secondary infections; these include the type of HA and NA surface antigen. As mentioned previously, HA mediates virion binding to the host cell via sialic acid receptors. Binding is followed by endocytosis and the movement of the virion into the host cell within an endosome (Samji, 2009). HA binds to sialylated glycans found on the surface of human epithelial cells; traditionally seasonal influenza A virus binds to α2-6 sialylated glycans on cells in the upper respiratory tract whereas the highly pathogenic avian H5N1 strain binds to α2-3 sialylated glycans on type 2 pneumocytes lining lung alveoli (Shinya et al., 2006). Clearly the type of HA impacts on the site and development of infection. The low pH in the endosome causes a conformational change to the HA allowing it to be cleaved, an important step in penetrating into the host cell. Therefore HA and the availability of appropriate host proteases are determinants of infectivity (Steinhauer, 1999; Samji, 2009). Interestingly non-pathogenic and mammalian influenza HA undergoes cleavage outside of the host cell where as highly pathogenic strains are cleaved inside host cells (Steinhauer, 1999). Another example of how the type of HA can make a difference to infection, and therefore the impact of an epi- or pandemic, is that traditionally trypsin-like protease cleaves influenza HA; however, some HA types (i.e., types 5 and 7) have the ability to acquire insertional mutations at the cleavage site which changes their recognition site in such a way that specificity is broadened so more proteases are recognized (Kash and Taubenberger, 2015).

Neuraminidase enables the release of newly formed progeny virions; by hydrolysing the sialic acid and detaching it from the HA the virion becomes liberated from the host cell (Zambon, 2001). To be truly effective the NA must be complementary and share the same receptor specificity as HA, so if the viral HA binds to α2-3 sialic acid then the NA should hydrolyse α2-3 sialic acid (Baum and Paulson, 1991).

The production of viral toxins that impact host cell integrity is another important factor in the development of co/secondary bacterial infection. Influenza A virus can produce a viral cytotoxin PB1-F2 (Conenello and Palese, 2007; Iverson et al., 2011) which plays a role in increasing inflammation and therefore host cell damage and bacterial adherence, increasing mortality and morbidity (Lee et al., 2016). It also helps reduce bacterial clearance, increasing the occurrence and severity of co/secondary bacterial infection, by causing cell death in host monocytes (Conenello and Palese, 2007; Iverson et al., 2011).

Molecular Co-pathogenesis

Following bacterial colonization, disease develops due to specific characteristics of viral infection that facilitate bacterial adhesion and penetration (Selinger et al., 1981). Influenza produces NA, which increases adhesion of some bacterial species by removing sialic acid to expose host cell receptors (McCullers and Bartmess, 2003; Peltola and McCullers, 2004). Alternatively some bacteria, i.e., group B Streptococci, contain sialic acid which allows for direct binding to the viral HA expressed by influenza infected host cells (Okamoto et al., 2003; Peltola and McCullers, 2004). Damaged host cells, whether damaged directly by the virus or by inflammation and immune cell responses, provide additional adhesion sites allowing for increased bacterial adhesion. For example the exposure of apical receptors like integrins permit the adhesion of bacteria such as *S. aureus* and *P. aeruginosa* (Sanford et al., 1978; Davison and Sanford, 1981; Bucior et al., 2012; Smith and McCullers, 2014). In response to viral infection, host inflammatory responses may cause an up-regulation in the expression of host receptor molecules and other molecules that bacteria can use as a receptors (Hakansson et al., 1994; Peltola and McCullers, 2004). For example Cundell et al. (1995) showed an increased presentation of G-protein-coupled platelet-activating factor (PAF) receptor, which certain bacteria, i.e., *S. pneumoniae*, can utilize for cell attachment and colonization in endothelial cells (Cundell et al., 1995; van der Sluijs et al., 2010). In contrast it has been suggested that the PAF receptor does not affect initial bacterial adherence and colonization but is more involved with assisting bacterial transition/spread into the blood and thus the development of invasive disease (McCullers et al., 2008).

Influenza infection appears to prime the host airways for bacterial infection, whilst modifying and impairing immune responses in a number of ways (Joseph et al., 2013). Viral induced immunosuppression can allow for a bacterial super infection, as host immune responses can be suppressed when immunologic cells are impaired during influenza infection and immune cell dysfunction can reduce the host's ability to fight bacteria (Peltola and McCullers, 2004; Brundage, 2006; Wu et al., 2011). Many studies involving animal models have shown that influenza infection increases and prolongs bacterial growth, due to reduced macrophage accumulation and decreased bacterial clearance due to reduced phagocytic activity (Kleinerman et al., 1976; Wyde et al., 1989; Sun and Metzger, 2008). Additionally it has recently been shown that *S. pneumoniae* and influenza co-infection results in a reduction in the number of local alveolar macrophages, this due to increased death of these macrophages by apoptosis and necrosis (Sharma-Chawla et al., 2016). This reduction is likely to hinder bacterial clearance, hence the increased bacterial load found during co-infection during this study, and results in prolonged inflammatory response increasing morbidity. Even after influenza is cleared, *S. pneumoniae* bacterial clearance is affected. This study has highlighted some serotype dependent differences, suggesting different treatment programs would be beneficial for different serotypes. Evidence that it is worth further looking into co-infection of influenza with different serotypes of *S. pneumoniae* and other bacteria of interest (Sharma-Chawla et al., 2016). Impaired neutrophils have been shown to correlate with secondary bacterial infection in Chinchillas, due to the importance of phagocytosis during innate immunity (Abramson et al., 1981). Influenza infection is known to result in the production of IFN; pulmonary IFNγ pro-inflammatory cytokines are produced by natural killer (NK) cells as part of innate immunity and by CD4 and CD8 NK T cells as part of adaptive immunity (Schoenborn and Wilson, 2007). They increase macrophage activation during innate immunity (Scott et al., 2004) however, during T cell responses to viral infection they have been shown to inhibit bacterial clearance from the respiratory system by macrophages. It is thought that as they assist in the induction of specific anti-influenza adaptive immunity they down regulate innate immunity. The resulting suppression of phagocytosis paves the way for successful bacterial infection (Sun and Metzger, 2008). Additionally, Type I IFNs inhibit interleukin 23 (IL-23) dependent induction of T helper cell 17 (Th17) immunity, and therefore there is a reduction in the levels of CD4+ T cells and gamma delta T cells and hence a reduction in the production of IL-17 and IL-22, preventing the clearance of bacteria (Shahangian et al., 2009; Kudva et al., 2011; Mulcahy and McLoughlin, 2016). Robinson et al. (2013) have also shown that influenza A infection significantly decreased IL-1β production; IL-1β has been shown to play a role in Th17 polarization, therefore further hindering this pathway of immunity (Robinson et al., 2013). During co/secondary *S. pneumoniae* infection, type I IFNs have been shown to inhibit the production of specific chemokines (KC/CXCL1 and Mip2/CXCL2) resulting in an attenuated neutrophil response (Shahangian et al., 2009). Viral and bacterial co-infection of monocyte derived macrophages synergistically induces a pro-inflammatory response related to the type-I IFN and JAK-STAT signaling pathways (Hoffmann et al., 2016). Inflammation causes tissue damage, revealing more attachment sites for increased/developed bacterial infection. Co-infection also results in a synergistic increase in type II IFN (IFNγ) when compared to individual infection of influenza or *S. pneumoniae*, CXCL10 (aka IFNγ-induced protein 10/IP-10) is secreted in response. IP-10 attracts various immune cells including activated T cells, monocytes and macrophages, therefore causing inflammation (Dufour et al., 2002; Hoffmann et al., 2016). Patients suffering severe pneumonia show significantly higher levels of IP-10 than those with minor cases of pneumonia. IP-10 is increased during *H. influenzae* and *S. aureus* co-infection as well, and like with *S. pneumoniae*, correlates with/highlights pneumonia etiology (Hoffmann et al., 2016). In addition, when *S. pneumoniae* successively co-infects with influenza it leads to severe clinical complications; partly due to an increase in apoptosis of dendritic cells, which therefore reduces T cell priming impairing the development of adaptive immunity. Influenza and *S. pneumoniae* infections can also lead to synergistic and non-synergistic dysregulation of cytokine responses (Wu et al., 2011).

As previously described influenza infection results in a reduction in the production of IL-17 and 22. IL-17 is important in the clearance of *S. aureus* by neutrophils (Archer et al., 2013). IL-22 in involved in controlling the production of antimicrobial peptides as well as staphylococcal ligand expression (Robinson et al., 2014; Mulcahy et al., 2016). In addition to this, influenza positively affects the colonization of *S. aureus* by causing increased type III- IFN expression, which alters the IL-22 responses impairing host expression of antimicrobial peptides (Mulcahy and McLoughlin, 2016). Distress signals, such as ATP and norepinephrine, produced by damaged influenza also have several effects on *S. aureus*; namely the instigation of biofilm dispersal which helps the spread of *S. aureus* into the lungs, assisting in the development of pneumonia and invasive disease (Mulcahy and McLoughlin, 2016).

Of course viral infection doesn't just benefit bacteria; several mechanisms of synergism between viruses and bacteria have been suggested. A number of studies have documented an increase in viral load as viral clearance is reduced during bacterial co-infection. It is, however, unclear whether this is from bacterial and viral cooperation/interactions or simply from bacteria burdening the host immune system resulting in the reduction of viral eradication. Therefore further research is required to develop our understanding of the interaction between bacteria and viruses within co-infection (Peltola and McCullers, 2004; Iverson et al., 2011; Smith and McCullers, 2014).

CONCLUSION

Viral infection aids bacterial infection in a number of ways, including unveiling/providing more sites for adhesion, impairing immune responses and causing cell and tissue destruction allowing for the spread of bacteria and development of invasive infection. Bacterial infection is then able to worsen clinical outcome and the severity of disease. Of course viral and bacterial co-infection can be mutually beneficial, further helping viral infection, which is bad news for public health. Although antibiotics can help reduce the impact of co/secondary bacterial infection, we still need to better understand the interactions between viruses, bacteria and their host, and to fully understand all mechanisms of disease. Particularly in light of increased antibiotic resistance and the ability of microbes to adapt and evade vaccine induced immunity.

The aim of this review was to emphasize the historical and continuing threat of influenza and to highlight the risk of bacterial co/secondary infection. Vaccines and antibiotics are readily available, however, with antibiotic resistance at an all-time high, vaccination is becoming even more vital in the fight against influenza epidemics and pandemics and the bacterial co/secondary infections commonly associated. It is important to examine the strains and types of bacteria and viruses being spread amongst and transmitted throughout the general public (or continue to in the case of influenza) to inform clinical treatment and development, particularly in the setting of an influenza epidemic or pandemic. As the threat from influenza is ever changing, we need to ensure we know what strains are circulating, which could cause issue and how they interact with other potential pathogens. This preparation also entails monitoring the changing epidemiology of bacterial pathogens associated with secondary infection, such as capsule switching which help *S. pneumoniae* evade immunity (Pai et al., 2005a,b).

AUTHOR CONTRIBUTIONS

DM, DC, and SC designed, planned and wrote the manuscript.

REFERENCES

Abdullahi, O., Nyiro, J., Lewa, P., Slack, M., and Scott, J. A. (2008). The descriptive epidemiology of *Streptococcus pneumoniae* and *Haemophilus influenzae* nasopharyngeal carriage in children and adults in Kilifi district, Kenya. *Pediatr. Infect. Dis. J.* 27, 59–64. doi: 10.1097/INF.0b013e31814da70c

Abrahams, A., Hallows, N., and French, H. (1919). A further investigation into influenzo-pneumococcal and influenzo-streptococcal septicæmia: epidemic influenzal "pneumonia" of highly fatal type and its relation to "purulent bronchitis" *Lancet* 193, 1–11. doi: 10.1016/S0140-6736(01)22115-1

Abramson, J. S., Giebink, G. S., Mills, E. L., and Quie, P. G. (1981). Polymorphonuclear leukocyte dysfunction during influenza virus infection in chinchillas. *J. Infect. Dis.* 143, 836–845. doi: 10.1093/infdis/143.6.836

Aebi, T., Weisser, M., Bucher, E., Hirsch, H. H., Marsch, S., and Siegemund, M. (2010). Co-infection of influenza B and *Streptococci* causing severe pneumonia and septic shock in healthy women. *BMC Infect. Dis.* 10:308. doi: 10.1186/1471-2334-10-308

Archer, N. K., Harro, J. M., and Shirtliff, M. E. (2013). Clearance of *Staphylococcus aureus* nasal carriage is T cell dependent and mediated through interleukin-17A expression and neutrophil influx. *Infect. Immun.* 81, 2070–2075. doi: 10.1128/IAI.00084-13

Baum, L. G., and Paulson, J. C. (1991). The N2 neuraminidase of human influenza virus has acquired a substrate specificity complementary to the hemagglutinin receptor specificity. *Virology* 180, 10–15. doi: 10.1016/0042-6822(91)90003-T

Bentley, S. D., Aanensen, D. M., Mavroidi, A., Saunders, D., Rabbinowitsch, E., Collins, M., et al. (2006). Genetic analysis of the capsular biosynthetic locus from all 90 pneumococcal serotypes. *PLoS Genet.* 2:e31. doi: 10.1371/journal.pgen.0020031

Berendt, R. F., Long, G. G., and Walker, J. S. (1975). Influenza alone and in sequence with pneumonia due to *Streptococcus pneumoniae* in the squirrel monkey. *J. Infect. Dis.* 132, 689–693. doi: 10.1093/infdis/132.6.689

Berg, R. A., and Bartley, D. L. (1987). Pneumonia associated with *Branhamella catarrhalis* in infants. *Pediatr. Infect. Dis. J.* 6, 569–573. doi: 10.1097/00006454-198706000-00017

Biggerstaff, M., Cauchemez, S., Reed, C., Gambhir, M., and Finelli, L. (2014). Estimates of the reproduction number for seasonal, pandemic, and zoonotic influenza: a systematic review of the literature. *BMC Infect. Dis.* 14:480. doi: 10.1186/1471-2334-14-480

Blanton, W. B., and Irons, E. E. (1918). A recent epidemic of acute respiratory infection at camp custer, mich: preliminary laboratory report. *J. Am. Med. Assoc.* 71, 1988–1991. doi: 10.1001/jama.1918.26020500014006d

Bluestone, C. D. (1986). Otitis media and sinusitis in children. Role of Branhamella catarrhalis. *Drugs* 31(Suppl. 3), 132–141.

Brem, W. V., Bolling, G. E., and Casper, E. J. (1918). Pandemic "influenza" and secondary pneumonia at camp fremont, Calif. *J. Am. Med. Assoc.* 71, 2138–2144. doi: 10.1001/jama.1918.26020520007010b

Bridy-Pappas, A. E., Margolis, M. B., Center, K. J., and Isaacman, D. J. (2005). *Streptococcus pneumoniae*: description of the pathogen, disease epidemiology, treatment, and prevention. *Pharmacotherapy* 25, 1193–1212. doi: 10.1592/phco.2005.25.9.1193

Brouwer, M. C., Tunkel, A. R., and van de Beek, D. (2010). Epidemiology, diagnosis, and antimicrobial treatment of acute bacterial meningitis. *Clin. Microbiol. Rev.* 23, 467–492. doi: 10.1128/CMR.00070-09

Brundage, J. F. (2006). Interactions between influenza and bacterial respiratory pathogens: implications for pandemic preparedness. *Lancet Infect. Dis.* 6, 303–312. doi: 10.1016/S1473-3099(06)70466-2

Brundage, J. F., and Shanks, G. D. (2008). Deaths from bacterial pneumonia during 1918-19 influenza pandemic. *Emerg. Infect. Dis.* 14, 1193–1199. doi: 10.3201/eid1408.071313

Bucior, I., Pielage, J. F., and Engel, J. N. (2012). *Pseudomonas aeruginosa* pili and flagella mediate distinct binding and signaling events at the apical and basolateral surface of airway epithelium. *PLoS Pathog.* 8:e1002616. doi: 10.1371/journal.ppat.1002616

Burnet, F. M., and Clark, E. (1942). *Influenza: A Survey of the Last 50 Years in the Light of Modern Work on the Virus of Epidemic Influenza*. London: Macmillan and co ltd.

Calix, J. J., and Nahm, M. H. (2010). A new pneumococcal serotype, 11E, has a variably inactivated wcjE gene. *J. Infect. Dis.* 202, 29–38. doi: 10.1086/653123

Calix, J. J., Porambo, R. J., Brady, A. M., Larson, T. R., Yother, J., Abeygunwardana, C., et al. (2012). Biochemical, genetic, and serological characterization of two capsule subtypes among *Streptococcus pneumoniae* serotype 20 strains: discovery of a new pneumococcal serotype. *J. Biol. Chem.* 287, 27885–27894. doi: 10.1074/jbc.M112.380451

Catlin, B. W. (1990). *Branhamella catarrhalis*: an organism gaining respect as a pathogen. *Clin. Microbiol. Rev.* 3, 293–320. doi: 10.1128/CMR.3.4.293

Cauley, L. S., and Vella, A. T. (2015). Why is coinfection with influenza virus and bacteria so difficult to control? *Discov. Med.* 19, 33–40.

Centers for Disease Control and Prevention (2009). Bacterial co-infection in lung tissue specimens from fatal cases of 2009 pandemic influenza A (H1N1) - United States, May-August 2009. *Morb. Mortal. Wkly. Rep.* 58, 1071–1074.

Chen, R., and Holmes, E. C. (2006). Avian influenza virus exhibits rapid evolutionary dynamics. *Mol. Biol. Evol.* 23, 2336–2341. doi: 10.1093/molbev/msl102

Chertow, D. S., and Memoli, M. J. (2013). Bacterial coinfection in influenza: a grand rounds review. *JAMA* 309, 275–282. doi: 10.1001/jama.2012.194139

Chien, Y. W., Klugman, K. P., and Morens, D. M. (2009). Bacterial pathogens and death during the 1918 influenza pandemic. *N. Engl. J. Med.* 361, 2582–2583. doi: 10.1056/NEJMc0908216

Chin, C. L., Manzel, L. J., Lehman, E. E., Humlicek, A. L., Shi, L., Starner, T. D., et al. (2005). *Haemophilus influenzae* from patients with chronic obstructive pulmonary disease exacerbation induce more inflammation than colonizers. *Am. J. Respir. Crit. Care Med.* 172, 85–91. doi: 10.1164/rccm.200412-1687OC

Cockburn, W. C., Delon, P. J., and Ferreira, W. (1969). Origin and progress of the 1968-69 Hong Kong influenza epidemic. *Bull. World Health Organ.* 41, 345–348.

Collazos, J., de Miguel, J., and Ayarza, R. (1992). *Moraxella catarrhalis* bacteremic pneumonia in adults: two cases and review of the literature. *Eur. J. Clin. Microbiol. Infect. Dis.* 11, 237–240. doi: 10.1007/BF02098086

Conenello, G. M., and Palese, P. (2007). Influenza A virus PB1-F2: a small protein with a big punch. *Cell Host Microbe* 2, 207–209. doi: 10.1016/j.chom.2007.09.010

Couch, R. B., Kasel, J. A., Gerin, J. L., Schulman, J. L., and Kilbourne, E. D. (1974). Induction of partial immunity to influenza by a neuraminidase-specific vaccine. *J. Infect. Dis.* 129, 411–420. doi: 10.1093/infdis/129.4.411

Cox, N. J., and Subbarao, K. (2000). Global epidemiology of influenza: past and present. *Annu. Rev. Med.* 51, 407–421. doi: 10.1146/annurev.med.51.1.407

Cundell, D. R., Gerard, N. P., Gerard, C., Idanpaan-Heikkila, I., and Tuomanen, E. I. (1995). *Streptococcus pneumoniae* anchor to activated human cells by the receptor for platelet-activating factor. *Nature* 377, 435–438. doi: 10.1038/377435a0

Cunha, B. A., Pherez, F. M., and Schoch, P. (2009). Diagnostic importance of relative lymphopenia as a marker of swine influenza (H1N1) in adults. *Clin. Infect. Dis.* 49, 1454–1456. doi: 10.1086/644496

Daszak, P. (2012). Anatomy of a pandemic. *Lancet* 380, 1883–1884. doi: 10.1016/S0140-6736(12)61887-X

Davison, V. E., and Sanford, B. A. (1981). Adherence of *Staphylococcus aureus* to influenza A virus-infected madin-darby canine kidney cell cultures. *Infect. Immun.* 32, 118–126.

de Jong, M. D., Simmons, C. P., Thanh, T. T., Hien, V. M., Smith, G. J., Chau, T. N., et al. (2006). Fatal outcome of human influenza A (H5N1) is associated with high viral load and hypercytokinemia. *Nat. Med.* 12, 1203–1207. doi: 10.1038/nm1477

Dela Cruz, C. S., and Wunderink, R. G. (2017). *Respiratory Viral and Atypical Pneumonia in Adults*. Amsterdam: Elsevier.

DeLeo, F. R., and Musser, J. M. (2010). Axis of coinfection evil. *J. Infect. Dis.* 201, 488–490. doi: 10.1086/650304

Dennis Shanks, G., Mackenzie, A., Waller, M., and Brundage, J. F. (2012). Relationship between "purulent bronchitis" in military populations in Europe prior to 1918 and the 1918-1919 influenza pandemic. *Influenza Other Respir. Viruses* 6, 235–239. doi: 10.1111/j.1750-2659.2011.00309.x

Dhanoa, A., Fang, N. C., Hassan, S. S., Kaniappan, P., and Rajasekaram, G. (2011). Epidemiology and clinical characteristics of hospitalized patients with pandemic influenza A (H1N1) 2009 infections: the effects of bacterial coinfection. *Virol. J.* 8:501. doi: 10.1186/1743-422X-8-501

Dormitzer, P. R., Galli, G., Castellino, F., Golding, H., Khurana, S., Del Giudice, G., et al. (2011). Influenza vaccine immunology. *Immunol. Rev.* 239, 167–177. doi: 10.1111/j.1600-065X.2010.00974.x

Dufour, J. H., Dziejman, M., Liu, M. T., Leung, J. H., Lane, T. E., and Luster, A. D. (2002). IFN-gamma-inducible protein 10 (IP-10; CXCL10)-deficient mice reveal a role for IP-10 in effector T cell generation and trafficking. *J. Immunol.* 168, 3195–3204. doi: 10.4049/jimmunol.168.7.3195

Dunn, F. L. (1958). Pandemic influenza in 1957; review of international spread of new Asian strain. *J. Am. Med. Assoc.* 166, 1140–1148. doi: 10.1001/jama.1958.02990100028006

Dupont, D., Mahjoub-Messai, F., Francois, M., Doit, C., Mariani-Kurkdjian, P., Bidet, P., et al. (2010). Evolving microbiology of complicated acute otitis media before and after introduction of the pneumococcal conjugate vaccine in France. *Diagn. Microbiol. Infect. Dis.* 68, 89–92. doi: 10.1016/j.diagmicrobio.2010.04.012

Eldin, C., Melenotte, C., Mediannikov, O., Ghigo, E., Million, M., Edouard, S., et al. (2017). From Q fever to *Coxiella burnetii* infection: a paradigm change. *Clin. Microbiol. Rev* 30, 115–190. doi: 10.1128/CMR.00045-16

Eom, S.-H., Lee, D.-S., Jung, Y.-J., Park, J.-H., Choi, J.-I., Yim, M.-J., et al. (2014). The mechanism of antibacterial activity of phlorofucofuroeckol-A against methicillin-resistant *Staphylococcus aureus*. *Appl. Microbiol. Biotechnol.* 98, 9795–9804. doi: 10.1007/s00253-014-6041-8

Faden, H., Harabuchi, Y., and Hong, J. J. (1994). Epidemiology of *Moraxella catarrhalis* in children during the first 2 years of life: relationship to otitis media. *J. Infect. Dis.* 169, 1312–1317. doi: 10.1093/infdis/169.6.1312

Fainstein, V., Musher, D. M., and Cate, T. R. (1980). Bacterial adherence to pharyngeal cells during viral infection. *J. Infect. Dis.* 141, 172–176. doi: 10.1093/infdis/141.2.172

Feder, H. M. Jr., and Garibaldi, R. A. (1984). The significance of nongonococcal, nonmeningococcal *Neisseria* isolates from blood cultures. *Rev. Infect. Dis.* 6, 181–188. doi: 10.1093/clinids/6.2.181

Fedson, D. S., Wajda, A., Nicol, J. P., Hammond, G. W., Kaiser, D. L., and Roos, L. L. (1993). Clinical effectiveness of influenza vaccination in Manitoba. *JAMA* 270, 1956–1961. doi: 10.1001/jama.1993.03510160074032

Finelli, L., Fiore, A., Dhara, R., Brammer, L., Shay, D. K., Kamimoto, L., et al. (2008). Influenza-associated pediatric mortality in the United States: increase of *Staphylococcus aureus* coinfection. *Pediatrics* 122, 805–811. doi: 10.1542/peds.2008-1336

Fishovitz, J., Rojas-Altuve, A., Otero, L. H., Dawley, M., Carrasco-López, C., Chang, M., et al. (2014). Disruption of allosteric response as an unprecedented mechanism of resistance to antibiotics. *J. Am. Chem. Soc.* 136, 9814–9817. doi: 10.1021/ja5030657

Fukumi, H. (1959). Summary report on the Asian influenza epidemic in Japan, 1957. *Bull. World Health Organ.* 20, 187–198.

Gatherer, D. (2009). The 2009 H1N1 influenza outbreak in its historical context. *J. Clin. Virol.* 45, 174–178. doi: 10.1016/j.jcv.2009.06.004

Gill, J. R., Sheng, Z. M., Ely, S. F., Guinee, D. G., Beasley, M. B., Suh, J., et al. (2010). Pulmonary pathologic findings of fatal 2009 pandemic influenza A/H1N1 viral infections. *Arch. Pathol. Lab. Med.* 134, 235–243. doi: 10.1043/1543-2165-134.2.235

Glezen, W. P. (1996). Emerging infections: pandemic influenza. *Epidemiol. Rev.* 18, 64–76. doi: 10.1093/oxfordjournals.epirev.a017917

Grijalva, C. G., Griffin, M. R., Edwards, K. M., Williams, J. V., Gil, A. I., Verastegui, H., et al. (2014). The role of influenza and parainfluenza infections in nasopharyngeal pneumococcal acquisition among young children. *Clin. Infect. Dis.* 58, 1369–1376. doi: 10.1093/cid/ciu148

Gupta, R. K., George, R., and Nguyen-Van-Tam, J. S. (2008). Bacterial pneumonia and pandemic influenza planning. *Emerg. Infect. Dis.* 14, 1187–1192. doi: 10.3201/eid1408.070751

Hager, H., Verghese, A., Alvarez, S., and Berk, S. L. (1987). Branhamella catarrhalis respiratory infections. *Rev. Infect. Dis.* 9, 1140–1149. doi: 10.1093/clinids/9.6.1140

Hakansson, A., Kidd, A., Wadell, G., Sabharwal, H., and Svanborg, C. (1994). Adenovirus infection enhances in vitro adherence of *Streptococcus pneumoniae*. *Infect. Immun.* 62, 2707–2714.

Hall, J. N., Stone, M. C., and Simpson, J. C. (1918). The epidemic of pneumonia following influenza at Camp Logan, Texas: preliminary report. *J. Am. Med. Assoc.* 71, 1986–1987. doi: 10.1001/jama.1918.26020500012006c

Hament, J.-M., Kimpen, J. L. L., Fleer, A., and Wolfs, T. F. W. (1999). Respiratory viral infection predisposing for bacterial disease: a concise review. *FEMS Immunol. Med. Microbiol.* 26, 189–195. doi: 10.1111/j.1574-695X.1999.tb01389.x

Hampson, A. W., and Mackenzie, J. S. (2006). The influenza viruses. *Med. J. Aust.* 185, S39–S43.

Hers, J. F., Masurel, N., and Mulder, J. (1958). Bacteriology and histopathology of the respiratory tract and lungs in fatal Asian influenza. *Lancet* 2, 1141–1143. doi: 10.1016/S0140-6736(58)92404-8

Hilleman, M. R. (2002). Realities and enigmas of human viral influenza: pathogenesis, epidemiology and control. *Vaccine* 20, 3068–3087. doi: 10.1016/S0264-410X(02)00254-2

Hoffmann, J., Machado, D., Terrier, O., Pouzol, S., Messaoudi, M., Basualdo, W., et al. (2016). Viral and bacterial co-infection in severe pneumonia triggers innate immune responses and specifically enhances IP-10: a translational study. *Sci. Rep.* 6:38532. doi: 10.1038/srep38532

Holland, J., Spindler, K., Horodyski, F., Grabau, E., Nichol, S., and VandePol, S. (1982). Rapid evolution of RNA genomes. *Science* 215, 1577–1585. doi: 10.1126/science.7041255

Hooper, L. V., Littman, D. R., and Macpherson, A. J. (2012). Interactions between the microbiota and the immune system. *Science* 336, 1268–1273. doi: 10.1126/science.1223490

Houser, K., and Subbarao, K. (2015). Influenza vaccines: challenges and solutions. *Cell Host Microbe* 17, 295–300. doi: 10.1016/j.chom.2015.02.012

Huang, S. S., Johnson, K. M., Ray, G. T., Wroe, P., Lieu, T. A., Moore, M. R., et al. (2011). Healthcare utilization and cost of pneumococcal disease in the United States. *Vaccine* 29, 3398–3412. doi: 10.1016/j.vaccine.2011.02.088

Ioannidis, J. P., Worthington, M., Griffiths, J. K., and Snydman, D. R. (1995). Spectrum and significance of bacteremia due to *Moraxella catarrhalis*. *Clin. Infect. Dis.* 21, 390–397. doi: 10.1093/clinids/21.2.390

Iverson, A. R., Boyd, K. L., McAuley, J. L., Plano, L. R., Hart, M. E., and McCullers, J. A. (2011). Influenza virus primes mice for pneumonia from *Staphylococcus aureus*. *J. Infect. Dis.* 203, 880–888. doi: 10.1093/infdis/jiq113

Jackson, C., Mangtani, P., Hawker, J., Olowokure, B., and Vynnycky, E. (2014). The effects of school closures on influenza outbreaks and pandemics: systematic review of simulation studies. *PLoS ONE* 9:e97297. doi: 10.1371/journal.pone.0097297

Jacoby, P., Watson, K., Bowman, J., Taylor, A., Riley, T. V., Smith, D. W., et al. (2007). Modelling the co-occurrence of *Streptococcus pneumoniae* with other bacterial and viral pathogens in the upper respiratory tract. *Vaccine* 25, 2458–2464. doi: 10.1016/j.vaccine.2006.09.020

Jakeman, K. J., Tisdale, M., Russell, S., Leone, A., and Sweet, C. (1994). Efficacy of 2'-deoxy-2'-fluororibosides against influenza A and B viruses in ferrets. *Antimicrob. Agents Chemother.* 38, 1864–1867. doi: 10.1128/AAC.38.8.1864

Jin, P., Kong, F., Xiao, M., Oftadeh, S., Zhou, F., Liu, C., et al. (2009). First report of putative *Streptococcus pneumoniae* serotype 6D among nasopharyngeal isolates from Fijian children. *J. Infect. Dis.* 200, 1375–1380. doi: 10.1086/606118

Johnson, B. (2016). *The Spanish Flu Pandemic of 1918*. Available at: http://www.historic-uk.com/HistoryUK/HistoryofBritain/The-Spanish-Flu-pandemic-of-1918/

Johnson, M. A., Drew, W. L., and Roberts, M. (1981). Branhamella (*Neisseria*) catarrhalis-a lower respiratory tract pathogen? *J. Clin. Microbiol.* 13, 1066–1069.

Jones, K. E., Patel, N. G., Levy, M. A., Storeygard, A., Balk, D., Gittleman, J. L., et al. (2008). Global trends in emerging infectious diseases. *Nature* 451, 990–993. doi: 10.1038/nature06536

Joseph, C., Togawa, Y., and Shindo, N. (2013). Bacterial and viral infections associated with influenza. *Influenza Other Respir. Viruses* 7(Suppl. 2), 105–113. doi: 10.1111/irv.12089

Karesh, W. B., Dobson, A., Lloyd-Smith, J. O., Lubroth, J., Dixon, M. A., Bennett, M., et al. (2012). Ecology of zoonoses: natural and unnatural histories. *Lancet* 380, 1936–1945. doi: 10.1016/S0140-6736(12)61678-X

Kash, J. C., and Taubenberger, J. K. (2015). The role of viral, host, and secondary bacterial factors in influenza pathogenesis. *Am. J. Pathol.* 185, 1528–1536. doi: 10.1016/j.ajpath.2014.08.030

Kash, J. C., Tumpey, T. M., Proll, S. C., Carter, V., Perwitasari, O., Thomas, M. J., et al. (2006). Genomic analysis of increased host immune and cell death responses induced by 1918 influenza virus. *Nature* 443, 578–581. doi: 10.1038/nature05181

Kash, J. C., Walters, K. A., Davis, A. S., Sandouk, A., Schwartzman, L. M., Jagger, B. W., et al. (2011). Lethal synergism of 2009 pandemic H1N1 influenza virus and *Streptococcus pneumoniae* coinfection is associated with loss of murine lung repair responses. *MBio* 2:e00172-11. doi: 10.1128/mBio.00172-11

Kash, J. C., Xiao, Y., Davis, A. S., Walters, K. A., Chertow, D. S., Easterbrook, J. D., et al. (2014). Treatment with the reactive oxygen species scavenger EUK-207 reduces lung damage and increases survival during 1918 influenza virus infection in mice. *Free Radic. Biol. Med.* 67, 235–247. doi: 10.1016/j.freeradbiomed.2013.10.014

Kilpi, T., Herva, E., Kaijalainen, T., Syrjanen, R., and Takala, A. K. (2001). Bacteriology of acute otitis media in a cohort of Finnish children followed for the first two years of life. *Pediatr. Infect. Dis. J.* 20, 654–662. doi: 10.1097/00006454-200107000-00004

Klein, E. Y., Monteforte, B., Gupta, A., Jiang, W., May, L., Hsieh, Y. H., et al. (2016). The frequency of influenza and bacterial coinfection: a systematic review and meta-analysis. *Influenza Other Respir. Viruses* 10, 394–403. doi: 10.1111/irv.12398

Kleinerman, E. S., Daniels, C. A., Polisson, R. P., and Snyderman, R. (1976). Effect of virus infection on the inflammatory response. Depression of macrophage accumulation in influenza-infected mice. *Am. J. Pathol.* 85, 373–382.

Klimov, A., Simonsen, L., Fukuda, K., and Cox, N. (1999). Surveillance and impact of influenza in the United States. *Vaccine* 17(Suppl. 1), S42–S46. doi: 10.1016/S0264-410X(99)00104-8

Kobasa, D., Jones, S. M., Shinya, K., Kash, J. C., Copps, J., Ebihara, H., et al. (2007). Aberrant innate immune response in lethal infection of macaques with the 1918 influenza virus. *Nature* 445, 319–323. doi: 10.1038/nature05495

Kollef, M. H., Shorr, A., Tabak, Y. P., Gupta, V., Liu, L. Z., and Johannes, R. S. (2005). Epidemiology and outcomes of health-care-associated pneumonia: results from a large US database of culture-positive pneumonia. *Chest* 128, 3854–3862. doi: 10.1378/chest.128.6.3854

Kudva, A., Scheller, E. V., Robinson, K. M., Crowe, C. R., Choi, S. M., Slight, S. R., et al. (2011). Influenza A inhibits Th17-mediated host defense against bacterial pneumonia in mice. *J. Immunol.* 186, 1666–1674. doi: 10.4049/jimmunol.1002194

Lamkanfi, M., and Dixit, V. M. (2010). Manipulation of Host Cell Death Pathways during Microbial Infections. *Cell Host Microbe* 8, 44–54. doi: 10.1016/j.chom.2010.06.007

Langereis, J. D., and de Jonge, M. I. (2015). Invasive disease caused by nontypeable *Haemophilus influenzae*. *Emerg. Infect. Dis.* 21, 1711–1718. doi: 10.3201/eid2110.150004

Lee, E. H., Wu, C., Lee, E. U., Stoute, A., Hanson, H., Cook, H. A., et al. (2010). Fatalities associated with the 2009 H1N1 influenza A virus in New York city. *Clin. Infect. Dis.* 50, 1498–1504. doi: 10.1086/652446

Lee, L. N., Dias, P., Han, D., Yoon, S., Shea, A., Zakharov, V., et al. (2010). A mouse model of lethal synergism between influenza virus and *Haemophilus influenzae*. *Am. J. Pathol.* 176, 800–811. doi: 10.2353/ajpath.2010.090596

Lee, M. H., Arrecubieta, C., Martin, F. J., Prince, A., Borczuk, A. C., and Lowy, F. D. (2010). A postinfluenza model of *Staphylococcus aureus* pneumonia. *J. Infect. Dis.* 201, 508–515. doi: 10.1086/650204

Lee, P. H., Bird, N., MacKenzie-Kludas, C., Mansell, A., Kedzierska, K., Brown, L., et al. (2016). Induction of memory cytotoxic T cells to influenza A virus and subsequent viral clearance is not modulated by PB1-F2-dependent inflammasome activation. *Immunol. Cell Biol.* 94, 439–446. doi: 10.1038/icb.2015.115

Lee, S. E., Eick, A., Bloom, M. S., and Brundage, J. F. (2008). Influenza immunization and subsequent diagnoses of group A streptococcus-illnesses among U.S. Army trainees, 2002-2006. *Vaccine* 26, 3383–3386. doi: 10.1016/j.vaccine.2008.04.041

Li, Q., Sun, X., Li, Z., Liu, Y., Vavricka, C. J., Qi, J., et al. (2012). Structural and functional characterization of neuraminidase-like molecule N10 derived from bat influenza A virus. *Proc. Natl. Acad. Sci. U.S.A.* 109, 18897–18902. doi: 10.1073/pnas.1211037109

Lindsay, M. I. Jr., Herrmann, E. C. Jr., Morrow, G. W. Jr., and Brown, A. L. Jr. (1970). Hong kong influenza: clinical, microbiologic, and pathologic features in 127 cases. *JAMA* 214, 1825–1832. doi: 10.1001/jama.1970.03180100019004

Lozano, R., Naghavi, M., Foreman, K., Lim, S., Shibuya, K., Aboyans, V., et al. (2012). Global and regional mortality from 235 causes of death for 20 age groups in 1990 and 2010: a systematic analysis for the Global Burden of Disease Study 2010. *Lancet* 380, 2095–2128. doi: 10.1016/S0140-6736(12)61728-0

Lucas, S. (2010). Predictive clinicopathological features derived from systematic autopsy examination of patients who died with A/H1N1 influenza infection in the UK 2009-10 pandemic. *Health Technol. Assess.* 14, 83–114. doi: 10.3310/hta14550-02

Madhi, S. A., Adrian, P., Kuwanda, L., Cutland, C., Albrich, W. C., and Klugman, K. P. (2007). Long-term effect of pneumococcal conjugate vaccine on nasopharyngeal colonization by *Streptococcus pneumoniae* and associated interactions with *Staphylococcus aureus* and *Haemophilus influenzae* colonization in HIV-Infected and HIV-uninfected children. *J. Infect. Dis.* 196, 1662–1666. doi: 10.1086/522164

Marchant, C. D. (1990). Spectrum of disease due to *Branhamella catarrhalis* in children with particular reference to acute otitis media. *Am. J. Med.* 88, 15S–19S. doi: 10.1016/0002-9343(90)90255-c

Matsuzaki, Y., Sugawara, K., Mizuta, K., Tsuchiya, E., Muraki, Y., Hongo, S., et al. (2002). Antigenic and genetic characterization of influenza C viruses which caused two outbreaks in Yamagata City, Japan, in 1996 and 1998. *J. Clin. Microbiol.* 40, 422–429. doi: 10.1128/JCM.40.2.422-429.2002

McCullers, J. A., and Bartmess, K. C. (2003). Role of neuraminidase in lethal synergism between influenza virus and *Streptococcus pneumoniae*. *J. Infect. Dis.* 187, 1000–1009. doi: 10.1086/368163

McCullers, J. A., Iverson, A. R., McKeon, R., and Murray, P. J. (2008). The platelet activating factor receptor is not required for exacerbation of bacterial pneumonia following influenza. *Scand. J. Infect. Dis.* 40, 11–17. doi: 10.1080/00365540701477568

McCullers, J. A., McAuley, J. L., Browall, S., Iverson, A. R., Boyd, K. L., and Henriques Normark, B. (2010). Influenza enhances susceptibility to natural acquisition of and disease due to *Streptococcus pneumoniae* in ferrets. *J. Infect. Dis.* 202, 1287–1295. doi: 10.1086/656333

McCullers, J. A., and Rehg, J. E. (2002). Lethal synergism between influenza virus and *Streptococcus pneumoniae*: characterization of a mouse model and the role of platelet-activating factor receptor. *J. Infect. Dis.* 186, 341–350. doi: 10.1086/341462

McLeod, D. T., Ahmad, F., Power, J. T., Calder, M. A., and Seaton, A. (1983). Bronchopulmonary infection due to *Branhamella catarrhalis*. *Br. Med. J. (Clin. Res. Ed)* 287, 1446–1447. doi: 10.1136/bmj.287.6403.1446

McNeely, D. J., Kitchens, C. S., and Kluge, R. M. (1976). Fatal *Neisseria (Branhamella)* catarrhalis pneumonia in an immunodeficient host. *Am. Rev. Respir. Dis.* 114, 399–402.

Metersky, M. L., Masterton, R. G., Lode, H., File, T. M. Jr., and Babinchak, T. (2012). Epidemiology, microbiology, and treatment considerations for bacterial pneumonia complicating influenza. *Int. J. Infect. Dis.* 16, e321-31. doi: 10.1016/j.ijid.2012.01.003

Michael, M. Jr. (1942). *Staphylococcus aureus* pneumonia: with special reference to its occurrence as a complication of influenza. *J. Am. Med. Assoc.* 118, 869–874. doi: 10.1001/jama.1942.02830110011004

Michaelis, M., Doerr, H. W., and Cinatl, J. Jr. (2009). Novel swine-origin influenza A virus in humans: another pandemic knocking at the door. *Med. Microbiol. Immunol.* 198, 175–183. doi: 10.1007/s00430-009-0118-5

Michaels, R. H., Myerowitz, R. L., and Klaw, R. (1977). Potentiation of experimental meningitis due to *Haemophilus influenzae* by influenza A virus. *J. Infect. Dis.* 135, 641–645. doi: 10.1093/infdis/135.4.641

Morens, D. M., Taubenberger, J. K., and Fauci, A. S. (2008). Predominant role of bacterial pneumonia as a cause of death in pandemic influenza: implications for pandemic influenza preparedness. *J. Infect. Dis.* 198, 962–970. doi: 10.1086/591708

Morse, S. S., Mazet, J. A., Woolhouse, M., Parrish, C. R., Carroll, D., Karesh, W. B., et al. (2012). Prediction and prevention of the next pandemic zoonosis. *Lancet* 380, 1956–1965. doi: 10.1016/S0140-6736(12)61684-5

Mulcahy, M. E., Leech, J. M., Renauld, J. C., Mills, K. H., and McLoughlin, R. M. (2016). Interleukin-22 regulates antimicrobial peptide expression and keratinocyte differentiation to control *Staphylococcus aureus* colonization of the nasal mucosa. *Mucosal Immunol.* 9, 1429–1441. doi: 10.1038/mi.2016.24

Mulcahy, M. E., and McLoughlin, R. M. (2016). *Staphylococcus aureus* and Influenza A Virus: partners in coinfection. *MBio* 7:e2068-16. doi: 10.1128/mBio.02068-16

Murphy, F. A. (1998). Emerging zoonoses. *Emerg. Infect. Dis.* 4, 429–435. doi: 10.3201/eid0403.980324

Murphy, T. F. (2003). Respiratory infections caused by non-typeable *Haemophilus influenzae*. *Curr. Opin. Infect. Dis.* 16, 129–134. doi: 10.1097/00001432-200304000-00009

Murphy, T. F., Brauer, A. L., Grant, B. J., and Sethi, S. (2005). *Moraxella catarrhalis* in chronic obstructive pulmonary disease: burden of disease and immune response. *Am. J. Respir. Crit. Care Med.* 172, 195–199. doi: 10.1164/rccm.200412-1747OC

Murray, R. J., Robinson, J. O., White, J. N., Hughes, F., Coombs, G. W., Pearson, J. C., et al. (2010). Community-acquired pneumonia due to pandemic A(H1N1)2009 influenzavirus and methicillin resistant *Staphylococcus aureus* co-infection. *PLoS ONE* 5:e8705. doi: 10.1371/journal.pone.0008705

Nakajima, K., Desselberger, U., and Palese, P. (1978). Recent human influenza A (H1N1) viruses are closely related genetically to strains isolated in 1950. *Nature* 274, 334–339. doi: 10.1038/274334a0

Nayak, D. P., Balogun, R. A., Yamada, H., Zhou, Z. H., and Barman, S. (2009). Influenza virus morphogenesis and budding. *Virus Res.* 143, 147–161. doi: 10.1016/j.virusres.2009.05.010

Newing, W. J., and Christie, R. (1947). Meningitis; isolation of an organism resembling *Neisseria catarrhalis* from cerebro-spinal fluid; report of a case. *Med. J. Aust.* 1, 306.

Nguyen, T., Kyle, U. G., Jaimon, N., Tcharmtchi, M. H., Coss-Bu, J. A., Lam, F., et al. (2012). Coinfection with *Staphylococcus aureus* increases risk of severe coagulopathy in critically ill children with influenza A (H1N1) virus infection. *Crit. Care Med.* 40, 3246–3250. doi: 10.1097/CCM.0b013e318260c7f8

Nichol, K. L., Wuorenma, J., and von Sternberg, T. (1998). Benefits of influenza vaccination for low-, intermediate-, and high-risk senior citizens. *Arch. Intern. Med.* 158, 1769–1776. doi: 10.1001/archinte.158.16.1769

Okamoto, S., Kawabata, S., Nakagawa, I., Okuno, Y., Goto, T., Sano, K., et al. (2003). Influenza A virus-infected hosts boost an invasive type of *Streptococcus pyogenes* infection in mice. *J. Virol.* 77, 4104–4112. doi: 10.1128/JVI.77.7.4104-4112.2003

Olson, D. R., Simonsen, L., Edelson, P. J., and Morse, S. S. (2005). Epidemiological evidence of an early wave of the 1918 influenza pandemic in New York City. *Proc. Natl. Acad. Sci. U.S.A.* 102, 11059–11063. doi: 10.1073/pnas.0408290102

Onofrio, J. M., Shulkin, A. N., Heidbrink, P. J., Toews, G. B., and Pierce, A. K. (1981). Pulmonary clearance and phagocytic cell response to normal pharyngeal flora. *Am. Rev. Respir. Dis.* 123, 222–225.

Opie, E. L., Blake, F., Small, J. C., and Rivers, T. M. (1921). "The pneumonias and other infections of the respiratory tract accompanying influenza and measles. St. Louis: CV Mosby," in *Epidemic Respiratory Disease*, eds E. L. Opie, F. G. Blake, J. C. Small, and T. M. Rivers (New York, NY: Forgotten Books), 107–281.

Osterhaus, A. D., Rimmelzwaan, G. F., Martina, B. E., Bestebroer, T. M., and Fouchier, R. A. (2000). Influenza B virus in seals. *Science* 288, 1051–1053. doi: 10.1126/science.288.5468.1051

Oxford, J. S., Sefton, A., Jackson, R., Innes, W., Daniels, R. S., and Johnson, N. P. (2002). World war I may have allowed the emergence of "Spanish" influenza. *Lancet Infect. Dis.* 2, 111–114. doi: 10.1016/S1473-3099(02)00185-8

Pai, R., Gertz, R. E., Whitney, C. G., and Beall, B. (2005a). Clonal association between *Streptococcus pneumoniae* serotype 23A, circulating within the United States, and an internationally dispersed clone of serotype 23F. *J. Clin. Microbiol.* 43, 5440–5444.

Pai, R., Moore, M. R., Pilishvili, T., Gertz, R. E., Whitney, C. G., Beall, B., et al. (2005b). Postvaccine genetic structure of *Streptococcus pneumoniae* serotype 19A from children in the United States. *J. Infect. Dis.* 192, 1988–1995.

Palacios, G., Hornig, M., Cisterna, D., Savji, N., Bussetti, A. V., Kapoor, V., et al. (2009). *Streptococcus pneumoniae* coinfection is correlated with the severity of H1N1 pandemic influenza. *PLoS ONE* 4:e8540. doi: 10.1371/journal.pone.0008540

Park, I. H., Pritchard, D. G., Cartee, R., Brandao, A., Brandileone, M. C., and Nahm, M. H. (2007). Discovery of a new capsular serotype (6C) within serogroup 6 of *Streptococcus pneumoniae*. *J. Clin. Microbiol.* 45, 1225–1233. doi: 10.1128/JCM.02199-06

Parrish, C. R., Murcia, P. R., and Holmes, E. C. (2015). Influenza virus reservoirs and intermediate hosts: dogs, horses, and new possibilities for influenza virus exposure of humans. *J. Virol.* 89, 2990–2994. doi: 10.1128/JVI.03146-14

Payne, A. M. (1958). Some aspects of the epidemiology of the 1957 influenza pandemic. *Proc. R. Soc. Med.* 51, 1009–1015.

Peckham, R. (2016). *Epidemics in Modern Asia*. Cambridge: Cambridge University Press. doi: 10.1017/CBO9781316026939

Peiris, J. S., Poon, L. L., and Guan, Y. (2009). Emergence of a novel swine-origin influenza A virus (S-OIV) H1N1 virus in humans. *J. Clin. Virol.* 45, 169–173. doi: 10.1016/j.jcv.2009.06.006

Peltola, V. T., and McCullers, J. A. (2004). Respiratory viruses predisposing to bacterial infections: role of neuraminidase. *Pediatr. Infect. Dis. J.* 23, S87–S97. doi: 10.1097/01.inf.0000108197.81270.35

Peltola, V. T., Murti, K. G., and McCullers, J. A. (2005). Influenza virus neuraminidase contributes to secondary bacterial pneumonia. *J. Infect. Dis.* 192, 249–257. doi: 10.1086/430954

Petersdorf, R. G., Fusco, J. J., Harter, D. H., and Albrink, W. S. (1959). Pulmonary infections complicating Asian influenza. *AMA Arch. Intern. Med.* 103, 262–272. doi: 10.1001/archinte.1959.00270020090010

Pittet, L. F., and Posfay-Barbe, K. M. (2012). Pneumococcal vaccines for children: a global public health priority. *Clin. Microbiol. Infect.* 18(Suppl. 5), 25–36. doi: 10.1111/j.1469-0691.2012.03938.x

Plotkowski, M. C., Puchelle, E., Beck, G., Jacquot, J., and Hannoun, C. (1986). Adherence of type I *Streptococcus pneumoniae* to tracheal epithelium of mice infected with influenza A/PR8 virus. *Am. Rev. Respir. Dis.* 134, 1040–1044. doi: 10.1164/arrd.1986.134.5.1040

Potter, C. W. (2001). A history of influenza. *J. Appl. Microbiol.* 91, 572–579. doi: 10.1046/j.1365-2672.2001.01492.x

Racaniello, V. (2009). *The A, B, and C of Influenza Virus*. Available at: http://www.virology.ws/2009/09/22/the-a-b-and-c-of-influenza-virus/

Randolph, A. G., Vaughn, F., Sullivan, R., Rubinson, L., Thompson, B. T., Yoon, G., et al. (2011). Critically ill children during the 2009-2010 influenza pandemic in the United States. *Pediatrics* 128:e1450-8. doi: 10.1542/peds.2011-0774

Reid, A. H., Taubenberger, J. K., and Fanning, T. G. (2001). The 1918 Spanish influenza:integrating history and biology. *Microbes Infect.* 3, 81–87. doi: 10.1016/S1286-4579(00)01351-4

Rice, T. W., Rubinson, L., Uyeki, T. M., Vaughn, F. L., John, B. B., Miller, R. R., et al. (2012). Critical illness from 2009 pandemic influenza A virus and bacterial coinfection in the United States. *Crit. Care Med.* 40, 1487–1498. doi: 10.1097/CCM.0b013e3182416f23

Rizzo, C., Caporali, M. G., and Rota, M. C. (2010). Pandemic influenza and pneumonia due to *Legionella pneumophila*: a frequently underestimated coinfection. *Clin. Infect. Dis.* 51:115. doi: 10.1086/653444

Robertson, L., Caley, J. P., and Moore, J. (1958). Importance of *Staphylococcus aureus* in pneumonia in the 1957 epidemic of influenza A. *Lancet* 2, 233–236. doi: 10.1016/S0140-6736(58)90060-6

Robinson, K. M., Choi, S. M., McHugh, K. J., Mandalapu, S., Enelow, R. I., Kolls, J. K., et al. (2013). Influenza A exacerbates *Staphylococcus aureus* pneumonia by attenuating IL-1beta production in mice. *J. Immunol.* 191, 5153–5159. doi: 10.4049/jimmunol.1301237

Robinson, K. M., Kolls, J. K., and Alcorn, J. F. (2015). The immunology of influenza virus-associated bacterial pneumonia. *Curr. Opin. Immunol.* 34, 59–67. doi: 10.1016/j.coi.2015.02.002

Robinson, K. M., McHugh, K. J., Mandalapu, S., Clay, M. E., Lee, B., Scheller, E. V., et al. (2014). Influenza A virus exacerbates *Staphylococcus aureus* pneumonia in mice by attenuating antimicrobial peptide production. *J. Infect. Dis.* 209, 865–875. doi: 10.1093/infdis/jit527

Rock, K. L., and Kono, H. (2008). The inflammatory response to cell death. *Annu. Rev. Pathol.* 3, 99–126. doi: 10.1146/annurev.pathmechdis.3.121806.151456

Rosenstein, N. E., and Perkins, B. A. (2000). Update on *Haemophilus influenzae* serotype b and meningococcal vaccines. *Pediatr. Clin. North Am.* 47, 337–352. doi: 10.1016/S0031-3955(05)70210-8

Rupp, S., Ambata, P., Narat, V., and Giles-Vernick, T. (2016). Beyond the cut hunter: a historical epidemiology of HIV beginnings in central africa. *Ecohealth* 13, 661–671. doi: 10.1007/s10393-016-1189-6

Samji, T. (2009). Influenza A: understanding the viral life cycle. *Yale J. Biol. Med.* 82, 153–159.

Sanford, B. A., and Ramsay, M. A. (1987). Bacterial adherence to the upper respiratory tract of ferrets infected with influenza A virus. *Proc. Soc. Exp. Biol. Med.* 185, 120–128. doi: 10.3181/00379727-185-42525

Sanford, B. A., Shelokov, A., and Ramsay, M. A. (1978). Bacterial adherence to virus-infected cells: a cell culture model of bacterial superinfection. *J. Infect. Dis.* 137, 176–181. doi: 10.1093/infdis/137.2.176

Schoenborn, J. R., and Wilson, C. B. (2007). Regulation of interferon-gamma during innate and adaptive immune responses. *Adv. Immunol.* 96, 41–101. doi: 10.1016/S0065-2776(07)96002-2

Schwarzmann, S. W., Adler, J. L., Sullivan, R. J. Jr., and Marine, W. M. (1971). Bacterial pneumonia during the Hong Kong influenza epidemic of 1968-1969. *Arch. Intern. Med.* 127, 1037–1041. doi: 10.1001/archinte.1971.00310180053006

Scott, M. J., Hoth, J. J., Stagner, M. K., Gardner, S. A., Peyton, J. C., and Cheadle, W. G. (2004). CD40-CD154 interactions between macrophages and natural killer cells during sepsis are critical for macrophage activation and are not interferon gamma dependent. *Clin. Exp. Immunol.* 137, 469–477. doi: 10.1111/j.1365-2249.2004.02547.x

Seki, M., Yanagihara, K., Higashiyama, Y., Fukuda, Y., Kaneko, Y., Ohno, H., et al. (2004). Immunokinetics in severe pneumonia due to influenza virus and bacteria coinfection in mice. *Eur. Respir. J.* 24, 143–149. doi: 10.1183/09031936.04.00126103

Selinger, D. S., Reed, W. P., and McLaren, L. C. (1981). Model for studying bacterial adherence to epithelial cells infected with viruses. *Infect. Immun.* 32, 941–944.

Shahangian, A., Chow, E. K., Tian, X., Kang, J. R., Ghaffari, A., Liu, S. Y., et al. (2009). Type I IFNs mediate development of postinfluenza bacterial pneumonia in mice. *J. Clin. Invest.* 119, 1910–1920. doi: 10.1172/JCI35412

Sharma-Chawla, N., Sender, V., Kershaw, O., Gruber, A. D., Volckmar, J., Henriques-Normark, B., et al. (2016). Influenza A virus infection predisposes hosts to secondary infection with different *Streptococcus pneumoniae* serotypes with similar outcome but serotype-specific manifestation. *Infect. Immun.* 84, 3445–3457. doi: 10.1128/IAI.00422-16

Shinya, K., Ebina, M., Yamada, S., Ono, M., Kasai, N., and Kawaoka, Y. (2006). Avian flu: influenza virus receptors in the human airway. *Nature* 440, 435–436. doi: 10.1038/440435a

Shope, R. E. (1931). Swine Influenza: Iii. Filtration experiments and etiology. *J. Exp. Med.* 54, 373–385. doi: 10.1084/jem.54.3.373

Shrestha, S., Foxman, B., Weinberger, D. M., Steiner, C., Viboud, C., and Rohani, P. (2013). Identifying the interaction between influenza and pneumococcal pneumonia using incidence data. *Sci. Transl. Med.* 5:191ra184. doi: 10.1126/scitranslmed.3005982

Siegel, S. J., Roche, A. M., and Weiser, J. N. (2014). Influenza promotes pneumococcal growth during coinfection by providing host sialylated substrates as a nutrient source. *Cell Host Microbe* 16, 55–67. doi: 10.1016/j.chom.2014.06.005

Smith, A. M., and McCullers, J. A. (2014). Secondary bacterial infections in influenza virus infection pathogenesis. *Curr. Top. Microbiol. Immunol.* 385, 327–356. doi: 10.1007/82_2014_394

Smith, F. B. (1995). The Russian influenza in the United Kingdom, 1889-1894. *Soc. Hist. Med.* 8, 55–73. doi: 10.1093/shm/8.1.55

Spooner, L. H., Scott, J. M., and Heath, E. H. (1919). A bacteriologic study of the influenza epidemic at Camp Devens. *JAMA* 72, 155–159. doi: 10.1001/jama.1919.02610030001001

Steinhauer, D. A. (1999). Role of hemagglutinin cleavage for the pathogenicity of influenza virus. *Virology* 258, 1–20. doi: 10.1006/viro.1999.9716

Sullivan, K. M., Monto, A. S., and Longini, I. M. Jr. (1993). Estimates of the US health impact of influenza. *Am. J. Public Health* 83, 1712–1716. doi: 10.2105/AJPH.83.12.1712

Sun, K., and Metzger, D. W. (2008). Inhibition of pulmonary antibacterial defense by interferon-gamma during recovery from influenza infection. *Nat. Med.* 14, 558–564. doi: 10.1038/nm1765

Takase, H., Nitanai, H., Yamamura, E., and Otani, T. (1999). Facilitated expansion of pneumococcal colonization from the nose to the lower respiratory tract in mice preinfected with influenza virus. *Microbiol. Immunol.* 43, 905–907. doi: 10.1111/j.1348-0421.1999.tb01226.x

Tashiro, M., Ciborowski, P., Reinacher, M., Pulverer, G., Klenk, H. D., and Rott, R. (1987). Synergistic role of staphylococcal proteases in the induction of influenza virus pathogenicity. *Virology* 157, 421–430. doi: 10.1016/0042-6822(87)90284-4

Taubenberger, J. K., and Morens, D. M. (2008). The pathology of influenza virus infections. *Annu. Rev. Pathol.* 3, 499–522. doi: 10.1146/annurev.pathmechdis.3.121806.154316

Taubenberger, J. K., Reid, A. H., Janczewski, T. A., and Fanning, T. G. (2001). Integrating historical, clinical and molecular genetic data in order to explain the origin and virulence of the 1918 Spanish influenza virus. *Philos. Trans. R. Soc. Lond. B Biol. Sci.* 356, 1829–1839. doi: 10.1098/rstb.2001.1020

Tillett, H. E., Smith, J. W., and Gooch, C. D. (1983). Excess deaths attributable to influenza in England and Wales: age at death and certified cause. *Int. J. Epidemiol.* 12, 344–352. doi: 10.1093/ije/12.3.344

Tong, S., Li, Y., Rivailler, P., Conrardy, C., Castillo, D. A., Chen, L. M., et al. (2012). A distinct lineage of influenza A virus from bats. *Proc. Natl. Acad. Sci. U.S.A.* 109, 4269–4274. doi: 10.1073/pnas.1116200109

Tong, S. Y., Davis, J. S., Eichenberger, E., Holland, T. L., Fowler, V. G. Jr., et al. (2015). *Staphylococcus aureus* infections: epidemiology, pathophysiology, clinical manifestations, and management. *Clin. Microbiol. Rev.* 28, 603–661. doi: 10.1128/CMR.00134-14

Topham, N. J., and Hewitt, E. W. (2009). Natural killer cell cytotoxicity: how do they pull the trigger? *Immunology* 128, 7–15. doi: 10.1111/j.1365-2567.2009.03123.x

Trilla, A., Trilla, G., and Daer, C. (2008). The 1918 "Spanish Flu" in Spain. *Clin. Infect. Dis.* 47, 668–673. doi: 10.1086/590567

Valleron, A. J., Cori, A., Valtat, S., Meurisse, S., Carrat, F., and Boelle, P. Y. (2010). Transmissibility and geographic spread of the 1889 influenza pandemic. *Proc. Natl. Acad. Sci. U.S.A.* 107, 8778–8781. doi: 10.1073/pnas.1000886107

van der Sluijs, K. F., van der Poll, T., Lutter, R., Juffermans, N. P., and Schultz, M. J. (2010). Bench-to-bedside review: bacterial pneumonia with influenza - pathogenesis and clinical implications. *Crit. Care* 14:219. doi: 10.1186/cc8893

Verduin, C. M., Hol, C., Fleer, A., van Dijk, H., and van Belkum, A. (2002). *Moraxella catarrhalis*: from emerging to established pathogen. *Clin. Microbiol. Rev.* 15, 125–144. doi: 10.1128/CMR.15.1.125-144.2002

Wadowsky, R. M., Mietzner, S. M., Skoner, D. P., Doyle, W. J., and Fireman, P. (1995). Effect of experimental influenza A virus infection on isolation of *Streptococcus pneumoniae* and other aerobic bacteria from the oropharynges of allergic and nonallergic adult subjects. *Infect. Immun.* 63, 1153–1157.

Wang, T. T., and Palese, P. (2009). Unraveling the mystery of swine influenza virus. *Cell* 137, 983–985. doi: 10.1016/j.cell.2009.05.032

Webster, R. G., Bean, W. J., Gorman, O. T., Chambers, T. M., and Kawaoka, Y. (1992). Evolution and ecology of influenza A viruses. *Microbiol. Rev.* 56, 152–179.

Weinberger, D. M., Simonsen, L., Jordan, R., Steiner, C., Miller, M., and Viboud, C. (2012). Impact of the 2009 influenza pandemic on pneumococcal pneumonia hospitalizations in the United States. *J. Infect. Dis.* 205, 458–465. doi: 10.1093/infdis/jir749

Wenger, J. D. (1998). Epidemiology of Haemophilus influenzae type b disease and impact of Haemophilus influenzae type b conjugate vaccines in the United States and Canada. *Pediatr. Infect. Dis. J.* 17(9 Suppl.), S132–S136.

Wertheim, H. F., Melles, D. C., Vos, M. C., van Leeuwen, W., van Belkum, A., Verbrugh, H. A., et al. (2005). The role of nasal carriage in *Staphylococcus aureus* infections. *Lancet Infect. Dis.* 5, 751–762. doi: 10.1016/S1473-3099(05)70295-4

World Health Organization (2016). *Influenza (Seasonal) Factsheet*. Available at: http://www.who.int/mediacentre/factsheets/fs211/en/

World Health Organization (2017). *Smallpox*. Available at: http://web.archive.org/web/20070921235036/http://www.who.int/mediacentre/factsheets/smallpox/en/

Wu, S. W., de Lencastre, H., and Tomasz, A. (2005). Expression of high-level methicillin resistance in *Staphylococcus aureus* from the Staphylococcus sciuri mec A homologue: role of mutation(s) in the genetic background and in the coding region of mec A. *Microb. Drug Resist.* 11, 215–224. doi: 10.1089/mdr.2005.11.215

Wu, Y., Mao, H., Ling, M. T., Chow, K. H., Ho, P. L., Tu, W., et al. (2011). Successive influenza virus infection and *Streptococcus pneumoniae* stimulation alter human dendritic cell function. *BMC Infect. Dis.* 11:201. doi: 10.1186/1471-2334-11-201

Wu, Y., Wu, Y., Tefsen, B., Shi, Y., and Gao, G. F. (2014). Bat-derived influenza-like viruses H17N10 and H18N11. *Trends Microbiol.* 22, 183–191. doi: 10.1016/j.tim.2014.01.010

Wyde, P. R., Six, H. R., Ambrose, M. W., and Throop, B. J. (1989). Influenza virus infection and bacterial clearance in young adult and aged mice. *J. Gerontol.* 44, B118–B124. doi: 10.1093/geronj/44.5.b118

Yamashita, M., Krystal, M., Fitch, W. M., and Palese, P. (1988). Influenza B virus evolution: co-circulating lineages and comparison of evolutionary pattern with those of influenza A and C viruses. *Virology* 163, 112–122. doi: 10.1016/0042-6822(88)90238-3

Yang, M., Gao, H., Chen, J., Xu, X., Tang, L., Yang, Y., et al. (2016a). Bacterial coinfection is associated with severity of avian influenza A (H7N9), and procalcitonin is a useful marker for early diagnosis. *Diagn. Microbiol. Infect. Dis.* 84, 165–169. doi: 10.1016/j.diagmicrobio.2015.10.018

Yang, R., Cui, Y., and Bi, Y. (2016b). Perspectives on *Yersinia pestis*: a model for studying zoonotic pathogens. *Adv. Exp. Med. Biol.* 918, 377–391.

Youzbashi, E., Marschall, M., Chaloupka, I., and Meier-Ewert, H. (1996). [Distribution of influenza C virus infection in dogs and pigs in Bavaria]. *Tierarztl. Prax.* 24, 337–342.

Zambon, M. C. (2001). The pathogenesis of influenza in humans. *Rev. Med. Virol.* 11, 227–241. doi: 10.1002/rmv.319

Zumla, A., Hui, D. S., Al-Tawfiq, J. A., Gautret, P., McCloskey, B., and Memish, Z. A. (2014). Emerging respiratory tract infections. *Lancet Infect. Dis.* 14, 910–911. doi: 10.1016/S1473-3099(14)70899-0

18

The Pivotal Role of DNA Repair in Infection-Mediated Inflammation and Cancer

Ayse Z. Sahan[1], *Tapas K. Hazra*[2] *and Soumita Das*[1]*

[1] *Department of Pathology, University of California, San Diego, San Diego, CA, United States,* [2] *Department of Internal Medicine, University of Texas Medical Branch, Galveston, TX, United States*

Pathogenic and commensal microbes induce various levels of inflammation and metabolic disease in the host. Inflammation caused by infection leads to increased production of reactive oxygen species (ROS) and subsequent oxidative DNA damage. These in turn cause further inflammation and exacerbation of DNA damage, and pose a risk for cancer development. *Helicobacter pylori*-mediated inflammation has been implicated in gastric cancer in many previously established studies, and *Fusobacterium nucleatum* presence has been observed with greater intensity in colorectal cancer patients. Despite ambiguity in the exact mechanism, infection-mediated inflammation may have a link to cancer development through an accumulation of potentially mutagenic DNA damage in surrounding cells. The multiple DNA repair pathways such as base excision, nucleotide excision, and mismatch repair that are employed by cells are vital in the abatement of accumulated mutations that can lead to carcinogenesis. For this reason, understanding the role of DNA repair as an important cellular mechanism in combatting the development of cancer will be essential to characterizing the effect of infection on DNA repair proteins and to identifying early cancer biomarkers that may be targeted for cancer therapies and treatments.

Keywords: bacterial infection, commensal bacteria, DNA damage, inflammation and cancer, *Fusobacterium nucleatum*, *Helicobacter pylori*

Edited by:
Chew Chieng Yeo,
Sultan Zainal Abidin University,
Malaysia

Reviewed by:
M. Gabriela Kramer,
University of the Republic, Uruguay
Arijit Dutta,
Yale University, United States

***Correspondence:**
Soumita Das
sodas@ucsd.edu

INTRODUCTION

The significance of cancer as a disease that affects a large percentage of the world population is undeniable. It is one of the leading causes of death worldwide and according to World Health Organization, it causes 8–9 million deaths/year. In the United States alone, it is projected that 39.6% of the population will have some type of cancer during their life. Consequently, there are enormous expenditures in the field of cancer care. The expenditures for cancer care in the United States were nearly $125 billion in 2010 and could reach $156 billion in 2020; as mentioned in the website of National Cancer Institute. Infection and infection-associated inflammation is the major threat of cancer. Chronic inflammation from infection causes abnormal immune response, obesity, DNA damage and cancer. The best example is the inflammatory bowel disease where Ulcerative colitis and Crohn's disease develop colon cancer.

In 1863, Rudolf Virchow was the first scientist to link inflammation with cancer. He mentioned that the origin of cancer was at sites of chronic inflammation and a group of irritants with tissue injury causes cell proliferation (Balkwill and Mantovani, 2001). How does inflammation initiate malignancies? The possible answer to this question is the infection associated with chronic inflammation. Approximately 20% of cancer worldwide is caused by infection (Kuper et al., 2000; Parkin, 2006). The last 20 years of research shows that microbial infection is associated with cancer and can induce cancer progression. According to the estimate of the International Agency for Research on Cancer (IARC) approximately 18% of cancer are associated to infectious diseases that are caused by bacteria, viruses, and parasites. The well-known examples are human papilloma viruses (HPV; causing anogenital cancers), *Helicobacter pylori* (gastric cancers), hepatitis B and C viruses (hepatic cancers), and *Fusobacterium* (Colon cancer). The detailed list of microbes that have been researched in relation to their effect on certain DNA repair proteins are added in the **Table 1**. The gaps in knowledge in this field are reflected in the table, where there are many pathogen-associated cancers that have not been studied in relation to DNA repair proteins.

As cited in the press release of the Nobel Assembly (Marshall and Warren, 1984, 2005): "Many diseases in humans such as Crohn's disease, ulcerative colitis, rheumatoid arthritis, and atherosclerosis are due to chronic inflammation. The discovery that one of the most common diseases of mankind, peptic ulcer disease, has a microbial cause has stimulated the search for microbes as possible causes of other chronic inflammatory conditions. Even though no definite answers are at hand, recent data clearly suggest that a dysfunction in the recognition of microbial products by the human immune system can result in disease development. The discovery of *Helicobacter pylori* has led to an increased understanding of the connection between chronic infection, inflammation, and cancer."

Chronic infection generates a milieu of inflammatory cytokines that leads to inflammatory microenvironment, a critical modulator of carcinogenesis. The persistent infection and chronic inflammation changes somatic cells by the influence of associated microbes and epigenetic factors (Fernandes et al., 2015). Hanahan et al. showed that genome instability and inflammation are the emerging hallmarks associated with cancer (Hanahan and Weinberg, 2011). **Figure 1** shows the responsive elements that can trigger carcinogenesis. Bacterial infections increase cancer risk through either an extrinsic pathway, linked to induction of chronic inflammatory diseases that can increase cancer risk, or an intrinsic pathway, which is the accrual of genetic mutations that cause inflammation and transformation (Mantovani et al., 2008). Chronic inflammation has been associated with multiple types of cancer to the extent that inflammation period has been linked to increased risk of carcinogenesis (Shacter and Weitzman, 2002). Chronic inflammation is able to adjust the tumor microenvironment with cells such as tumor associated macrophages and various inflammatory agents such as chemokines, to regulate both tumor growth and angiogenesis (Coussens and Werb, 2002). Inflammation is also able to induce growth factors that serve several roles in carcinogenesis and tumorigenesis (Hanahan and Weinberg, 2011). The intrinsic pathway of genome alterations caused by infection is often linked to inflammation-mediated reactive oxygen species (ROS) production, which can increase the rate of genetic mutations that can accumulate to cause cancer (Hanahan and Weinberg, 2011).

Bacterial infection causes inflammatory response and the ROS generated by bacterial infection often results in genomic instability (Chumduri et al., 2016). This implicates a bacterial infection in compromising or at least impeding some of the several cellular mechanisms for maintaining genetic integrity and repairing mutations. As genomic instability is an underlying factor in almost all cancer cells, the link between infection and cancer development and progression is a significant one. This review will focus on the mechanism of inflammation and ROS production post-infection, and then elaborate on the genomic instability induced by infection/inflammation by discussing the effect on various DNA repair pathways. As a major focus, we will bring the link of *Helicobacter pylori* with gastric cancer and microbial infection associated colorectal cancer under the scope of DNA damage repair following infection.

MICROBIAL INFECTION-MEDIATED INFLAMMATION LINKED TO CANCER

Innate and adaptive immune responses are important to protect self from pathogenic microbial attack. Understanding of the infection process is important as bacterial and viral infection induces the inflammation that increases cancer risk (de Martel and Franceschi, 2009). The innate and adaptive immune responses will be discussed in the next section.

Innate Immune Response

The immune response following recognition and invasion of microbes such as bacteria and viruses split into the innate and adaptive responses. Recognition and the initial precautionary actions are taken by the pattern recognition receptors (PRRs) of innate immune system (Pasare and Medzhitov, 2004; Akira et al., 2006). PRRs are found on the surfaces of epithelial cells and several immune cells to recognize structurally conserved pathogen-associated microbial products (PAMPs). For example, Toll-like receptor 4 (TLR4) is a PRR that binds lipopolysaccharide (LPS) on the outer membrane of gram-negative bacteria; TLR2 binds bacterial lipoproteins and lipoteichoic acids. The cytosolic sensor; NOD-like receptors (NLRs) detect intracellular pathogens. The innate immune response is the immediate mechanism by which the host attempts to clear an infection. Most of the PRRs lead to activation of MYD88-dependent pathways that eventually lead to NF-κB activation. This leads to further production of inflammatory cytokines and chemokines as well as antimicrobial peptides. Like TLRs, NLRs lead to an inflammatory response from both this signaling cascade and through the activation of caspases that act on cytokines that mediate the rest of the inflammatory response (Zarember and Godowski, 2002; Barton and Medzhitov, 2003; Basset et al., 2003; Tsung et al., 2007; Church et al., 2008; Martinon et al., 2009; West et al., 2011b). The PRRs mediated inflammatory response is initiated by various inflammatory cytokines and chemokines, which draw macrophages and mast cells that

TABLE 1 | A compilation of bacteria, virus, and parasite-associated cancers with some of the available information on their link to BER, NER, and MMR protein expression or mutations.

Bacteria-associated cancers	Gastric cancer	*Helicobacter pylori*		
		BER	NER	MMR
		APE1 (Ding et al., 2004)		MSH2 and MLH1 (Kim et al., 2002)
				MLH1, MSH2, MSH3, and MSH6 (Santos et al., 2017)
	Colon cancer	Enteropathogenic *Escherichia coli*		
		BER	NER	MMR
				MSH2 and MLH1 (Maddocks et al., 2013)
		Streptococcus bovis (Ellmerich et al., 2000)		
		BER	NER	MMR
		Fusobacterium nucleatum (Repass et al., 2016)		
		BER	NER	MMR
	Lung cancer	*Mycobacterium tuberculosis* (Wu et al., 2011)		
		BER	NER	MMR
		Chlamydia pneumoniae (Koyi et al., 2001)		
		BER	NER	MMR
	Bladder cancer	*Schistosoma hematobium* (Jemal et al., 2011)		
		BER	NER	MMR
		8-oxo-dG (Salim et al., 2008)		
		APE1 (Salim et al., 2008)		
		Salmonella enterica serovar Typhi (Dutta et al., 2000)		
	Ovarian cancer	*Chlamydia trachomatis* (Xie et al., 2017)		
		BER	NER	MMR
		Mycoplasma genitalium		
		BER	NER	MMR
Virus-associated cancers	Cervical cancer	Human Papilloma Virus (HPV) (Bosch et al., 1995)		
		BER	NER	MMR
		XRCC1 (Bajpai et al., 2016)	ERCC2 (Bajpai et al., 2016)	hMLH1 (Ciavattini et al., 2005)
		APE-1 (Shekari et al., 2008)	ERCC4 (Bajpai et al., 2016)	hMSH2 (Ciavattini et al., 2005)
			ERCC5(Joo et al., 2016)	hMLH3 (Ye et al., 2014)
	Head and neck (Oropharyngeal) Squamous cell cancer	Human Papilloma Virus (HPV) (Hajek et al., 2017)		
		BER	NER	MMR
		XRCC1 (Nickson et al., 2017)	ATR (Maddocks et al., 2013)	
		DNA polymerase β (Nickson et al., 2017)	ERCC1 (Langer, 2012)	

(Continued)

TABLE 1 | Continued

		PNKP (Nickson et al., 2017)		
		PARP-1 (Nickson et al., 2017)		
	Hepatocellular Carcinoma	Hepatitis B Virus (HBV) (El-Serag, 2012; Lu et al., 2017)		
		BER	NER	MMR
		XRCC1 (Almeida Pereira Leite et al., 2013; Bose et al., 2013)	TFIIH (Arbuthnot and Kew, 2001; Qadri et al., 2011)	
		OGG1 (Bose et al., 2013)	TFIIB (Arbuthnot and Kew, 2001)	
		TDG (van de Klundert et al., 2012)	XPB (Jia et al., 1999; Arbuthnot and Kew, 2001)	
			XPD (Jia et al., 1999; Arbuthnot and Kew, 2001)	
			XAP-1 (DDB1) (Sohn et al., 2000)	
		Hepatitis C Virus (HCV) (El-Serag, 2012; Lu et al., 2017)		
		BER	NER	MMR
		Neil1 (Higgs et al., 2014)	XPD (Gulnaz et al., 2013)	hMSH2 (Helal et al., 2010)
		XRCC1 (Gulnaz et al., 2013)		hMLH1 (Helal et al., 2010)
		XRCC3 (Gulnaz et al., 2013)		
	Nasopharyngeal Cancer	Epstein-Barr Virus (EBV) (Huang et al., 2017)		
		BER	NER	MMR
			XPA (Fu et al., 2015)	
	T Cell leukemia	Human T Lymphotropic Virus Type I (HTLV-1) (Liao, 2006)		
		BER	NER	MMR
		XRCC5 (Ng et al., 2001)	ERCC5 (Ng et al., 2001)	
	Kaposi sarcoma	Kaposi sarcoma-associated Herpes Virus (KSHV or HHV8) (Liao, 2006)		
		BER	NER	MMR
		APE1 (Zhong et al., 2017)		
		UNG2 (Verma et al., 2006)		
Parasite-associated cancer	Urinary bladder cancer	*Schistosoma haematobium* (Kawanishi et al., 2016)		
		BER	NER	MMR
	Cholangio carcinoma	*Opisthorchis Viverrini* (Kawanishi et al., 2016)		
		BER	NER	MMR
				hMSH2 (Liengswangwong et al., 2006)
				hMLH1 (Liengswangwong et al., 2006)

There are many gaps in knowledge in this field, particularly in the field of bacteria-associated cancers and regulation of various DNA repair pathway proteins.

release inflammatory mediators to recruit neutrophils and plasma proteins. Neutrophils phagocytose the pathogen and surrounding debris. Phagolysosomes form through the fusion of phagosomes with granules consisting of enzymes and ROS that can kill the phagocytosed pathogen. The neutrophils can release these toxic granules and cause collateral damage to the surrounding tissue (Segal, 2005; Medzhitov, 2008; Serhan et al., 2008).

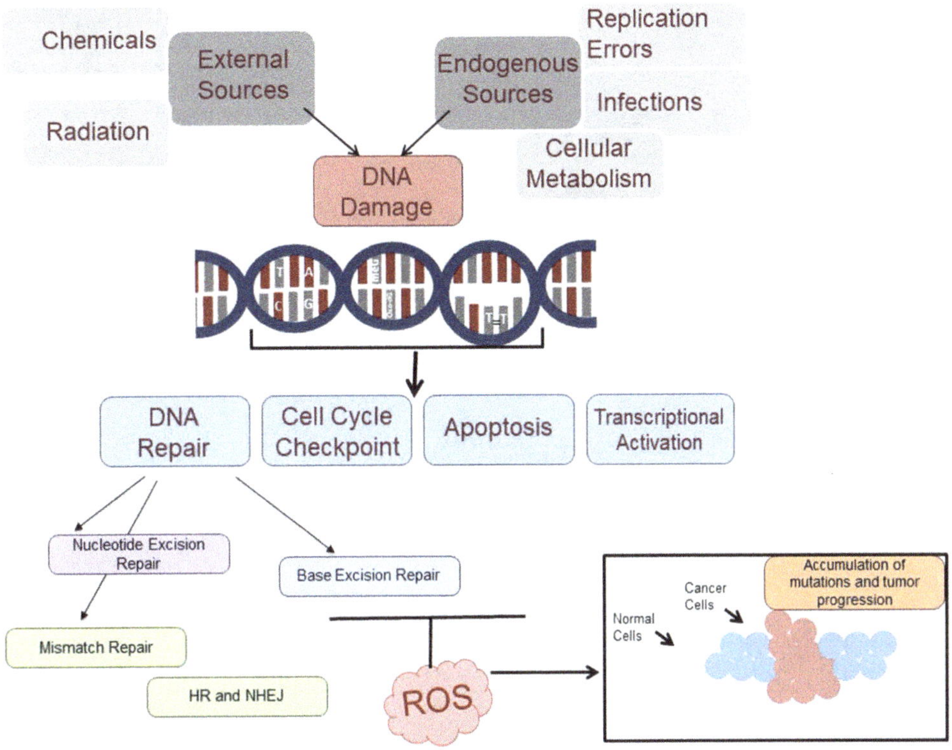

FIGURE 1 | DNA damage response and the potential role of ROS in inhibition of DNA repair. DNA damage is induced by external (various environmental pollutants, chemicals, and radiation), and internal resources (infection, cellular metabolism and replication errors). These DNA damages affect the cell cycle check point, apoptosis, transcriptional activation and cancer. Part of the DNA damages (DNA base adducts, mismatch bases, damaged bases, and double strand breaks) are repaired by the nucleotide excision repair, mismatch repair, base excision repair, and homologous recombination (HR) and non-homologous end joining (NHEJ) pathways, respectively. Many of the DNA damages are ROS-induced and recognized by the BER pathway which excises and repairs the lesions. However, ROS may potentially inhibit repair through down-regulation of certain initial proteins in the BER pathway, which can cause a buildup of carcinogenic mutations and ultimately lead to tumor progression.

The increase in the pro-inflammatory cytokines will clear the bacteria and the release of pro-inflammatory agents will be halted (**Figure 2**), when the tissue repair phase will be started. The tissue damage and debris resulting from the neutrophil activity may prevent impedance of inflammatory response and lead to continued chronic inflammation (Nathan, 2002). Various non-degradable components of the eliminated pathogen also contribute to a lasting inflammatory response even after the threat of the invading pathogen has been erased (Medzhitov, 2008).

Adaptive Immune Response

If the innate immune response is not adequate to kill the pathogen, the adaptive immune response is activated by continued inflammation at the site of infection (Zarember and Godowski, 2002; Martinon et al., 2009). The microbes must then be cleared by specialized lymphocyte (B and T-cell)-mediated mechanisms (Medzhitov, 2008). For this reason, the adaptive immune response requires time to be initiated. Prolonged infection can lead to continuous tissue damage and a chronic inflammatory response that results in various diseases. For instance, in the intestinal tract, inflammatory bowel diseases such as Crohn's disease and ulcerative colitis may manifest due to conditions of chronic inflammation that result from excessive immune activation exacerbating the initial inflammatory response to a foreign or commensal microbe (Cong et al., 2002; Macdonald and Monteleone, 2005). Specifically in Crohn's disease, the adaptive immune response plays a role through an extreme CD4 T helper type I response to overexpression of various inflammatory cytokines (Macdonald and Monteleone, 2005).

MECHANISMS THAT LINK INFLAMMATION TO CANCER PROGRESSION

The cytokines and chemokines from innate and adaptive immune cells direct the progression of the tumor microenvironment. Downstream effectors NF-κB, AP-1, STAT control the inflammatory milieu either by affecting tumor survival, growth, and tumor progression by signaling for molecules such as IL-6, IL-23 or by anti-tumor immunity (IFN-γ, IL-12). The inflammatory cytokines have been reported to generate ROS and reactive nitrogen intermediates (RNI) using NADPH oxidase in phagocytic cells and epithelial cells (Yang et al., 2007). These

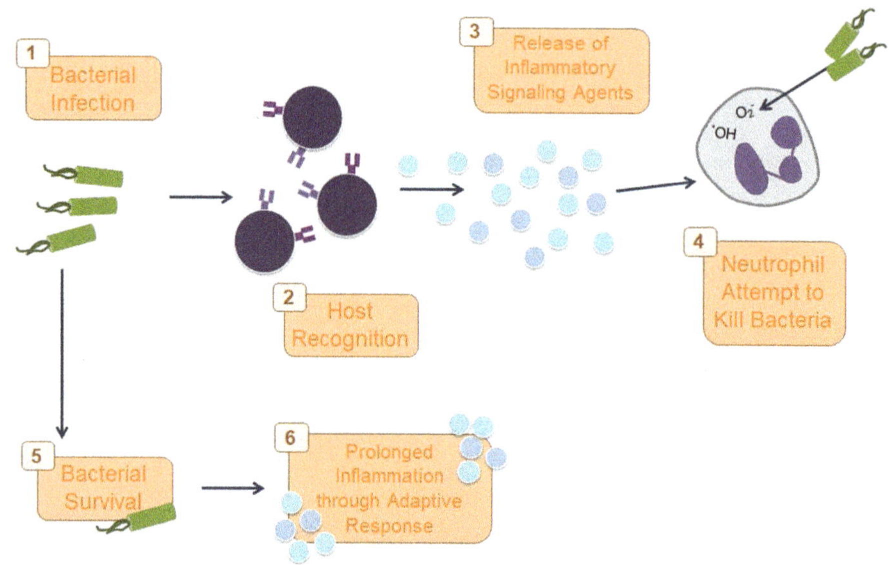

FIGURE 2 | Immune response to bacterial infection. Once a bacterial infection is recognized by PRRs on host cells as part of the innate immune response, inflammatory cytokines and chemokines are released to draw neutrophils to the site of infection. Neutrophils phagocytose and kill the bacteria with ROS. If this initial response does not kill the bacteria, the adaptive immune response kicks in and may result in a chronic inflammatory response.

ROS are the major source of damage to nucleic acid, proteins and lipids.

Reactive Oxygen Species (ROS)

ROS refer to various, highly reactive and partially reduced metabolites of oxygen such as H_2O_2 that are essential signaling molecules in the human immune system (Martindale and Holbrook, 2002; West et al., 2011b). ROS includes oxygen radicals (superoxide, hydroxyl, peroxyl and alkoxyl) and certain nonradicals that are either oxidizing agents and/or are easily converted into radicals, such as HOCl, ozone, peroxynitrite, singlet oxygen and hydrogen peroxide. ROS initiates DNA base oxidation which, if not repaired properly, may lead to induce a mutation.

Inflammatory cytokine signaling also depend on ROS. ROS is also crucial for inflammasome signaling and increased mitochondrial ROS activate the NLRP3 inflammasome (Zhou et al., 2011). Additionally, the post-translational modifications that ROS are associated with are often linked to modifications in protein cysteine residues that are generally associated with either Ca^{2+} mediated signaling or tyrosine phosphorylation. This implicates ROS in cell motility, mitosis, differentiation, and immune response or regulation (van der Vliet, 2008). The immediate effect of excessive ROS presence in the host is a chronic inflammatory state that exacerbates both inflammation and, as a result, ROS production. Also, cancer cells utilize mROS to constitutively activate proliferation pathways to promote tumor growth (Cairns et al., 2011). Inability of cellular antioxidants to curtail ROS leads to oxidative stress on host cells that can lead to many adverse effects including induction of DNA damage (Ernst, 1999). Therefore, following oxidative stress, the cell must survive by either adapting to the induced stress or by repairing the damage (Martindale and Holbrook, 2002). Inability to repair or adjust to the damage will lead to chronic conditions such as cancer, diabetes, and various neurological or cardiovascular diseases. The specific mechanisms of DNA damage repair that diminish the oxidative damage done by ROS will be discussed later in this review.

ROS in immune cells are important to kill extracellular pathogens by using the NADPH oxidase in phagosomes. Despite the various positive roles of ROS in immunity and other processes, they are associated with conditions such as diabetes, hypertension, cancer, and autoimmune diseases due to their ability to change and damage cellular proteins, lipids, and DNA (Zimmerman and Cerutti, 1984; Guzik et al., 2007). ROS is acquired through both exogenous and endogenous sources. Exogenous sources of ROS include carcinogen induced or generated ROS; for example, xenobiotics, chlorinated compounds, and radiation are associated to oxidative stress. The main endogenous source of ROS in the human body is through mitochondrial respiration or Ox-Phos (Oxidative phosphorylation) system. The electron transport chain uses mitochondrial oxidative phosphorylation complexes that lead to generation of ROS that can potentially harm cellular components such as proteins and nucleic acids through post-translational modifications and oxidation (Molina-Cruz et al., 2008). ROS produced by Ox-Phos pathway participate in immune signaling with TLR and cytosolic RIG-I like receptors (RLRs). West et al. reported that stimulation of cell-surface TLRs (TLR1, TLR2, and TLR4), but not endosomal TLRs (TLR3, TLR7, TLR8, and TLR9), leads to an increase in mROS (Mitochondrial Reactive Oxygen Species) production through TRAF6 and ECSIT signaling (West et al., 2011a). Mitochondrial ROS enhance RLR signaling in autophagy-dependent Atg5-depleted cells that indicate the importance of autophagy in innate antiviral defense (Tal et al., 2009).

The most significant damages caused to DNA due to high concentrations of ROS include double and single strand breaks, oxidized DNA bases, and aberrant DNA cross-linking. Each type of ROS plays a different role in inflicting damage. For example, hydrogen peroxide (H_2O_2) directly induces DNA damage. The versatility of different ROS is reflected in the wide array of DNA damage that they can cause. In general, the hydroxyl molecule (OH^-) is the most damaging form of ROS, but other forms like the oxygen molecule (O_2), RO_2, and RO are also capable of different types of damage. For instance, while OH^- can react with all of the bases and the deoxyribose backbone of DNA, O_2 preferentially targets guanine residues (Wiseman and Halliwell, 1996; Valko et al., 2007; Imlay, 2008). In a wider scope, ROS are generally likely to react with and damage DNA through oxidation, methylation, depurination, and deamination. The lesion most often found due to oxidative DNA damage, 8-hydroxyguanine (8-OHdG) or the nucleoside 8-hydroxydeoxyguanosine, is generally considered as marker of oxidative damage incurred by a cell. For instance, many studies examined for increased production of ROS culminating in an increased level of oxidative DNA damage by measuring 8-hydroxyguanine (Dandona et al., 1996; Farinati et al., 1998). In another study, it was observed that P53 acts as a defense against ROS-mediated DNA oxidation in various experimental conditions by detecting the presence of 8-oxodeoxyguanosine in the DNA (Sablina et al., 2005). The 8-hydroxydeoxyguanosine and other oxidized DNA lesions (8-oxo-adenine, thymine glycol, 5-hydroxy-deoxycytidine) have also been observed in many mutated oncogenes and tumor suppressor genes and these lesions are able to induce further neoplastic mutations in the DNA. The presence of high levels of 8-oxoguanine lesions, along with many other oxidative lesions, was shown in the DNA of tumorous tissues from many patients with different types of cancers (Klaunig and Kamendulis, 2004). This indicates a serious implication of oxidative stress and oxidative DNA damage in carcinogenesis (**Figure 1**).

INVOLVEMENT OF MICROBES IN ROS-LINKED CARCINOGENESIS

The gut microbiota supplement a significant portion of the human metabolism, and the composition and activity of the microbiota plays a large role in susceptibility to metabolic diseases such as hyperglycemia, hyperlipidemia, insulin resistance, and obesity (Vijay-Kumar et al., 2010; Spencer et al., 2011). Microbes protect themselves from the ROS generated by host using an enzyme called superoxide dismutase (Sod), which is abundant in cells throughout the body. This enzyme attaches (binds) to molecules of copper and zinc to break down toxic, charged oxygen molecules called superoxide radicals. Interestingly, bacteria has protective proteins such as SodA, SodB, SodC, AhpCF, KatG, KatE to detoxify ROS and proteins to counter damage (e.g., SoxRS, OxyRS, and SOS regulons) (Imlay, 2008). On the host side, Nox2, present in the NADPH complex is responsible for the generation of ROS (Lambeth, 2004). The generation of ROS is important for host defense as patients with chronic granulomatous disease (CGD) with deficiencies in NOX2 components are susceptible to infection (Cross et al., 2000).

Transcription factors vital to cellular processes such as inflammation, cell cycle regulation, motility, and growth like NF-κB, p53, HIF-1α, β-catenin/Wnt can be activated by the oxidative stress caused by an imbalance in the presence of oxidative ROS and the countering antioxidants. By mediating activity of genes and proteins related to oxidative stress, ROS are able to affect further cellular properties, such as cell growth, differentiation, and apoptosis, which can induce transformation (Reuter et al., 2010). For instance, ROS produced as a result of exposure of mouse mammary epithelial cells to MMP-3, a stromal enzyme that is linked to inducing epithelial-mesenchymal (EMT) transition and malignant transformation, caused activation of the transcription factor SNAIL, which induced oxidative DNA damage and EMT (Radisky et al., 2005). ROS-induced oxidative stress can also affect gene expression through direct DNA (de)methylation, which is an epigenetic method of silencing and activating certain genes by changing the physical accessibility of certain genes (Klaunig and Kamendulis, 2004). Chromosomal alterations induced by ROS can also lead to cellular damage. The genomic instability and transcriptional changes that accompany ROS and oxidative stress can therefore lead to carcinogenesis.

Post-infection inflammation-mediated mechanisms that assist in tumor formation and progression are an indirect method of bacterial infection leading to cancer. There are also several factors in a bacterial infection that can directly induce DNA damage or alter cell-signaling pathways that can lead to carcinogenesis. For instance, several bacteria such as *Escherichia coli* and *Shigella dysenteriae* are able to produce genotoxins (colibactin and shiga toxin, respectively) that inflict damage on the host DNA such as DNA strand breaks that may affect tumor suppressors or oncogenes (Gagnaire et al., 2017).

DNA REPAIR MECHANISMS

Cellular DNA is altered either during replication or by external mutagens. Misincorporation of DNA bases can occur during replication; however, it is combatted through the proofreading activity of DNA polymerases. There may however be errors during DNA replication that are not recognized by the polymerases. There is also a great wealth of mutagens that can cause extensive changes in the sequence of human DNA. Of these, just oxidative DNA damage is estimated to arise about 10^5 times in 1 day due to ROS-induced damages (Lengauer et al., 1998). Accumulation of all these mutations would severely inhibit the ability of cells to survive and/or maintain proper cellular functions in almost all cases. Fortunately, the cell has several mechanisms that are specialized to recognize and repair

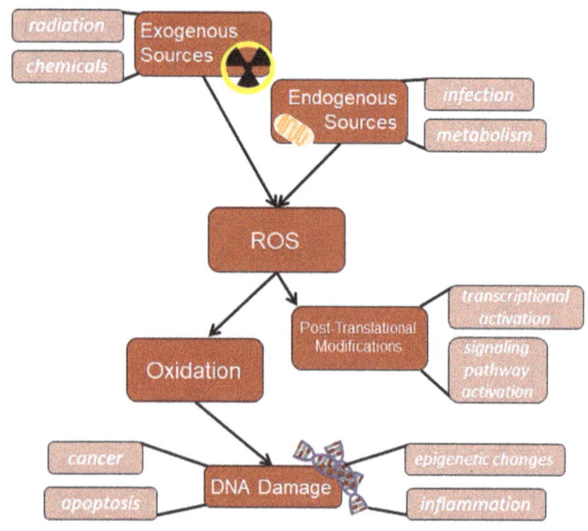

FIGURE 3 | ROS flowchart. ROS produced by either exogenous sources, such as radiation, or endogenous sources, such as through cellular mitochondria, can induce DNA damage through oxidation or cause post-translational modifications on cellular proteins.

different types of DNA damage. Of these, this review will describe the mechanisms that address alterations in DNA sequence and their link to various cancers, including mismatch repair (MMR), nucleotide excision repair (NER), base excision repair (BER), and homologous recombination (HR) and non-homologous end joining (NHEJ). Most importantly, MMR, BER, HR, and NHEJ alterations have been linked to chronic inflammatory states (**Figure 3**). Therefore, the focus will be on genetic instabilities induced through deregulation of these repair pathways, in some cases due to inflammation, that lead to genetic instabilities that may contribute to cancer formation.

DNA repair pathways recognize and correct mismatches present in the DNA, abnormal bases, single-stranded and double-stranded DNA breaks (DSBs). The MMR, BER, and NER pathways respond to specific lesions in DNA residues. DSBs are particularly dangerous lesion and are repaired by two principal pathways: NHEJ pathway functions in all phases of the cell cycle, while the high-fidelity HR pathway requires a template for repair and utilizes available sister chromatids during the S and G2 phases of the cell cycle.

Nucleotide Excision Repair

Nucleotide excision repair, or NER, is a versatile repair pathway that recognizes bulky adducts and general base lesions that cause a distortion of the double-helix structure of DNA. The major sources of these types of DNA damage are ultraviolet radiation and various types of genotoxic chemicals (Hoeijmakers, 2001). These lead to lesions such as pyrimidine dimers, cyclobutane pyrimidine dimers (CPD), and 6–4 photoproducts. NER-mediated pathway employs several different proteins to carry out a multi-step "cut-and-patch"-like pathway (Shuck et al., 2008). The defect in NER generates human genetic disorders and the bulky adducts targeted by NER mechanism can block replication and/or transcription which can lead to apoptosis or necrosis (Sancar et al., 2004).

The NER mechanism is divided into two pathways that have different methods of lesion recognition. These are the global-genome (GG-NER) and transcription-coupled repair (TC-NER) pathways. In GG-NER, DNA damage is removed from the whole genome while TC-NER is primarily involved in repairing the damage on the coding strand of actively transcribed genes (Hoeijmakers, 2001; Jackson and Bartek, 2009). Both pathways differ in the initial recognition steps. In GG-NER, the major proteins involved in the recognition are the XPC/HR23B/CEN2 (XP complementation group C/Rad23 homolog B/Centrin-2) protein complex. TC-NER is important to protect the cells from UV-light-induced apoptosis. In TC-NER, the damage is recognized and RNA Pol II stalls, Cockayne syndrome protein CSB transiently interact with RNA Pol II and the other associated proteins can take care of the damage to repair. In patients with Cockayne syndrome have defective TC-NER (Sancar et al., 2004; Fousteri and Mullenders, 2008; Hanawalt and Spivak, 2008).

Mismatch Repair

Mismatch repair, or MMR, fixes mismatched base pairs and insertion-deletion loops that are generally a product of incorrect genomic DNA replication. MMRs have the ability to inflict serious damage on the cell without killing it since they may go unnoticed and accumulate (Kunkel and Erie, 2005). MMR deficient cells can display a mutator phenotype, characterized by microsatellite instability and an elevated mutation frequency. The germline mutations in MMR genes can lead to a variety of cancers, including the non-polyposis colon cancer/ Lynch syndrome (Peltomäki, 2001). MMR involves three steps: a recognition step for identifying the mispaired bases, an excision step to remove the error-containing and the synthesis step, where the gap is filled-in by DNA polymerases. Therefore, the MMR pathway is very important to prevent cancer.

Several MMR proteins can be regulated upon chronic inflammation through the activation of HIF-1α by inflammatory cytokines and ROS (Colotta et al., 2009). The inability of MMR to repair single base pair or small-scale mutations results in microsatellite instabilities such as poly CA repeats, present in various cancers (Lengauer et al., 1998).

Base Excision Repair

Base-excision repair or BER recognizes a wide variety of damaged bases including those that underwent oxidation, alkylation, methylation, deamination, and hydroxylation (Hoeijmakers, 2001). It also repairs ROS induced strand breaks that consist of sugar fragments or 3′phosphate ends that are non-ligatable (Hegde et al., 2008). Due to its recognition and repair of an extensive range of ROS induced damages, BER is the major defense against accumulation of mutations caused by ROS. Additionally, the lesions targeted by BER are generally small-DNA base adducts. These types of lesions are more likely to be kept within the genome without cell apoptosis and can result in continued mutations in tumor suppressor and oncogenes that can lead to cancer (Hoeijmakers, 2001). Especially through this framework, BER earns greater importance in preventing cellular transformation and cancer.

Single strand break repair pathway (SSBR) is now considered a specialized sub-pathway of BER. They share several common proteins including APE1, Polβ, LIGIIIα, along with the nick sensor poly (ADP-ribose) polymerase 1 (PARP1) and the scaffold protein X-ray cross-complementation group 1 (XRCC1) (Caldecott, 2008). ROS generate 8-oxoguanine (8-oxoG), ring-opened purines (formamidopyrimidines or Fapys), and other oxidized DNA base lesions that are repaired via the DNA BER pathway. BER begins with the recognition of an altered base by a DNA glycosylase. There are two classes of DNA glycosylases, the first group with the enzymes OGG1 and NTH1 utilize an internal Lys residue as the active site nucleophile. The second group comprising of NEIL1, NEIL2, and NEIL3 use N-terminal Pro or Val as the active site (Hazra et al., 2002, 2007; Sancar et al., 2004). The first and second group have distinct structural features and reaction mechanism but have overlapping substrate specificities. The NEIL proteins are able to preferentially target single stranded DNA and also lesions from a DNA bubble while NTH1 and OGG1 only excise lesions from double-stranded DNA, as they use the second strand as a template for repair (Hegde et al., 2008). Therefore, the NEIL proteins can be functional for replication/transcription errors (Banerjee et al., 2011).

These oxidized base-specific mammalian DNA glycosylases are bifunctional. Monofunctional DNA glycosylases excise the altered base in a way that leaves behind an AP site that needs to be processed by an AP endonuclease (3′-OH and 5′dRP are generated by APE1). On the other hand, the mammalian bifunctional DNA glycosylases have an associated AP lyase activity generating 3′dRP (OGG1, NTH1, and NEIL3) or 3′-P (NEIL1 and 2) and 5′-P. The DNA polymerase then fills in the gap using the template DNA strand and finally, DNA ligase seals the nick by completing the repair of the DNA duplex (Hitomi et al., 2007; Hegde et al., 2008; Maynard et al., 2009). The essential component of BER is the DNA glycosylase that recognizes and removes the oxidized base. Inhibition of the BER proteins may lead to an accumulation of oxidized DNA damage induced by ROS and the possible escalation of mutation rate upon an inflammatory response to bacterial infection that can contribute to carcinogenesis (Maynard et al., 2009).

The fact that defects in various components of NER and MMR have been shown to contribute to certain cancers indicates that the mutations that arise from inhibition of these repair mechanisms are certainly an important causative factor to cancer. For this reason, although BER has not been studied as thoroughly as NER and MMR in the context of cancer, it may have a significant role in the perpetuation of bacterial infection and inflammation mediated cancers such as gastric cancer and, to some degree, colorectal cancer (Wallace et al., 2012; Leguisamo et al., 2017). Various human DNA glycosylases have already been implicated in various types of cancer. NEIL2 was shown to protect against the oxidative damage that is induced by secondhand smoke in human lung cells, and lower levels of NEIL2 are associated with development of lung tumors (Sarker et al., 2014). It has also been shown that knock out of NEIL2 increased the accumulation of spontaneous mutations, and a variant of NEIL2 was observed in lung cancer samples (Dey et al., 2012). Furthermore, functional variants of NEIL2 have been linked to greater risk of squamous cell carcinoma in the oral cavity and oropharynx, making it a possible marker for risk to and progression of squamous cell carcinoma in the oral cavity and oropharynx (Zhai et al., 2008). In a study of human non-small cell lung cancer, it was found that carriers who were positive for silencing of the DNA glycosylase OGG1 via methylation had a 2.25-fold higher risk of developing non-small cell lung cancer than carriers who did not exhibit OGG1 methylation (Qin et al., 2017). In another study, an OGG1 variant (Ser326Cys polymorphism) was found to increase lung cancer risk by 24% in an analysis of seven studies totaling over 3,000 cases and controls. Both the mRNA and protein expression of MUTYH, a human DNA glycosylase that repairs the foremost oxidative DNA damage in prostate cancer (8-hydroxyguanine), was down-regulated in about two thirds of prostate cancers compared to the non-cancerous prostate tissue data presented in two separate publicly available databases (Shinmura et al., 2017). The apparent links between DNA glycosylase inhibition or silencing and various cancers indicates that: (1) Oxidative damage is a major contributor to the accumulation of genetic mutations that can lead to carcinogenesis, and (2) BER plays a significant role in repressing the accumulation of ROS-induced mutations and therefore inhibition of certain BER proteins may contribute to carcinogenesis. The roles of various bacterial infections in the development of gastric and colon cancer will be discussed later to explain the relationship of bacterial infection to DNA damage and repair inhibition.

Homologous Recombination and Non-homologous End Joining

HR and NHEJ are mechanisms to repair double strand breaks in DNA (Jackson and Bartek, 2009). These repair processes are important since double strand breaks can be extremely harmful to the genome and are largely considered the most lethal type of DNA lesions since both strands of the DNA are affected (Helleday et al., 2008). Double strand breaks can be induced by X-rays, genotoxic chemicals, during replication of single strand breaks, or by ROS (Hoeijmakers, 2001). Due to the possibility of ROS inducing double strand breaks in the DNA, bacterial infection and the resulting inflammation are implicated in this type of DNA damage. HR is present during DNA replication in the S and G2 phase of the DNA cycle while NHEJ combats direct double strand breaks that can be induced by the other factors listed above, such as X-ray or ROS exposure, and is predominant in the G1 phase of the cell cycle (Hoeijmakers, 2001; Jackson and Bartek, 2009).

In HR, the MRN sensor complex; containing MRE11 (meiotic recombination 11), RAD50, and NBS1 recruits ATM (ataxia telangiectasia mutated), generates DNA breaks followed by phosphorylation of histone H2AX (generating γH2AX) that amplifies the damage signal (Blackwood et al., 2013). Classical non-homologous end-joining (C-NHEJ) is the major pathway for DNA double strand break repair. Depletion of C-NHEJ factors significantly abrogates double strand break repair in transcribed but not in non-transcribed genes (Chakraborty et al., 2016). In NHEJ, the Ku70-Ku80 initiates NHEJ, which are the sensor proteins that recruit DNA–PK (DNA-dependent protein kinase) and end-processing proteins, followed by ligation of the breaks by a complex consisting of DNA ligase IV/XRCC4; all the end-processing enzymes (Jackson and Bartek, 2009). Mutations in proteins that repair double strand breaks are linked to higher risk for different types of cancers, predominantly lymphomas (Hoeijmakers, 2001). Interestingly, HR occurs only in cycling cells while NHEJ occurs in all cells.

NON-ROS LINKED INFLAMMATION-ASSOCIATED MECHANISMS FOR PROMOTING CARCINOGENESIS

Cancer related inflammation has been observed in most neoplastic tissues through observations of white blood cell and tumor-associated macrophage infiltration into the tumor microenvironment as well as pro-inflammatory cytokine and chemokine presence. These inflammatory factors have been linked to increased tissue remodeling and angiogenesis and are therefore stated to lead to cancer related inflammation (Balkwill and Mantovani, 2001; Colotta et al., 2009; Ostrand-Rosenberg and Sinha, 2009). Other than the intrinsic effect of cancer stimulating genomic instability induced by chronic inflammation, there are many components of an extrinsic inflammatory pathway that accelerate the further development of cancer. In this case, various pro-inflammatory cytokines, chemokines, and other inflammation-linked factors assist in tissue remodeling and angiogenesis.

For instance, inflammation activates NF-κB, which in turn activates various other inflammatory cytokines as well as angiogenic factors (Colotta et al., 2009). In the case of an acute inflammatory response of the innate immune system, the activation of NF- κB is not significant to the extent that it will affect angiogenesis and cancer progression. However, in the case of chronic inflammation, as is common in various bacterial infections such as *H. pylori* infection, the activation of these cytokines and factors will be consistent and significant enough to contribute to cancer development (Mantovani et al., 2008). For instance, the mediators that are downstream of NF-κB will help in neoplasia assisted by inflammation. Therefore, prolonged activation of NF-κB will promote tumor cell survival, initiation and progression of tumor tissue formation (Karin, 2006; Bollrath and Greten, 2009). In addition to NF-κB, various other inflammation-associated molecules such as IL-6 and TNF will aid in cancer progression through various mechanisms (Colotta et al., 2009). IL-6 specifically assists in tumor cell survival and growth. IL-6 is produced by myeloid-derived suppressor cells that, as suggested by the name, suppress T-cell activation. These cells are recruited to areas with chronic inflammation by pro-inflammatory mediators and are extremely influential in promoting cancer survival by allowing them to evade the immune system's attacks via T-cell activation. TNF, on the other hand, mediates inflammation and can promote the growth of a tumor by assisting in angiogenesis, epithelial to mesenchymal transition, and other mechanisms. This pro-inflammatory cytokine is associated with tumor-associated macrophages, which are also found in areas of chronic inflammation. When tumor-associated macrophages secrete TNF, the activation of Wnt/ β-catenin signaling pathway is promoted and this leads to greater tumor development (Colotta et al., 2009). The Wnt/ β-catenin pathway is a significant signaling mechanism that controls transcription for proteins involved in cell proliferation and cell fate determination (MacDonald et al., 2009). Moreover, in both T cells and ECs, an upregulation of the Wnt/β-Catenin pathway occurs upon infection, a pathway usually associated with changes in the cellular turnover rate, tissue regeneration and cellular metabolism (Karin and Clevers, 2016). The activation of microbe-sensing pathways by ECs, associated with similar gene expression changes in both ECs and IELs in immune-response-related and metabolic pathways, pointed to ECs as potential primary microbe-responding cells that could prompt neighboring IELs. Lastly, NF- κB activation leads to the production of proangiogenic factors like vascular endothelial growth factor (VEGF), which allow for the tumor to grow and spread to distal sites to result in metastasis (Ellis and Hicklin, 2008). Of the countless links between inflammation and cancer progression, only a few have been discussed here. Ultimately, however, it is clear that cancer related inflammation is a significant factor in both the initiation of cancer as well as the consecutive growth and metastasis of a tumor.

Bacterial Toxins Can Cause DNA Damage

Bacteria not only generates DNA damage but also interacts with the host DDR pathways so damage cannot be efficiently repaired. Some of the pathogenic strains of bacteria produce toxins such as Cytolethal distending toxin (cdt) and Colibactin. Cdt is present in *Campylobacter jejuni*, *Haemophilus ducreyi*, *Actinobacillus actinomycetemcomitans*, *Shigella dysenteriae*, *Helicobacter cinaedi*, *Helicobacter hepaticus*, *Salmonella* species. Cdts recruit the MRN complex (MRE11/Rad50/NBS1) and generate DSBs that ultimately progress to gastro-intestinal cancer (Taieb et al., 2016). Another toxin, Colibactin is a polyketide nonribosomal peptide produced by several species of *Enterobacteriaceae;* for example in some of the *E. coli* strain with pks, *Klebsiella pneumoniae* and *Enterobacter aerogenes.* Colibactin is responsible for alkylation and interstrand crosslinks of DNA followed by generation of DSBs (Nougayrède et al., 2006).

There are multiple mechanisms for bacterial infections to induce cancer, and we focused gastric cancer and colon cancer here. Further studies in this arena will give more in depth insight to the mechanisms of association between specific bacterial infections and cancers, but in the next section we will discuss the well-known and characterized bacterial-infection associated cancers especially through the lens of ROS induced DNA damage and alteration of DNA repair pathways.

BACTERIAL INFECTION-ASSOCIATED WITH CANCERS

The cumulative effect of the various processes that take place after a bacterial infection, either through the direct action of the bacterium or indirectly through bacteria-induced pathways, is linked to carcinogenesis in various tissues. Many infectious agents have previously been linked to cancers, and are implicated in about 20% of human tumors (de Martel et al., 2012). For instance, respiratory tract and lung cancers have been linked to pulmonary infections caused by *Chlamydia pneumonia* and *Mycobacterium tuberculosis* (Chaturvedi et al., 2010). *Chlamydia trachomatis* and *Neisseria gonorrhoeae* infections have been associated with genitourinary cancers, and so on (Smith et al., 2004). Of the many bacterial species that have been implicated in cancer formation, *Helicobacter pylori* has been shown to play a significant role in contributing to the global gastric cancer burden and has therefore been studied more thoroughly than many other species linked to cancers. Despite a clear, demonstrated link between *H. pylori* infection and gastric cancer incidence, the exact mechanism of carcinogenesis has yet to be discovered and characterized in depth. *Fusobacterium nucleatum*, on the other hand, has been observed in large amounts in the intestinal tissues of colorectal cancer patients, but a link to cancer has not been concretely established as of yet (Gagnaire et al., 2017). Although there are some contributions available on the possible mechanisms with which it may be linked to colorectal cancer, they have not been properly defined either.

Microbial Infection-Associated Mechanisms to Promote Cancer

In mammalian cells, all the above-mentioned repair mechanisms can repair damage using the DNA damage responses (DDRs) and failure in these bring DNA damage/mutation and genomic instability. Microbial infection is one of the major reasons of the failure of DDRs.

Microbes such as bacteria, virus and parasites are able to activate or alter various signaling pathways that may lead to either activation of oncogenes or down-regulation of tumor suppressor genes in a contribution to cancer progression (Francescone et al., 2014; Sheflin et al., 2014). Examples of these modifications and pathways are discussed in this section. They also modulate repair pathways as mentioned in the table that can generate mutations linked to cancer.

Helicobacter pylori infection leads to activation of PI3K-AKT pathway, which ultimately leads to degradation of tumor suppressor p53. It also contributes to cell transformation and growth by both preventing the degradation of and activating β-catenin through various bacterial effectors such as VacA (vacuolating cytotoxin A) (Tabassam et al., 2009). Similarly, a virulence factor of *Fusobacterium nucleatum* called *Fusobacterium* adhesin A (FadA) is able to bind E-cadherin to induce greater β-catenin release and ultimately activate WNT signaling, which is oncogenic (Rubinstein et al., 2013). *Helicobacter pylori* and *Salmonella enterica* serovar Typhimurium both activate MAPK and AKT signaling pathways upon infection leading to altered mediation of cell growth, proliferation, migration, and other important processes relevant to cellular transformation (Sokolova et al., 2008; Gagnaire et al., 2017).

The bacterial toxins which cells are exposed when infected can alter the cell cycle and ultimately affect some of the processes, which are implicated in carcinogenesis: proliferation, apoptosis, and differentiation (Mager, 2006). Certain bacteria, which are classified as cyclomodulins, are able to change host cell cycle patterns with cell-cycle inhibitors, such as cytolethal distending toxins and cycle inhibiting factor, and cell cycle stimulators such as cytotoxic necrotizing factor (Nougayrède et al., 2005).

Bacteria of the microbiota are also implicated in cancer through their ability to construct biofilm. Biofilms form when bacteria aggregate and secrete a substance that allows them to stick to surfaces that generally have a mucosal lining (Johnson et al., 2016). In addition to being linked to inflammatory bowel conditions, bacterial biofilms have been seen on colorectal cancers, preferentially in proximal colon cancers, which have a higher mortality rate, than distal colon cancers (Dejea et al., 2014).

H. pylori Associated Gastric Cancer

Helicobacter pylori is a gram-negative and spiral shaped bacterium. It has flagella that assist in movement and is able to survive at very low pH, making it a main colonizer of the stomach. Infection with *H. pylori* is associated with greater susceptibility to further infections, diarrhea, and chronic gastritis (Tomb et al., 1997; Crew and Neugut, 2006). While eradication of *H. pylori* infection is possible, there is high probability of relapse as well as antimicrobial resistance in many strains of the bacteria. The standard triple antibiotic treatment of *H. pylori* cures up to 70% of infected patients since there is a growing resistance to clarithromycin. Additionally, a very large portion of the world, about half of the total population, is exposed to or infected by *H. pylori* (Parsonnet et al., 1994; Malfertheiner et al., 2012). As *H. pylori* is the greatest risk factor for gastric cancer development, it is very important to study mechanisms of carcinogenesis upon infection in order to develop possible therapeutic and treatment options that address the specific pathways altered by *H. pylori*. Studies have shown that eradicating *H. pylori* infection decreases gastric cancer development in patients without premalignant tumors and prevents malignant transformation in patients with premalignant tumors (Wong et al., 2004; Malfertheiner et al., 2012). This further reinforces the link between infection and gastric cancer.

Non-inflammatory Pathways for Cellular Transformation Upon *H. pylori* Infection

H. pylori has multiple bacterial effectors that are able to alter cellular signaling pathways in favor of carcinogenesis. For instance, vacuolating cytotoxin A (VacA) and outer inflammatory protein A (OipA) are involved in epidermal growth factor receptor activation which leads to PI3K-AKT signaling and ultimately activates β-catenin (Suzuki et al., 2009; Wroblewski et al., 2010). This signaling cascade leads to transcriptional activation for cell growth. If the infection cannot be cleared, this becomes constitutively active and can then cause cellular transformation. This is just one example of how bacterial effectors can contribute to carcinogenesis in a non-inflammatory pathway. However, these effectors such as OipA are also able to induce proinflammatory cytokine expression along with other oncogenic proteins that have a significant effect on cellular transformation. Bacterial oncoproteins such as CagA are additional features of the bacteria that contribute to carcinogenesis (same reference as on line 573). The cag pathogenicity island, present in cag^+ strains of *H. pylori* has genes which encode for type IV bacterial secretion system, commonly known as T4SS, that is able to export bacterial proteins such as CagA upon bacterial attachment to host cells. Once it enters a host epithelial cell, this protein can be activated via phosphorylation to mitigate apoptosis and induce greater cell proliferation to contribute to carcinogenesis (Polk and Peek, 2010). Despite their active role in altering many cancer-associated cellular pathways, bacterial oncoproteins and effectors are only able to increase gastric cancer risk. The ultimate development of gastric cancer also relies on chronic inflammatory response to *H. pylori* infection, which is accompanied by a wide variety of consequences that contribute to cellular transformation (Lamb and Chen, 2013).

Inflammation-Associated Pathways for Cellular Transformation Upon *H. pylori* Infection

The inflammatory response to *H. pylori* infection can significantly alter cellular signaling and activity to induce transformation through some of the many pathways already discussed in this review (Figueiredo et al., 2002). The focus here will be on DNA damage induced by *H. pylori* mediated inflammation as well as obstruction of DNA repair pathways by the bacterium.

H. pylori colonization is characterized by recurring infections even after assumed eradication of the bacteria as well as chronic gastritis, which signifies a continuous inflammatory response that is started as a host response to infection but is continued with the aid of the bacteria by effectors and oncoproteins that alter chemokine and cytokine release in the infected host cells (Miftahussurur et al., 2017). One such oncoprotein is Tipα, a membrane protein secreted by *H. pylori* that is associated with epithelial to mesenchymal transition by activating IL-6 cytokine-dependent STAT3 signaling and greatly impacts cancer cell invasiveness. Similar to IL-6, many other inflammatory cytokines play a role in extrcellular matrix degradation to promote cell motility and angiogenesis (Chen et al., 2017).

An accumulation of DNA damage caused by chronic inflammation and resulting ROS has the potential to induce cellular transformation. In a normally functioning cell, the multiple DNA repair pathways are able to curb the accumulation of DNA mutations by addressing the damage as it occurs. In a cell infected by *H. pylori*, it is shown that mismatch repair (MMR), the major pathway that repairs small-scale mutations such as single base pair mismatches, is inhibited (Kim et al., 2002; Santos et al., 2017). This potentially allows for seemingly minute mutations to accumulate in oncogenic or tumor suppressor genes and lead to cancer development. Two vital MMR proteins, MSH2 and MLH1, are directly affected by *H. pylori* infection. Additional MMR proteins MLH1, MSH3, MSH6, PMS1, and PMS2 have also been shown to be down-regulated upon *H. pylori* infection of gastric cell lines AGS and BG (Machado et al., 2013; Strickertsson et al., 2014; Santos et al., 2017). The various oxidative damages induced to DNA by *H. pylori* infection are supposed to be repaired by the BER pathway in a normally functioning cell. In a cell infected with *H. pylori,* there is decreased expression of vital BER proteins. APE-1, an AP endonuclease, and YB-1, an early-stage repair protein of BER, are down-regulated upon *H. pylori* infection (Machado et al., 2013). Down-regulation of the DNA glycosylase OGG1, which is very important for recognition and removal of abasic sites induced by bacterial and host ROS, has been observed in gastric epithelial cells upon infection as well. Abasic sites that are targeted by OGG1, such as 8oxodG lesions, are induced at a greater frequency upon *H. pylori* infection. Reduced expression of OGG1 allows for accumulation of abasic sites that would not be repaired by the other DNA repair pathways and lead to carcinogenic mutation-build up and cellular transformation (Kidane et al., 2014).

Bacteria Associated Colon Cancer

The association between bacterial infection and colon cancer has not been elucidated to the extent that *H. pylori* induced gastric cancer has been. Certain pathogenic species such as enterotoxigenic *Bacteroides fragilis* and *Escherichia coli* strain NC101 have been linked to colitis-associated colon cancer (Wu et al., 2009; Arthur et al., 2012), but no bacterial species has been proven to be a major causative agent in colorectal carcinogenesis. Rather, a collection of gram-negative and anaerobic bacteria has been observed in colorectal tumor tissues and may serve as a marker of cancer. The colon is home to commensal bacterial species that play various supporting and valuable roles in processes such as metabolism in the host. These microbiota have also been implicated in carcinogenesis and tumor formation in the colon, potentially through bacterial dysbiosis in the gut (McCoy et al., 2013; Warren et al., 2013).

One area of interest in this field is the presence of *Fusobacterium nucleatum* in abundance in colorectal tumor tissues (Kostic et al., 2013). *F. nucleatum* is a gram-negative microbe more often associated with the oral cavity. However, the bacterial species have been identified in the early stages of cancer in colorectal adenomas as well as carcinoma samples (McCoy et al., 2013). Introduction of *F. nucleatum* to mice was shown to speed up colonic tumorigenesis and induce a pro-inflammatory state through NF-κB signaling (Kostic et al., 2013). *F. nucleatum* has also been positively correlated with mortality linked to colorectal cancer, meaning that greater abundance of the bacteria more likely resulted in mortality due to the cancer (Mima et al., 2016). There is still much to study concerning whether *F. nucleatum* is a causative agent in colorectal carcinogenesis, but the preliminary studies have indicated the involvement of bacteria in the induction of cancer (Ray, 2011; Kostic et al., 2013; Gagnaire et al., 2017).

In the oral cavity, *F. nucleatum* is a very invasive bacterial species due to its ability to adhere well to mucous surfaces (McCoy et al., 2013). In the intestines, *F. nucleatum* can therefore act similar to *H. pylori* and adhere to host cell surfaces to alter cellular pathways with its bacterial proteins. Once *F. nucleatum* adheres to host cells, bacterial FadA adhesin binds E-cadherin to activate β-catenin signaling to increase cell growth and proliferation as well as to regulate inflammatory response of the cell (Rubinstein et al., 2013). *F. nucleatum* has also been found to activate TLR4 and ultimately lead to NF-κB activation (Yang et al., 2017). Many studies have linked *F. nucleatum* to a pro-inflammatory state (Kostic et al., 2013; Rubinstein et al., 2013). Inflammatory cytokine gene expression such as IL-10 and TNF- α has a positive association to abundance of *F. nucleatum* in the colon (Rubinstein et al., 2013). This also poses the possibility of *F. nucleatum* contributing to carcinogenesis and tumorigenesis by inducing genetic mutations as a result of prolonged inflammation and potentially the down-regulation of various DNA repair pathways, similar to what occurs in *H. pylori*-mediated inflammation leading to gastric cancer.

Other bacteria associated cancers: Other that *H. pylori* and *Fusobacterium*, *Salmonella typhi* infection has been associated with the development of gallbladder cancer (Mager, 2006; Di Domenico et al., 2017). All other bacteria associated with cancer is mentioned in the **Table 1**.

BACTERIAL INFECTION IN CANCER PREVENTION AND THERAPY

While a number of bacterial infections have been shown to increase the potential for carcinogenesis, recent developments in the field have provided evidence for a positive role of certain bacteria and toxins in cancer prevention and therapy (Mager, 2006). For example in one case-control study, researchers found that *Helicobacter pylori* infection correlated to a lower risk for esophageal cancer development (de Martel et al., 2005). Alternatively, the introduction of bacteria or its toxins to treat cancer has become a method of interest for many types of cancer. Dr. William Coley started to treat end stage cancers with a vaccine made of killed *Streptococcus pyogenes* and *Serratia marcescens* in the late 1800's to induce an initial fever followed by treatment for many different types of cancers (de Martel et al., 2005). More recently, researchers have shown that a vaccine with live attenuated *Salmonella enterica* serovar Typhi reduced tumor growth and enhanced survival in mice (Vendrell et al., 2011). The Bacillus Calmette-Guérin vaccine, which has a strain of *Mycobacterium bovis*, is used clinically for the treatment of high-risk urinary bladder cancer (Kucerova and Cervinkova, 2016). Finally, a number of bacteria species has been tried as anti-tumor agents in experimental models of cancer (Ryan et al., 2006).

FUTURE RESEARCH DIRECTIONS

About 20% of the global cancer burden is linked to infectious agents including, but not limited to, *H. pylori*, *Hepatitis B* and *C* virus, and *Human papilloma virus* (Mantovani et al., 2008; Gagnaire et al., 2017). By studying how these infectious agents can lead to and exacerbate cancer states, we may be able to prevent certain cancers from forming or advancing and identify cancer markers or therapeutic targets in the treatment of cancer. It is already known that a chronic inflammatory state is able to induce DNA damage through oxidative stress induced by ROS production. As it is a hallmark of cancer, DNA damage must be repaired properly through the multiple machineries present within the cell (Colotta et al., 2009; Hanahan and Weinberg, 2011). However, along with inducing a chronic inflammatory state, bacterial infection may affect function and/or the level of DNA repair proteins leading to a buildup of genetic mutations in potential oncogenic and tumor suppressor genes that are major contributors to carcinogenesis. Studying the effect of infectious agents that are already linked to cancers, such as *F. nucleatum* and colorectal cancer, on the functionality of the various DNA repair pathways, can lead to novel identification of various markers for early cancer detection as well as more effective therapies and treatments that can combat the loss in DNA repair function in host cells.

AUTHOR CONTRIBUTIONS

All authors listed, have made substantial, direct and intellectual contribution to the work, and approved it for publication.

ACKNOWLEDGMENTS

This work was supported, in whole or in part, by National Institute of Health Grants: DK107585 and DK099275 (to SD); R01 NS073976 (to TH).

REFERENCES

Akira, S., Uematsu, S., and Takeuchi, O. (2006). Pathogen recognition and innate immunity. *Cell* 124, 783–801. doi: 10.1016/j.cell.2006.02.015

Almeida Pereira Leite, S. T., Marques-Guimarães, N., Silva-Oliveira, J. C., Dutra-Souto, F. J., Alves-dos-Santos, R., and Bassi-Branco, C. L. (2013). The X-ray repair cross complementing protein 1 (XRCC1) rs25487 polymorphism and susceptibility to cirrhosis in Brazilian patients with chronic viral hepatitis. *Ann. Hepatol.* 12, 733–739.

Arbuthnot, P., and Kew, M. (2001). Hepatitis B virus and hepatocellular carcinoma. *Int. J. Exp. Pathol.* 82, 77–100. doi: 10.1111/j.1365-2613.2001.iep178.x

Arthur, J. C., Perez-Chanona, E., Mühlbauer, M., Tomkovich, S., Uronis, J. M., Fan, T. J., et al. (2012). Intestinal inflammation targets cancer-inducing activity of the microbiota. *Science* 338, 120–123. doi: 10.1126/science.1224820

Bajpai, D., Banerjee, A., Pathak, S., Thakur, B., Jain, S. K., and Singh, N. (2016). Single nucleotide polymorphisms in the DNA repair genes in HPV-positive cervical cancer. *Eur. J. Cancer Prev.* 25, 224–231. doi: 10.1097/CEJ.0000000000000159

Balkwill, F., and Mantovani, A. (2001). Inflammation and cancer: back to Virchow? *Lancet* 357, 539–545. doi: 10.1016/S0140-6736(00)04046-0

Banerjee, D., Mandal, S. M., Das, A., Hegde, M. L., Das, S., Bhakat, K. K., et al. (2011). Preferential repair of oxidized base damage in the transcribed genes of mammalian cells. *J. Biol. Chem.* 286, 6006–6016. doi: 10.1074/jbc.M110.198796

Barton, G. M., and Medzhitov, R. (2003). Toll-like receptor signaling pathways. *Science* 300, 1524–1525. doi: 10.1126/science.1085536

Basset, C., Holton, J., O'Mahony, R., and Roitt, I. (2003). Innate immunity and pathogen-host interaction. *Vaccine* 21(Suppl. 2), S12–S23. doi: 10.1016/S0264-410X(03)00195-6

Blackwood, J. K., Rzechorzek, N. J., Bray, S. M., Maman, J. D., Pellegrini, L., and Robinson, N. P. (2013). End-resection at DNA double-strand breaks in the three domains of life. *Biochem. Soc. Trans.* 41, 314–320. doi: 10.1042/BST20120307

Bollrath, J., and Greten, F. R. (2009). IKK/NF-kappaB and STAT3 pathways: central signalling hubs in inflammation-mediated tumour promotion and metastasis. *EMBO Rep.* 10, 1314–1319. doi: 10.1038/embor.2009.243

Bosch, F. X., Manos, M. M., Munoz, N., Sherman, M., Jansen, A. M., Peto, J., et al. (1995). Prevalence of human papillomavirus in cervical cancer: a worldwide perspective. International biological study on cervical cancer (IBSCC) Study Group. *J. Natl. Cancer Inst.* 87, 796–802. doi: 10.1097/00006254-199510000-00015

Bose, S., Tripathi, D. M., Sukriti, Sakhuja, P., Kazim, S. N., and Sarin, S. K. (2013). Genetic polymorphisms of CYP2E1 and DNA repair genes HOGG1 and XRCC1: association with hepatitis B related advanced liver disease and cancer. *Gene* 519, 231–237. doi: 10.1016/j.gene.2013.02.025

Cairns, R. A., Harris, I. S., and Mak, T. W. (2011). Regulation of cancer cell metabolism. *Nat. Rev. Cancer* 11, 85–95. doi: 10.1038/nrc2981

Caldecott, K. W. (2008). Single-strand break repair and genetic disease. *Nat. Rev. Genet.* 9, 619–631. doi: 10.1038/nrg2380

Chakraborty, A., Tapryal, N., Venkova, T., Horikoshi, N., Pandita, R. K., Sarker, A. H., et al. (2016). Classical non-homologous end-joining pathway utilizes nascent RNA for error-free double-strand break repair of transcribed genes. *Nat. Commun.* 7:13049. doi: 10.1038/ncomms13049

Chaturvedi, A. K., Gaydos, C. A., Agreda, P., Holden, J. P., Chatterjee, N., Goedert, J. J., et al. (2010). *Chlamydia pneumoniae* infection and risk for lung cancer. *Cancer Epidemiol. Biomarkers Prev.* 19, 1498–1505. doi: 10.1158/1055-9965.EPI-09-1261

Chen, G., Tang, N., Wang, C., Xiao, L., Yu, M., Zhao, L., et al. (2017). TNF-alpha-inducing protein of *Helicobacter pylori* induces epithelial-mesenchymal transition (EMT) in gastric cancer cells through activation of IL-6/STAT3 signaling pathway. *Biochem. Biophys. Res. Commun.* 484, 311–317. doi: 10.1016/j.bbrc.2017.01.110

Chumduri, C., Gurumurthy, R. K., Zietlow, R., and Meyer, T. F. (2016). Subversion of host genome integrity by bacterial pathogens. *Nat. Rev. Mol. Cell Biol.* 17, 659–673. doi: 10.1038/nrm.2016.100

Church, L. D., Cook, G. P., and McDermott, M. F. (2008). Primer: inflammasomes and interleukin 1beta in inflammatory disorders. *Nat. Clin. Pract. Rheumatol.* 4, 34–42. doi: 10.1038/ncprheum0681

Ciavattini, A., Piccioni, M., Tranquilli, A. L., Filosa, A., Pieramici, T., and Goteri, G. (2005). Immunohistochemical expression of DNA mismatch repair (MMR) system proteins (hMLH1, hMSH2) in cervical preinvasive and invasive lesions. *Pathol. Res. Pract.* 201, 21–25. doi: 10.1016/j.prp.2004.09.012

Colotta, F., Allavena, P., Sica, A., Garlanda, C., and Mantovani, A. (2009). Cancer-related inflammation, the seventh hallmark of cancer: links to genetic instability. *Carcinogenesis* 30, 1073–1081. doi: 10.1093/carcin/bgp127

Cong, Y., Weaver, C. T., Lazenby, A., and Elson, C. O. (2002). Bacterial-reactive T regulatory cells inhibit pathogenic immune responses to the enteric flora. *J. Immunol.* 169, 6112–6119. doi: 10.4049/jimmunol.169.11.6112

Coussens, L. M., and Werb, Z. (2002). Inflammation and cancer. *Nature* 420, 860–867. doi: 10.1038/nature01322

Crew, K. D., and Neugut, A. I. (2006). Epidemiology of gastric cancer. *World J. Gastroenterol.* 12, 354–362. doi: 10.3748/wjg.v12.i3.354

Cross, A. R., Noack, D., Rae, J., Curnutte, J. T., and Heyworth, P. G. (2000). Hematologically important mutations: the autosomal recessive forms of chronic granulomatous disease (first update). *Blood Cells Mol. Dis.* 26, 561–565. doi: 10.1006/bcmd.2000.0333

Dandona, P., Thusu, K., Cook, S., Snyder, B., Makowski, J., Armstrong, D., et al. (1996). Oxidative damage to DNA in diabetes mellitus. *Lancet* 347, 444–445. doi: 10.1016/S0140-6736(96)90013-6

Dejea, C. M., Wick, E. C., Hechenbleikner, E. M., White, J. R., Mark Welch, J. L., Rossetti, B. J., et al. (2014). Microbiota organization is a distinct feature of proximal colorectal cancers. *Proc. Natl. Acad. Sci. U.S.A.* 111, 18321–18326. doi: 10.1073/pnas.1406199111

de Martel, C., Ferlay, J., Franceschi, S., Vignat, J., Bray, F., Forman, D., et al. (2012). Global burden of cancers attributable to infections in 2008: a review and synthetic analysis. *Lancet Oncol.* 13, 607–615. doi: 10.1016/S1470-2045(12)70137-7

de Martel, C., and Franceschi, S. (2009). Infections and cancer: established associations and new hypotheses. *Crit. Rev. Oncol. Hematol.* 70, 183–194. doi: 10.1016/j.critrevonc.2008.07.021

de Martel, C., Llosa, A. E., Farr, S. M., Friedman, G. D., Vogelman, J. H., Orentreich, N., et al. (2005). *Helicobacter pylori* infection and the risk of development of esophageal adenocarcinoma. *J. Infect. Dis.* 191, 761–767. doi: 10.1086/427659

Dey, S., Maiti, A. K., Hegde, M. L., Hegde, P. M., Boldogh, I., Sarkar, P. S., et al. (2012). Increased risk of lung cancer associated with a functionally impaired polymorphic variant of the human DNA glycosylase NEIL2. *DNA Repair* 11, 570–578. doi: 10.1016/j.dnarep.2012.03.005

Di Domenico, E. G., Cavallo, I., Pontone, M., Toma, L., and Ensoli, F. (2017). Biofilm producing *Salmonella typhi*: chronic colonization and development of gallbladder cancer. *Int. J. Mol. Sci.* 18:1887. doi: 10.3390/ijms18091887

Ding, S. Z., O'Hara, A. M., Denning, T. L., Dirden-Kramer, B., Mifflin, R. C., Reyes, V. E., et al. (2004). *Helicobacter pylori* and H2O2 increase AP endonuclease-1/redox factor-1 expression in human gastric epithelial cells. *Gastroenterology* 127, 845–858. doi: 10.1053/j.gastro.2004.06.017

Dutta, U., Garg, P. K., Kumar, R., and Tandon, R. K. (2000). Typhoid carriers among patients with gallstones are at increased risk for carcinoma of the gallbladder. *Am. J. Gastroenterol.* 95, 784–787. doi: 10.1111/j.1572-0241.2000.01860.x

Ellis, L. M., and Hicklin, D. J. (2008). VEGF-targeted therapy: mechanisms of anti-tumour activity. *Nat. Rev. Cancer* 8, 579–591. doi: 10.1038/nrc2403

Ellmerich, S., Schöller, M., Duranton, B., Gossé, F., Galluser, M., Klein, J. P., et al. (2000). Promotion of intestinal carcinogenesis by *Streptococcus bovis*. *Carcinogenesis* 21, 753–756. doi: 10.1093/carcin/21.4.753

El-Serag, H. B. (2012). Epidemiology of viral hepatitis and hepatocellular carcinoma. *Gastroenterology* 142, 1264.e1–1273.e1. doi: 10.1053/j.gastro.2011.12.061

Ernst, P. (1999). Review article: the role of inflammation in the pathogenesis of gastric cancer. *Aliment Pharmacol. Ther.* 13(Suppl. 1), 13–18. doi: 10.1046/j.1365-2036.1999.00003.x

Farinati, F., Cardin, R., Degan, P., Rugge, M., Mario, F. D., Bonvicini, P., et al. (1998). Oxidative DNA damage accumulation in gastric carcinogenesis. *Gut* 42, 351–356. doi: 10.1136/gut.42.3.351

Fernandes, J. V., DE Medeiros Fernandes, T. A., DE Azevedo, J. C., Cobucci, R. N., DE Carvalho, M. G., Andrade, V. S., et al. (2015). Link between chronic inflammation and human papillomavirus-induced carcinogenesis (Review). *Oncol. Lett.* 9, 1015–1026. doi: 10.3892/ol.2015.2884

Figueiredo, C., Machado, J. C., Pharoah, P., Seruca, R., Sousa, S., Carvalho, R., et al. (2002). *Helicobacter pylori* and interleukin 1 genotyping: an opportunity to identify high-risk individuals for gastric carcinoma. *J. Natl. Cancer Inst.* 94, 1680–1687. doi: 10.1093/jnci/94.22.1680

Fousteri, M., and Mullenders, L. H. (2008). Transcription-coupled nucleotide excision repair in mammalian cells: molecular mechanisms and biological effects. *Cell Res.* 18, 73–84. doi: 10.1038/cr.2008.6

Francescone, R., Hou, V., and Grivennikov, S. I. (2014). Microbiome, inflammation, and cancer. *Cancer J.* 20, 181–189. doi: 10.1097/PPO.0000000000000048

Fu, X., Hu, J., Han, H. Y., Hua, Y. J., Zhou, L., Shuai, W. D., et al. (2015). High expression of XPA confers poor prognosis for nasopharyngeal carcinoma patients treated with platinum-based chemoradiotherapy. *Oncotarget* 6, 28478–28490. doi: 10.18632/oncotarget.4424

Gagnaire, A., Nadel, B., Raoult, D., Neefjes, J., and Gorvel, J. P. (2017). Collateral damage: insights into bacterial mechanisms that predispose host cells to cancer. *Nat. Rev. Microbiol.* 15, 109–128. doi: 10.1038/nrmicro.2016.171

Gulnaz, A., Sayyed, A. H., Amin, F., Khan, A., Aslam, M. A., Shaikh, R. S., et al. (2013). Association of XRCC1, XRCC3, and XPD genetic polymorphism with an increased risk of hepatocellular carcinoma because of the hepatitis B and C virus. *Eur. J. Gastroenterol. Hepatol.* 25, 166–179. doi: 10.1097/MEG.0b013e328359a775

Guzik, T. J., Hoch, N. E., Brown, K. A., McCann, L. A., Rahman, A., Dikalov, S., et al. (2007). Role of the T cell in the genesis of angiotensin II induced hypertension and vascular dysfunction. *J. Exp. Med.* 204, 2449–2460. doi: 10.1084/jem.20070657

Hajek, M., Sewell, A., Kaech, S., Burtness, B., Yarbrough, W. G., and Issaeva, N. (2017). TRAF3/CYLD mutations identify a distinct subset of human papillomavirus-associated head and neck squamous cell carcinoma. *Cancer* 123, 1778–1790. doi: 10.1002/cncr.30570

Hanahan, D., and Weinberg, R. A. (2011). Hallmarks of cancer: the next generation. *Cell* 144, 646–674. doi: 10.1016/j.cell.2011.02.013

Hanawalt, P. C., and Spivak, G. (2008). Transcription-coupled DNA repair: two decades of progress and surprises. *Nat. Rev. Mol. Cell Biol.* 9, 958–970. doi: 10.1038/nrm2549

Hazra, T. K., Das, A., Das, S., Choudhury, S., Kow, Y. W., and Roy, R. (2007). Oxidative DNA damage repair in mammalian cells: a new perspective. *DNA Repair* 6, 470–480. doi: 10.1016/j.dnarep.2006.10.011

Hazra, T. K., Izumi, T., Boldogh, I., Imhoff, B., Kow, Y. W., Jaruga, P., et al. (2002). Identification and characterization of a human DNA glycosylase for repair of modified bases in oxidatively damaged DNA. *Proc. Natl. Acad. Sci. U.S.A.* 99, 3523–3528. doi: 10.1073/pnas.062053799

Hegde, M. L., Hazra, T. K., and Mitra, S. (2008). Early steps in the DNA base excision/single-strand interruption repair pathway in mammalian cells. *Cell Res.* 18, 27–47. doi: 10.1038/cr.2008.8

Helal, T. E., Khamis, N. S., El-Sharkawy, T. M., Nada, O. H., and Radwan, N. A. (2010). Immunohistochemical expression of mismatch repair genes (hMSH2 and hMLH1) in hepatocellular carcinoma in Egypt. *APMIS* 118, 934–940. doi: 10.1111/j.1600-0463.2010.02658.x

Helleday, T., Petermann, E., Lundin, C., Hodgson, B., and Sharma, R. A. (2008). DNA repair pathways as targets for cancer therapy. *Nat. Rev. Cancer* 8, 193–204. doi: 10.1038/nrc2342

Higgs, M. R., Chouteau, P., and Lerat, H. (2014). "Liver let die": oxidative DNA damage and hepatotropic viruses. *J. Gen. Virol.* 95, 991–1004. doi: 10.1099/vir.0.059485-0

Hitomi, K., Iwai, S., and Tainer, J. A. (2007). The intricate structural chemistry of base excision repair machinery: implications for DNA damage recognition, removal, and repair. *DNA Repair* 6, 410–428. doi: 10.1016/j.dnarep.2006.10.004

Hoeijmakers, J. H. (2001). Genome maintenance mechanisms for preventing cancer. *Nature* 411, 366–374. doi: 10.1038/35077232

Huang, J., Fogg, M., Wirth, L. J., Daley, H., Ritz, J., Posner, M. R., et al. (2017). Epstein-Barr virus-specific adoptive immunotherapy for recurrent, metastatic nasopharyngeal carcinoma. *Cancer* 123, 2642–2650. doi: 10.1002/cncr.30541

Imlay, J. A. (2008). Cellular defenses against superoxide and hydrogen peroxide. *Annu. Rev. Biochem.* 77, 755–776. doi: 10.1146/annurev.biochem.77.061606.161055

Jackson, S. P., and Bartek, J. (2009). The DNA-damage response in human biology and disease. *Nature* 461, 1071–1078. doi: 10.1038/nature08467

Jemal, A., Bray, F., Center, M. M., Ferlay, J., Ward, E., and Forman, D. (2011). Global cancer statistics. *CA Cancer J. Clin.* 61, 69–90. doi: 10.3322/caac.20107

Jia, L., Wang, X. W., and Harris, C. C. (1999). Hepatitis B virus X protein inhibits nucleotide excision repair. *Int. J. Cancer* 80, 875–879. doi: 10.1002/(SICI)1097-0215(19990315)80:6<875::AID-IJC13>3.0.CO;2-Z

Johnson, C. H., Spilker, M. E., Goetz, L., Peterson, S. N., and Siuzdak, G. (2016). Metabolite and microbiome interplay in cancer immunotherapy. *Cancer Res.* 76, 6146–6152. doi: 10.1158/0008-5472.CAN-16-0309

Joo, J., Yoon, K. A., Hayashi, T., Kong, S. Y., Shin, H. J., Park, B., et al. (2016). Nucleotide excision repair gene ERCC2 and ERCC5 variants increase risk of uterine cervical cancer. *Cancer Res. Treat.* 48, 708–714. doi: 10.4143/crt.2015.098

Karin, M. (2006). Nuclear factor-kappaB in cancer development and progression. *Nature* 441, 431–436. doi: 10.1038/nature04870

Karin, M., and Clevers, H. (2016). Reparative inflammation takes charge of tissue regeneration. *Nature* 529, 307–315. doi: 10.1038/nature17039

Kawanishi, S., Ohnishi, S., Ma, N., Hiraku, Y., Oikawa, S., and Murata, M. (2016). Nitrative and oxidative DNA damage in infection-related carcinogenesis in relation to cancer stem cells. *Genes Environ.* 38:26. doi: 10.1186/s41021-016-0055-7

Kidane, D., Murphy, D. L., and Sweasy, J. B. (2014). Accumulation of abasic sites induces genomic instability in normal human gastric epithelial cells during *Helicobacter pylori* infection. *Oncogenesis* 3:e128. doi: 10.1038/oncsis.2014.42

Kim, J. J., Tao, H., Carloni, E., Leung, W. K., Graham, D. Y., and Sepulveda, A. R. (2002). *Helicobacter pylori* impairs DNA mismatch repair in gastric epithelial cells. *Gastroenterology* 123, 542–553. doi: 10.1053/gast.2002.34751

Klaunig, J. E., and Kamendulis, L. M. (2004). The role of oxidative stress in carcinogenesis. *Annu. Rev. Pharmacol. Toxicol.* 44, 239–267. doi: 10.1146/annurev.pharmtox.44.101802.121851

Kostic, A. D., Chun, E., Robertson, L., Glickman, J. N., Gallini, C. A., Michaud, M., et al. (2013). *Fusobacterium nucleatum* potentiates intestinal tumorigenesis and modulates the tumor-immune microenvironment. *Cell Host Microbe* 14, 207–215. doi: 10.1016/j.chom.2013.07.007

Koyi, H., Brandén, E., Gnarpe, J., Gnarpe, H., and Steen, B. (2001). An association between chronic infection with *Chlamydia pneumoniae* and lung cancer. A prospective 2-year study. *APMIS* 109, 572–580. doi: 10.1034/j.1600-0463.2001.d01-177.x

Kucerova, P., and Cervinkova, M. (2016). Spontaneous regression of tumour and the role of microbial infection–possibilities for cancer treatment. *Anticancer. Drugs* 27, 269–277. doi: 10.1097/CAD.0000000000000337

Kunkel, T. A., and Erie, D. A. (2005). DNA mismatch repair. *Annu. Rev. Biochem.* 74, 681–710. doi: 10.1146/annurev.biochem.74.082803.133243

Kuper, H., Adami, H. O., and Trichopoulos, D. (2000). Infections as a major preventable cause of human cancer. *J. Intern. Med.* 248, 171–183. doi: 10.1046/j.1365-2796.2000.00742.x

Lamb, A., and Chen, L. F. (2013). Role of the *Helicobacter pylori*-induced inflammatory response in the development of gastric cancer. *J. Cell. Biochem.* 114, 491–497. doi: 10.1002/jcb.24389

Lambeth, J. D. (2004). NOX enzymes and the biology of reactive oxygen. *Nat. Rev. Immunol.* 4, 181–189. doi: 10.1038/nri1312

Langer, C. J. (2012). Exploring biomarkers in head and neck cancer. *Cancer* 118, 3882–3892. doi: 10.1002/cncr.26718

Leguisamo, N. M., Gloria, H. C., Kalil, A. N., Martins, T. V., Azambuja, D. B., Meira, L. B., et al. (2017). Base excision repair imbalance in colorectal cancer has prognostic value and modulates response to chemotherapy. *Oncotarget* 8, 54199–54214. doi: 10.18632/oncotarget.14909

Lengauer, C., Kinzler, K. W., and Vogelstein, B. (1998). Genetic instabilities in human cancers. *Nature* 396, 643–649. doi: 10.1038/25292

Liao, J. B. (2006). Viruses and human cancer. *Yale J. Biol. Med.* 79, 115–122.

Liengswangwong, U., Karalak, A., Morishita, Y., Noguchi, M., Khuhaprema, T., Srivatanakul, P., et al. (2006). Immunohistochemical expression of mismatch repair genes: a screening tool for predicting mutator phenotype in liver fluke infection-associated intrahepatic cholangiocarcinoma. *World J. Gastroenterol.* 12, 3740–3745. doi: 10.3748/wjg.v12.i23.3740

Lu, R., Zhang, Y. G., and Sun, J. (2017). STAT3 activation in infection and infection-associated cancer. *Mol. Cell. Endocrinol.* 451, 80–87. doi: 10.1016/j.mce.2017.02.023

MacDonald, B. T., Tamai, K., and He, X. (2009). Wnt/beta-catenin signaling: components, mechanisms, and diseases. *Dev. Cell* 17, 9–26. doi: 10.1016/j.devcel.2009.06.016

Macdonald, T. T., and Monteleone, G. (2005). Immunity, inflammation, and allergy in the gut. *Science* 307, 1920–1925. doi: 10.1126/science.1106442

Machado, A. M., Desler, C., Bøggild, S., Strickertsson, J. A., Friis-Hansen, L., Figueiredo, C., et al. (2013). *Helicobacter pylori* infection affects mitochondrial function and DNA repair, thus, mediating genetic instability in gastric cells. *Mech. Ageing Dev.* 134, 460–466. doi: 10.1016/j.mad.2013.08.004

Maddocks, O. D., Scanlon, K. M., and Donnenberg, M. S. (2013). An *Escherichia coli* effector protein promotes host mutation via depletion of DNA mismatch repair proteins. *MBio* 4:e00152-13. doi: 10.1128/mBio.00152-13

Mager, D. L. (2006). Bacteria and cancer: cause, coincidence or cure? A review. *J. Transl. Med.* 4:14. doi: 10.1186/1479-5876-4-14

Malfertheiner, P., Megraud, F., O'Morain, C. A., Atherton, J., Axon, A. T., Bazzoli, F., et al. (2012). Management of *Helicobacter pylori* infection–the Maastricht IV/ Florence consensus report. *Gut* 61, 646–664. doi: 10.1136/gutjnl-2012-302084

Mantovani, A., Allavena, P., Sica, A., and Balkwill, F. (2008). Cancer-related inflammation. *Nature* 454, 436–444. doi: 10.1038/nature07205

Marshall, B. J., and Warren, J. R. (1984). Unidentified curved bacilli in the stomach of patients with gastritis and peptic ulceration. *Lancet* 1, 1311–1315. doi: 10.1016/S0140-6736(84)91816-6

Marshall, B. J., and Warren, J. R. (2005). *The Bacterium Helicobacter pylori and Its Role in Gastritis and Peptic ulcer Disease*. The Nobel Prize in Physiology or Medicine, Press Release.

Martindale, J. L., and Holbrook, N. J. (2002). Cellular response to oxidative stress: signaling for suicide and survival. *J. Cell. Physiol.* 192, 1–15. doi: 10.1002/jcp.10119

Martinon, F., Mayor, A., and Tschopp, J. (2009). The inflammasomes: guardians of the body. *Annu. Rev. Immunol.* 27, 229–265. doi: 10.1146/annurev.immunol.021908.132715

Maynard, S., Schurman, S. H., Harboe, C., de Souza-Pinto, N. C., and Bohr, V. A. (2009). Base excision repair of oxidative DNA damage and association with cancer and aging. *Carcinogenesis* 30, 2–10. doi: 10.1093/carcin/bgn250

McCoy, A. N., Araujo-Perez, F., Azcarate-Peril, A., Yeh, J. J., Sandler, R. S., and Keku, T. O. (2013). *Fusobacterium* is associated with colorectal adenomas. *PLoS ONE* 8:e53653. doi: 10.1371/journal.pone.0053653

Medzhitov, R. (2008). Origin and physiological roles of inflammation. *Nature* 454, 428–435. doi: 10.1038/nature07201

Miftahussurur, M., Yamaoka, Y., and Graham, D. Y. (2017). *Helicobacter pylori* as an oncogenic pathogen, revisited. *Expert Rev. Mol. Med.* 19:e4. doi: 10.1017/erm.2017.4

Mima, K., Nishihara, R., Qian, Z. R., Cao, Y., Sukawa, Y., Nowak, J. A., et al. (2016). *Fusobacterium nucleatum* in colorectal carcinoma tissue and patient prognosis. *Gut* 65, 1973–1980. doi: 10.1136/gutjnl-2015-310101

Molina-Cruz, A., DeJong, R. J., Charles, B., Gupta, L., Kumar, S., Jaramillo-Gutierrez, G., et al. (2008). Reactive oxygen species modulate *Anopheles gambiae* immunity against bacteria and Plasmodium. *J. Biol. Chem.* 283, 3217–3223. doi: 10.1074/jbc.M705873200

Nathan, C. (2002). Points of control in inflammation. *Nature* 420, 846–852. doi: 10.1038/nature01320

Ng, P. W., Iha, H., Iwanaga, Y., Bittner, M., Chen, Y., Jiang, Y., et al. (2001). Genome-wide expression changes induced by HTLV-1 Tax: evidence for MLK-3 mixed lineage kinase involvement in tax-mediated NF-kappaB activation. *Oncogene* 20, 4484–4496. doi: 10.1038/sj.onc.1204513

Nickson, C. M., Moori, P., Carter, R. J., Rubbi, C. P., and Parsons, J. L. (2017). Misregulation of DNA damage repair pathways in HPV-positive head and neck squamous cell carcinoma contributes to cellular radiosensitivity. *Oncotarget* 8, 29963–29975. doi: 10.18632/oncotarget.16265

Nougayrède, J. P., Homburg, S., Taieb, F., Boury, M., Brzuszkiewicz, E., Gottschalk, G., et al. (2006). *Escherichia coli* induces DNA double-strand breaks in eukaryotic cells. *Science* 313, 848–851. doi: 10.1126/science.1127059

Nougayrède, J. P., Taieb, F., De Rycke, J., and Oswald, E. (2005). Cyclomodulins: bacterial effectors that modulate the eukaryotic cell cycle. *Trends Microbiol.* 13, 103–110. doi: 10.1016/j.tim.2005.01.002

Ostrand-Rosenberg, S., and Sinha, P. (2009). Myeloid-derived suppressor cells: linking inflammation and cancer. *J. Immunol.* 182, 4499–4506. doi: 10.4049/jimmunol.0802740

Parkin, D. M. (2006). The global health burden of infection-associated cancers in the year 2002. *Int. J. Cancer* 118, 3030–3044. doi: 10.1002/ijc.21731

Parsonnet, J., Hansen, S., Rodriguez, L., Gelb, A. B., Warnke, R. A., Jellum, E., et al. (1994). *Helicobacter pylori* infection and gastric lymphoma. *N. Engl. J. Med.* 330, 1267–1271. doi: 10.1056/NEJM199405053301803

Pasare, C., and Medzhitov, R. (2004). Toll-like receptors and acquired immunity. *Semin. Immunol.* 16, 23–26. doi: 10.1016/j.smim.2003.10.006

Peltomäki, P. (2001). Deficient DNA mismatch repair: a common etiologic factor for colon cancer. *Hum. Mol. Genet.* 10, 735–740. doi: 10.1093/hmg/10.7.735

Polk, D. B., and Peek, R. M. Jr. (2010). *Helicobacter pylori*: gastric cancer and beyond. *Nat. Rev. Cancer* 10, 403–414. doi: 10.1038/nrc2857

Qadri, I., Fatima, K., and Abdel Hafiz, H. (2011). Hepatitis B virus X protein impedes the DNA repair via its association with transcription factor, TFIIH. *BMC Microbiol.* 11:48. doi: 10.1186/1471-2180-11-48

Qin, H., Zhu, J., Zeng, Y., Du, W., Shen, D., Lei, Z., et al. (2017). Aberrant promoter methylation of hOGG1 may be associated with increased risk of non-small cell lung cancer. *Oncotarget* 8, 8330–8341. doi: 10.18632/oncotarget.14177

Radisky, D. C., Levy, D. D., Littlepage, L. E., Liu, H., Nelson, C. M., Fata, J. E., et al. (2005). Rac1b and reactive oxygen species mediate MMP-3-induced EMT and genomic instability. *Nature* 436, 123–127. doi: 10.1038/nature03688

Ray, K. (2011). Colorectal cancer: *Fusobacterium nucleatum* found in colon cancer tissue–could an infection cause colorectal cancer? *Nat. Rev. Gastroenterol. Hepatol.* 8:662. doi: 10.1038/nrgastro.2011.208

Repass, J., Maherali, N., Owen, K., and Reproducibility Project: Cancer Biology. (2016). Registered report: *Fusobacterium nucleatum* infection is prevalent in human colorectal carcinoma. *Elife* 5:e10012. doi: 10.7554/eLife.10012

Reuter, S., Gupta, S. C., Chaturvedi, M. M., and Aggarwal, B. B. (2010). Oxidative stress, inflammation, and cancer: how are they linked? *Free Radic. Biol. Med.* 49, 1603–1616. doi: 10.1016/j.freeradbiomed.2010.09.006

Rubinstein, M. R., Wang, X., Liu, W., Hao, Y., Cai, G., and Han, Y. W. (2013). *Fusobacterium nucleatum* promotes colorectal carcinogenesis by modulating E-cadherin/beta-catenin signaling via its FadA adhesin. *Cell Host Microbe* 14, 195–206. doi: 10.1016/j.chom.2013.07.012

Ryan, R. M., Green, J., and Lewis, C. E. (2006). Use of bacteria in anti-cancer therapies. *Bioessays* 28, 84–94. doi: 10.1002/bies.20336

Sablina, A. A., Budanov, A. V., Ilyinskaya, G. V., Agapova, L. S., Kravchenko, J. E., and Chumakov, P. M. (2005). The antioxidant function of the p53 tumor suppressor. *Nat. Med.* 11, 1306–1313. doi: 10.1038/nm1320

Salim, E. I., Morimura, K., Menesi, A., El-Lity, M., Fukushima, S., and Wanibuchi, H. (2008). Elevated oxidative stress and DNA damage and repair levels in urinary bladder carcinomas associated with schistosomiasis. *Int. J. Cancer* 123, 601–608. doi: 10.1002/ijc.23547

Sancar, A., Lindsey-Boltz, L. A., Unsal-Kaçmaz, K., and Linn, S. (2004). Molecular mechanisms of mammalian DNA repair and the DNA damage checkpoints. *Annu. Rev. Biochem.* 73, 39–85. doi: 10.1146/annurev.biochem.73.011303.073723

Santos, J. C., Brianti, M. T., Almeida, V. R., Ortega, M. M., Fischer, W., Haas, R., et al. (2017). *Helicobacter pylori* infection modulates the expression of miRNAs associated with DNA mismatch repair pathway. *Mol. Carcinog.* 56, 1372–1379. doi: 10.1002/mc.22590

Sarker, A. H., Chatterjee, A., Williams, M., Lin, S., Havel, C., Jacob, P. 3rd., et al. (2014). NEIL2 protects against oxidative DNA damage induced by sidestream smoke in human cells. *PLoS ONE* 9:e90261. doi: 10.1371/journal.pone.0090261

Segal, A. W. (2005). How neutrophils kill microbes. *Annu. Rev. Immunol.* 23, 197–223. doi: 10.1146/annurev.immunol.23.021704.115653

Serhan, C. N., Chiang, N., and van Dyke, T. E. (2008). Resolving inflammation: dual anti-inflammatory and pro-resolution lipid mediators. *Nat. Rev. Immunol.* 8, 349–361. doi: 10.1038/nri2294

Shacter, E., and Weitzman, S. A. (2002). Chronic inflammation and cancer. *Oncology* 16, 217–226, 229. discussion: 230–232.

Sheflin, A. M., Whitney, A. K., and Weir, T. L. (2014). Cancer-promoting effects of microbial dysbiosis. *Curr. Oncol. Rep.* 16:406. doi: 10.1007/s11912-014-0406-0

Shekari, M., Sobti, R. C., Tamandani, D. M., Malekzadeh, K., Kaur, P., and Suri, V. (2008). Association of genetic polymorphism of the DNA base excision repair gene (APE-1 Asp/148 Glu) and HPV type (16/18) with the risk of cervix cancer in north Indian population. *Cancer Biomark.* 4, 63–71. doi: 10.3233/CBM-2008-4202

Shinmura, K., Kato, H., Kawanishi, Y., Yoshimura, K., Igarashi, H., Goto, M., et al. (2017). Reduced expression of the DNA glycosylase gene MUTYH is associated with an increased number of somatic mutations via a reduction in the DNA repair capacity in prostate adenocarcinoma. *Mol. Carcinog.* 56, 781–788. doi: 10.1002/mc.22509

Shuck, S. C., Short, E. A., and Turchi, J. J. (2008). Eukaryotic nucleotide excision repair: from understanding mechanisms to influencing biology. *Cell Res.* 18, 64–72. doi: 10.1038/cr.2008.2

Smith, J. S., Bosetti, C., Muñoz, N., Herrero, R., Bosch, F. X., Eluf-Neto, J., et al. (2004). *Chlamydia trachomatis* and invasive cervical cancer: a pooled analysis of the IARC multicentric case-control study. *Int. J. Cancer* 111, 431–439. doi: 10.1002/ijc.20257

Sohn, S., Jaitovitch-Groisman, I., Benlimame, N., Galipeau, J., Batist, G., and Alaoui-Jamali, M. A. (2000). Retroviral expression of the hepatitis B virus x gene promotes liver cell susceptibility to carcinogen-induced site specific mutagenesis. *Mutat. Res.* 460, 17–28. doi: 10.1016/S0921-8777(00)00010-0

Sokolova, O., Bozko, P. M., and Naumann, M. (2008). *Helicobacter pylori* suppresses glycogen synthase kinase 3beta to promote beta-catenin activity. *J. Biol. Chem.* 283, 29367–29374. doi: 10.1074/jbc.M801818200

Spencer, M. D., Hamp, T. J., Reid, R. W., Fischer, L. M., Zeisel, S. H., and Fodor, A. A. (2011). Association between composition of the human gastrointestinal microbiome and development of fatty liver with choline deficiency. *Gastroenterology* 140, 976–986. doi: 10.1053/j.gastro.2010.11.049

Strickertsson, J. A., Desler, C., and Rasmussen, L. J. (2014). Impact of bacterial infections on aging and cancer: impairment of DNA repair and mitochondrial function of host cells. *Exp. Gerontol.* 56, 164–174. doi: 10.1016/j.exger.2014.03.024

Suzuki, M., Mimuro, H., Kiga, K., Fukumatsu, M., Ishijima, N., Morikawa, H., et al. (2009). *Helicobacter pylori* CagA phosphorylation-independent function in epithelial proliferation and inflammation. *Cell Host Microbe* 5, 23–34. doi: 10.1016/j.chom.2008.11.010

Tabassam, F. H., Graham, D. Y., and Yamaoka, Y. (2009). *Helicobacter pylori* activate epidermal growth factor receptor- and phosphatidylinositol 3-OH kinase-dependent Akt and glycogen synthase kinase 3beta phosphorylation. *Cell. Microbiol.* 11, 70–82. doi: 10.1111/j.1462-5822.2008.01237.x

Taieb, F., Petit, C., Nougayrède, J. P., and Oswald, E. (2016). The enterobacterial genotoxins: cytolethal distending toxin and colibactin. *EcoSal Plus* 7, 1–21. doi: 10.1128/ecosalplus.ESP-0008-2016

Tal, M. C., Sasai, M., Lee, H. K., Yordy, B., Shadel, G. S., and Iwasaki, A. (2009). Absence of autophagy results in reactive oxygen species-dependent amplification of RLR signaling. *Proc. Natl. Acad. Sci. U.S.A.* 106, 2770–2775. doi: 10.1073/pnas.0807694106

Tomb, J. F., White, O., Kerlavage, A. R., Clayton, R. A., Sutton, G. G., Fleischmann, R. D., et al. (1997). The complete genome sequence of the gastric pathogen *Helicobacter pylori*. *Nature* 388, 539–547. doi: 10.1038/41483

Tsung, A., Klune, J. R., Zhang, X., Jeyabalan, G., Cao, Z., Peng, X., et al. (2007). HMGB1 release induced by liver ischemia involves Toll-like receptor 4 dependent reactive oxygen species production and calcium-mediated signaling. *J. Exp. Med.* 204, 2913–2923. doi: 10.1084/jem.20070247

Valko, M., Leibfritz, D., Moncol, J., Cronin, M. T., Mazur, M., and Telser, J. (2007). Free radicals and antioxidants in normal physiological functions and human disease. *Int. J. Biochem. Cell Biol.* 39, 44–84. doi: 10.1016/j.biocel.2006.07.001

van de Klundert, M. A., van Hemert, F. J., Zaaijer, H. L., and Kootstra, N. A. (2012). The hepatitis B virus x protein inhibits thymine DNA glycosylase initiated base excision repair. *PLoS ONE* 7:e48940. doi: 10.1371/journal.pone.0048940

van der Vliet, A. (2008). NADPH oxidases in lung biology and pathology: host defense enzymes, and more. *Free Radic. Biol. Med.* 44, 938–955. doi: 10.1016/j.freeradbiomed.2007.11.016

Vendrell, A., Gravisaco, M. J., Pasetti, M. F., Croci, M., Colombo, L., Rodriguez, C., et al. (2011). A novel *Salmonella typhi*-based immunotherapy promotes tumor killing via an antitumor Th1-type cellular immune response and neutrophil activation in a mouse model of breast cancer. *Vaccine* 29, 728–736. doi: 10.1016/j.vaccine.2010.11.017

Verma, S. C., Bajaj, B. G., Cai, Q., Si, H., Seelhammer, T., and Robertson, E. S. (2006). Latency-associated nuclear antigen of Kaposi's sarcoma-associated herpesvirus recruits uracil DNA glycosylase 2 at the terminal repeats and is important for latent persistence of the virus. *J. Virol.* 80, 11178–11190. doi: 10.1128/JVI.01334-06

Vijay-Kumar, M., Aitken, J. D., Carvalho, F. A., Cullender, T. C., Mwangi, S., Srinivasan, S., et al. (2010). Metabolic syndrome and altered gut microbiota in mice lacking Toll-like receptor 5. *Science* 328, 228–231. doi: 10.1126/science.1179721

Wallace, S. S., Murphy, D. L., and Sweasy, J. B. (2012). Base excision repair and cancer. *Cancer Lett.* 327, 73–89. doi: 10.1016/j.canlet.2011.12.038

Warren, R. L., Freeman, D. J., Pleasance, S., Watson, P., Moore, R. A., Cochrane, K., et al. (2013). Co-occurrence of anaerobic bacteria in colorectal carcinomas. *Microbiome* 1:16. doi: 10.1186/2049-2618-1-16

West, A. P., Brodsky, I. E., Rahner, C., Woo, D. K., Erdjument-Bromage, H., Tempst, P., et al. (2011a). TLR signalling augments macrophage bactericidal activity through mitochondrial ROS. *Nature* 472, 476–480. doi: 10.1038/nature09973

West, A. P., Shadel, G. S., and Ghosh, S. (2011b). Mitochondria in innate immune responses. *Nat. Rev. Immunol.* 11, 389–402. doi: 10.1038/nri2975

Wiseman, H., and Halliwell, B. (1996). Damage to DNA by reactive oxygen and nitrogen species: role in inflammatory disease and progression to cancer. *Biochem. J.* 313 (Pt 1), 17–29. doi: 10.1042/bj3130017

Wong, B. C., Lam, S. K., Wong, W. M., Chen, J. S., Zheng, T. T., Feng, R. E., et al. (2004). *Helicobacter pylori* eradication to prevent gastric cancer in a high-risk region of China: a randomized controlled trial. *JAMA* 291, 187–194. doi: 10.1001/jama.291.2.187

Wroblewski, L. E., Peek, R. M. Jr., and Wilson, K. T. (2010). *Helicobacter pylori* and gastric cancer: factors that modulate disease risk. *Clin. Microbiol. Rev.* 23, 713–739. doi: 10.1128/CMR.00011-10

Wu, C. Y., Hu, H. Y., Pu, C. Y., Huang, N., Shen, H. C., Li, C. P., et al. (2011). Pulmonary tuberculosis increases the risk of lung cancer: a population-based cohort study. *Cancer* 117, 618–624. doi: 10.1002/cncr.25616

Wu, S., Rhee, K. J., Albesiano, E., Rabizadeh, S., Wu, X., Yen, H. R., et al. (2009). A human colonic commensal promotes colon tumorigenesis via activation of T helper type 17 T cell responses. *Nat. Med.* 15, 1016–1022. doi: 10.1038/nm.2015

Xie, X., Yang, M., Ding, Y., and Chen, J. (2017). Microbial infection, inflammation and epithelial ovarian cancer. *Oncol. Lett.* 14, 1911–1919. doi: 10.3892/ol.2017.6388

Yang, D., Elner, S. G., Bian, Z. M., Till, G. O., Petty, H. R., and Elner, V. M. (2007). Pro-inflammatory cytokines increase reactive oxygen species through mitochondria and NADPH oxidase in cultured RPE cells. *Exp. Eye Res.* 85, 462–472. doi: 10.1016/j.exer.2007.06.013

Yang, Y., Weng, W., Peng, J., Hong, L., Yang, L., Toiyama, Y., et al. (2017). *Fusobacterium nucleatum* increases proliferation of colorectal cancer cells and tumor development in mice by activating toll-like receptor 4 signaling to nuclear factor-kappaB, and up-regulating expression of microRNA-21. *Gastroenterology* 152, 851.e24–866.e24. doi: 10.1053/j.gastro.2016.11.018

Ye, F., Cheng, Q., Shen, J., Zhou, C., and Chen, H. (2014). Mismatch repair gene MLH3 Pro844Leu and Thr942Ile polymorphisms and the susceptibility to cervical carcinoma and HPV infection: a case-control study in a Chinese population. *PLoS ONE* 9:e96224. doi: 10.1371/journal.pone.0096224

Zarember, K. A., and Godowski, P. J. (2002). Tissue expression of human Toll-like receptors and differential regulation of Toll-like receptor mRNAs in leukocytes in response to microbes, their products, and cytokines. *J. Immunol.* 168, 554–561. doi: 10.4049/jimmunol.168.2.554

Zhai, X., Zhao, H., Liu, Z., Wang, L. E., El-Naggar, A. K., Sturgis, E. M., et al. (2008). Functional variants of the NEIL1 and NEIL2 genes and risk and progression of squamous cell carcinoma of the oral cavity and oropharynx. *Clin. Cancer Res.* 14, 4345–4352. doi: 10.1158/1078-0432.CCR-07-5282

Zhong, C., Xu, M., Wang, Y., Xu, J., and Yuan, Y. (2017). An APE1 inhibitor reveals critical roles of the redox function of APE1 in KSHV replication and pathogenic phenotypes. *PLoS Pathog.* 13:e1006289. doi: 10.1371/journal.ppat.1006289

Zhou, R., Yazdi, A. S., Menu, P., and Tschopp, J. (2011). A role for mitochondria in NLRP3 inflammasome activation. *Nature* 469, 221–225. doi: 10.1038/nature09663

Zimmerman, R., and Cerutti, P. (1984). Active oxygen acts as a promoter of transformation in mouse embryo C3H/10T1/2/C18 fibroblasts. *Proc. Natl. Acad. Sci. U.S.A.* 81, 2085–2087. doi: 10.1073/pnas.81.7.2085

19

Conjugation Inhibitors and their Potential Use to Prevent Dissemination of Antibiotic Resistance Genes in Bacteria

Elena Cabezón, Fernando de la Cruz and Ignacio Arechaga**

Instituto de Biomedicina y Biotecnología de Cantabria (IBBTEC), CSIC-Universidad de Cantabria and Departamento de Biología Molecular, Universidad de Cantabria, Santander, Spain

Edited by:
Chew Chieng Yeo, Universiti Sultan Zainal Abidin, Malaysia

Reviewed by:
Charles Martin Dozois, Institut National de la Recherche Scientifique (INRS), Canada
Irene Wagner-Doebler, Helmholtz-Zentrum für Infektionsforschung (HZI), Germany
Radoslaw Pluta, International Institute of Molecular and Cell Biology in Warsaw (IIMCB), Poland

***Correspondence:**
Elena Cabezón
cabezone@unican.es
Ignacio Arechaga
arechagai@unican.es

Antibiotic resistance has become one of the most challenging problems in health care. Bacteria conjugation is one of the main mechanisms whereby bacteria become resistant to antibiotics. Therefore, the search for specific conjugation inhibitors (COINs) is of interest in the fight against the spread of antibiotic resistances in a variety of laboratory and natural environments. Several compounds, discovered as COINs, are promising candidates in the fight against plasmid dissemination. In this review, we survey the effectiveness and toxicity of the most relevant compounds. Particular emphasis has been placed on unsaturated fatty acid derivatives, as they have been shown to be efficient in preventing plasmid invasiveness in bacterial populations. Biochemical and structural studies have provided insights concerning their potential molecular targets and inhibitory mechanisms. These findings open a new avenue in the search of new and more effective synthetic inhibitors. In this pursuit, the use of structure-based drug design methods will be of great importance for the screening of ligands and binding sites of putative targets.

Keywords: antibiotic resistance, type IV secretion systems, inhibitors, fatty acids, bacterial conjugation

INTRODUCTION

Antibiotic resistance is becoming a major threat for human health (WHO, 2014). Widespread abuse of antibiotics in human health and food production are threating our current health systems and challenge future care (Giske et al., 2008; Boucher et al., 2009; Rice, 2009; Chang et al., 2015). However, despite these threats, few new antibiotics are becoming available to fight against multi-resistant bugs (Gould and Bal, 2013; Hede, 2014). Bacterial conjugation is one of the main mechanisms whereby bacteria become resistant to antibiotics (Mazel and Davies, 1999; Waters, 1999). Thus, the search for specific conjugation inhibitors (COINs) is a foremost concern in the fight against the spread of antibiotic resistance genes (Smith and Romesberg, 2007; Baquero et al., 2011; Baym et al., 2016). In this pursuit, several compounds were reported to inhibit bacterial conjugation specifically, although most turned out to be unspecific growth inhibitors (Michel-Briand and Laporte, 1985; Hooper et al., 1989; Conter et al., 2002; Lujan et al., 2007; Nash et al., 2012).

Bacterial conjugation is a mechanism by which DNA is transferred between two bacterial cells. The process consists of two steps. In a first stage, DNA is mobilized by a set of proteins, encoded by MOB genes. In a second step, DNA is transported across a secretion system [type IV secretion system (T4SS)] (de la Cruz et al., 2010; Cabezón et al., 2015). T4SS is a complex formed by proteins encoded by another set of genes, also known as MPF (mating pore formation) genes (Fernandez-Lopez et al., 2006). Conjugative Gram negative bacteria usually contain three MOB genes encoding proteins involved in DNA processing (Smillie et al., 2010). The most ubiquitous of these genes encodes a relaxase protein, which cleaves one of the plasmid strands at the origin of transfer (oriT) (Garcillán-Barcia et al., 2009) (Supplementary Figure S1A). Upon this, the relaxase protein remains covalently bound to the DNA at the 5′ end. This nucleoprotein complex is recruited at the secretion channel (T4SS) with the assistance of the coupling protein, an ATPase present in most conjugative plasmids (Llosa et al., 2002; Tato et al., 2005) (Supplementary Figure S1B). A third, accessory protein is usually involved to ensure the correct DNA folding for the relaxase action (Moncalian and de la Cruz, 2004).

Conjugative T4SS are large multi-subunit complexes involved in substrate transport and pilus biogenesis. The simplest T4SS consists of 11 proteins, named VirB1 to VirB11, after *Agrobaterium tumefaciens* T4SS (Christie et al., 2005, 2014). This macromolecular complex spans across the inner and outer membranes and the periplasm in between. T4SS architecture is well-preserved in most conjugative bacteria, consisting of four distinct sections: the pilus, the core channel complex, the inner membrane platform and the hexameric ATPases that provide the energy for substrate transport and pilus biogenesis (Cabezón et al., 2015). One of them, the traffic ATPase VirB11, was shown to be the target for inhibition by unsaturated fatty acids (Ripoll-Rozada et al., 2016). Here, we will analyze the progress on the different strategies to inhibit the VirB11 ATPase and the rest of the T4SS machinery. The impact of these results on the fight against the spread of antibiotic resistance genes is discussed.

Strategies for the Identification of Conjugation Inhibitors

Bacterial conjugation has been reported to be inhibited by a variety of compounds. Indeed, chemicals such as heterocyclic compounds, intercalators, acridine dyes, or quinolones were reported to inhibit conjugation (Hahn and Ciak, 1976; Michel-Briand and Laporte, 1985; Molnar et al., 1992; Mazel and Davies, 1999; Nash et al., 2012). However, posterior revisions showed that these molecules were unspecific, mainly affecting bacterial growth or DNA synthesis. Plants are a rich source of bioactive compounds, such as phenolics, which are able to modify bacterial resistances (Oyedemi et al., 2016). Therefore, a current approach consists of isolating molecules from different parts of medicinal plants to discover new inhibitors. By using this approach, two new drugs: rottlerin [5,7-dihydroxy-2,2-dimethyl-6-(2,4,6-trihydroxy-3-methyl-5-acetylbenzyl)-8-cinnamoyl-1,2-chromene] and the red compound (8-cinnamoyl-5,7-dihydroxy-2,2,6-trimethylchromene) were identified as potent antibacterial chemicals against Gram-positive bacteria. These compounds did not hamper Gram-negative bacteria growth but inhibited conjugal transfer of plasmids pKM101, TP114, pUB307, and R6K (Oyedemi et al., 2016). The planar structure of the compounds suggests that the target of these inhibitors might be the DNA replication system but further studies are required to elucidate the mode of inhibition of these agents.

Alternative attempts to inhibit bacterial conjugation have been based on bottom up strategies, targeting essential compounds of the secretion machinery. One study focused on targeting the conjugative relaxase protein, which is the protein that initiates conjugation upon nicking plasmid DNA at the origin of transfer. Due to its key role in plasmid conjugation, relaxases have been considered as potential targets for inhibitors. Some of these potential relaxase-specific inhibitors belong to the bisphosphonates family of compounds, such as etidronate (Didronel) and clodronate (Bonefos) (Lujan et al., 2007). These compounds were reported to be efficient in restraining conjugative DNA transfer. However, these results turned out to be misleading, as these putative inhibitors were found to work as unspecific chelating agents (Nash et al., 2012). An alternative method to inhibit specifically the conjugative relaxase consisted of the expression of specific single chain Fv antibodies (intrabodies) against the relaxase TrwC of conjugative plasmid R388 (Garcillan-Barcia et al., 2007). Expression of these intrabodies in the recipient cell prevented the accretion of the conjugative plasmid. However, the usefulness of intrabodies in practical clinical care is hampered by the need of a transgenic recipient population expressing them. Besides, each intrabody would be specific only against its cognate plasmid.

VirB8 is an essential assembly protein of bacterial T4SS that also acts as molecular target of small-molecule inhibitors (Smith et al., 2012). A high throughput assay based on the restoration of interactions between two split domains of the *Brucella* VirB8 protein allowed the identification of several compounds that inhibited protein-protein interactions (Paschos et al., 2011). One of the most efficient molecules, B8I-2, is a salicylidene acylhydrazide derivative, also known to inhibit T3SS (Keyser et al., 2008). Posterior analysis by X-ray crystallography and *in silico* docking of several of these compounds allowed the determination of VirB8 binding site (Smith et al., 2012). Recently, it has been reported that these small molecules also bind TraE, the VirB8 homolog of the conjugative plasmid pKM101, and some of them inhibit plasmid transfer (Casu et al., 2016). Although some of these molecules displayed a low Kd value in *in vitro* binding experiments, no significant impact was observed on plasmid transfer frequencies, with a 10-fold reduction as the strongest effect. Moreover, none of these molecules had an effect on the conjugation of the unrelated plasmid RP4, diminishing the effectiveness of these compounds to overcome antibiotic resistance.

Other alternatives to develop specific inhibitors focused on the conjugative pilus. These appendages are targeted by a variety of bacteriophages that, upon binding to them, enter inside the bacteria cytoplasm. Some of these bacteriophages are specific to conjugative pili and, therefore, have the potential of discriminating between different bacterial species. For instance, filamentous bacteriophages, such as M13, display high affinity for F-type pilus. This interaction is mediated by the phage

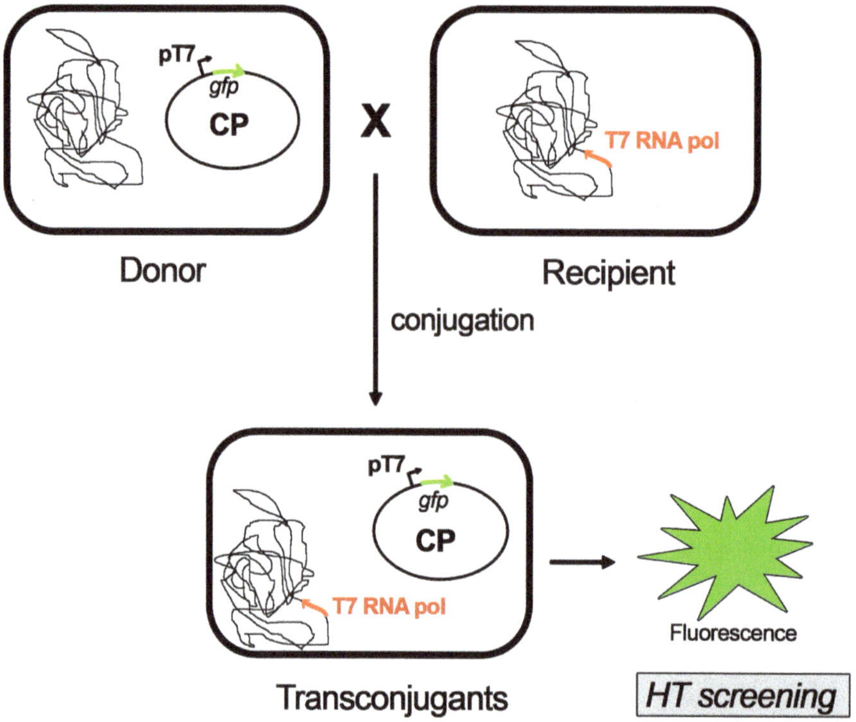

FIGURE 1 | High-throughput conjugation (HTC) assay. A whole-cell automated assay for conjugation based on fluorescence emission of transconjugant cells. A fluorescent protein (GFP) is cloned in the conjugative plasmid (CP) under the control of pT7 promoter and transformed in donor cells. Recipient cells contains a T7 RNA polymerase gene cloned in the bacteria chromosome. Upon mating, the conjugative plasmid is transferred into the recipient cell. The resultant transconjugants will express the GFP after addition of IPTG (isopropyl-β-D-thiogalactopyranoside). Upon induction, the emission fluorescence of GFP is measured, along with the optical densities (ODs) of the cultures. Conjugation frequencies are estimated as a ratio between the fluorescence emitted by transconjugant cells and the total number of cells measured by the OD (for further details on this method, please see Getino et al., 2015).

coat protein g3p. Addition of the soluble N-terminal domain of g3p to F-plasmid containing bacteria resulted in inhibition of conjugation (Lin et al., 2011). Considering that conjugative pili are needed for bacterial cell contact (Anthony et al., 1994), a step required for the spread of antibiotic resistance genes, the use of specific compounds that inhibit pilus formation could result in a powerful strategy to fight against this problem.

In that sense, peptidomimetic small molecules, such as C10 and KSK85, have been found to disrupt T4SS-dependent transport of pathogenic factors, as well as DNA transfer in conjugative *Escherichia coli*. KSK85 acts impeding biogenesis of the pilus appendage, whereas C10 disrupts T4SS activity without affecting pilus assembly (Shaffer et al., 2016). In this case, authors have used a phenotypic screen to identify these two inhibitors. Both compounds have been tested in plasmids pKM101 (IncN) and R1-16 (IncF) but conjugation efficiency is only decreased to 25% and a high concentration of inhibitor is required (150 μM). Therefore, although these compounds are promising scaffolds, new derivatives need to be found to inhibit more effectively the conjugative process.

A novel approach in the pursuit of specific COINs is the screening of potential compounds in whole-cell assays (**Figure 1**). This methodology should be designed in a way that discriminates COINs from false positives affecting cell growth. Thus, the development of luminescence-based high-throughput conjugation (HTC) assays has been shown to be effective in the identification of potential hits. By using HTC assays, unsaturated fatty acids were found to inhibit conjugation of IncF and IncW plasmids without affecting cell growth (Fernández-Lopez et al., 2005). To this end, a library consisting of more than 12,000 natural compounds (NatChem library) was tested. The most effective COIN found in this screening was dehydrocrepnynic acid (DHCA) (Fernández-Lopez et al., 2005). Considering that this compound was extracted from tropical plant seeds (Gussoni et al., 1994) the viability of using it without vast downstream process improvement is limited.

Nonetheless, by using DHCA structure as a chemical template, new synthetic compounds were developed as specific COINs. In particular, synthetic 2-hexadecynoic acid (2-HDA) and other 2-alkynoic fatty acids (2-AFAs) were found to be specific inhibitors of a wide range of conjugative plasmids in different bacteria, including the highly infective and prevalent IncF plasmids (Getino et al., 2015). Furthermore, due to the effect of plasmid burden on host fitness, 2-AFAs could in fact eliminate transmissible plasmids, such as IncF, from bacterial populations. However, other plasmid groups, such as IncN and IncP, were not affected (Fernández-Lopez et al., 2005; Getino et al., 2015).

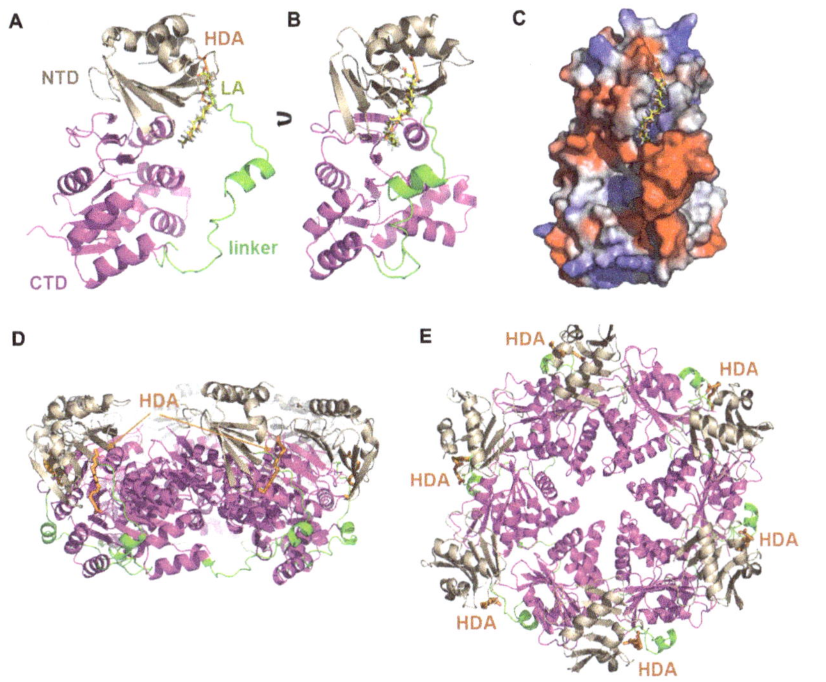

FIGURE 2 | Blind docking of fatty acids in TrwD. Docking predictions between a molecular model of TrwD, the traffic ATPase of the conjugative plasmid R388, and linoleic acid and 2-HDA are shown **(A,B)** (Ripoll-Rozada et al., 2016). These unsaturated fatty acids bind to a pocket formed by the end of the N-terminal domain (NTD, *wheat*) and the linker region (*green*) that connects it to the C-terminal domain (CTD, *purple*). An electrostatic map showing these interactions is included in **(C)**. Inhibitor binding sites in the hexameric TrwD protein are shown in **(D,E)**, corresponding to side and top views, respectively. Docking was performed using the EADock dihedral spacing sampling engine of the Swiss-dock server (Grosdidier et al., 2011).

In addition to 2-AFAs compounds, another family of bioactive compounds was tested in the search for COINs. This collection, named AQUAc, contains more than 1,600 natural compounds. AQUAc was evaluated in order to find potential COINs (Getino et al., 2016). As a result, new COINs were found. Among them, tanzawaic acids A and B were identified as best hits. They specifically inhibited IncW and IncFII conjugative plasmids. The advantage of these compounds is their lower toxicity to animal cells, in comparison to other synthetic COINs. Unsaturated fatty acids (oleic and linoleic acids), 2-HDA, 2-alkynoic fatty acids and tanzawaic acids, all share similar chemical characteristics: a carboxylic group, a long unsaturated carbon chain and the presence of double or triple bonds. These compounds present a 100-fold reduction in plasmid transfer frequencies and, although higher inhibition rates must be achieved to maximize their effectiveness, they constitute key scaffold structures on which to develop more potent and specific COINs. In that respect, knowing the molecular target of these compounds is extremely important, since the use of structure-based drug design (SBDD) methods will allow the design of modified synthetic compounds with higher binding affinities.

Unsaturated Fatty Acids As Specific Inhibitors of Conjugative Traffic ATPases

Bacterial conjugation is driven by a group of ATPases that empowers almost every step in the conjugative process: DNA unwinding, DNA transport, pilus biogenesis and protein transport (Cabezón et al., 2015). Each of these steps is catalyzed by a specific ATPase, that in the conjugative plasmid R388 are named TrwC, TrwB, TrwK, and TrwD, respectively. TrwC is a protein that nicks the DNA, thus relaxing the conjugative plasmid in an ATP-independent manner (Llosa et al., 1995, 1996). In addition, TrwC displays a DNA helicase activity that results in nucleic acid unwinding (Grandoso et al., 1994; Llosa et al., 1996). TrwB is a DNA-dependent ATPase involved in DNA transfer to the secretion channel (Tato et al., 2005, 2007; Matilla et al., 2010). TrwK is a hexameric ATPase (Arechaga et al., 2008; Peña et al., 2011) that participates in the transport of the pilin molecules from the inner to the outer membranes during pilus biogenesis (Kerr and Christie, 2010). Finally, TrwD is a traffic ATPase that contributes to pilus biogenesis and to DNA translocation (Atmakuri et al., 2004), thus working as a molecular switch between pilus synthesis and substrate transport (Ripoll-Rozada et al., 2013).

Each of these ATPases has been purified to homogeneity and their enzymatic activities have been characterized (Grandoso et al., 1994; Tato et al., 2005; Arechaga et al., 2008; Ripoll-Rozada et al., 2012). The kinetic parameters of each ATPase were analyzed in the presence of the unsaturated fatty acids shown to be efficient COINs (Ripoll-Rozada et al., 2016). Interestingly, only the traffic ATPase TrwD was inhibited by unsaturated fatty acids, such a linoleic acid, and 2-AFAs like 2-HDA. The kinetic parameters for TrwD ATPase inhibition by these fatty acids were determined, indicating that in all cases it was a non-competitive inhibition (Ripoll-Rozada et al., 2016). In contrast, saturated fatty acids, such as palmitic acid, showed no inhibitory effect in conjugation experiments and in ATPase assays.

TrwD belongs to the secretion ATPase superfamily, which also includes members of Type II secretion, Type IV pilus and flagellar biogenesis machineries (Planet et al., 2001). All members of this superfamily are hexameric ATPases, in which each monomer is formed by two domains at the N- and C-termini (NTD and CTD, respectively), connected by a flexible linker of a variable length (Planet et al., 2001; Peña and Arechaga, 2013). ATPase catalysis is driven by swapping this linker over the NTD and CTD (Savvides et al., 2003; Hare et al., 2006). Blind docking predictions (Grosdidier et al., 2011) suggested a putative binding site for uFAs and 2-aFAs located at the end of the NTD and beginning of the linker region that connects it to the CTD where the nucleotide binding site is located (**Figure 2**) (Ripoll-Rozada et al., 2016). These predictions were compatible with a model in which the mode of action of the inhibitors consisted in preventing the swapping movements between the N- and C-terminal domains along the linker region that are required in the catalytic cycle of the protein.

In summary, the discovery of traffic ATPases as potential targets of bacterial COINs opens a promising avenue for the development of new and more potent compounds capable to impair the dissemination of antibiotic resistance genes. However, despite these promising data, the efficacy of these inhibitors to prevent the general spread of antibiotic resistance in naturally occurring environments (Martinez, 2008; von Wintersdorff et al., 2016), hospitals, wastewater systems (Hocquet et al., 2016), agriculture (Capita and Alonso-Calleja, 2013; Economou and Gousia, 2015) and farming settings (Wegener, 2003) needs to be tested. Even today, the main reservoirs where the antibiotic resistance genes arise from and how these genes are rapidly acquired by human pathogens are a matter of debate. COINs could be effective in the discovery of these natural reservoirs. Experiments in controlled microcosms (e.g., freshwater microcosms) and/or experimental animals (e.g., mice gut) should be most instructive. Not only COIN-related experiments will help to identify reservoirs, they will also serve to quantify the dynamics of plasmids in those experiments and the rate at which they can be mobilized to recipient strains that are potential human pathogens. These are some of the exciting goals of ongoing COINs research.

OUTLOOK

The battle against antibiotic resistance is a challenging problem which is likely to become a progressively increasing burden to our health systems. Several approaches are currently envisaged to fight back against antibiotic resistant bugs. Among them, a promising alternative consists of preventing the propagation of antibiotic resistance genes. Bacterial conjugation is the main mechanism for the wide spread dissemination of these genes. Hence, the search of compounds able to specifically inhibit bacterial conjugation is a preeminent undertaking in the global war against antibiotic resistant bugs. Here, we have reviewed several compounds that are competent in this pursuit. Among these, unsaturated fatty acids and derivatives have been proved to be the most efficient specific COINs. Moreover, the identification of the traffic ATPase TrwD as the molecular target of these uFAs enables the future development of more efficient inhibitors directed against this essential protein of T4SS.

AUTHOR CONTRIBUTIONS

All authors listed have made a substantial, direct and intellectual contribution to the work, and approved it for publication.

REFERENCES

Anthony, K. G., Sherburne, C., Sherburne, R., and Frost, L. S. (1994). The role of the pilus in recipient cell recognition during bacterial conjugation mediated by F-like plasmids. *Mol. Microbiol.* 13, 939–953. doi: 10.1111/j.1365-2958.1994.tb00486.x

Arechaga, I., Peña, A., Zunzunegui, S., del Carmen Fernandez-Alonso, M., Rivas, G., and de la Cruz, F. (2008). ATPase activity and oligomeric state of TrwK, the VirB4 homologue of the plasmid R388 type IV secretion system. *J. Bacteriol.* 190, 5472–5479. doi: 10.1128/JB.00321-08

Atmakuri, K., Cascales, E., and Christie, P. J. (2004).). Energetic components VirD4, VirB11 and VirB4 mediate early DNA transfer reactions required for bacterial type IV secretion. *Mol. Microbiol.* 54, 1199–1211. doi: 10.1111/j.1365-2958.2004.04345.x

Baquero, F., Coque, T. M., and de la Cruz, F. (2011). Ecology and evolution as targets: the need for novel eco-evo drugs and strategies to fight antibiotic resistance. *Antimicrob. Agents Chemother.* 55, 3649–3660. doi: 10.1128/AAC.00013-11

Baym, M., Stone, L. K., and Kishony, R. (2016). Multidrug evolutionary strategies to reverse antibiotic resistance. *Science* 351:aad3292. doi: 10.1126/science.aad3292

Boucher, H. W., Talbot, G. H., Bradley, J. S., Edwards, J. E., Gilbert, D., Rice, L. B., et al. (2009). Bad bugs, no drugs: no ESKAPE! An update from the infectious diseases society of America. *Clin. Infect. Dis.* 48, 1–12. doi: 10.1086/595011

Cabezón, E., Ripoll-Rozada, J., Peña, A., de la Cruz, F., and Arechaga, I. (2015). Towards an integrated model of bacterial conjugation. *FEMS Microbiol. Rev.* 39, 81–95. doi: 10.1111/1574-6976.12085

Capita, R., and Alonso-Calleja, C. (2013). Antibiotic-resistant bacteria: a challenge for the food industry. *Crit. Rev. Food Sci. Nutr.* 53, 11–48. doi: 10.1080/10408398.2010.519837

Casu, B., Smart, J., Hancock, M. A., Smith, M., Sygusch, J., and Baron, C. (2016). Structural analysis and inhibition of TraE from the pKM101 type IV secretion system. *J. Biol. Chem.* 291, 23817–23829. doi: 10.1074/jbc.M116.753327

Chang, Q., Wang, W., Regev-Yochay, G., Lipsitch, M., and Hanage, W. P. (2015). Antibiotics in agriculture and the risk to human health: how worried should we be? *Evol. Appl.* 8, 240–247. doi: 10.1111/eva.12185

Christie, P. J., Atmakuri, K., Krishnamoorthy, V., Jakubowski, S., and Cascales, E. (2005). Biogenesis, architecture, and function of bacterial type IV secretion systems. *Annu. Rev. Microbiol.* 59, 451–485. doi: 10.1146/annurev.micro.58.030603.123630

Christie, P. J., Whitaker, N., and Gonzalez-Rivera, C. (2014). Mechanism and structure of the bacterial type IV secretion systems. *Biochim. Biophys. Acta* 1843, 1578–1591. doi: 10.1016/j.bbamcr.2013.12.019

Conter, A., Sturny, R., Gutierrez, C., and Cam, K. (2002). The RcsCB His-Asp phosphorelay system is essential to overcome chlorpromazine-induced stress in *Escherichia coli*. *J. Bacteriol.* 184, 2850–2853. doi: 10.1128/JB.184.10.2850-2853.2002

de la Cruz, F., Frost, L. S., Meyer, R. J., and Zechner, E. L. (2010). Conjugative DNA metabolism in Gram-negative bacteria. *FEMS Microbiol. Rev.* 34, 18–40. doi: 10.1111/j.1574-6976.2009.00195.x

Economou, V., and Gousia, P. (2015). Agriculture and food animals as a source of antimicrobial-resistant bacteria. *Infect. Drug Resist.* 8, 49–61. doi: 10.2147/IDR.S55778

Fernandez-Lopez, R., Garcillan-Barcia, M. P., Revilla, C., Lazaro, M., Vielva, L., and de la Cruz, F. (2006). Dynamics of the IncW genetic backbone imply general trends in conjugative plasmid evolution. *FEMS Microbiol. Rev.* 30, 942–966. doi: 10.1111/j.1574-6976.2006.00042.x

Fernández-Lopez, R., Machon, C., Longshaw, C. M., Martin, S., Molin, S., Zechner, E. L., et al. (2005). Unsaturated fatty acids are inhibitors of bacterial conjugation. *Microbiology* 151(Pt 11), 3517–3526. doi: 10.1099/mic.0.28216-0

Garcillán-Barcia, M. P., Francia, M. V., and de la Cruz, F. (2009). The diversity of conjugative relaxases and its application in plasmid classification. *FEMS Microbiol. Rev.* 33, 657–687. doi: 10.1111/j.1574-6976.2009.00168.x

Garcillan-Barcia, M. P., Jurado, P., Gonzalez-Perez, B., Moncalian, G., Fernandez, L. A., and de la Cruz, F. (2007). Conjugative transfer can be inhibited by blocking relaxase activity within recipient cells with intrabodies. *Mol. Microbiol.* 63, 404–416. doi: 10.1111/j.1365-2958.2006.05523.x

Getino, M., Sanabria-Ríos, D. J., Fernández-López, R., Campos-Gómez, J., Sánchez-López, J. M., Fernández, A., et al. (2015). Synthetic fatty acids prevent plasmid-mediated horizontal gene transfer. *mbio* 6:e01032-15. doi: 10.1128/mBio.01032-15

Getino, M., Fernandez-Lopez, R., Palencia-Gandara, C., Campos-Gomez, J., Sanchez-Lopez, J. M., Martinez, M., et al. (2016). Tanzawaic acids, a chemically novel set of bacterial conjugation inhibitors. *PLOS ONE* 11:e0148098. doi: 10.1371/journal.pone.0148098

Giske, C. G., Monnet, D. L., Cars, O., and Carmeli, Y. (2008). Clinical and economic impact of common multidrug-resistant gram-negative bacilli. *Antimicrob. Agents Chemother.* 52, 813–821. doi: 10.1128/AAC.01169-07

Gould, I. M., and Bal, A. M. (2013). New antibiotic agents in the pipeline and how they can help overcome microbial resistance. *Virulence* 4, 185–191. doi: 10.4161/viru.22507

Grandoso, G., Llosa, M., Zabala, J. C., and de la Cruz, F. (1994). Purification and biochemical characterization of TrwC, the helicase involved in plasmid R388 conjugal DNA transfer. *Eur. J. Biochem.* 226, 403–412. doi: 10.1111/j.1432-1033.1994.tb20065.x

Grosdidier, A., Zoete, V., and Michielin, O. (2011). SwissDock, a protein-small molecule docking web service based on EADock DSS. *Nucleic Acids Res.* 39, W270–W277. doi: 10.1093/nar/gkr366

Gussoni, M., Greco, F., Pegna, M., Bianchi, G., and Zetta, L. (1994). Solid state and microscopy NMR study of the chemical constituents of Afzelia cuanzensis seeds. *Magn. Reson. Imaging* 12, 477–486. doi: 10.1016/0730-725X(94)92542-9

Hahn, F. E., and Ciak, J. (1976). Elimination of resistance determinants from R-factor R1 by intercalative compounds. *Antimicrob. Agents Chemother.* 9, 77–80. doi: 10.1128/AAC.9.1.77

Hare, S., Bayliss, R., Baron, C., and Waksman, G. (2006). A large domain swap in the VirB11 ATPase of *Brucella suis* leaves the hexameric assembly intact. *J. Mol. Biol.* 360, 56–66. doi: 10.1016/j.jmb.2006.04.060

Hede, K. (2014). Antibiotic resistance: An infectious arms race. *Nature* 509, S2–S3. doi: 10.1038/509S2a

Hocquet, D., Muller, A., and Bertrand, X. (2016). What happens in hospitals does not stay in hospitals: antibiotic-resistant bacteria in hospital waste water systems. *J. Hosp. Inf.* 93, 395–402. doi: 10.1016/j.jhin.2016.01.010

Hooper, D. C., Wolfson, J. S., Tung, C., Souza, K. S., and Swartz, M. N. (1989). Effects of inhibition of the B subunit of DNA gyrase on conjugation in *Escherichia coli*. *J. Bacteriol.* 171, 2235–2237. doi: 10.1128/jb.171.4.2235-2237.1989

Kerr, J. E., and Christie, P. J. (2010). Evidence for VirB4-mediated dislocation of membrane-integrated VirB2 pilin during biogenesis of the *Agrobacterium* VirB/VirD4 type IV secretion system. *J. Bacteriol.* 192, 4923–4934. doi: 10.1128/JB.00557-10

Keyser, P., Elofsson, M., Rosell, S., and Wolf-Watz, H. (2008). Virulence blockers as alternatives to antibiotics: type III secretion inhibitors against gram-negative bacteria. *J. Intern. Med.* 264, 17–29. doi: 10.1111/j.1365-2796.2008.01941.x

Lin, A., Jimenez, J., Derr, J., Vera, P., Manapat, M. L., Esvelt, K. M., et al. (2011). Inhibition of bacterial conjugation by phage M13 and its protein g3p: quantitative analysis and model. *PLOS ONE* 6:e19991. doi: 10.1371/journal.pone.0019991

Llosa, M., Gomis-Ruth, F. X., Coll, M., and de la Cruz Fd, F. (2002). Bacterial conjugation: a two-step mechanism for DNA transport. *Mol. Microbiol.* 45, 1–8. doi: 10.1046/j.1365-2958.2002.03014.x

Llosa, M., Grandoso, G., and de la Cruz, F. (1995). Nicking activity of TrwC directed against the origin of transfer of the IncW plasmid R388. *J. Mol. Biol.* 246, 54–62. doi: 10.1006/jmbi.1994.0065

Llosa, M., Grandoso, G., Hernando, M. A., and de la Cruz, F. (1996). Functional domains in protein TrwC of plasmid R388: dissected DNA strand transferase and DNA helicase activities reconstitute protein function. *J. Mol. Biol.* 264, 56–67. doi: 10.1006/jmbi.1996.0623

Lujan, S. A., Guogas, L. M., Ragonese, H., Matson, S. W., and Redinbo, M. R. (2007). Disrupting antibiotic resistance propagation by inhibiting the conjugative DNA relaxase. *Proc. Natl. Acad. Sci. U.S.A.* 104, 12282–12287. doi: 10.1073/pnas.0702760104

Martinez, J. L. (2008). Antibiotics and antibiotic resistance genes in natural environments. *Science* 321, 365–367. doi: 10.1126/science.1159483

Matilla, I., Alfonso, C., Rivas, G., Bolt, E. L., de la Cruz, F., and Cabezon, E. (2010). The conjugative DNA translocase TrwB is a structure-specific DNA-binding protein. *J. Biol. Chem.* 285, 17537–17544. doi: 10.1074/jbc.M109.084137

Mazel, D., and Davies, J. (1999). Antibiotic resistance in microbes. *Cell Mol. Life Sci.* 56, 742–754. doi: 10.1007/s000180050021

Michel-Briand, Y., and Laporte, J. M. (1985). Inhibition of conjugal transfer of R plasmids by nitrofurans. *J. Gen. Microbiol.* 131, 2281–2284. doi: 10.1099/00221287-131-9-2281

Molnar, J., Fischer, J., and Nakamura, M. J. (1992). Mechanism of chlorpromazine binding by gram-positive and gram-negative bacteria. *Antonie Van Leeuwenhoek* 62, 309–314. doi: 10.1007/BF00572599

Moncalian, G., and de la Cruz, F. (2004). DNA binding properties of protein TrwA, a possible structural variant of the Arc repressor superfamily. *Biochim. Biophys. Acta* 1701, 15–23. doi: 10.1016/j.bbapap.2004.05.009

Nash, R. P., McNamara, D. E., Ballentine, W. K. III., Matson, S. W., and Redinbo, M. R. (2012). Investigating the impact of bisphosphonates and structurally related compounds on bacteria containing conjugative plasmids. *Biochem. Biophys. Res. Commun.* 424, 697–703. doi: 10.1016/j.bbrc.2012.07.012

Oyedemi, B. O., Shinde, V., Shinde, K., Kakalou, D., Stapleton, P. D., and Gibbons, S. (2016). Novel R-plasmid conjugal transfer inhibitory and antibacterial activities of phenolic compounds from *Mallotus philippensis* (Lam.) Mull. Arg. *J. Glob. Antimicrob. Resist.* 5, 15–21. doi: 10.1016/j.jgar.2016.01.011

Paschos, A., den Hartigh, A., Smith, M. A., Atluri, V. L., Sivanesan, D., Tsolis, R. M., et al. (2011). An *in vivo* high-throughput screening approach targeting the type IV secretion system component VirB8 identified inhibitors of *Brucella abortus* 2308 proliferation. *Infect. Immun.* 79, 1033–1043. doi: 10.1128/IAI.00993-10

Peña, A., and Arechaga, I. (2013). Molecular motors in bacterial secretion. *J. Mol. Microbiol. Biotechnol.* 23, 357–369. doi: 10.1159/000351360

Peña, A., Ripoll-Rozada, J., Zunzunegui, S., Cabezon, E., de la Cruz, F., and Arechaga, I. (2011). Autoinhibitory regulation of TrwK, an essential VirB4 ATPase in type IV secretion systems. *J. Biol. Chem.* 286, 17376–17382. doi: 10.1074/jbc.M110.208942

Planet, P. J., Kachlany, S. C., DeSalle, R., and Figurski, D. H. (2001). Phylogeny of genes for secretion NTPases: identification of the widespread tadA subfamily and development of a diagnostic key for gene classification. *Proc. Natl. Acad. Sci. U.S.A.* 98, 2503–2508. doi: 10.1073/pnas.051436598

Rice, L. B. (2009). The clinical consequences of antimicrobial resistance. *Curr. Opin. Microbiol.* 12, 476–481. doi: 10.1016/j.mib.2009.08.001

Ripoll-Rozada, J., Garcia-Cazorla, Y., Getino, M., Machon, C., Sanabria-Rios, D., de la Cruz, F., et al. (2016). Type IV traffic ATPase TrwD as molecular target to inhibit bacterial conjugation. *Mol. Microbiol.* 100, 912–921. doi: 10.1111/mmi.13359

Ripoll-Rozada, J., Pena, A., Rivas, S., Moro, F., de la Cruz, F., Cabezon, E., et al. (2012). Regulation of the type IV secretion ATPase TrwD by magnesium: implications for catalytic mechanism of the secretion ATPase superfamily. *J. Biol. Chem.* 287, 17408–17414. doi: 10.1074/jbc.M112.357905

Ripoll-Rozada, J., Zunzunegui, S., de la Cruz, F., Arechaga, I., and Cabezon, E. (2013). Functional interactions of VirB11 traffic ATPases with VirB4 and VirD4 molecular motors in type IV secretion systems. *J. Bacteriol.* 195, 4195–4201. doi: 10.1128/JB.00437-13

Savvides, S. N., Yeo, H. J., Beck, M. R., Blaesing, F., Lurz, R., Lanka, E., et al. (2003). VirB11 ATPases are dynamic hexameric assemblies: new insights into bacterial type IV secretion. *EMBO J.* 22, 1969–1980. doi: 10.1093/emboj/cdg223

Shaffer, C. L., Good, J. A. D., Kumar, S., Krishnan, K. S., Gaddy, J. A., Loh, J. T., et al. (2016). Peptidomimetic small molecules disrupt type IV secretion activity in diverse bacterial pathogens. *mBio* 7:e221-16. doi: 10.1128/mBio.00221-16

Smillie, C., Garcillan-Barcia, M. P., Francia, M. V., Rocha, E. P., and de la Cruz, F. (2010). Mobility of plasmids. *Microbiol. Mol. Biol. Rev.* 74, 434–452. doi: 10.1128/MMBR.00020-10

Smith, M. A., Coincon, M., Paschos, A., Jolicoeur, B., Lavallee, P., Sygusch, J., et al. (2012). Identification of the binding site of *Brucella* VirB8 interaction inhibitors. *Chem. Biol.* 19, 1041–1048. doi: 10.1016/j.chembiol.2012.07.007

Smith, P. A., and Romesberg, F. E. (2007). Combating bacteria and drug resistance by inhibiting mechanisms of persistence and adaptation. *Nat. Chem. Biol.* 3, 549–556. doi: 10.1038/nchembio.2007.27

Tato, I., Matilla, I., Arechaga, I., Zunzunegui, S., de la Cruz, F., and Cabezon, E. (2007). The ATPase activity of the DNA transporter TrwB is modulated by protein TrwA: implications for a common assembly mechanism of DNA translocating motors. *J. Biol. Chem.* 282, 25569–25576. doi: 10.1074/jbc.M703464200

Tato, I., Zunzunegui, S., de la Cruz, F., and Cabezón, E. (2005). TrwB, the coupling protein involved in DNA transport during bacterial conjugation, is a DNA-dependent ATPase. *Proc. Natl. Acad. Sci. U.S.A.* 102, 8156–8161. doi: 10.1073/pnas.0503402102

von Wintersdorff, C., Penders, J., van Niekerk, J. M., Mills, N. D., Majumder, S., van Alphen, L. B., et al. (2016). Dissemination of antimicrobial resistance in microbial ecosystems through horizontal gene transfer. *Front. Microbiol.* 7:173. doi: 10.3389/fmicb.2016.00173

Waters, V. L. (1999). Conjugative transfer in the dissemination of beta-lactam and aminoglycoside resistance. *Front. Biosci.* 4, D433–D456.

Wegener, H. C. (2003). Antibiotics in animal feed and their role in resistance development. *Curr. Opin. Microbiol.* 6, 439–445. doi: 10.1016/j.mib.2003.09.009

WHO (2014). *WHO: Antimicrobial Resistance. Global Report on Surveillance.* Geneva: WHO.

20

Targeting the Bacterial Orisome in the Search for New Antibiotics

Julia E. Grimwade and Alan C. Leonard*

Department of Biological Sciences, Florida Institute of Technology, Melbourne, FL, United States

There is an urgent need for new antibiotics to combat drug resistant bacteria. Existing antibiotics act on only a small number of proteins and pathways in bacterial cells, and it seems logical that expansion of the target set could lead to development of novel antimicrobial agents. One essential process, not yet exploited for antibiotic discovery, is the initiation stage of chromosome replication, mediated by the bacterial orisome. In all bacteria, orisomes assemble when the initiator protein, DnaA, as well as accessory proteins, bind to a DNA scaffold called the origin of replication (*oriC*). Orisomes perform the essential tasks of unwinding *oriC* and loading the replicative helicase, and orisome assembly is tightly regulated in the cell cycle to ensure chromosome replication begins only once. Only a few bacterial orisomes have been fully characterized, and while this lack of information complicates identification of all features that could be targeted, examination of assembly stages and orisome regulatory mechanisms may provide direction for some effective inhibitory strategies. In this perspective, we review current knowledge about orisome assembly and regulation, and identify potential targets that, when inhibited pharmacologically, would prevent bacterial chromosome replication.

Keywords: antibiotic discovery, orisome, *oriC*, DnaA, initiation of bacterial DNA replication

Edited by:
Tatiana Venkova,
Fox Chase Cancer Center,
United States

Reviewed by:
Dhruba Chattoraj,
National Institutes of Health (NIH),
United States
Anders Løbner-Olesen,
University of Copenhagen, Denmark

***Correspondence:**
Julia E. Grimwade
grimwade@fit.edu

INTRODUCTION

The increase in life-threatening infections caused by multi-drug resistant bacteria has caused an urgent need for new antibiotics. Prevalence of drug-resistant bacteria can be partly attributed to over-use of antibiotics, both clinically and agriculturally (Ventola, 2015), but antibiotic resistance is an ancient phenomenon (D'Costa et al., 2011), and selection of resistant organisms is a predictable and inevitable consequence of antibiotic use. Complicating the problem is lack of diversity in current antibiotic targets; of the approximately 200 essential genes identified in bacteria, only a handful are currently targeted (Lewis, 2013). Because recent drug discovery efforts have focused largely on modifying existing scaffolds, any new drug that acts on molecular targets in the few exploited processes risks encountering pre-selected, resistance-causing mutations (Barker, 1999). Therefore, one logical way to combat antibiotic resistance is to expand the set of targeted essential processes and proteins. One unexploited process is assembly of the orisome, the nucleoprotein complex that mediates initiation of bacterial chromosome replication, a critical event in the bacterial cell cycle (Leonard and Grimwade, 2015). In this perspective, we review orisome assembly, and address whether or not orisomes contain molecular targets that are not only novel, but which might also lead to the development of clinically useful antibiotics.

ORISOME ASSEMBLY

All bacteria must duplicate their genomes before they divide into two identical daughter cells. With a few exceptions, all bacteria share fundamental molecular machinery responsible for triggering new rounds of DNA synthesis, comprising a unique chromosomal replication origin, *oriC*, and the conserved initiator protein, DnaA, a member of the AAA+ family of ATPases. The nucleoprotein complex formed by these two components is termed the orisome, which, when fully assembled, unwinds *oriC* DNA, and recruits replicative helicase, preparing the origin for the two replisomes required to bi-directionally replicate the circular genome (Wolański et al., 2014; Leonard and Grimwade, 2015).

The model for orisome assembly (**Figure 1**) is based largely on studies using *Escherichia coli* (Leonard and Grimwade, 2011, 2015). The orisome assembles from a persistent scaffold comprising three molecules of DnaA, interacting with three high affinity recognition boxes (R1, R2, and R4) (Cassler et al., 1995). The scaffold (stage 1) establishes a conformation of *oriC* that prevents premature unwinding and allows negative regulation by the DNA-bending protein Fis (Kaur et al., 2014). This scaffold also recruits and positions additional DnaA molecules for the next assembly stage (stage 2) (Miller et al., 2009). In stage 2, the N-terminal domain of DnaA bound to the high affinity R1 or R4 sites recruits DnaA to the proximal low affinity site (R5M or C1), followed by progressive binding of DnaA to the remaining lower affinity (non-consensus) binding sites; these sites preferentially bind DnaA-ATP (McGarry et al., 2004; Rozgaja et al., 2011). In the left region of *oriC*, DNA bending, assisted by the IHF protein, brings R1 and R5M into proximity to facilitate the cooperative DnaA site filling in *oriC*'s left half (Grimwade et al., 2000). Occupation of low affinity sites is required for the final stage (stage 3), when AT-rich DNA in a DNA Unwinding Element (DUE) is unwound, and DnaA-ATP associates with the single-stranded region (Yung and Kornberg, 1989; Speck and Messer, 2001), either in the form of a compact filament, or through interactions between ssDNA and domain III of DnaA bound to the left array of sites (Duderstadt et al., 2010; Ozaki and Katayama, 2012). DnaA in the DUE then recruits the replicative helicase and the helicase loader (DnaB and DnaC, respectively, in *E. coli*) (Sutton et al., 1998; Mott et al., 2008).

The instructions for orisome assembly are carried in all bacterial *oriC*'s in the form of precisely positioned recognition sites that direct DnaA binding (Rozgaja et al., 2011). DnaA is highly conserved and the consensus DnaA recognition motif in *E. coli* (5′-TTATCCACA) is also utilized by most bacteria (Schaper and Messer, 1995; Speck et al., 1997). However, there can be significant differences in the affinity each DnaA has for recognition sequences, particularly those that diverge from consensus (Zawilak-Pawlik et al., 2005; Ozaki et al., 2006). In addition, a database (DoriC[1]) (Gao et al., 2013) of over 1000 bacterial replication origins reveals a surprising variation in the arrangement, orientation and number of consensus or near consensus DnaA recognition sites among the *oriC*s of different bacterial types. Thus, although all orisomes contain a conserved protein (DnaA) and all perform the same essential function of origin activation, there is little obvious similarity in the set of instructions used to assemble them. How this diversity influences individual assembly stages and the transitions between those stages is not yet clear, and this lack of information could hamper identification of some conserved features essential for the mechanical aspects of origin activation that could be used as targets in antibiotic screens. Studies on orisomes outside of *E. coli* are ongoing, and the reader is referred to recent reviews discussing orisome assembly in different bacterial types (Wolański et al., 2014), as well as a review that includes strategies for rapid comparative analyses of diverse orisomes (Leonard and Grimwade, 2015).

ORISOME REGULATORY MECHANISMS: A POTENTIAL GUIDE TO EFFECTIVE DRUG TARGETS?

Because more research is required before there is a unified paradigm for how orisomes trigger initiation, the best current strategy for identifying orisome targets may be to examine molecular mechanisms that regulate assembly. Logically, conserved mechanisms that inhibit orisome assembly will prevent initiation, and should provide "proof of principle" to justify targets as appropriate for pharmacological inhibition.

All orisomes are tightly regulated so that they trigger initiation of chromosome replication once, only once, and at the correct time in the cell cycle (Skarstad and Katayama, 2013). Delayed, or under-initiation leads to eventual chromosome loss, while re-initiation from the same origin can result in replication fork collapse and genome instability (Simmons et al., 2004). Like orisome assembly, regulation is best understood in *E. coli*, where two non-competing mechanisms, regulation of DnaA/*oriC* interactions, and regulation of cellular DnaA-ATP levels, predominate. Below, we review these two mechanisms and evaluate their possible utility as drug targets.

Orisome Regulation by Controlling DnaA-*oriC* Interactions

In *E. coli*, DnaA binding to *oriC* is controlled both before and immediately after initiation by mechanisms that prevent completion of orisome assembly stages 2 and 3 (Leonard and Grimwade, 2005). Before initiation, the DNA bending protein Fis helps maintain the origin in a conformation that reduces DnaA's ability to bind low affinity sites, until levels of DnaA increase enough to displace Fis from its recognition site (Ryan et al., 2004; Kaur et al., 2014). Since *E. coli oriC* contains multiple low affinity DnaA binding sites that preferentially bind DnaA-ATP (McGarry et al., 2004; Kawakami et al., 2005), orisome assembly cannot be completed until DnaA-ATP levels rise to a critical level. (Regulation of DnaA-ATP levels is discussed below.) After initiation, the SeqA protein binds hemimethylated GATC motifs in *oriC*, several of which are inside or overlap low affinity DnaA recognition sites (Lu et al., 1994; Skarstad et al., 2001). SeqA

[1] http://tubic.tju.edu.cn/doric/index.php

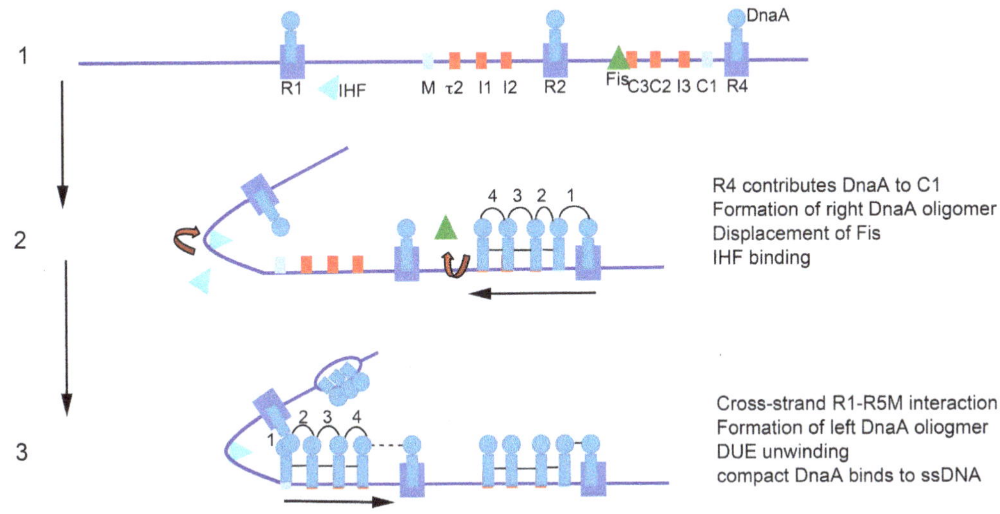

FIGURE 1 | Model of staged orisome assembly. Stage 1: Immediately after initiation of chromosome replication, DnaA rebinds to high affinity R1, R2 and R4 sites. Fis is also bound, but IHF is not. Low affinity sites are unoccupied. Stage 2: DnaA bound to R4 recruits DnaA for binding to its proximal site, and DnaA then progressively fills the remaining arrayed sites. DnaA displaces Fis, and this allows IHF to bind to its cognate site. Stage 3: The bend induced by IHF binding allows DnaA, recruited by R1, to bind to R5M, and form a cross-strand DnaA interaction. DnaA then progressively fills the sites between R5M and R2. Coincident with completion of this stage, *oriC* DNA is unwound in the DUE, and DnaA binds to the ssDNA (Figure from Leonard and Grimwade, 2015).

blocks DnaA-ATP from re-occupying low affinity sites and the DUE region for approximately one third of the cell cycle (Nievera et al., 2006).

It isn't known how many bacterial origins contain low affinity recognition sites with preference for DnaA-ATP, and not all bacteria use Fis or SeqA to regulate orisome assembly (Brézellec et al., 2006; Madiraju et al., 2006). Regardless, the basic paradigm of controlling DnaA's access to *oriC* as a way of regulating orisome assembly can be found in many bacterial types. For example, response regulators CtrA, MtrA, and HP1021 inhibit DnaA occupation of *oriC* in *Caulobacter crescentus*, *Mycobacteria tuberculosis*, and *Helicobacter pylori*, respectively, and by doing so, help prevent untimely initiations (Taylor et al., 2011; Donczew et al., 2015; Purushotham et al., 2015). *H. pylori* also uses DNA topology to regulate DnaA/*oriC* interactions (Donczew et al., 2014). In *Bacillus subtilis*, several proteins have been identified that negatively regulate initiation by inhibiting cooperative binding of DnaA at *oriC*; these include YabA (Merrikh and Grossman, 2011; Scholefield and Murray, 2013), DnaD (Bonilla and Grossman, 2012; Scholefield and Murray, 2013), and Soj (Scholefield et al., 2012). In several systems, orisome assembly is also controlled by positive regulators that increase DnaA binding to low affinity sites. In *E. coli* and *Caulobacter crescentus*, low affinity DnaA binding is stimulated by the DNA bending protein IHF (Grimwade et al., 2000; Siam et al., 2003). Additionally, the *E. coli* DiaA protein (Ishida et al., 2004), and its homolog in *H. pylori*, HobA (Natrajan et al., 2007; Zawilak-Pawlik et al., 2007, 2011), bind to DnaA's domain I and increase weak site occupation.

The studies described above suggest that several different regions of DnaA could be targeted to inhibit DnaA binding. Obviously, blocking the DNA binding domain (domain IV) should inhibit all stages of orisome formation. Although protein–DNA interactions have not traditionally been considered to be "druggable" targets, recent studies have reported success in identifying inhibitors of DNA binding (Huang et al., 2016; Grimley et al., 2017). Further, inhibition of the self-oligomerization regions of DnaA in domains I and III should prevent cooperative binding and thus assembly stages 2 and 3 (Kawakami et al., 2005; Miller et al., 2009; Duderstadt et al., 2010; Scholefield and Murray, 2013). Like protein–DNA interactions, protein–protein interactions have not traditionally been favored as drug targets, but recent reports raise optimism that targeting DnaA oligomerization could be successful (Marceau et al., 2013; Voter et al., 2017).

Several other must be resolved before inhibition of DnaA's access to *oriC* can be determined to be a practical antimicrobial strategy. First, it is not yet clear how much binding must be prevented to inhibit replication. All origins contain multiple DnaA binding sites (Leonard and Mechali, 2013), and studies that removed or inactivated DnaA recognition sites in *E. coli* chromosomal *oriC* revealed a tremendous plasticity in orisome assembly. Remarkably, deletion of the entire right region of *oriC* is tolerated in slow growing cells (Stepankiw et al., 2009). Additionally, directed mutations that knocked out binding to individual chromosomal *oriC* sites had little effect on viability (Weigel et al., 2001; Riber et al., 2009; Kaur et al., 2014). However, eliminating binding to more than one high affinity site did cause loss of viability (Kaur et al., 2014). Similar plasticity was noted in SeqA regulation of the number of occupied DnaA sites; even though loss of SeqA binding would be expected to allow DnaA re-binding at some *oriC* sites after initiation, mutating individual GATCs had little effect on initiation synchrony (Jha and Chattoraj, 2016). In *Bacillus*, some individual chromosomal *oriC* DnaA binding sites were shown to be essential, but others were not (Richardson et al., 2016). These studies, although by no means comprehensive, suggest that any pharmacological strategy should aim to inhibit DnaA binding at a majority of *oriC* sites, at least until future orisome studies reveal which sites are needed to assemble sub-complexes that carry out the essential mechanical functions of origin activation. Additionally, several studies suggest that assays used to screen for inhibitors of DnaA binding should be based on inhibiting chromosomal *oriC* rather than cloned origins, since inactivating individual sites is much more detrimental to plasmid *oriC* function (Weigel et al., 2001). Also, given the diversity in bacterial origin configurations (Leonard and Mechali, 2013), screens using a single bacterial type might not be sufficient to identify agents that act against a broad spectrum of bacteria. It might be necessary to utilize multiple types of bacteria, unless methodology is developed that allows the function of any chromosomal origin to be examined in an easily cultured strain. One strategy, involving heterologous origin transplantation, was described in a recent review (Leonard and Grimwade, 2015).

Orisome Regulation by Controlling DnaA-ATP Levels

Based on seminal studies of *in vitro E. coli* DNA replication by the Kornberg lab (Sekimizu et al., 1987), DnaA-ATP is the active initiator form, and it is widely accepted that all bacteria share the requirement for DnaA-ATP in origin activation. In *E. coli*, DnaA-ATP levels are tightly regulated during the cell cycle to ensure precise initiation timing. Prior to the initiation step, DnaA-ATP levels rise due to new synthesis and a combination of recharging systems that include the DARS loci and acidic phospholipids in the membrane, reviewed in Skarstad and Katayama (2013). After initiation, the synthesis of DnaA-ATP is repressed for 1/3 of the cell cycle by SeqA, which binds to hemi-methylated GATC motifs in the *dnaA* promoter (Campbell and Kleckner, 1990). To inactivate DnaA-ATP, DnaA's intrinsic ATPase activity is stimulated by the Hda protein associated with the β-clamp (DnaN) (Su'etsugu et al., 2004; Kim et al., 2017). Excess DnaA-ATP can also bind to a high capacity locus, termed *datA* (Kitagawa et al., 1998), which also stimulates DnaA-ATP hydrolysis (Kasho and Katayama, 2013).

The critical importance of mechanisms regulating DnaA-ATP levels in *E. coli* is demonstrated by the lethality observed in mutants, such as *dnaA*(cos) and *hda* null, that have lost the ability to inactivate DnaA-ATP by hydrolysis (Nishida et al., 2002; Felczak and Kaguni, 2009). DnaA(cos) carries two amino acid substitutions, one that prevents nucleotide binding (A184V), and another that stabilizes the mutated form (Y271H) (Simmons and Kaguni, 2003). Cells harboring *dnaA*(cos) are non-viable at 30°C, most likely due to over-initiation that results in co-directional replication fork collisions at stalled forks, leading to catastrophic double-stranded breaks (Simmons et al., 2004). A similar lethal phenotype is seen when Hda is inactivated, unless suppressor mutations arise (Riber et al., 2009; Charbon et al., 2011). Interestingly, although diverse suppressor mutations have been identified (Charbon et al., 2011), they all seem to cause tolerance of over-initiation by decreasing the chance of fork collisions, either by reducing initiation frequencies, or by preventing DNA lesions, such as oxidative DNA damage, that would slow forks (Charbon et al., 2014, 2017).

There are several aspects of DnaA inactivation mutants that are relevant to identifying antibiotic targets. First, lethality is caused by increasing, rather than decreasing the initiation frequency (Simmons et al., 2004). The run-away replication observed in DnaA(cos) mutants correlates with the inability to bind adenine nucleotide (Simmons et al., 2003), although it is not clear why loss of nucleotide binding leads to over-replication rather than orisome inactivation. Second, it is not yet known how many other bacterial types use regulation of DnaA-ATP levels as a regulatory mechanism. While some bacteria, such as *Caulobacter* and most enterobacteria, appear to have homologs of *hda* (Wargachuk and Marczynski, 2015), others, such as *Bacillus*, *Staphylococcus*, and *H. pylori*, do not (Katayama et al., 2010). DnaA in *B. subtilis* and *S. aureus* exchange bound ADP for ADP much more rapidly than *E. coli* does (Kurokawa et al., 2009; Bonilla and Grossman, 2012), and negative regulation of orisomes in these bacteria is focused mainly on DnaA-DNA interactions. Thus, screens to identify stimulators of DnaA hydrolysis may be ineffective in identifying broad-spectrum antimicrobials. In contrast, the growth inhibition/lethality caused when DnaA can't hydrolyze ATP suggests that identification of inhibitors of ATP binding or ATPase activity, causing lethality by over-initiation, may be more fruitful. While targeting of the ATPase of AAA+ proteins is still in its infancy there are reports of successful inhibition of this protein class (Chou et al., 2011; Firestone et al., 2012). Targeting of DnaA's ATPase, however, could generate suppressor mutations that reduce fork collisions (Charbon et al., 2017) and thus be prone to rapid resistance development. Possibly, this could be resolved by combination with an agent that inhibits DNA repair to counteract the actions of suppressor mutations (Simmons et al., 2004; Sutera and Lovett, 2006).

ADDITIONAL CONSIDERATIONS IN TARGETING ORISOME FUNCTION

Obviously, any antibiotic acting on the orisome must enter the bacterial cell. This presents a problem with all bacteria, but particularly Gram negative bacteria, where the relatively impermeable outer membrane presents a potential barrier to drug delivery (Lewis, 2013; Brown, 2016). Until more is known about transport across the outer membrane, successful platforms to discover drugs affecting orisomes or any other intracellular target are likely to require living cells to augment or replace *in vitro* biochemical assays. While screen development is beyond the scope of this Perspective, we note that one cell-based assay, to identify agents that allow *dnaA*(cos) cells to grow at non-permissive temperature, has been described (Fossum et al., 2008), but failed to identify any small molecule inhibitors of DnaA function in a limited trial screen, although it is possible that lead compounds could be identified by screening a much larger library.

Of greater concern is generation of intra- or extra-genic suppressors, particularly if a new drug causes over-replication. Unfortunately, bacteria are adept in their ability to survive initiation perturbation. In cases where rapid development of resistance is expected, hybrid antibiotics or combination chemotherapy, where orisome inhibitors are combined with drugs that act on different pathways, should be considered. Alternatively, it might be useful to target features within DnaA that are shared by other proteins, since the majority of currently used successful antibiotics delay resistance development by attacking more than one target (Silver, 2011; Brown and Wright, 2016). One possible shared motif is the AAA+ domain, since the AAA+ domain of DnaC is quite similar to that of DnaA (Mott et al., 2008). Interestingly, hydrolysis of the ATP bound to DnaC is required before DnaB helicase can function (Mott et al., 2008), and it may be possible to identify inhibitors of DnaA's intrinsic ATPase that also inhibit DnaB activation.

It is interesting that no natural product that inhibits orisome function has been identified in many years of antibiotic screening. This may be because the assays are not designed to identify drugs inhibiting this essential process, or that targeting the orisome is an inherently risky competition strategy for any bacteria, and so it rarely evolves. Regardless, the orisome appears to have potential as a novel and effective drug target, and its usefulness in antibiotic discovery should increase as more studies reveal conserved and non-conserved features of orisome assembly among bacterial types.

AUTHOR CONTRIBUTIONS

All authors listed have made a substantial, direct and intellectual contribution to the work, and approved it for publication.

ACKNOWLEDGMENT

The work in our laboratories was supported by Public Health Service grant GM54042. Publication of this article was funded in part by the Open Access Subvention Fund and the Florida Tech Libraries.

REFERENCES

Barker, K. F. (1999). Antibiotic resistance: a current perspective. *Br. J. Clin. Pharmacol.* 48, 109–124. doi: 10.1046/j.1365-2125.1999.00997.x

Bonilla, C. Y., and Grossman, A. D. (2012). The primosomal protein DnaD inhibits cooperative DNA binding by the replication initiator DnaA in *Bacillus subtilis*. *J. Bacteriol.* 194, 5110–5117. doi: 10.1128/JB.00958-12

Brézellec, P., Hoebeke, M., Hiet, M. S., Pasek, S., and Ferat, J. L. (2006). DomainSieve: a protein domain-based screen that led to the identification of dam-associated genes with potential link to DNA maintenance. *Bioinformatics* 22, 1935–1941. doi: 10.1093/bioinformatics/btl336

Brown, D. G. (2016). Drug discovery strategies to outer membrane targets in Gram-negative pathogens. *Bioorg. Med. Chem.* 24, 6320–6331. doi: 10.1016/j.bmc.2016.05.004

Brown, E. D., and Wright, G. D. (2016). Antibacterial drug discovery in the resistance era. *Nature* 529, 336–343. doi: 10.1038/nature17042

Campbell, J. L., and Kleckner, N. (1990). *E. coli* oriC and the dnaA gene promoter are sequestered from dam methyltransferase following the passage of the chromosomal replication fork. *Cell* 62, 967–979. doi: 10.1016/0092-8674(90)90271-F

Cassler, M. R., Grimwade, J. E., and Leonard, A. C. (1995). Cell cycle-specific changes in nucleoprotein complexes at a chromosomal replication origin. *EMBO J.* 14, 5833–5841.

Charbon, G., Bjørn, L., Mendoza-Chamizo, B., Frimodt-Møller, J., and Løbner-Olesen, A. (2014). Oxidative DNA damage is instrumental in hyperreplication stress-induced inviability of *Escherichia coli*. *Nucleic Acids Res.* 42, 13228–13241. doi: 10.1093/nar/gku1149

Charbon, G., Riber, L., Cohen, M., Skovgaard, O., Fujimitsu, K., Katayama, T., et al. (2011). Suppressors of DnaA(ATP) imposed overinitiation in *Escherichia coli*. *Mol. Microbiol.* 79, 914–928. doi: 10.1111/j.1365-2958.2010.07493.x

Charbon, G., Riber, L., and Løbner-Olesen, A. (2017). Countermeasures to survive excessive chromosome replication in *Escherichia coli*. *Curr. Genet.* doi: 10.1007/s00294-017-0725-4 [Epub ahead of print].

Chou, T. F., Brown, S. J., Minond, D., Nordin, B. E., Li, K., Jones, A. C., et al. (2011). Reversible inhibitor of p97, DBeQ, impairs both ubiquitin-dependent and autophagic protein clearance pathways. *Proc. Natl. Acad. Sci. U.S.A.* 108, 4834–4839. doi: 10.1073/pnas.1015312108

D'Costa, V. M., King, C. E., Kalan, L., Morar, M., Sung, W. W., Schwarz, C., et al. (2011). Antibiotic resistance is ancient. *Nature* 477, 457–461. doi: 10.1038/nature10388

Donczew, R., Makowski, Ł., Jaworski, P., Bezulska, M., Nowaczyk, M., Zakrzewska-Czerwińska, J., et al. (2015). The atypical response regulator HP1021 controls formation of the *Helicobacter pylori* replication initiation complex. *Mol. Microbiol.* 95, 297–312. doi: 10.1111/mmi.12866

Donczew, R., Mielke, T., Jaworski, P., Zakrzewska-Czerwińska, J., and Zawilak-Pawlik, A. (2014). Assembly of *Helicobacter pylori* initiation complex is determined by sequence-specific and topology-sensitive DnaA-oriC interactions. *J. Mol. Biol.* 426, 2769–2782. doi: 10.1016/j.jmb.2014.05.018

Duderstadt, K. E., Mott, M. L., Crisona, N. J., Chuang, K., Yang, H., and Berger, J. M. (2010). Origin remodeling and opening in bacteria rely on distinct assembly states of the DnaA initiator. *J. Biol. Chem.* 285, 28229–28239. doi: 10.1074/jbc.M110.147975

Felczak, M. M., and Kaguni, J. M. (2009). DnaAcos hyperinitiates by circumventing regulatory pathways that control the frequency of initiation in *Escherichia coli*. *Mol. Microbiol.* 72, 1348–1363. doi: 10.1111/j.1365-2958.2009.06724.x

Firestone, A. J., Weinger, J. S., Maldonado, M., Barlan, K., Langston, L. D., O'Donnell, M., et al. (2012). Small-molecule inhibitors of the AAA+ ATPase motor cytoplasmic dynein. *Nature* 484, 125–129. doi: 10.1038/nature10936

Fossum, S., De Pascale, G., Weigel, C., Messer, W., Donadio, S., and Skarstad, K. (2008). A robust screen for novel antibiotics: specific knockout of the initiator of bacterial DNA replication. *FEMS Microbiol. Lett.* 281, 210–214. doi: 10.1111/j.1574-6968.2008.01103.x

Gao, F., Luo, H., and Zhang, C. T. (2013). DoriC 5.0: an updated database of oriC regions in both bacterial and archaeal genomes. *Nucleic Acids Res.* 41, D90–D93. doi: 10.1093/nar/gks990

Grimley, E., Liao, C., Ranghini, E. J., Nikolovska-Coleska, Z., and Dressler, G. R. (2017). Inhibition of Pax2 transcription activation with a small molecule that targets the DNA binding domain. *ACS Chem. Biol.* 12, 724–734. doi: 10.1021/acschembio.6b00782

Grimwade, J. E., Ryan, V. T., and Leonard, A. C. (2000). IHF redistributes bound initiator protein, DnaA, on supercoiled oriC of *Escherichia coli*. *Mol. Microbiol.* 35, 835–844. doi: 10.1046/j.1365-2958.2000.01755.x

Huang, W., Dong, Z., Chen, Y., Wang, F., Wang, C. J., Peng, H., et al. (2016). Small-molecule inhibitors targeting the DNA-binding domain of STAT3 suppress tumor growth, metastasis and STAT3 target gene expression in vivo. *Oncogene* 35, 783–792. doi: 10.1038/onc.2015.215

Ishida, T., Akimitsu, N., Kashioka, T., Hatano, M., Kubota, T., Ogata, Y., et al. (2004). DiaA, a novel DnaA-binding protein, ensures the timely initiation of *Escherichia coli* chromosome replication. *J. Biol. Chem.* 279, 45546–45555. doi: 10.1074/jbc.M402762200

Jha, J. K., and Chattoraj, D. K. (2016). Inactivation of individual SeqA binding sites of the *E. coli* origin reveals robustness of replication initiation synchrony. *PLOS ONE* 11:e0166722. doi: 10.1371/journal.pone.0166722

Kasho, K., and Katayama, T. (2013). DnaA binding locus datA promotes DnaA-ATP hydrolysis to enable cell cycle-coordinated replication initiation. *Proc. Natl. Acad. Sci. U.S.A.* 110, 936–941. doi: 10.1073/pnas.1212070110

Katayama, T., Ozaki, S., Keyamura, K., and Fujimitsu, K. (2010). Regulation of the replication cycle: conserved and diverse regulatory systems for DnaA and oriC. *Nat. Rev. Microbiol.* 8, 163–170. doi: 10.1038/nrmicro2314

Kaur, G., Vora, M. P., Czerwonka, C. A., Rozgaja, T. A., Grimwade, J. E., and Leonard, A. C. (2014). Building the bacterial orisome: high-affinity DnaA recognition plays a role in setting the conformation of oriC DNA. *Mol. Microbiol.* 91, 1148–1163. doi: 10.1111/mmi.12525

Kawakami, H., Keyamura, K., and Katayama, T. (2005). Formation of an ATP-DnaA-specific initiation complex requires DnaA Arginine 285, a conserved motif in the AAA+ protein family. *J. Biol. Chem.* 280, 27420–27430. doi: 10.1074/jbc.M502764200

Kim, J. S., Nanfara, M. T., Chodavarapu, S., Jin, K. S., Babu, V. M. P., Ghazy, M. A., et al. (2017). Dynamic assembly of Hda and the sliding clamp in the regulation of replication licensing. *Nucleic Acids Res.* 45, 3888–3905. doi: 10.1093/nar/gkx081

Kitagawa, R., Ozaki, T., Moriya, S., and Ogawa, T. (1998). Negative control of replication initiation by a novel chromosomal locus exhibiting exceptional affinity for *Escherichia coli* DnaA protein. *Genes Dev.* 12, 3032–3043. doi: 10.1101/gad.12.19.3032

Kurokawa, K., Mizumura, H., Takaki, T., Ishii, Y., Ichihashi, N., Lee, B. L., et al. (2009). Rapid exchange of bound ADP on the *Staphylococcus aureus* replication initiation protein DnaA. *J. Biol. Chem.* 284, 34201–34210. doi: 10.1074/jbc.M109.060681

Leonard, A. C., and Grimwade, J. E. (2005). Building a bacterial orisome: emergence of new regulatory features for replication origin unwinding. *Mol. Microbiol.* 55, 978–985. doi: 10.1111/j.1365-2958.2004.04467.x

Leonard, A. C., and Grimwade, J. E. (2011). Regulation of DnaA assembly and activity: taking directions from the genome. *Annu. Rev. Microbiol.* 65, 19–35. doi: 10.1146/annurev-micro-090110-102934

Leonard, A. C., and Grimwade, J. E. (2015). The orisome: structure and function. *Front. Microbiol.* 6:545. doi: 10.3389/fmicb.2015.00545

Leonard, A. C., and Mechali, M. (2013). DNA replication origins. *Cold Spring Harb. Perspect. Biol.* 5:a010116. doi: 10.1101/cshperspect.a010116

Lewis, K. (2013). Platforms for antibiotic discovery. *Nat. Rev. Drug Discov.* 12, 371–387. doi: 10.1038/nrd3975

Lu, M., Campbell, J. L., Boye, E., and Kleckner, N. (1994). SeqA: a negative modulator of replication initiation in *E. coli*. *Cell* 77, 413–426. doi: /10.1016/0092-8674(94)90156-2

Madiraju, M. V., Moomey, M., Neuenschwander, P. F., Muniruzzaman, S., Yamamoto, K., Grimwade, J. E., et al. (2006). The intrinsic ATPase activity of *Mycobacterium tuberculosis* DnaA promotes rapid oligomerization of DnaA on oriC. *Mol. Microbiol.* 59, 1876–1890. doi: 10.1111/j.1365-2958.2006.05068.x

Marceau, A. H., Bernstein, D. A., Walsh, B. W., Shapiro, W., Simmons, L. A., and Keck, J. L. (2013). Protein interactions in genome maintenance as novel antibacterial targets. *PLOS ONE* 8:e58765. doi: 10.1371/journal.pone.0058765

McGarry, K. C., Ryan, V. T., Grimwade, J. E., and Leonard, A. C. (2004). Two discriminatory binding sites in the *Escherichia coli* replication origin are required for DNA strand opening by initiator DnaA-ATP. *Proc. Natl. Acad. Sci. U.S.A.* 101, 2811–2816. doi: 10.1073/pnas.0400340101

Merrikh, H., and Grossman, A. D. (2011). Control of the replication initiator DnaA by an anti-cooperativity factor. *Mol. Microbiol.* 82, 434–446. doi: 10.1111/j.1365-2958.2011.07821.x

Miller, D. T., Grimwade, J. E., Betteridge, T., Rozgaja, T., Torgue, J. J., and Leonard, A. C. (2009). Bacterial origin recognition complexes direct assembly of higher-order DnaA oligomeric structures. *Proc. Natl. Acad. Sci. U.S.A.* 106, 18479–18484. doi: 10.1073/pnas.0909472106

Mott, M. L., Erzberger, J. P., Coons, M. M., and Berger, J. M. (2008). Structural synergy and molecular crosstalk between bacterial helicase loaders and replication initiators. *Cell* 135, 623–634. doi: 10.1016/j.cell.2008.09.058

Natrajan, G., Hall, D. R., Thompson, A. C., Gutsche, I., and Terradot, L. (2007). Structural similarity between the DnaA-binding proteins HobA (HP1230) from *Helicobacter pylori* and DiaA from *Escherichia coli*. *Mol. Microbiol.* 65, 995–1005. doi: 10.1111/j.1365-2958.2007.05843.x

Nievera, C., Tor, J. J., Grimwade, J. E., and Leonard, A. C. (2006). SeqA blocking of DnaA-oriC interactions ensures staged assembly of the *E. coli* pre-RC. *Mol. Cell* 24, 581–592. doi: 10.1016/j.molcel.2006.09.016

Nishida, S., Fujimitsu, K., Sekimizu, K., Ohmura, T., Ueda, T., and Katayama, T. (2002). A nucleotide switch in the *Escherichia coli* DnaA protein initiates chromosomal replication: evidence from a mutant DnaA protein defective in regulatory ATP hydrolysis in vitro and in vivo. *J. Biol. Chem.* 277, 14986–14995. doi: 10.1074/jbc.M108303200

Ozaki, S., Fujimitsu, K., Kurumizaka, H., and Katayama, T. (2006). The DnaA homolog of the hyperthermophilic eubacterium *Thermotoga maritima* forms an open complex with a minimal 149-bp origin region in an ATP-dependent manner. *Genes Cells* 11, 425–438. doi: 10.1111/j.1365-2443.2006.00950.x

Ozaki, S., and Katayama, T. (2012). Highly organized DnaA-oriC complexes recruit the single-stranded DNA for replication initiation. *Nucleic Acids Res.* 40, 1648–1665. doi: 10.1093/nar/gkr832

Purushotham, G., Sarva, K. B., Blaszczyk, E., Rajagopalan, M., and Madiraju, M. V. (2015). *Mycobacterium tuberculosis* oriC sequestration by MtrA response regulator. *Mol. Microbiol.* 98, 586–604. doi: 10.1111/mmi.13144

Riber, L., Fujimitsu, K., Katayama, T., and Løbner-Olesen, A. (2009). Loss of Hda activity stimulates replication initiation from I-box, but not R4 mutant origins in *Escherichia coli*. *Mol. Microbiol.* 71, 107–122. doi: 10.1111/j.1365-2958.2008.06516.x

Richardson, T. T., Harran, O., and Murray, H. (2016). The bacterial DnaA-trio replication origin element specifies single-stranded DNA initiator binding. *Nature* 534, 412–416. doi: 10.1038/nature17962

Rozgaja, T. A., Grimwade, J. E., Iqbal, M., Czerwonka, C., Vora, M., and Leonard, A. C. (2011). Two oppositely oriented arrays of low-affinity recognition sites in oriC guide progressive binding of DnaA during *Escherichia coli* pre-RC assembly. *Mol. Microbiol.* 82, 475–488. doi: 10.1111/j.1365-2958.2011.07827.x

Ryan, V. T., Grimwade, J. E., Camara, J. E., Crooke, E., and Leonard, A. C. (2004). *Escherichia coli* prereplication complex assembly is regulated by dynamic interplay among Fis. IHF and DnaA. *Mol. Microbiol.* 51, 1347–1359. doi: 10.1046/j.1365-2958.2003.03906.x

Schaper, S., and Messer, W. (1995). Interaction of the initiator protein DnaA of *Escherichia coli* with its DNA target. *J. Biol. Chem.* 270, 17622–17626. doi: 10.1074/jbc.270.29.17622

Scholefield, G., Errington, J., and Murray, H. (2012). Soj/ParA stalls DNA replication by inhibiting helix formation of the initiator protein DnaA. *EMBO J.* 31, 1542–1555. doi: 10.1038/emboj.2012.6

Scholefield, G., and Murray, H. (2013). YabA and DnaD inhibit helix assembly of the DNA replication initiation protein DnaA. *Mol. Microbiol.* 90, 147–159. doi: 10.1111/mmi.12353

Sekimizu, K., Bramhill, D., and Kornberg, A. (1987). ATP activates dnaA protein in initiating replication of plasmids bearing the origin of the *E. coli* chromosome. *Cell* 50, 259–265. doi: 10.1016/0092-8674(87)90221-2

Siam, R., Brassinga, A. K., and Marczynski, G. T. (2003). A dual binding site for integration host factor and the response regulator CtrA inside the *Caulobacter crescentus* replication origin. *J. Bacteriol.* 185, 5563–5572. doi: 10.1128/JB.185.18.5563-5572.2003

Silver, L. L. (2011). Challenges of antibacterial discovery. *Clin. Microbiol. Rev.* 24, 71–109. doi: 10.1128/CMR.00030-10

Simmons, L. A., Breier, A. M., Cozzarelli, N. R., and Kaguni, J. M. (2004). Hyperinitiation of DNA replication in *Escherichia coli* leads to replication fork collapse and inviability. *Mol. Microbiol.* 51, 349–358. doi: 10.1046/j.1365-2958.2003.03842.x

Simmons, L. A., Felczak, M., and Kaguni, J. M. (2003). DnaA Protein of *Escherichia coli*: oligomerization at the *E. coli* chromosomal origin is required for initiation and involves specific N-terminal amino acids. *Mol. Microbiol.* 49, 849–858. doi: 10.1046/j.1365-2958.2003.03603.x

Simmons, L. A., and Kaguni, J. M. (2003). The DnaAcos allele of *Escherichia coli*: hyperactive initiation is caused by substitution of A184V and Y271H, resulting in defective ATP binding and aberrant DNA replication control. *Mol. Microbiol.* 47, 755–765. doi: 10.1046/j.1365-2958.2003.03333.x

Skarstad, K., and Katayama, T. (2013). Regulating DNA replication in bacteria. *Cold Spring Harb. Perspect. Biol.* 5:a012922. doi: 10.1101/cshperspect.a012922

Skarstad, K., Torheim, N., Wold, S., Lurz, R., Messer, W., Fossum, S., et al. (2001). The *Escherichia coli* SeqA protein binds specifically to two sites in fully and hemimethylated oriC and has the capacity to inhibit DNA replication and affect chromosome topology. *Biochimie* 83, 49–51. doi: 10.1016/S0300-9084(00)01207-4

Speck, C., and Messer, W. (2001). Mechanism of origin unwinding: sequential binding of DnaA to double- and single-stranded DNA. *EMBO J.* 20, 1469–1476. doi: 10.1093/emboj/20.6.1469

Speck, C., Weigel, C., and Messer, W. (1997). From footprint to toeprint: a close-up of the DnaA box, the binding site for the bacterial initiator protein DnaA. *Nucleic Acids Res.* 25, 3242–3247. doi: 10.1093/nar/25.16.3242

Stepankiw, N., Kaidow, A., Boye, E., and Bates, D. (2009). The right half of the *Escherichia coli* replication origin is not essential for viability, but facilitates multi-forked replication. *Mol. Microbiol.* 74, 467–479. doi: 10.1111/j.1365-2958.2009.06877.x

Su'etsugu, M., Takata, M., Kubota, T., Matsuda, Y., and Katayama, T. (2004). Molecular mechanism of DNA replication-coupled inactivation of the initiator protein in *Escherichia coli*: interaction of DnaA with the sliding clamp-loaded DNA and the sliding clamp-Hda complex. *Genes Cells* 9, 509–522. doi: 10.1111/j.1356-9597.2004.00741.x

Sutera, V. A., and Lovett, S. T. (2006). The role of replication initiation control in promoting survival of replication fork damage. *Mol. Microbiol.* 60, 229–239. doi: 10.1111/j.1365-2958.2006.05093.x

Sutton, M. D., Carr, K. M., Vicente, M., and Kaguni, J. M. (1998). *Escherichia coli* DnaA protein. The N-terminal domain and loading of DnaB helicase at the *E. coli* chromosomal origin. *J. Biol. Chem.* 273, 34255–34262. doi: 10.1074/jbc.273.51.34255

Taylor, J. A., Ouimet, M. C., Wargachuk, R., and Marczynski, G. T. (2011). The *Caulobacter crescentus* chromosome replication origin evolved two classes of weak DnaA binding sites. *Mol. Microbiol.* 82, 312–326. doi: 10.1111/j.1365-2958.2011.07785.x

Ventola, C. L. (2015). The antibiotic resistance crisis: part 1: causes and threats. *P T* 40, 277–283.

Voter, A. F., Killoran, M. P., Ananiev, G. E., Wildman, S. A., Hoffmann, F. M., and Keck, J. L. (2017). A high-throughput screening strategy to identify inhibitors of SSB protein-protein interactions in an academic screening facility. *SLAS Discov.* doi: 10.1177/2472555217712001 [Epub ahead of print].

Wargachuk, R., and Marczynski, G. T. (2015). The *Caulobacter crescentus* homolog of DnaA (HdaA) also regulates the proteolysis of the replication initiator protein DnaA. *J. Bacteriol.* 197, 3521–3532. doi: 10.1128/JB.00460-15

Weigel, C., Messer, W., Preiss, S., Welzeck, M., Morigen, and Boye, E. (2001). The sequence requirements for a functional *Escherichia coli* replication origin are different for the chromosome and a minichromosome. *Mol. Microbiol.* 40, 498–507. doi: 10.1046/j.1365-2958.2001.02409.x

Wolański, M., Donczew, R., Zawilak-Pawlik, A., and Zakrzewska-Czerwińska, J. (2014). *oriC*-encoded instructions for the initiation of bacterial chromosome replication. *Front. Microbiol.* 5:735. doi: 10.3389/fmicb.2014.00735

Yung, B. Y., and Kornberg, A. (1989). The dnaA initiator protein binds separate domains in the replication origin of *Escherichia coli*. *J. Biol. Chem.* 264, 6146–6150.

Zawilak-Pawlik, A., Donczew, R., Szafrański, S., Mackiewicz, P., Terradot, L., and Zakrzewska-Czerwińska, J. (2011). DiaA/HobA and DnaA: a pair of proteins co-evolved to cooperate during bacterial orisome assembly. *J. Mol. Biol.* 408, 238–251. doi: 10.1016/j.jmb.2011.02.045

Zawilak-Pawlik, A., Kois, A., Majka, J., Jakimowicz, D., Smulczyk-Krawczyszyn, A., Messer, W., et al. (2005). Architecture of bacterial replication initiation complexes: orisomes from four unrelated bacteria. *Biochem. J.* 389, 471–481. doi: 10.1042/BJ20050143

Zawilak-Pawlik, A., Kois, A., Stingl, K., Boneca, I. G., Skrobuk, P., Piotr, J., et al. (2007). HobA–a novel protein involved in initiation of chromosomal replication in *Helicobacter pylori*. *Mol. Microbiol.* 65, 979–994. doi: 10.1111/j.1365-2958.2007.05853.x

Permissions

All chapters in this book were first published in TGTBTUMRBHL, by Frontiers; hereby published with permission under the Creative Commons Attribution License or equivalent. Every chapter published in this book has been scrutinized by our experts. Their significance has been extensively debated. The topics covered herein carry significant findings which will fuel the growth of the discipline. They may even be implemented as practical applications or may be referred to as a beginning point for another development.

The contributors of this book come from diverse backgrounds, making this book a truly international effort. This book will bring forth new frontiers with its revolutionizing research information and detailed analysis of the nascent developments around the world.

We would like to thank all the contributing authors for lending their expertise to make the book truly unique. They have played a crucial role in the development of this book. Without their invaluable contributions this book wouldn't have been possible. They have made vital efforts to compile up to date information on the varied aspects of this subject to make this book a valuable addition to the collection of many professionals and students.

This book was conceptualized with the vision of imparting up-to-date information and advanced data in this field. To ensure the same, a matchless editorial board was set up. Every individual on the board went through rigorous rounds of assessment to prove their worth. After which they invested a large part of their time researching and compiling the most relevant data for our readers.

The editorial board has been involved in producing this book since its inception. They have spent rigorous hours researching and exploring the diverse topics which have resulted in the successful publishing of this book. They have passed on their knowledge of decades through this book. To expedite this challenging task, the publisher supported the team at every step. A small team of assistant editors was also appointed to further simplify the editing procedure and attain best results for the readers.

Apart from the editorial board, the designing team has also invested a significant amount of their time in understanding the subject and creating the most relevant covers. They scrutinized every image to scout for the most suitable representation of the subject and create an appropriate cover for the book.

The publishing team has been an ardent support to the editorial, designing and production team. Their endless efforts to recruit the best for this project, has resulted in the accomplishment of this book. They are a veteran in the field of academics and their pool of knowledge is as vast as their experience in printing. Their expertise and guidance has proved useful at every step. Their uncompromising quality standards have made this book an exceptional effort. Their encouragement from time to time has been an inspiration for everyone.

The publisher and the editorial board hope that this book will prove to be a valuable piece of knowledge for researchers, students, practitioners and scholars across the globe.

List of Contributors

Martín Masner
Department of Biotechnology, Instituto de Higiene, Facultad de Medicina, Universidad de la República, Montevideo, Uruguay

Fernando A. Ferreira and M. Gabriela Kramer
Department of Biotechnology, Instituto de Higiene, Facultad de Medicina, Universidad de la República, Montevideo, Uruguay
Laboratory of Carbohydrates and Glycoconjugates, Department of Organic Chemistry, Facultad de Química,Universidad de la República, Montevideo, Uruguay

Robert M. Hoffman
AntiCancer, Inc., San Diego, CA, United States
Department of Surgery, University of California, San Diego, San Diego, CA, United States

Sandra Águila-Arcos, Itxaso Álvarez-Rodríguez, Olatz Garaiyurrebaso and Itziar Alkorta
Instituto Biofisika (UPV/EHU, CSIC), Department of Biochemistry and Molecular Biology, University of the Basque Country, Bilbao, Spain

Carlos Garbisu,
Department of Conservation of Natural Resources, Soil Microbial Ecology Group, NEIKER-Tecnalia, Derio, Spain

Elisabeth Grohmann
Life Sciences and Technology, Beuth University of Applied Sciences, Berlin, Germany

Saskia Neuert, Claire Jenkins, Gauri Godbole and Timothy J. Dallman
National Institute for Health Research Health Protection Research Unit in Gastrointestinal Infections, University of Liverpool, Liverpool, United Kingdom
Bacteriology Reference Department, National Infection Service, Public Health England, London, United Kindom

Satheesh Nair, Martin R. Day, Michel Doumith, Philip M. Ashton, Katie L. Hopkins, Neil Woodford and Elizabeth de Pinna
Bacteriology Reference Department, National Infection Service, Public Health England, London, United Kindom

Kate C. Mellor
Department of Pathobiology and Population Sciences, Royal Veterinary College, London, United Kingdom, London School of Hygiene & Tropical Medicine, London, United Kingdom

Marta Herranz, Laura Pérez-Lago and Darío García-de-Viedma, Paloma Gijón and María Jesús Ruiz Serrano
Servicio Microbiología Clínica y Enfermedades Infecciosas, Hospital General Universitario Gregorio Marañón, Madrid, Spain
Instituto de Investigación Sanitaria Gregorio Marañón, Madrid, Spain
CIBER Enfermedades Respiratorias (CIBERES), Madrid, Spain

Ilva Pole
Childhood Tuberculosis Department, Centre of Tuberculosis and Lung Diseases, Riga East University Hospital, Riga, Latvia
Latvian Biomedical Research and Study Centre, Riga, Latvia

Iveta Ozere
Childhood Tuberculosis Department, Centre of Tuberculosis and Lung Diseases, Riga East University Hospital, Riga, Latvia
Department of Infectology and Dermatology, Riga Stradinš University, Riga, Latvia

Álvaro Chiner-Oms
Unidad Mixta Genómica y Salud, Centro Superior de Investigación en Salud Pública (FISABIO)-Universitat de València, Valencia, Spain

Miguel Martínez-Lirola
Servicio de Microbiología, Complejo Hospitalario Torrecárdenas, Almería, Spain

Laura Clotet Romero
Servei de Vigilància Epidemiològica i Resposta a Emergències de Salut Pública al Vallès Occidental i Vallès Oriental, Subdirecció General de Vigilància i Resposta a Emergències de Salut Pública, Agència de Salut Pública de Catalunya, Barcelona, Spain

Oscar Cuevas
Servicio de Laboratorio, Institut d'Investigació i Innovació Parc Taulí, I3PT Parc Taulí Hospital Universitari, Universitat Autònoma de Barcelona, Barcelona, Spain

Iñaki Comas
Instituto de Biomedicina de Valencia, Consejo Superior de Investigaciones Científicas, Valencia, Spain
CIBER en Epidemiología y Salud Pública, Madrid, Spain

Emilio Bouza and Felipe Pérez-García
Servicio Microbiología Clínica y Enfermedades Infecciosas, Hospital General Universitario Gregorio Marañón, Madrid, Spain
Instituto de Investigación Sanitaria Gregorio Marañón, Madrid, Spain
CIBER Enfermedades Respiratorias (CIBERES), Madrid, Spain
Departamento de Medicina, Facultad de Medicina, Universidad Complutense de Madrid, Madrid, Spain

Stephen M. Kwong and Neville Firth
School of Life and Environmental Sciences, University of Sydney, Sydney, NSW, Australia

Joshua P. Ramsay
School of Biomedical Sciences, Curtin University, Perth, WA, Australia

Slade O. Jensen
Antimicrobial Resistance and Mobile Elements Group, Ingham Institute for Applied Medical Research, Sydney, NSW, Australia

Richard R. Rodrigues, Xiaoxi Dong, Karen N. DSouza, Andrey Morgun
Department of Pharmaceutical Sciences, Oregon State University, Corvallis, OR, United States

Renee L. Greer, Manoj Gurung, Jia Y. Wu and Natalia Shulzhenko
Department of Biomedical Sciences, Oregon State University, Corvallis, OR, United States

Manuel Ares-Arroyo, Cristina Bernabe-Balas, Alfonso Santos-Lopez, Dolores Cid, Carmen Martin-Espada and Bruno Gonzalez-Zorn
Departamento de Sanidad Animal and Centro de Vigilancia Sanitaria Veterinaria (VISAVET), Facultad de Veterinaria, Universidad Complutense de Madrid, Madrid, Spain

Maria R. Baquero
Departamento de Microbiología, Facultad de Veterinaria, Universidad Alfonso X el Sabio, Madrid, Spain

Kashi N. Prasad
Department of Microbiology, Sanjay Gandhi Postgraduate Institute of Medical Sciences, Lucknow, India

Alvaro San Millan
Servicio de Microbiología Hospital Universitario Ramón y Cajal, Instituto de Investigación Sanitaria (IRYCIS), Madrid, Spain

María López-Sanz, Alvaro Quevedo-Olmos and Silvia Ayora
Department of Microbial Biotechnology, Centro Nacional de Biotecnología, Consejo Superior de Investigaciones Científicas, Madrid, Spain

Ana Valero-Rello
Department of Microbial Biotechnology, Centro Nacional de Biotecnología, Consejo Superior de Investigaciones Científicas, Madrid, Spain
Micalis Institute, INRA, AgroParisTech, Universite Paris-Saclay, Jouy-en-Josas, France

Alexei Sorokin
Micalis Institute, INRA, AgroParisTech, Universite Paris-Saclay, Jouy-en-Josas, France

Martina Pasqua, Maria Letizia Di Martino, Milena Grossi, Bianca Colonna and Gianni Prosseda
Istituto Pasteur Italia, Department of Biology and Biotechnology "C. Darwin", Sapienza Università di Roma, Rome, Italy

Valeria Michelacci, Rosangela Tozzoli and Stefano Morabito
European Union Reference Laboratory for Escherichia coli, Department of Veterinary Public Health and Food Safety, Istituto Superiore di Sanità, Rome, Italy

Illiassou Hamidou Soumana
Department of Infectious Diseases, University of Georgia, Athens, GA, USA
Center for Vaccines and Immunology, University of Georgia, Athens, GA, USA

Bodo Linz
Center for Vaccines and Immunology, University of Georgia, Athens, GA, USA
Department of Veterinary and Biomedical Sciences, Pennsylvania State University,University Park, PA, USA

Eric T. Harvill
Department of Infectious Diseases, University of Georgia, Athens, GA, USA
Center for Vaccines and Immunology, University of Georgia, Athens, GA, USA
Department of Veterinary and Biomedical Sciences, Pennsylvania State University,University Park, PA, USA

Darren J. Wight
Institute of Virology, Free University of Berlin, Berlin, Germany

Yurena Navarro
Microbiología Clínica y Enfermedades Infecciosas, Hospital General Universitario Gregorio Marañón, Madrid, Spain
Instituto de Investigación Sanitaria Gregorio Marañón, Madrid, Spain
CIBER Enfermedades Respiratorias, CIBERES,Madrid, Spain
CEI Campus Moncloa, UCM-UPM, Madrid, Spain
Centro de Vigilancia Sanitaria Veterinaria, Universidad Complutense Madrid, Madrid, Spain

Laura Pérez-Lago and Marta Herranz
Microbiología Clínica y Enfermedades Infecciosas, Hospital General Universitario Gregorio Marañón, Madrid, Spain
Instituto de Investigación Sanitaria Gregorio Marañón, Madrid, Spain
CIBER Enfermedades Respiratorias, CIBERES,Madrid, Spain

Olalla Sierra
Microbiología Clínica y Enfermedades Infecciosas, Hospital General Universitario Gregorio Marañón, Madrid, Spain
Instituto de Investigación Sanitaria Gregorio Marañón, Madrid, Spain

Iñaki Comas
Unidad Mixta Genómica y Salud, Centro Superior de Investigación en Salud Pública (FISABIO)-Universitat de València, Valencia, Spain
CIBER en Epidemiología y Salud Pública, Madrid, Spain

Javier Sicilia
Instituto de Investigación Sanitaria Gregorio Marañón, Madrid, Spain
Unidad de Medicina y Cirugía Experimental, Hospital General Universitario Gregorio Marañón, Madrid, Spain

Emilio Bouza
Microbiología Clínica y Enfermedades Infecciosas, Hospital General Universitario Gregorio Marañón, Madrid, Spain
Instituto de Investigación Sanitaria Gregorio Marañón, Madrid, Spain
CIBER Enfermedades Respiratorias, CIBERES,Madrid, Spain
Departamento de Medicina, Facultad de Medicina, Universidad Complutense de Madrid, Madrid, Spain

Darío García de Viedma
Microbiología Clínica y Enfermedades Infecciosas, Hospital General Universitario Gregorio Marañón, Madrid, Spain
Instituto de Investigación Sanitaria Gregorio Marañón, Madrid, Spain
CIBER Enfermedades Respiratorias, CIBERES, Madrid, Spain
CEI Campus Moncloa, UCM-UPM, Madrid, Spain

Ankita Vaishampayan
Life Sciences and Technology, Beuth University of Applied Sciences Berlin, Berlin, Germany

Anne de Jong and Jan Kok
Department of Molecular Genetics, University of Groningen, Groningen, Netherlands

Elisabeth Grohmann
Life Sciences and Technology, Beuth University of Applied Sciences Berlin, Berlin, Germany
Division of Infectious Diseases, University Medical Center Freiburg, Freiburg, Germany

Ferran Navarro
Servei de Microbiologia, Hospital de la Santa Creu i Sant Pau, Institut d'Investigació Biomèdica Sant Pau, Barcelona, Spain

Maite Muniesa
Department of Microbiology, University of Barcelona, Barcelona, Spain

Felipe Lira and José L. Martínez
Centro Nacional de Biotecnología, Consejo Superior de Investigaciones Científicas, Madrid, Spain

Gabriele Berg
Institute of Environmental Biotechnology, Graz University of Technology, Graz, Austria

Mirian Domenech and Ernesto García
Departamento de Microbiología Molecular y Biología de las Infecciones, Centro de Investigaciones Biológicas, Consejo Superior de Investigaciones Científicas, Madrid, Spain

Dolores L. Guzmán-Herrador, Anabel Alperi, Coral González-Prieto and Matxalen Llosa
Departamento de Biología Molecular, Universidad de Cantabria (UC), Instituto de Biomedicina y Biotecnología de Cantabria(IBBTEC, UC-CSIC-SODERCAN), Santander, Spain

Samuel Steiner and Craig R. Roy
Department of Microbial Pathogenesis, Boyer Center for Molecular Medicine, Yale University School of Medicine, New Haven, CT, United States

Denise E. Morris and David W. Cleary
Infectious Disease Epidemiology Group, Academic Unit of Clinical and Experimental Sciences, Faculty of Medicine, Institute for Life Sciences, University of Southampton, University Hospital Southampton Foundation NHS Trust, Southampton, United Kingdom

Stuart C. Clarke
Infectious Disease Epidemiology Group, Academic Unit of Clinical and Experimental Sciences, Faculty of Medicine, Institute for Life Sciences, University of Southampton, University Hospital Southampton Foundation NHS Trust, Southampton, United Kingdom
Global Health Research Institute, University of Southampton, Southampton, United Kingdom
NIHR Southampton Respiratory Biomedical Research Unit, Southampton, United Kingdom

Ayse Z. Sahan and Soumita Das
Department of Pathology, University of California, San Diego, San Diego, CA, United States

Tapas K. Hazra
Department of Internal Medicine, University of Texas Medical Branch, Galveston, TX, United States

Elena Cabezón, Fernando de la Cruz and Ignacio Arechaga
Instituto de Biomedicina y Biotecnología de Cantabria (IBBTEC), CSIC-Universidad de Cantabria and Departamento de Biología Molecular, Universidad de Cantabria, Santander, Spain

Julia E. Grimwade and Alan C. Leonard
Department of Biological Sciences, Florida Institute of Technology, Melbourne, FL, United States

Index

www.ingramcontent.com/pod-product-compliance
Lightning Source LLC
LaVergne TN
LVHW061512220926
848485LV00017B/879